DATE DUE

NO 9 05			

DEMCO 38-296

OXFORD MONOGRAPHS ON GEOLOGY AND GEOPHYSICS NO. 36

OXFORD MONOGRAPHS ON GEOLOGY AND GEOPHYSICS

1. DeVerle P. Harris: *Mineral resources appraisal: mineral endowment, resources, and potential supply: concepts, methods, and cases*
2. J. J. Veevers (ed.): *Phanerozoic earth history of Australia*
3. Yang Zunyi, Cheng Yuqi, and Wang Hongzhen (eds.): *The geology of China*
4. Lin-gun Liu and William A. Bassett: *Elements, oxides, and silicates: high-pressure phases with implications for the earth's interior*
5. Antoni Hoffman and Matthew H. Nitecki (eds.): *Problematic fossil taxa*
6. S. Mahmood Naqvi and John J. W. Rogers: *Precambrian geology of India*
7. Chih-Pei Chang and T. N. Krishnamurti (eds.): *Monsoon meteorology*
8. Zvi Ben-Avraham (ed.): *The evolution of the Pacific Ocean margins*
9. Ian McDougall and T. Mark Harrison: *Geochronology and thermochronology by the $^{40}Ar/^{39}Ar$ method*
10. Walter C. Sweet: *The conodonta: morphology, taxonomy, paleoecology, and evolutionary history of a long-extinct animal phylum*
11. H. J. Melosh: *Impact cratering: a geologic process*
12. J. W. Cowie and M. D. Brasier (eds.): *The Precambrian-Cambrian boundary*
13. C. S. Hutchison: *Geological evolution of south-east Asia*
14. Anthony J. Naldrett: *Magmatic sulfide deposits*
15. D. R. Prothero and R. M. Schoch (eds.): *The evolution of perissodactyls*
16. M. Menzies (ed.): *Continental mantle*
17. R. J. Tingey (ed.): *Geology of the Antarctic*
18. Thomas J. Crowley and Gerald R. North: Paleoclimatology
19. Gregory J. Retallack: *Miocene paleosols and ape habitats in Pakistan and Kenya*
20. Kuo-Nan Liou: *Radiation and cloud processes in the atmosphere: theory, observation, and modeling*
21. Brian Bayly: *Chemical change in deforming materials*
22. A. K. Gibbs and C. N. Barron: *The geology of the Guiane Shield*
23. Peter J. Ortoleva: *Geochemical self-organization*
24. Robert G. Coleman: *Geologic evolution of the Red Sea*
25. Richard W. Spinrad, Kendall L. Carder, and Mary Jane Perry: *Ocean optics*
26. Clinton M. Case: *Physical principles of flow in unsaturated porous media*
27. Eric B. Kraus and Joost A. Businger: *Atmosphere-ocean interaction*, second edition
28. M. Solomon and D. I. Groves: *The geology and origins of Australia's mineral deposits*
29. R. L. Stanton: *Ore elements in arc lavas*
30. P. B. Wignall: *Black shales*
31. Orson L. Anderson: *Equations of state for solids in geophysics and ceramic science*
32. J. Alan Holman: *Pleistocene amphibians and reptiles in North America*
33. P. Janvier: *Early vertebrates*
34. David S. O'Hanley: *Serpentinites: records of tectonic and petrological history*
35. C. S. Hutchison: *South-east Asia oil, gas, coal, and mineral deposits*
36. Tina M. Niemi, Zvi Ben-Avraham, and Joel R. Gat (eds.): *The Dead Sea: the lake and its setting*

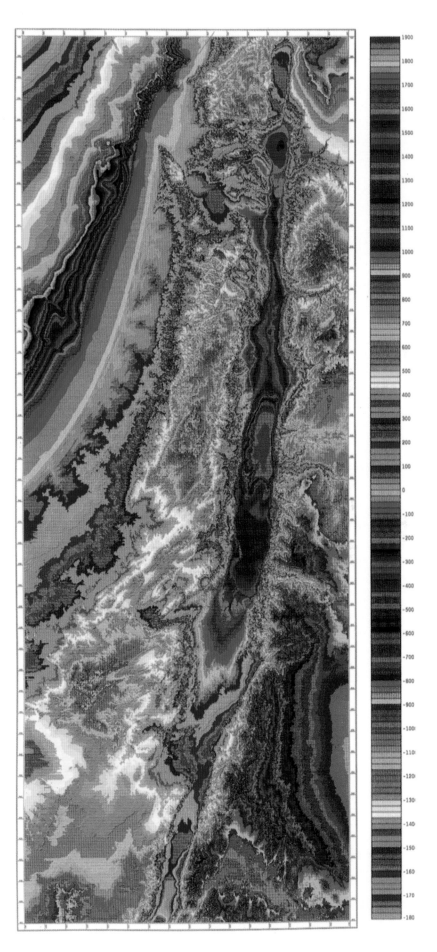

Hypsometric map of the Dead Sea depression, color coded at 25 m intervals. See chapter 2 by J. K. Hall for details of the construction of this map.

Satellite image of the Dead Sea produced by merging sequential LANDSAT 5 thematic mapper color scenes from January 18, 1987 with 10 m panchromatic imagery of Israel from the French SPOT satellite (© C.N.E.S., Historical Productions Ltd., Educational Programs, Inc., 1990).

THE
DEAD SEA

The Lake and Its Setting

Edited by

TINA M. NIEMI

ZVI BEN-AVRAHAM

JOEL R. GAT

New York   Oxford • Oxford University Press   1997

Oxford University Press

Oxford   New York

Athens   Auckland   Bangkok   Bogota   Bombay   Buenos Aires
Calcutta   Cape Town   Dar es Salaam   Delhi   Florence   Hong Kong
Istanbul   Karachi   Kuala Lumpur   Madras   Madrid   Melbourne
Mexico City   Nairobi   Paris   Singapore   Taipei   Tokyo   Toronto   Warsaw

and associated companies in
Berlin   Ibadan

Copyright © 1997 by Oxford University Press, Inc.

Published by Oxford University Press, Inc.
198 Madison Avenue, New York, New York 10016

Oxford is a registered trademark of Oxford University Press

Library of Congress Cataloging-in-Publication Data
The Dead Sea : the lake and its setting / edited by Tina M. Niemi, Zvi Ben-Avraham, Joel R. Gat.
p.   cm.
Includes bibliographical references and index.
ISBN 0-19-508703-8
1. Limnology—Dead Sea (Israel and Jordan)—Congresses.
I. Niemi, Tina M.   II.   Ben-Avraham, Zvi.   III.   Gat, Joel, 1926–
GB1759.D4D43   1996
551.48'2'095694—dc20   95-45027

1 3 5 7 9 8 6 4 2

Printed in the United States of America
on acid-free paper

PREFACE

This multidisciplinary book is a compilation of research on the tectonic and geologic history of the Dead Sea, physical, chemical, and biological aspects of the water body, and on environmental changes from the Pleistocene to the present. The book developed from a December 1992 conference in Tel Aviv entitled "The Dead Sea—A Summary of Recent Research" that was sponsored by the Dead Sea Research Center, Tel Aviv University. The meeting was organized to bring together researchers working in diverse fields to discuss recent advances and the state of our knowledge of various aspects of the Dead Sea and its environs. Because the lake spans political boundaries, most of the work describes only the western portion of the lake and its western shores. Although some of the findings are characteristic of the whole water body, many features must be supplemented by measurements taken on its eastern side. We hope that this volume will be a springboard for future collaborative research encompassing the entire extent of the Dead Sea.

The book is intended for both the general reader and scientists in various disciplines. Given the timely issues of resource management and water use and the political changes occurring in the region, this book on the dynamics of the Dead Sea should be of great interest to a wide variety of readers ranging from geologists, geochemists, and hydrologists to historians and political scientists.

One might ask, why study the Dead Sea when it is so extraordinarily unique? Not only does it lie more than 400 meters below sea level at the lowest continental elevation, but the 320-m-deep water of the Dead Sea is the most saline ever encountered in a large lake. There are numerous answers, ranging from economic assessments to basic research. Geologically, the Dead Sea is an active tectonic feature that provides us with modern analog for ancient pull-apart basins, the generation of hydrocarbon, and seismicity of active faults. It is a model for evaporite deposition from deep water and is a laboratory of measuring the changes in physical and chemical properties of the evolving sea. The life-forms in the water provide us with interesting insight on the adaptive abilities that indigenous species have developed to tolerate the composition of this hypersaline natural brine and survive. The waters of the Dead Sea are enriched with vast mineral resources. The same hot evaporitic days used to extract potash also create the unique climate that draws tourists to the Dead Sea's therapeutic brine and hot springs. Development of the region and uses of the water will only increase in the future, signaling the importance of monitoring our impact on this unique spot on Earth.

We have tried to standardize the geographic coordinates and names used in this book to the English edition of the 1:250,000 geographic map of Israel (Survey of Israel). The Palestine grid is a coordinate system established in the region by the British Mandate government that is still used today, as can be seen in many of the figures in this book. Place names in the Middle East have many spellings because of variations in transliteration. Furthermore, physiographic features and places have both common Arabic and Hebrew names (e.g., wadi and nahal mean river bed, arvat and ghor are words for playa), and both are used. However, multiple spellings are an unavoidable pitfall of the region. We hope that the cross-reference of these names in the index will prevent any confusion.

We would like to acknowledge the Minerva Dead Sea Research Center at Tel Aviv University, the meeting organizers, and all the participants and speakers at the conference. We extend thanks to the contributors to this volume and to the many people who helped by reviewing manuscripts for this book, including Rivka Amit, Yehouda Enzel, Jonathan Erez, Akiva Flexer, M. Gloor, Aharon Horowitz, Deiter Imboden, Yehoshua Kolodny, David Neev, A. Simon, Uri ten-Brink, Mina Weinstein-Evron, Alfred Wüest, and many of the contributors. The frontispiece was generously provided by John Hall. The efforts of graphic assistant Sara Lodigensky, word processing by Darlene Wallbillich, desktop publishing by Mark A. Decker, and the financial support for the final product of this book by the University of Missouri–Kansas City are gratefully acknowledged.

Kansas City, Missouri   T. M. N.
Tel Aviv, Israel   Z. B-A
Rehovot, Israel   J. R. G.
May 1996

CONTENTS

CONTRIBUTORS

ERGA ALONI Department of Botany, The George S. Wise Faculty of Life Sciences, Tel Aviv University, Tel Aviv 69978, Israel

DAVID A. ANATI Institute of Earth Science, The Hebrew University of Jerusalem, Givat Ram, Jerusalem 91904, Israel

ITZHAQ BEIT-ARIEH Institute of Archaeology, Tel Aviv University, Ramat Aviv 69978, Israel

ZVI BEN-AVRAHAM Department of Geophysics and Planetary Sciences, Tel Aviv University, Ramat Aviv 69978, Israel

DAN BOWMAN Department of Geography and Environmental Development, Ben-Gurion University of the Negev, Beer Sheva 84105, Israel

AMRAM ESHEL Department of Botany, The George S. Wise Faculty of Life Sciences, Tel Aviv University, Tel Aviv 69978, Israel

HEINZ-HERMANN ESSEN Institut für Meereskunde, Universität Hamburg, Troplowitzstraße 7, D-2000 Hamburg, Germany

TAL EZER Departmentof Atmospheric and Oceanic Sciences, Princeton University, P.O.B. CN710, Princeton, NJ 08544-0710, U.S.A.

AMOS FRUMKIN Department of Geography, The Hebrew University of Jerusalem, Jerusalem, 91905, Israel

MICHAEL GARDOSH The Israel National Oil Company, P.O. Box 50199, Tel Aviv 65100, Israel

ZVI GARFUNKEL Institute of Earth Science, The Hebrew University of Jerusalem, Givat Ram, Jerusalem 91904, Israel

JOEL R. GAT Department of Environmental Sciences and Energy Research, Weizmann Institute of Science, Rehovot 76100, Israel

ITTAI GAVRIELI Geological Survey of Israel, 30 Malkhe Yisrael Street, Jerusalem 95501, Israel

KLAUS-WERNER GURGEL Institut für Meereskunde, Universität Hamburg, Troplowitzstraße 7, D-22529 Hamburg, Germany

JOHN K. HALL Geological Survey of Israel, 30 Malkhe Yisrael Street, Jerusalem 95501, Israel

ARTUR HECHT Oceanographic and Limnological Research, Ltd., Tel Shikmona, P.O. Box 8030, Haifa 31080, Israel

AVRAHAM HUSS Department of Atmospheric Sciences, The Hebrew University of Jerusalem, Givat Ram, Jerusalem 91904, Israel

ELIEZER KASHAI The Israel National Oil Company, P.O. Box 50199, Tel Aviv 65100, Israel

PERLA KAUSHANSKY Department of Environmental Sciences, Weizmann Institute of Science, Rehovot 76100, Israel

BORIS S. KRUMGALZ Oceanographic and Limnological Research, Ltd., Tel Shikmona, P.O. Box 8030, Haifa 31080, Israel

BOAZ LUZ Institute of Earth Sciences, The Hebrew University of Jerusalem, Givat Ram, Jerusalem 91904, Israel

EMANUEL MAZOR Department of Environmental Sciences and Energy Research, Weizmann Institute of Science, Rehovot 76100, Israel

TINA M. NIEMI Department of Geosciences, University of Missouri–Kansas City, Kansas City, MO 64110-2499, U.S.A.

AMI NISHRI Oceanographic and Limnological Research, Ltd., Kinneret Laboratory, Tiberias 14133, Israel

AHARON OREN Division of Microbial and Molecular Ecology, The Insititue of Life Sciences, The Hebrew University of Jerusalem, Givat Ram, Jerusalem 91904, Israel

SHALOM SALHOV The Israel National Oil Company, P.O. Box 50199, Tel Aviv 65100, Israel

FLORIAN SCHIRMER Institut für Meereskunde, Universität Hamburg, Troplowitzstraße 7, D-22529 Hamburg, Germany

AVI SHAPIRA The Seismological Division, Institute for Petroleum Research and Geophysics, P.O. Box 2286, Holon 58122, Israel

AVIV SHAPIRA Oceanographic and Limnological Research, Ltd., Tel Shikmona, P.O. Box 8030, Haifa 31080, Israel

HAIM SHULMAN The Israel National Oil Company, P.O. Box 50199, Tel Aviv 65100, Israel

ZIV SIRKES Center for Ocean and Atmospheric Modeling, University of Southern Mississippi, Stennis Space Center, MS 39529, U.S.A

ILANA STEINHORN 204 Shrineview Avenue, Boalsburg, PA 16827, U.S.A.

MARIANA STILLER Department of Environmental Sciences and Energy Research, Weizmann Institute of Science, Rehovot 76100, Israel

A. SIEP TALMA Division of Water, Environment and Forestry Technology, Council of Scientific and Industrial Research, P.O. Box 395, Pretoria 0001, South Africa

ELI TANNENBAUM Kimron Oil and Minerals, 21 Yona Hanavi Street, Tel Aviv 63302, Israel

JOHN C. VOGEL Quaternary Dating Research Unit, Council of Scientific and Industrial Research, P.O. Box 395, Pretoria 0001, South Africa

YOAV WAISEL Department of Botany, The George S. Wise Faculty of Life Sciences, Tel Aviv University, Tel Aviv 69978, Israel

YOSEPH YECHIELI Geological Survey of Israel, 30 Malkhe Yisrael Street, Jerusalem 95501, Israel

ISRAEL ZAK Institute of Earth Sciences, The Hebrew University of Jerusalem, Givat Ram, Jerusalem 91904, Israel

THE
DEAD SEA

1. DEAD SEA RESEARCH—AN INTRODUCTION

Tina M. Niemi, Zvi Ben-Avraham, and Joel R. Gat

The Dead Sea is one of a series of intracontinental, terminal lakes within the East African–Syrian rift system (Fig. 1-1). The tectonic control on the Dead Sea's unique properties is undeniable. The Dead Sea is the lowest continental depression on Earth, with its surface standing at -410 m mean sea level (msl). This extreme negative elevation, in combination with the tectonically elevated mountains that flank the basin, promotes a very arid environment. Water that enters the Dead Sea has no outlet in the actively subsiding basin. These conditions control the physical, chemical, and biological properties of the inland lake. The hypersaline Dead Sea water is dense (1.234 g/l), with 30% dissolved solids—one of the highest average salinities of any lake on Earth (276 g/kg). The Dead Sea is inhabited only by highly specialized green algae and red archaeobacteria. The Dead Sea has influenced the course of human history because it is situated on the critical land bridge between the continents of Africa and Asia.

The 40,000-km² drainage basin of the Dead Sea (Bentor, 1961) spans six countries and territories: Israel, Jordan, Syria, Lebanon, the West Bank, and Egypt (Fig. 1-2a). The Jordan River and its intermediate basins (Hula, Sea of Galilee, and Beth Shean Basins; Fig. 1-2b) form the conduit that supplies the majority of the inflow to the Dead Sea. However, freshwater discharge from the snow-covered Mt. Hermon and Jebel Druze catchment area is a valuable commodity not to be wasted in the brine of the Dead Sea. Canals have been constructed to divert water from the Sea of Galilee (Israeli National Water Carrier), and a dam across the Yarmouk River channels water into the Jordan Valley canal. South of the Sea of Galilee, the Lower Jordan River, carries only a trickle of water the 105 km to the Dead Sea. Drainage areas around the Dead Sea and to the south only deliver water to the basin during flash floods, except for the flow of the Wadi Hasa and Wadi Mujib (Fig. 1-2a).

The area of the Dead Sea is sparsely populated with only a few settlements and resort spas. The largest town is Safi which lies along the perennial stream of Wadi Hasa in Jordan. A satellite view of the Dead Sea (Frontispiece A) shows a long, narrow lake that is separated into two basins by a peninsula. Known in Arabic as *El-Lishan*, or "the tongue," the Lisan Peninsula sticks out from the east shore and points northward into the basin. Not long ago, when the level of the Dead Sea stood at - 400 m, the rectangular, deep north basin was connected to the shallow-water, tear-shaped south basin. Today, salt-rich water is pumped from the north basin into the commercial evaporation ponds of the Dead Sea Works and the Arab Potash Company.

During the time in which hydrographic records have been taken (from 1929 to the present), the level of the Dead Sea has fallen 20 m (Fig. 1-3). This drop has caused enormous changes in the lake's morphology, physical hydrography, and geochemistry. Whether by tectonic, climatic, hydrologic, or anthropogenic forces, the Dead Sea and its environment are in a state of dynamic change. This multidisciplinary book provides both a synthesis of new data and a summary of more than 25 years of detailed monitoring and research on the Dead Sea. In this chapter, we briefly review the history of Dead Sea research and present a preview of the contents of this book.

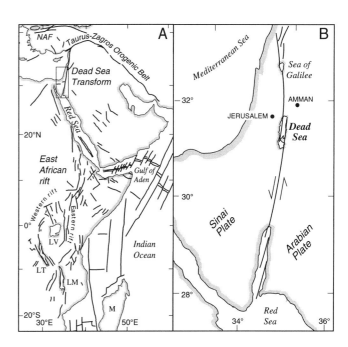

Figure 1-1 Regional maps of the Dead Sea transform (DST): A, The DST is part of the East African-Syrian rift system. LM=Lake Malawi, LT=Lake Tanganyika, LV=Lake Victoria, NAF=Northern Anatolian fault, M=Madagascar. (Modified after UNESCO, 1990); B, The Dead Sea basin formed by transtension along the transform plate boundary between the Arabian plate and the Sinai. (Map after Bartov, 1990)

HISTORY OF DEAD SEA RESEARCH

In the 19th and early 20th centuries, the structure of the linear valley that extends from the Gulf of Aqaba (Elat) northward to the Sea of Galilee was attributed to tectonic motion (Anderson, 1852; Hull, 1886; Lartet, 1869; Blanckenhorn, 1912, 1914). Picard (1943) called the depression the Jordan-Aqaba graben or rift valley because it was formed by downfaulting along border faults. The terms *rift valley* and *rift zone* are misnomers for, although large vertical motion is associated with the linear depression, there is overwhelming evidence that the "rift" is a continental transform plate boundary. An estimated 105 km of cumulative left-lateral, strike-slip offset (e.g., Quennell, 1959; Freund et al., 1970) has been measured across the transform that connects the spreading center of the Red Sea with the continental collision of the Taurus-Zagros mountain portion of the Alpine Orogenic belt in Turkey (Fig. 1-1). Although the name Dead Sea–Jordan transform more aptly fits the feature, the term Dead Sea rift is so embedded in the literature that its common usage is accepted.

Although information on the Dead Sea and its environs found in Biblical and historical records, as well as accounts of early travelers on Holy Land pilgrimages, are invaluable, their scientific merit is difficult to ascertain. Chemical analyses of Dead Sea water were made as early as the 18th century, with

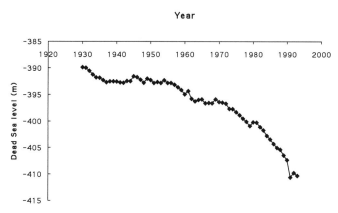

Figure 1-3 Hydrograph of the Dead Sea level since 1929

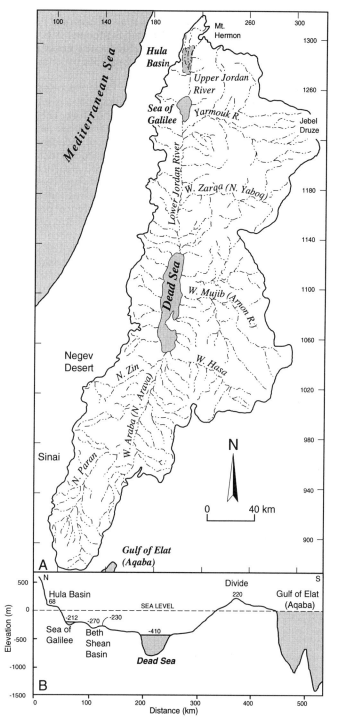

Figure 1-2 The Dead Sea and its drainage basin: a, Catchment area of the Dead Sea. Coordinates shown in the Palestine Grid (after Bentor, 1961); b, North–south longitudinal profile along the axis of the Dead Sea–Jordan transform valley. (Modified after Neev and Emery, 1967)

death of these explorers (Eriksen, 1989). After its illustrious beginnings in the 1830s and 1840s, scientific research was at a virtual standstill for one hundred years. Amazingly, there remain portions of the bathymetric map of the Dead Sea today that are based solely on the soundings of the 1848 data.

Through the pioneering effort of Moshe Novomeysky, the Palestine Potash Company received a concession from the British Mandate Government in 1929 to build and operate a plant to extract potash and other salts from the Dead Sea. Evaporation ponds were built on the north shore near Kalia, and a second plant was established near Mt. Sedom in 1934. The potash industry sparked new scientific research on the Dead Sea and monitoring of the Dead Sea levels. Geologic information was gleaned from water wells drilled in the vicinity of the plant (Picard, 1943). In the late 1930s to 1940s, several cores were taken in the Dead Sea by B. Elazari-Volcani (1943), who described the sediments and isolated the first live organism, a rather astonishing discovery for a lake with such an inhospitable name and nature. Presenting temperature and chemical data, Bloch et al. (1944) described an unusual phenomena in which the waters of the Dead Sea turned white (known as a whitening event).

After the 1948 war, the Dead Sea became divided between Israel and Jordan. The former Palestine Potash company became the Dead Sea Works and renewed potash production in the south plant in 1952. Oil exploration in the Dead Sea basin began in the 1950s resulting in a wealth of information on the subsurface lithology and structure of the transform valley.

Modern archaeological excavations began at Qumran after the chance discovery of the Dead Sea scrolls in nearby caves in 1948. These were followed by numerous excavations at Masada, Bab edh-Dhra, Numeira, and elsewhere.

Bentor (1961) presented theories on the evolution and age of the Dead Sea brine by comparing the inventory of salts in the lake to the known inputs. A similar approach had already been taken nearly 200 years before by Marcet in an avant-garde paper to the Royal Society of London (see Nissenbaum, 1970). But it was not until 1959–1960 that a comprehensive sampling and analysis of the water and sediment were conducted (Neev, 1964; Neev and Emery, 1967). This monumental work is the anchor point of all successive studies. In 1974, Neev and Hall (1976, 1979) collected seismic reflection and magnetic data from the Dead Sea, providing the first images of the deep structure and the faults along the basin's margin. In 1975, Ben-Avraham et al. (1978) measured the heat flow through the Dead Sea.

A regular hydrographic survey began in 1975 following the observation of Gad Assaf that the large density (salinity) gradi-

some analyses associated with illustrious names, such as Lavoisier, Klaproth, Gay-Lussac and Gmelin. These remarkable data were reviewed by Nissenbaum (1970). The first systematic scientific investigation of the lake was conducted in 1848 by the U.S. Navy under Lieutenant W. F. Lynch (Lynch, 1849). Data from two previous surveys by Irishman Christopher Costigan in 1835 and Lieutenant Thomas H. Molyneux of the British Royal Navy in 1845 were lost because of the tragic

tions were summarized in special sections of *Earth and Planetary Science Letters* in December 1984 and September 1987. This extraordinary limnological event heralded a new era of physiochemical and geologic research on the Dead Sea by a number of institutions.

During the late 1970s and early 1980s, Dead Sea research got a boost, both financially and in press coverage, as a result of the proposed Mediterranean-Dead Sea canal (Fig. 1-4). The project that proposed to simultaneously replenish the Dead Sea water and produce hydroelectricity captured global attention. The plight of the falling Dead Sea level was reported in articles such as "Better Med than Dead" (Rich, 1978) and others (Rich, 1979, 1980; Steinhorn and Gat, 1983). Scientists and engineers began to assess the feasibility of the canal with studies that encompassed a wide array of issues, including geotechnical problems, limnological and ecological impact, and geophysical concerns. A bibliographic reference to these works was compiled by Arad and Beyth (1990), with a historical review from 1837 of the many schemes of building a canal to connect the Dead Sea to the world's oceans (Vardi, 1990). Bibliographic information from the Dead Sea and its surroundings is also found in the compilation by Arad et al. (1984).

ORGANIZATION OF THIS BOOK

The book is organized into three parts—structure of the basin, properties of the water, and Quaternary environmental changes. A synopsis of Dead Sea research and directions for future research is provided in the final chapter.

The first part of the book covers the tectonic history of the Dead Sea from several different perspectives—geophysical, geological, petroleum exploration, and seismological. The book begins with a description of the bathymetry and topography of the Dead Sea–Jordan transform valley by Hall in chapter 2. This chapter sets the stage for chapters on the history and formation of the Dead Sea. Much of the data used in developing tectonic models are derived from geophysical studies. In chapter 3, Ben-Avraham reviews these data and models of the crustal structure and formation of the Dead Sea basin.

The Dead Sea is just the latest in a series of lakes that have filled the structural depression of the Dead Sea rift during the past 20 million years. Using stratigraphy and structural evidence mapped along the transform, Garfunkel (chapter 4) presents a pull-apart model based on basin floor extension of the entire 105-km Dead Sea transform slip. A summary of hydrocarbon exploration and interpretation of new seismic reflection and borehole data, including a new deep (approx. 6.5 km) well is presented by Gardosh et al. in chapter 5. These data on the complex subsurface lithology and structure of the Dead Sea's south basin illuminate the hidden potential of oil reservoirs.

The Dead Sea is one of the most seismically active sections of the transform. In chapter 6, Niemi and Ben-Avraham relate faults, slumps, and morphologic features mapped from high-resolution seismic reflection data to patterns of active faulting and tectonic subsidence. A review of earthquake parameters by Shapira is found in chapter 7.

The physical, chemical, and biological aspects of the Dead Sea water are largely a product of the deep sink of the Dead Sea, its interaction with residual brines, and evaporites of early lakes. These properties are interrelated and are the subject of the second part of this book. Freshwater flowing into the Dead Sea dilutes the surface water, creating a thermohaline stratification of the water column. As mentioned previously, the year 1979 is a noteworthy turning point in the history of the Dead Sea: It marks the time when this long-term stratification was destroyed in a complete overturn that brought oxygenated waters to the fossil deep water.

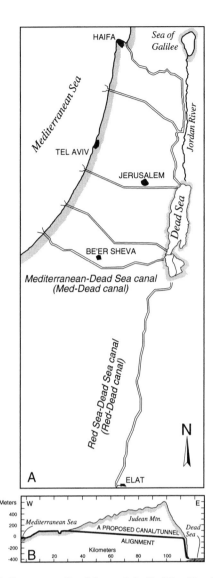

Figure 1-4 Mediterranean–Dead Sea and the Red Sea–Dead Sea canals: a, Proposed locations of possible canals and tunnels to connect the Mediterranean and Red Seas with the Dead Sea; b, Cross-sectional view of a Med–Dead canal alignment. (Modified after figures in Arad and Beyth, 1990)

ent between the upper and lower fossil water masses was greatly diminished and that an overturn of the water column was imminent. A team consisting of the Weizmann Institute, the Israel Geological Survey, and the U.S. laboratories of Prof. Harmon Craig (Scripps Institution of Oceanography) and Prof. G. M. Friedman (Rensselaer Polytechnic Institute) closely followed the exciting developments in the Dead Sea as the continuing negative water balance resulted in falling water levels, increasing salinity, and shrinking areal extent of the lake. In 1976, the southern basin became disconnected from the deeper, northern basin which still maintaining some stratification. However, the pycnocline descended in a series of steps from year to year, until finally the temperature and salinity gradient collapsed. In February 1979, complete mixing of the water column occurred in an overturn that was accompanied by far-reaching geochemical changes in the Dead Sea water, as described by Steinhorn et al. (1979). Many of these observa-

Three chapters cover the physical structure and features of the lake. In chapter 8, Anati summarizes the hydrography of the Dead Sea before the overturn and the subsequent seasonal stratification. The stability of the water column is a function of the amount of freshwater inflow and resultant salinity changes. Sirkes and coauthors (chapter 9) use radar to track the surface currents in the Dead Sea. Measurements of wave heights and periods are presented by Hecht et al. in chapter 10.

The importance of evaporation in the Dead Sea cannot be overstated, for it exerts the major control on the water balance and provides the energy for commercial extraction of potash, bromide, and other value minerals from the water. In chapter 11, Steinhorn discusses the rates of evaporation and the dependence of these calculations on salt concentration. The unique chemical composition of the Dead Sea and of subsurface brines encountered in the vicinity has aroused the curiosity of all researchers, as it contains clues about the genesis and evolution of the Dead Sea system. Zak reviews current ideas and hypotheses on this topic in chapter 12.

The multicomponent salinity of the Dead Sea brine poses challenges both in the measurement of the salinity and its modeling. The precipitation of evaporitic components, particularly halite, dominates the geochemical evolution of the brine. Each of three chapters on this subject takes a different approach to assessing halite solubility. Krumgalz uses Pitzer's thermodynamic approach to predict the ion activity of the waters under varying concentrations (chapter 13). In chapter 14, Gavrieli uses Harned's equation to calculate the saturation index, and by comparing it to the ion concentration in the Dead Sea, estimates the degree of supersaturation. Stiller et al. (chapter 15) attempted the direct experimental determination of the halite saturation curve as a function of temperature and salt composition. The actual rate of halite deposition was obained from sediment traps placed at various depths in the water column.

Detailed geochemical analyses of minor and trace elements of the Dead Sea and their isotopic composition are presented in three chapters. In chapter 16, Luz et al. measure the seasonal path of the total dissolved carbon dioxide and its stable carbon isotope composition. Radiocarbon analyses of the water before and after the 1979 mixing are presented in Talma et al. (chapter 17). Low ^{14}C values that do not show the impact of the bomb peak suggest that either the exchange between the water and the atmosphere is very slow or there is an input of low ^{14}C from an unknown source. The inventory of manganese and iron related to overturn and concurrent changes in rates of oxidation is discussed by Nishri and Stiller in chapter 18. Reductions in the concentration of trace elements are connected to their coprecipitation with halite following the termination of the long-term stratification.

The water composition of the Dead Sea differs from other saline lakes in its high concentration of divalent cations of magnesium and calcium. In chapter 19, Oren describes a unique, highly specialized endogenous community of microorganisms that are able to endure the hypersaline, Mg-enriched brines of the Dead Sea.

Simply stated, the level of a closed basin like the Dead Sea is the result of how much runoff from precipitation in the watershed reaches the lake less the amount of water lost from the surface of the lake by evaporation. However, in an actively subsiding basin, tectonic motion may also affect oscillation in the lake level. Three chapters deal with Pleistocene and Holocene lake-level fluctuations. Tectonic subsidence of the Dead Sea basin has occurred so rapidly that the streams draining into it have not had time to adjust and are marked by waterfalls and steep-walled canyons. Geomorphological changes of the Dead Sea erosional base level are summarized by Bowman in chapter 20. Levels of the Dead Sea precursor, Lake Lisan, are

an important indicator of climatic fluctuations in the late Pleistocene, as presented in chapter 21 by Niemi. In chapter 22, Frumkin describes a record of Holocene levels of the Dead Sea lake levels based on radiocarbon-dated cave outlets of the Mt. Sedom diapir, and he compares these data to geomorphic and archaeological indicators.

The shores of the Dead Sea are part of the fertile crescent, a unique land bridge that connects the three continents of Africa, Europe, and Asia. The Dead Sea region has played an important role in Near Eastern prehistory; archaeology; a summary of archaeological findings is presented in chapter 23 by Beit-Arieh. Periods of habitation correlate nicely with high stands of the lake during wetter climatic intervals.

Understanding the coastal environment, its water resources, and vegetation is important for future development of the Dead Sea area. The 20th century retreat of the shoreline caused by lowering levels of the Dead Sea exposes new land that is initially very saline and, therefore, an ideal environment to examine geochemical and botanical processes. Yechieli and Gat (chapter 24) examine the flux of salts between land and sea, as well as the rate of desalination of the newly exposed land. A survey of groundwater aquifers and freshwater and saline springs by Mazor in chapter 25 demonstrates the complexity and variability of the subsurface hydrological system. Aloni et al. (chapter 26) show that reduction of the substrate salinity by freshwater leaching (a process that takes 4–5 years) is the dominant factor in vegetating the emergent land exposed by the falling Dead Sea.

The final chapter of the book is a synopsis of material presented in this book and a projection of the possible directions of Dead Sea research in the 21st century.

REFERENCES

Anderson, H. J., 1852, Geological reconnaissance of part of the Holy Land, *in* Lynch, W. F., ed., *Official report of the U.S. Expedition to the River Jordan and the Dead Sea*: London, p. 75–206.

Arad, V., and Beyth, M., 1990, Mediterranean-Dead Sea Project bibliography: Geological Survey of Israel Report GSI/9/90, p. 1–29.

Arad, V., Beyth, M., and Bartov, Y., 1984, The Dead Sea and its surroundings: Bibliography of geological research: Geological Survey of Israel Special Publication 3, 111 p.

Bartov, Y., 1990, Geological photomap of Israel and adjacent areas: 1:750,000 scale map, Survey of Israel, 1 sheet.

Ben-Avraham, Z., Hänel, R., and Villinger, H., 1978, Heat flow through the Dead Sea rift: Marine Geology, v. 28, p. 253–269.

Bentor, Y., 1961, Some geochemical aspects of the Dead Sea and the question of its age: *Geochimica et Cosmochimica Acta*, v. 25, p. 239–260.

Blanckenhorn, M., 1912, *Naturwissenschaftliche Studien am Totem Meer und im Jordanthal*: Berlin, Friedländer & Sohn, 478 p.

Blanckenhorn, M., 1914, *Syrien, Arabien und Mesopotamien*: Heidelberg, Handbuch Region. Geol.

Bloch, R., Littman, H. Z., and Elazari-Volcani, B., 1944, Occasional whiteness of the Dead Sea: *Nature*, v. 154, p. 402–403.

Elazari-Volcani, B., 1943, Bacteria in the bottom sediments of the Dead Sea: *Nature*, v. 152, p. 274–275.

Eriksen, E. O., 1989, *Holy Land Explorers*: Jerusalem, Franciscan Printing Press, 171 p.

Freund, R., Garfunkel, Z., Zak, I., Goldberg, M., Weissbrod, T., and Derin, B., 1970, The shear along the Dead Sea rift: *Royal Society of London Philosophical Transactions*, v. A267, p. 107–130.

Hull, E., 1886, The survey of Western Palestine, memoir on the geology and geography of Arabia Petraea, Palestine and adjoining districts, with special reference to the mode of formation of the Jordan-Arabah Depression and the Dead Sea: London, R. Bentley & Sons.

Lartet, L., 1869, Essai sur la geologie de la Palestine et des contrées avoisinantes, telles que l'Egypte et l'Arabie: Paris, Masson, 292 p.

Lynch, W. F., 1849, *Narrative of the United States' Expedition to the River Jordan and the Dead Sea*: Lea and Blanchard, Philadelphia, 509 p.

Neev, D., 1964, The Dead Sea: Jerusalem, Geological Survey of Israel Report Q/2/64, 407 p.

Neev, D., and Emery, K. O., 1967, The Dead Sea: Depositonal processes and environments of evaporites: Jerusalem, Geological Survey of Israel Bulletin 41, 147 p.

Neev, D., and Hall, J. K., 1976, The Dead Sea geophysical survey, 19 July-1 August 1974, Final Report 2: Jerusalem, Geological Survey of Israel, Marine Geology Division, Report No. 6/76, 21 p.

Neev, D., and Hall, J. K., 1979, Geophysical investigations in the Dead Sea: *Sedimentary Geology*, v. 23, p. 209–238.

Nissenbaum, A., 1970, Chemical analyses of Dead Sea and Jordan River water 1778–1830: *Israel Journal of Chemistry*, v. 8, p. 281–287.

Picard, L., 1943, *Structure and evolution of Palestine*: Jerusalem, Bulletin of the Geology Department, The Hebrew University, v. 4, no. 2–4, 187 p.

Quennell, A. M., 1959, Tectonics of the Dead Sea rift: 20th International Geological Congress, Mexico, 1956, p. 385–405.

Rich, V., 1978, Better Med than Dead: *Nature*, v. 276, p. 746.

Rich, V., 1979, Drought could turn Dead Sea into a desert: *Nature*, v. 281, p. 516–517.

Rich, V., 1980, Dead Sea–Coming to life: *Nature*, v. 286, p. 436.

Steinhorn, I., Assaf, G., Gat, J. R., Nishry, A., Nissenbaum, A., Stiller, M., Beyth, M., Neev, D., Garber, R., Friedman, G. M., and Weiss, W., 1979, The Dead Sea: Deepening of the mixolimnion signifies the overture to overturn of the water column: *Science*, v. 206, p. 55–57.

Steinhorn, I., and Gat, J. R., 1983, The Dead Sea: *Scientific American*, v. 249, p. 102–109.

UNESCO, 1990, Geologic map of the world, 1:25,000,000 scale: Paris, Commission for the geological map of the World, CGMW and UNESCO.

Vardi, J., 1990, Historical review, *in* Arad, V., and Beyth, M., eds., Mediterranean-Dead Sea Project bibliography: Jerusalem, Geological Survey of Israel Report GSI/9/90, p. 31–50.

I. STRUCTURE AND TECTONICS OF THE DEAD SEA BASIN

2. TOPOGRAPHY AND BATHYMETRY OF THE DEAD SEA DEPRESSION

JOHN K. HALL

In 1987, the Geological Survey of Israel (GSI) undertook a project in conjunction with the Survey of Israel and Historical Productions Inc. to convert the 1:50,000-scale topographic maps of Israel and its surroundings into a digital terrain model (DTM) with elevations to decimeter resolution every 25 m on the local Israel geographic grid. The project was described by Hall et al. (1990), with more recent results given by Hall (1993). Altogether, some 203 20-x-20-km map sheets have been analyzed. The resulting DTM includes the bathymetry of the offshore Mediterranean, the Sea of Galilee, and the Dead Sea. The Dead Sea bathymetry is compiled from the surveys of Lynch (1849), Neev and Emery (1967), and Hall and Neev (1978), more recent surveys in the Lynch Straits, and the approximate heights and limits of the evaporation pans of the Dead Sea Works in Israel and of the Arab Potash Company in Jordan.

The DTM data offer a far better representation of the topography than the usual topographic maps. First, Figure 2-1 shows that it is possible to make essentially three-dimensional maps, comparable to actual satellite images (Frontispiece A, see also Cleave 1994, 1995) by using a shaded relief representation in which the surface is illuminated from a specific direction. Second, the hypsometric map (Frontispiece B) shows it is possible to build an image in which the height patterns of the topography onshore and bathymetry offshore are readily visible. Third, three-dimensional "fishnet" block diagrams of selected areas and topographic profiles can be made to get a more immediate appreciation of the topography (e.g., Figs. 2-4 a–e and 2-5). Finally, a number of mathematical operations can be carried out on the data, such as measuring the distribution of area as a function of elevation (hypsometry), mapping slopes, determining watersheds and drainage networks, and statistically characterizing the landscape.

THE SETTING OF THE DEAD SEA DEPRESSION

The Dead Sea occupies the lowest part of the Syrian-African rift or what Quennell (1987) called the Western Arabia Rift System. The depression enclosing the Dead Sea consists of the Jordan Valley and the Kinneret Basin to the north of the Dead Sea, and the Arava or Wadi Arava to the south. This feature has historically been termed the Dead Sea graben, the Dead Sea rift, or the Dead Sea basin. In this chapter, the term Dead Sea depression will be used for the closed basin including the Sea of Galilee and northern Arava, which is separated from the Mediterranean by a sill at an elevation of about 60.5 m and from the Red Sea by a sill at about 200 m.

Historically, the Dead Sea depression is the most famous of all the world's depressions, having figured prominently in the events of the Old Testament (Sodom and Gomorrah, Jericho, Moab, the cities of the plain, etc.). However, its position below sea level was only relatively recently discovered (Moore and Beke, 1837). Since then, the Dead Sea has been the source of many dreams and plans to connect the adjacent seas and use its great depth below sea level for power generation. An excellent historical review is provided by Vardi (1990). The Dead Sea itself is almost universally recognizable by its distinctive shape, its colorful visage, and its place of prominence in modern synoptic space imagery (Hall, 1984).

The Dead Sea depression is a very prominent feature. It is the lowest of all the continental depressions, with the water level of the Dead Sea currently standing at about -409 m. This is in marked contrast to Lake Assal in the Afar Triangle at -156 m, the Qattara depression in northwestern Egypt at -145 m, Death Valley in the western United States at -86 m, the Salina Gaulicho on the Valdes Peninsula in Argentina at -40 m, the Caspian Sea at -28 m, and South Australia's Lake Eyre at -16 m.

In places, the Dead Sea depression presents an abrupt asymmetric discontinuity (Figs. 2-1 and 2-5). The scarps on the west mark the abrupt drop of 300–500 m from the plateau of the Judean desert to the shores of the Dead Sea, and the wadis draining the Judean hills present precipices up to 350 m high. On the east, the plateaus of the mountains of Moab are considerably higher, with heights of 900–1,000 m; these heights progressively increase toward the south to 1736 m (-21.5N, 197E). The wadis on the eastern plateaus, namely, Wadi Zerka Ma'in, Wadi Mujib, and especially Wadi Hasa, are many times larger than those to the west; this is commensurate with the larger elevation change and increased watershed size. The overall elevation changes are large; for example, on the eastern side, only 22 km lie between the -700 m contour and Jebel Shihan (88.7N, 220E) at 1,065 m. The Dead Sea catchment area is 43,000 km^2 (Arkin, 1982). The location of all places mentioned in this chapter is given in Figure 2-2.

The topographic asymmetry, which is essentially present from the northern Dead Sea to the southern end of the Gulf of Elat (Aqaba), is not only apparent in the bathymetry of the Dead Sea (Fig. 2-3c) and the Gulf of Elat (Aqaba) (Ben-Avraham et al., 1979) but also in the heights of the bordering plateaus (Frontispiece B, Figs. 2-5). The plateau elevations increase 200 m between Jerusalem and Amman. An elevation difference of about 1,000 m between the Dead Sea depression and the plateaus exists in the southern Arava at the location of the 200-m high saddle (Fig. 2-4e) that defines the southern water divide in the valley. The lowest outlet or "spillway" from the Dead Sea depression lies in the northwest (222.7N, 178E), about 1 km southeast of Afula (Fig. 2-4d). At an elevation of approximately 60.5 m, this spillway lies about 7 km east-southeast of a northeast- to southwest-trending sill that acts as the divide between the Zebulun Plain and Haifa Bay to the northwest, and the Jezreel Valley to the southeast.

BATHYMETRY OF THE DEAD SEA

The earliest modern soundings of the deep Dead Sea were taken during an 8-day investigation of the Dead Sea by the Irish explorer Christopher Costigan and a Maltese sailor companion during August and September 1835. Unfortunately, Costigan was afflicted by heat stroke during the cruise and died soon afterward in Jerusalem (Abramowitz, 1978; Eriksen, 1985). The sounding data, known to consist of bottom contacts in all but one location, were lost. The northern cape of the Lisan Peninsula was named after Costigan.

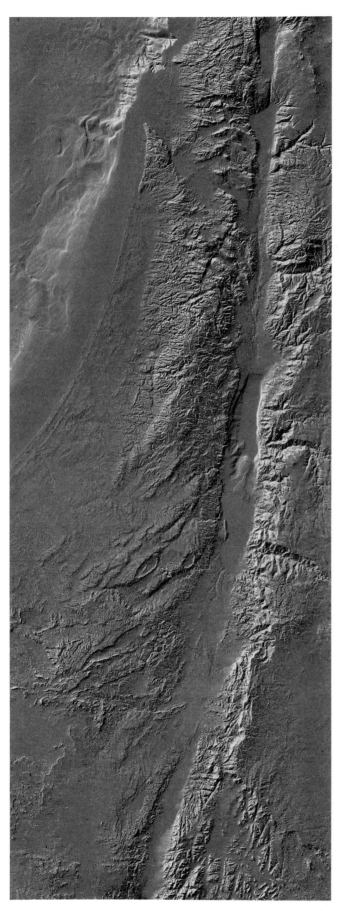

Exactly one decade later, an English naval officer, Lt. T. H. Molyneux, spent two and one half days studying the Dead Sea with several companions. Like Costigan, he died shortly afterward aboard the ship HMS *Spartan*, without having been able to transcribe his depth soundings (Abramowitz, 1978). The southwestern cape of the Lisan Peninsula was named after Molyneux.

Three years later, in 1848, the first systematic bathymetric survey of the Dead Sea was carried out by Lts. W. F. Lynch and J. B. Dale of the United States Navy in two copper rowboats named the *Fanny Skinner* and *Fanny Mason*. After navigating the Jordan River from the Sea of Galilee to the Dead Sea, these two officers and their crews surveyed the Dead Sea between April 20 and May 10, 1848. The deep northern basin was crossed eight times, and a number of inshore profiles were made. Altogether, 164 soundings were obtained with a linen sounding line, and a number of sediment samples were taken. A chart (Fig. 2-3a) at approximate scale of 1:146,000 was published in the cruise narrative by Lynch (1849). The now-dry straits between the northern and southern basins of the Dead Sea, west of the Lisan Peninsula, are named after Lynch.

From 1848 until the early 1940s, no additional systematic bathymetric information was collected. With the establishment in 1930 of the Palestine Potash Company Ltd. (PPC Ltd.), later to become the Dead Sea Works, Ltd., a flourishing commerce developed on the Dead Sea. Until the 1948 Arab-Israeli war, the potash and bromine extracted at the then-inaccessible Sodom plant in the south was transported by a very active shipping fleet to Kalia in the north. To support these operations, in the early 1940s, A. de Leeuw of the PPC Ltd. carried out a bathymetric survey to depths of 50 m in the Bay of Mazra'a east of the northern cape of the Lisan Peninsula (D. Neev, pers. com., 1994). Neev (1964) presents a chart of those earlier measurements, based on a 1944 contouring by the PPC Ltd. and supplemented by additional unspecified data.

In 1959, the first truly modern survey was conducted with echo sounder and precision visual navigation in the southwestern part of the southern Dead Sea between En Boqeq and En Gedi, in the area then within Israel. These results, based upon 2,000 soundings taken along 310 km of shiptrack, were published by Neev and Emery (1967), together with the Lynch soundings, in a colored chart (Fig. 2-3b) at a scale of 1:215,500. Inset in this chart is the first hypsometric curve giving the area and volume of the Dead Sea as a function of depth.

In 1974, with access to the entire northern basin following the 1967 Arab-Israeli war, a complete survey was carried out by the GSI, this time with electronic navigation, echo sounder, low-frequency seismic profiling equipment, and magnetometer (Hall and Neev, 1978; Neev and Hall, 1976, 1978, 1979). Between July

Figure 2-1 Shaded relief image of the area in Frontispiece B. The area encloses 67,200 km^2 (160 x 420 km). The image was produced by calculating the reflectance from the 107.52 million facets in the digital topographic surface, assuming Lambert's Law of diffuse reflection to apply. The surface is illuminated from the northwest (315°) by a light source at an altitude of 30°. The natural scale of this image is about 1:1.785 million. The original digital file was cleaned of DTM interpolation artifacts using a Scitex prepress system, and output on film for printing at 175 dots per inch. This image, at 1:500,000 scale, is available for sale from the library of the Geological Survey of Israel in Jerusalem. A number of interesting features are shown. The western border fault is clearly seen as a straight line extending from south of the isolated Mount Sedom, through the deep northern basin, and across the Jericho plain along the Jordan River. Minor faults and lineaments paralleling this trend are visible on the plateaus both east and west of the Dead Sea itself, and even away from the rift in the northwestern corner west of Jerusalem. Numerous other linears can be identified. Note the clarity with which faults and linears can be seen extending well beyond the limits usually attributed to them.

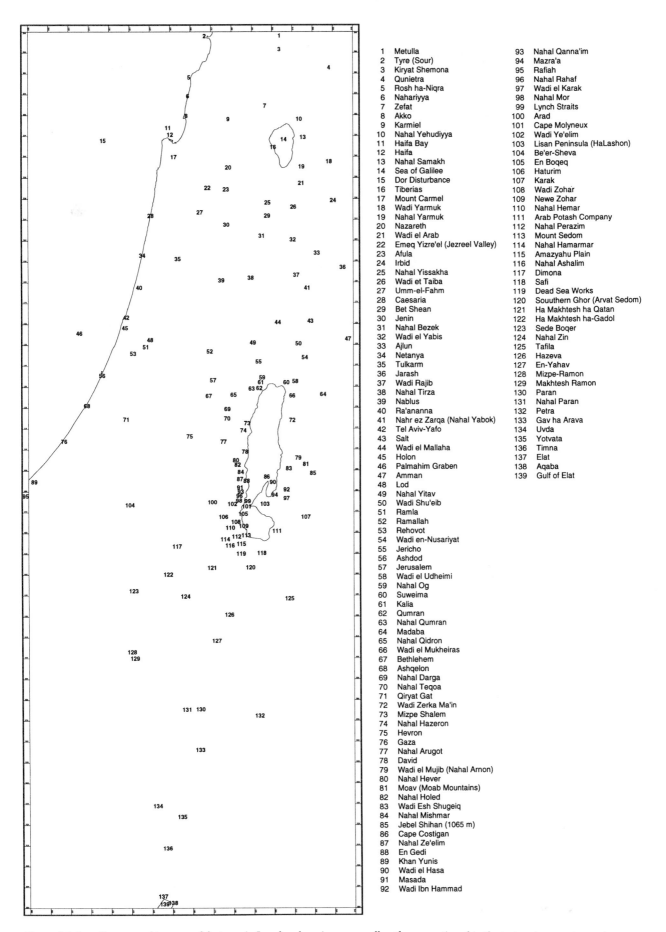

1	Metulla	93	Nahal Qanna'im
2	Tyre (Sour)	94	Mazra'a
3	Kiryat Shemona	95	Rafiah
4	Qunietra	96	Nahal Rahaf
5	Rosh ha-Niqra	97	Wadi el Karak
6	Nahariyya	98	Nahal Mor
7	Zefat	99	Lynch Straits
8	Akko	100	Arad
9	Karmiel	101	Cape Molyneux
10	Nahal Yehudiyya	102	Wadi Ye'elim
11	Haifa Bay	103	Lisan Peninsula (HaLashon)
12	Haifa	104	Be'er-Sheva
13	Nahal Samakh	105	En Boqeq
14	Sea of Galilee	106	Haturim
15	Dor Disturbance	107	Karak
16	Tiberias	108	Wadi Zohar
17	Mount Carmel	109	Newe Zohar
18	Wadi Yarmuk	110	Nahal Hemar
19	Nahal Yarmuk	111	Arab Potash Company
20	Nazareth	112	Nahal Perazim
21	Wadi el Arab	113	Mount Sedom
22	Emeq Yizre'el (Jezreel Valley)	114	Nahal Hamarmar
23	Afula	115	Amazyahu Plain
24	Irbid	116	Nahal Ashalim
25	Nahal Yissakha	117	Dimona
26	Wadi et Taiba	118	Safi
27	Umm-el-Fahm	119	Dead Sea Works
28	Caesaria	120	Soouthern Ghor (Arvat Sedom)
29	Bet Shean	121	Ha Makhtesh ha Qatan
30	Jenin	122	Ha Makhtesh ha-Gadol
31	Nahal Bezek	123	Sede Boqer
32	Wadi el Yabis	124	Nahal Zin
33	Ajlun	125	Tafila
34	Netanya	126	Hazeva
35	Tulkarm	127	En-Yahav
36	Jarash	128	Mizpe-Ramon
37	Wadi Rajib	129	Makhtesh Ramon
38	Nahal Tirza	130	Paran
39	Nablus	131	Nahal Paran
40	Ra'ananna	132	Petra
41	Nahr ez Zarqa (Nahal Yabok)	133	Gav ha Arava
42	Tel Aviv-Yafo	134	Uvda
43	Salt	135	Yotvata
44	Wadi el Mallaha	136	Timna
45	Holon	137	Elat
46	Palmahim Graben	138	Aqaba
47	Amman	139	Gulf of Elat
48	Lod		
49	Nahal Yitav		
50	Wadi Shu'eib		
51	Ramla		
52	Ramallah		
53	Rehovot		
54	Wadi en-Nusariyat		
55	Jericho		
56	Ashdod		
57	Jerusalem		
58	Wadi el Udheimi		
59	Nahal Og		
60	Suweima		
61	Kalia		
62	Qumran		
63	Nahal Qumran		
64	Madaba		
65	Nahal Qidron		
66	Wadi el Mukheiras		
67	Bethlehem		
68	Ashqelon		
69	Nahal Darga		
70	Nahal Teqoa		
71	Qiryat Gat		
72	Wadi Zerka Ma'in		
73	Mizpe Shalem		
74	Nahal Hazeron		
75	Hevron		
76	Gaza		
77	Nahal Arugot		
78	David		
79	Wadi el Mujib (Nahal Arnon)		
80	Nahal Hever		
81	Moav (Moab Mountains)		
82	Nahal Holed		
83	Wadi Esh Shugeiq		
84	Nahal Mishmar		
85	Jebel Shihan (1065 m)		
86	Cape Costigan		
87	Nahal Ze'elim		
88	En Gedi		
89	Khan Yunis		
90	Wadi el Hasa		
91	Masada		
92	Wadi Ibn Hammad		

Figure 2-2 Location map of towns and features in Israel and environs, as well as those mentioned in the text and appearing in the figures.

13

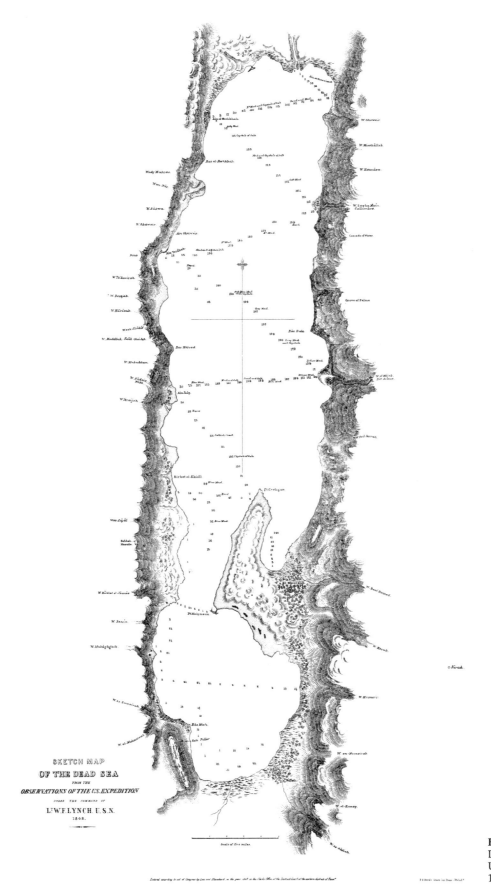

Figure 2-3 Bathymetric charts of the Dead Sea: a, Chart produced by the U.S. Expedition to the Dead Sea in 1848, from Lynch (1848).

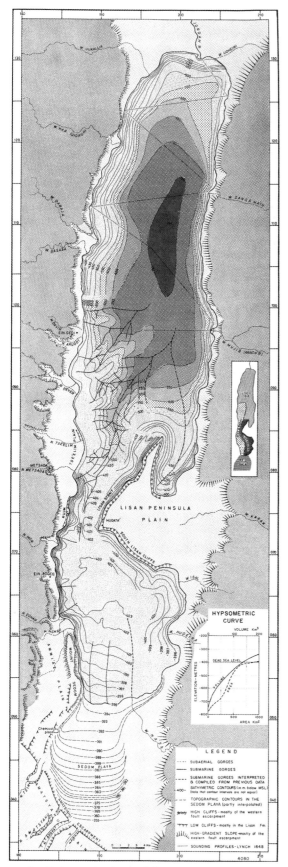

Figure 2-3b Bathymetric chart of the 1959 survey from Neev and Emery (1967). The inset shows the initial hypsometric calculations for the Dead Sea.

19 and August 1, measurements were made along 540 km of track, primarily in the deep northern basin. From these profiles, 5,148 depth soundings were digitized and used with previous data to prepare a revised bathymetric chart (Fig. 2-3b) at a 1:50,000 scale (Hall and Neev, 1978; Hall, 1979). The echo soundings were converted from two-way travel time to depth using an empirical relationship based on wire-measured depths obtained in 1959 at three core locations in deep flat areas. An average sound velocity of 1,770.6 m/sec was used; this velocity is about 18% greater than that of seawater. This work finally showed the general configuration of the Dead Sea to be like a flat-bottomed bathtub, rather than the V-shaped basin suggested by the Lynch soundings (Figs. 2-3a, b).

Results of this survey showed that the soundings of the 1848 Lynch expedition, indicating "abyssal" depths 24 to 27 m (and in one case 68 m) greater than those observed in 1974, were in error. Hall and Neev (1978) attribute these generally consistent differences to the oppressive conditions of observation that existed during the work in 1848, the possibility that currents produced a slant reading, and the likelihood that growth of salt crystals within the linen sounding line, which was premarked with indications of the length paid out, caused the line to shorten as the diameter increased. This shortening would have been compounded with each use. However, after the lines were washed in freshwater at the end of the expedition, this shortening disappeared.

Detailed surveys were carried out in the Lynch Straits to the west of the Lisan Peninsula in 1978 (Nir, 1980) and 1980 (Golik and Adler, 1980) to delineate the areas adjacent to the intakes and outlets of the Dead Sea Works Ltd. (Beyth and Olshina, 1983). Additional surveys were carried out in 1983 and 1984 (Ben-Avraham et al., 1985; Hall and Ben-Avraham, 1985) and in 1985 (Ben-Avraham et al., 1993), in which soundings were derived from high-resolution seismic profiles. The data from these last three surveys have not yet been used to upgrade the bathymetry, because we lack a careful analysis of the cross-over errors when compared with previous data, which would provide a sound velocity for the now overturned Dead Sea. However, only minor changes related to microtopography are anticipated. To date, with the exception of two profiles by Lynch (1849) and the work of de Leeuw in the early 1940s, there is poor coverage for the areas to the east of the northern cape of the Lisan Peninsula, and in the far northeastern corner of the northern basin. I am not aware of any Jordanian survey data.

BATHYMETRIC RESULTS

Figures 2-1, 2-3 a–c, 2-4 a–c, and 2-5 illustrate the bathymetry of the Dead Sea. The southern basin, which originally sloped gently down to central depths of slightly greater than -403 m, is now divided between the evaporation pans of the Dead Sea Works Ltd., and the Arab Potash Co. The pans are bounded by dykes made of the local silty sediment or, more recently, of salt removed from the pans. A channel carries the end-brines from the two extraction plants back to the northern basin.

The northern basin is generally shaped like a flat-bottomed bathtub. However, this relatively flat bottom, which overall exhibits up to 10 m of topography, does not appear to be related to the "salt-mirror" effect postulated by Bloch and Picard (1970). To the north, this basin is bounded by the gently sloping Jordan Delta (Fig. 2-4 a, c); to the east, by the steeply dipping eastern border fault; to the south, by the dissected surface of the Lisan and En Gedi diapirs (Fig. 2-4 a–b); and to the west, by the more gradually sloping diapir-related surface spanning the zone between several western boundary faults. The shaded relief presentation (Fig. 2-1) suggests the alignment of some of these features on the western margin with presumed faults to the

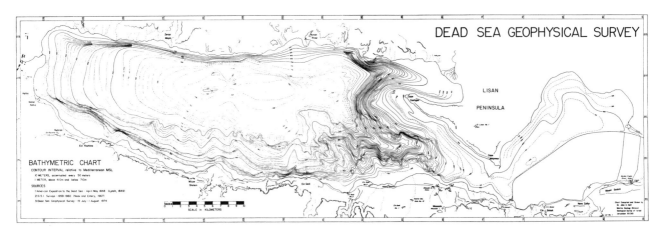

Figure 2-3c Bathymetric chart of the 1974 GSI survey from Hall and Neev (1978) and Hall (1979). Contour interval is 10 m to -720 m, and 1 m below that.

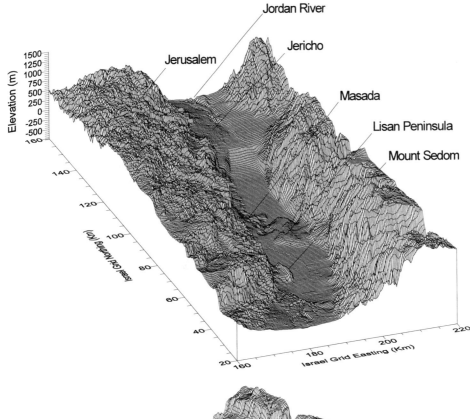

Figure 2-4a Three-dimensional "fishnet" block diagrams of the Dead Sea area, based upon the DTM data: The Dead Sea and environs, with a grid spacing of 300 x 700 m. View is toward N20°E from an elevation of 25°.

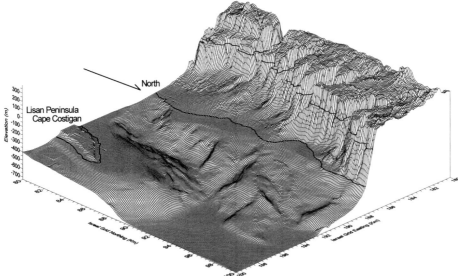

Figure 2-4b Three-dimensional "fishnet" block diagrams of the Dead Sea area, based upon the DTM data: The En Gedi area, at the southern end of the northern basin. Grid spacing is 100 m. View is toward 225° from an elevation of 25°. The -400 m msl shoreline and present Mediterranean sea level contours are indicated. The northward extension (Cape Costigan) of the Lisan Peninsula is shown, and the dissected relief of the offshore En Gedi diapir.

Figure 2-4c Three-dimensional "fishnet" block diagrams of the Dead Sea area, based upon the DTM data: The delta of the Jordan River at the north end of the Dead Sea northern basin. Grid spacing is 100 m. View is toward N25°E from an elevation of 25°. The -400 m msl shoreline and present Mediterranean sea level contours are indicated.

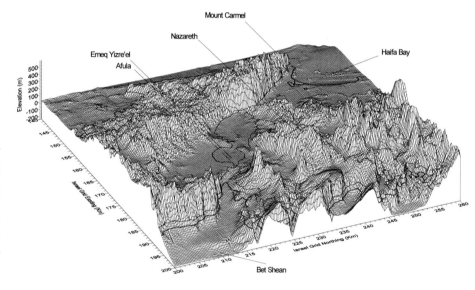

Figure 2-4d Three-dimensional "fishnet" block diagrams of the Dead Sea area, based upon the DTM data: The present area of the spillway of the Dead Sea depression into the Mediterranean. Grid spacing is 300 m. This view toward 295° from an elevation of 30° shows the area between Beth Shean and the Mediterranean. Sea level and the 60.5 m contour of the spillway are indicated.

Figure 2-4e The present area of the spillway of the Dead Sea depression into the Red Sea. This is in the central Arava at Gav Ha-Arava, midway between Lotan and Mishor Menuha, about 75 km north of Elat. Grid spacing is 100 m. The view is toward 170° from an elevation of 20°. The spillway elevation is about 200 m, at 955.9N, 165E. The 200-m contour is plotted on the block diagram. Because of artifacts in the original DTM from the widely spaced contours in the flat valley floor, a new DTM was calculated using Surfer® for Windows' inverse distance squared gridding routine on the more than three hundred thousand contour crossing points available to the DTM project from the Be'er Menuha and Arandal topographic sheets.

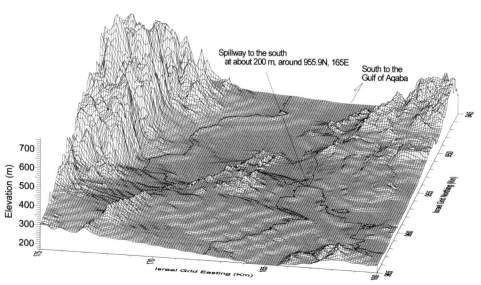

17

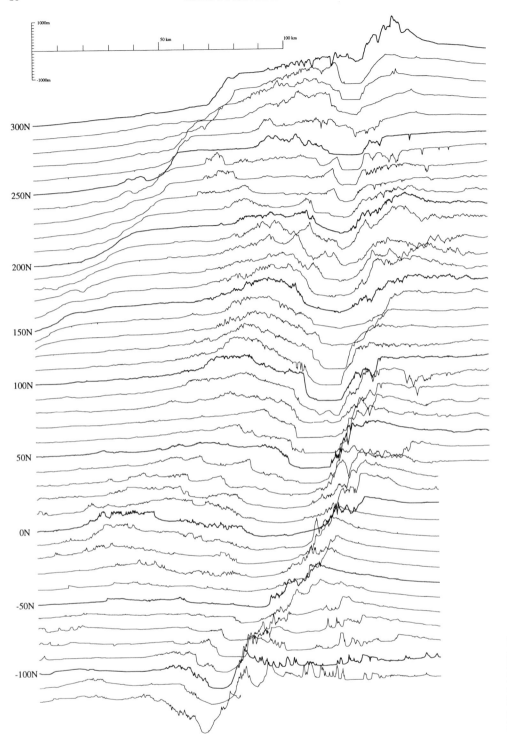

Figure 2-5 Topographic profiles across the Dead Sea depression. Heights are taken from the DTM at 100-m intervals and plotted as west-east profiles every 10 km. Note the asymmetry in elevation and steepness across most of the length of the Dead Sea depression.

north and south of the Dead Sea. The basin is asymmetrical in cross section, which Neev and Hall (1979) attributed to sinking of the underlying salt on the east and upward movement of salt on the west.

The maximum depths are found in two elliptical lows lying 4 km off the convex headland between Wadi Zerqa Ma'in and Wadi Arnon. The northern low (107–110.2N, 197.5–199.5E) is the deepest at slightly more than -730 m, whereas the southern low (98.5–106N, 197.3–199.5E) is between -729 and -730 m. The maximum depth is known only to ±2 m (Hall and Neev, 1978) because of errors in measurement related to rapid changes in the velocity of sound transmission that occurred during the survey. The velocity changes were apparently related to the events that

accompanied the overturn of the once stable water layers in the Dead Sea and were manifested in abrupt changes of up to 10°C at a thermistor suspended deep below a buoy at the south end of the northern basin (S. Serruya, pers. com., 1977). These changes may have been caused by longitudinal sloshing of an internal wave on the distinct density interface.

HYPSOGRAPHY OF THE DEAD SEA DEPRESSION AND ENVIRONS

Frontispiece B and Figure 2-1 are representations of the relief of Israel and its environs, including the Dead Sea depression from Metulla in the north to Elat on the Red Sea. Almost 3,600 m of topography are present, from the deep Mediterranean offshore to the heights on the Jordanian plateau in the south. The area encloses 67,200 km² (160 x 420 km). Frontispiece B is a colored hypsometric map using 72 colors that change with every 25 m in elevation. This type of representation is useful for comparing topography, for determining spot heights, and for seeing structural lineaments related to block movements. Figure 2-1 is the shaded relief image of the same area. The image was produced by assuming the sun to be in the northwest at an altitude of 30° and by calculating the reflectance from the more than 107 million surface elements in the DTM. Lambert's Law of diffuse reflection, which states that the reflected amplitude is proportional to the square of the sine of the angle of incidence, is assumed to apply. The natural scale of this image is about 1:1.785 million.

A number of interesting features may be noted. The western border fault is clearly seen as a straight line extending south of the isolated Mount Sedom through the deep northern basin across the Jericho plain along the Jordan River. Minor faults and lineaments paralleling this trend are visible on the plateaus both east and west of the Dead Sea itself, and even away from the rift in the northwestern corner west of Jerusalem. Numerous other linears and curvilinears can be identified. The faults and linears appear clearly, sometimes extending well beyond the limits usually attributed to them. Of speculative interest are indications of onshore extensions, even crossing the Dead Sea depression, of the bounding faults of the offshore Palmachim graben.

Hypsometric calculations for the Dead Sea Depression

Hypsometric curves giving the volume and surface area of the Dead Sea as a function of elevation were prepared following the surveys of Neev and Emery (1967) and Hall and Neev (1978). With the exception of a few point area-volume calculations for the Pleistocene Lake Lisan by Begin et al. (1974), I know of no curves that have been prepared for the entire enclosed basin of the Dead Sea depression including the Jordan Valley and Wadi Arava. This is interesting in light of the numerous proposed schemes to build a canal joining the Dead Sea with the Red Sea or the Mediterranean Sea (Vardi, 1990).

The hypsometric curves of Neev and Emery (1967) and Hall and Neev (1978) were made by computing the area within the bathymetric contours and then integrating these areas over the depth range. The existence of the new DTM permits a more accurate calculation to be made over the whole area. For each of the 63 topographic sheets within the watershed, a file was prepared giving the number of 25 m square facets whose center height lies within each meter of elevation. For one of these sheets (Nazareth) adjacent to the watershed spillway near Afula (Fig. 2-4d), it was necessary to block out the lower areas west of the spillway to prevent their being included in the tabulation.

The hypsometric curves for the Dead Sea depression giving the surface area and volume of the lake that would be formed by filling the depression, meter by meter, is shown in Figure 2-6. The solid line is the area, and the dashed line is the volume. The maximum area and volume at the 60.5-m height of the Afula spillway are 5,984.7 km² and 1,602.2 km³, respectively. This volume is more than 10 times the 147 km³ volume of the 1974 Dead Sea, and the area is 27% of the area of Israel (21,501 km; Wright, 1989). By comparison, this volume is only 8% of the estimated potential sea-filling volume (18,200 km³) of the Qattara depression (Newman and Fairbridge, 1987). Inflection points in the area curve occur at the -715 m transition from the sea's flat bottom to the steep sides, at the -385 m transition from the bathtub-like northern basin to the surrounding plain and southern basin, at -228 m with the step-up to the Kinneret Basin, and at the Mediterranean Sea level.

Figure 2-7 shows the area present at each meter of elevation from -732 m to 60.5 at the Afula spillway. This curve exhibits

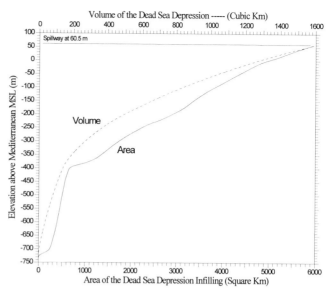

Figure 2-6 Hypsometric curve for the Dead Sea depression showing the surface area and volume of the water body that would be produced as the Dead Sea depression was filled up, meter by meter, to the Mediterranean spillway at 60.5-m elevation. The area and volume of this lake were 5,984.7 km² and 1,602.2 km³ respectively, the present area and volume of the Dead Sea at -409 m is approximately 638 km² and 132.6 km³.

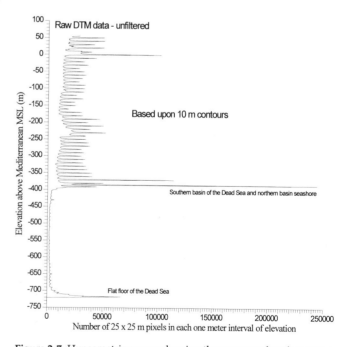

Figure 2-7 Hypsometric curves showing the amount of surface area as a function of elevation for the enclosed Dead Sea depression. The number of 25-x-25-m surface elements (pixels or picture elements) at each meter of elevation taken from the DTM. The spikes every 10 meters are a byproduct of the interpolation process in which bicubic splines are fit through the original 10-m contour data and then used to interpolate elevations every 25 m. Apparently, the splines preferentially flatten out as they pass through the contours, producing more surface area in their vicinity.

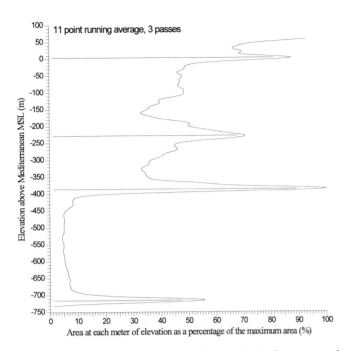

Figure 2-8 The same curve as shown in Figure 2.7 after three passes of a filter that averages the data over the 5 m above and below each 1-m level. Note the peaks at -715, -385, - 228, - 100 and +4 m msl.

two interesting characteristics. The first is pronounced maxima at the previously mentioned sites of the inflection points in the area curve. The second characteristic is secondary peaks every 10 m. The secondary peaks are apparently artifacts of the process by which DTM heights are interpolated. At each grid point in the DTM (every 25 m), an average height is computed from bicubic spline profiles passing both north-south and east-west through the original 10-m topographic contours, weighted inversely according to the distance to the nearest contour. The splines tend to flatten out as the contours are transited, thus producing more 25-x-25-m facets that are close to the 10-m contour values. Such "banding" has long been observed in slope maps and shaded relief displays of this and other DTM data (see Fig. 2-4e). However, since these measurements are conservative, that is, the number of facets exactly equals the area, it is possible to smooth out these secondary peaks by a running average of 11 heights extending over a span of 10 m.

Figure 2-8 shows the result after three passes of a smoothing filter have removed the peaks. The pronounced peak at +4 m around the Mediterranean mean sea level is of interest. This indicates a relatively large area at an elevation around sea level, which is a result either of the continental block isostatically preferring that level or of the depression having contained a relatively stable water body that eroded away material from above and deposited it in the depths below, thus producing a wide band about sea level. This curve bolsters the observation of various scientists (K. O. Emery and Yehuda Karmon, pers. com., 1971, 1993) that relict beaches at this level exist away from the present influence of the Mediterranean. These beaches presumably belonged to the lakes that have periodically occupied the depression since the Upper Pliocene (2 Ma) and during the more recent pluvial phases at about 20,000, 15,000, 9,000, and 4,000 yr B.P. (Neev and Hall, 1977). However, why this level should be just here, and not above or below, remains at present an intriguing question. A smaller and broader inflection is also observed at -100 m.

Acknowledgments

This work was carried out under Project 20743 (MEDMAP-DTM) of the Geological Survey of Israel, Ministry of Energy and Infrastructure. The support and encouragement of the Geological Survey directorate, the Dead Sea Works, Ltd., the Survey of Israel, and especially of Historical Productions Inc., is gratefully acknowledged. I thank David Neev for several historical insights and two anonymous reviewers for their critical comments.

Historical Productions (now Rohr Productions) has produced the following laminated natural-color high-resolution satellite images. Many are available from Shmuel Tal Publishing, 13 Rosanis Street, Tel Baruch, Tel Aviv 69 018, Tel: 972-3-649-9146, FAX 972-2-648-5655.

(1) 1988, Pictorial Archive LANDSAT 5 Satellite Map, 6 sheets with scale 1:150,000, LANDSAT TM data only)

(2) 1988, Pictorial Archive LANDSAT 5 Satellite Map, Levant: Ancient and Modern, Sheet 4 with ancient and modern names at scale 1:150,000, LANDSAT TM data only)

(3) 1988, Historical Productions - Untitled (Jerusalem and Jericho Area), Scale 1:37,500

(4) 1989, Historical Productions - Haifa & Mt. Carmel, Scale 1:40,000

(5) 1989, Historical Productions - Haifa & Mt. Carmel, Scale 1:25,000

(6) 1989, Historical Productions - Haifa & Mt. Carmel, Scale 1:60,000

(7) 1989, Historical Productions - STM (SPOT-LANDSAT TM Merge) Satellite Map of the Sea of Galilee and Upper Jordan Valley, Scale 1:100,000. Cassini-Soldner (Israel Grid) Projection

(8) 1989, Historical Productions - Hesban & the Madaba Plain, Scale 1:50,000

(9) 1990, Historical Productions - Jerusalem and Environs, Scale 1:50,000. Bible Lands Exhibit - Historical Geography of the Bible Lands

(10) 1990, Historical Productions - Judea and the Dead Sea, Scale 1:100,000

(11) 1990, Educational Programs, Inc., Israel Satellite Map, Scale 1:322,500

(12) 1991, Historical Productions - Satellite Map - Tel Aviv Area, Scale 1:50,000

(13) 1991, Historical Productions - Moab Archeological Survey, Scale 1:100,000 Satellite Image

(14) 1992, Historical Productions - Satellite map of Haifa Area, Scale 1:60,000. Includes place names in Hebrew.

REFERENCES

Abramowitz, L., 1978, Dead Sea explorers: *The Jerusalem Post Magazine*, Friday, October 6, 1978, p. 10–11.

Arkin, Y., compiler, 1982, Mediterranean-Dead Sea Project, Outline and Appraisal 1982: Mediterranean-Dead Sea Co., Ltd., Jerusalem, April 1982, 45 p.

Begin, Z. B., Ehrlich, A., and Natan Y., 1974, Lake Lisan: The Pleistocene precursor of the Dead Sea: Jerusalem, Geological Survey of Israel Bulletin 63, 30 p.

Ben-Avraham, Z., Garfunkel, Z., Almagor, G., and Hall, J. K., 1979, Continental Breakup by a Leaky Transform: The Gulf of Elat (Aqaba): *Science*, v. 206, p. 214–216.

Ben-Avraham, Z., Levy, Y., Hall, J. K., and Neev, D., 1985, The structure and composition of the uppermost layers in the Dead Sea: Tel Aviv University, Department of Geophysics and Planetary Sciences Report MGL TAU 1/85, 12 p., 15 figures.

Ben-Avraham, Z., Niemi, T. M., Neev, D., Hall, J. K., and Levy, Y., 1993, Distribution of Holocene sediments and neotectonics in the deep North Basin of the Dead Sea: *Marine Geology*, v. 113, p. 219–231.

Beyth, M., and Olshina, A., 1983, Mixing of end brines in the Lynch Straits, Dead Sea, 29 August–1 Sept. 1983: Israel Geological Survey Report MGG/7/83, Jerusalem, December 1983, 6 p.

Bloch, R., and Picard, L,. 1970, The Dead Sea—A sinkhole?, *in* Sonderheft Hydrogeologie und Hudrogeochemie: Stuttgart, Verlag Ferdinand Enke, *Zeitschrift der Deutschen Geologischen Gesellschaft*, p. 119–128.

Cleave, R. L. W. C., 1994, The Holy Land satellite atlas: Student map manual illustrated supplement, volume 1: POB 3312, Nocosia, Cyprus, Rohr Production, Ltd., 273 p.

Cleave, R. L. W. C., 1995, Satellite Revelations: New Views of the Holy Land: *National Geographic Magazine*, v. 187, no. 6, p. 88-105.

Eriksen, E. O., 1985, Dead Sea Explorer: *The Jerusalem Post*, Sept. 4, 1985.

Golik, A., and Adler, E., 1980, Bathymetric Chart—Dead Sea: Israel Oceanographic and Limnological Research Ltd., November 1980. Bathymetric chart of the area north of the Lynch Straits between 79N and 85N (Israel Grid), 1-m contour interval, original scale 1:10,000. Surveyed 10–11 November 1980.

Hall, J. K., 1979, Bathymetric Chart of the Dead Sea; prepared for the International Symposium on Rift Zones of the Earth: The Dead Sea: Jerusalem, September 10–20, 1979: One Sheet, 10- and 1-m contours, two colors, scale 1:100,000. Track chart of the survey is printed on the back.

Hall, J. K., 1984, LANDSAT Imagery Catalog for the Dead Sea and Arava (Including a primer on Remote Sensing for this area): Geological Survey of Israel Report GSI/8/84, Jerusalem, 86 p.

Hall, J. K., 1993, The GSI Digital Terrain Model (DTM) project completed: in Bogoch, R., and Eshet, Y., eds., *Geological Survey of Israel Current Research*, v. 8, p. 47–50.

Hall, J. K., and Ben-Avraham, Z., 1985, The Dead Sea shallow seismic survey of 5–9 June, 1983, *in* Bogoch, R., ed., *Geological Survey of Israel Current Research*, v. 5, p. 79–81.

Hall, J. K., and Neev, D., 1978, The Dead Sea Geophysical Survey 19 July–1 August 1974; Final Report No. 1 (previously MGD 2/75): Methods, Navigation, Bathymetry and Magnetics: Jerusalem, Geological Survey of Israel Marine Geology Division Report No. MG/1/78, 28 p.

Hall, J. K., Schwartz, E., and Cleave, R. L. W., 1990, The Israeli DTM (Digital Terrain Map) Project, *in* Hanley, J. T., and Merriam, D. F., eds., *Microcomputer Applications in Geology, II*: Oxford, England, Pergamon Press, p. 111–118.

Historical Publications, 1991a, *Iso Raamatun Tietosanakirja*, Ris-

tin Voitto ry., Raamatun Tietosanakirja, Osa 10, Atlas 2, 240 p. (in Finnish).

Historical Publications, 1991b, *Illustrerat Bibellexikon Bildatlas*, Studiebibelen a/s, Ski 1991, 128 p. (in Norwegian).

Lynch, W. F., 1849, *Narrative of the United States' Expedition to the River Jordan and the Dead Sea*: Philadelphia, Pennsylvania, Lea and Blanchard, 508 p.

Moore, H., and Beke, W. G., 1837. On the Dead Sea and some positions in Syria: *Royal Geographical Journal*, v. 7, p. 45–46.

Neev, D., 1964, The Dead Sea: Geological Survey of Israel, Quaternary and Recent Geology Division, Report Q/2/64, January 1964, 407 p. plus appendices and Hebrew abstract.

Neev, D., and Emery, K. O., 1967, The Dead Sea, depositional processes and environments of evaporites: Jerusalem, Geological Survey of Israel Bulletin 41, 147 p.

Neev, D., and Hall, J. K., 1976, The Dead Sea Geophysical Survey 19 July–1 August 1974. Final Report No. 2: Jerusalem, Geological Survey of Israel, Marine Geology Division Report No. 6/76, 21 p.

Neev, D., and Hall, J. K., 1977, Climatic fluctuations during the Holocene as reflected by the Dead Sea levels, *in* Greer, D. C., ed., *Deserti Terminal Lakes*, Proceedings from the International Conference on Desertic Terminal Lakes held at Weber State College, Ogden, Utah, May 2–5, 1977, Utah Water Research Laboratory, Utah State University, Logan, Utah, September 1977, p. 53–60.

Neev, D., and Hall, J. K., 1978, The Dead Sea Geophysical Survey 19 July–1 August 1974: Final Report No. 2 (previously MGD 6/76) Seismic Results and Interpretation; Jerusalem, Geological Survey of Israel, Marine Geology Division Report No. MG/1/78, 21 p.

Neev, D., and Hall, J. K., 1979, Geophysical Investigations in the Dead Sea: *Sedimentary Geology*, v. 23, p. 209–238.

Newman, W. S., and Fairbridge, R. W., 1987, Project Noah; regulating modern sealevel rise; Phase II; Jerusalem underground, *in* Carter, R. W., G., and Devoy, R. J. N., eds., *Progress in Oceanography* 18, Oxford, Pergamon Press, p. 61-78.

Nir, Y., 1980, The Dead Sea Lynch Straits Bathymetric Survey—6–11 August 1978: Geological Survey of Israel Report MG/1/80, 6 p., Appendix

Quennell, A. M., 1987, Rift valleys, *in* Seyfert, C. K., ed., *The encyclopedia of structural geology and plate tectonics*, Vol. X, Encyclopedia of Earth Sciences Series, Series Editor, R. W. Fairbridge: New York, Van Nostrand Reinhold Co., p. 671–688.

Vardi, J., 1990, Historical review: Jerusalem, Geological Survey of Israel Report GSI/9/90, p. 31–50.

Wright, J. W., ed., 1989. *The Universal Almanac 1990*: Kansas City, Missouri, Andrews and McMeel, 600 p.

3. GEOPHYSICAL FRAMEWORK OF THE DEAD SEA: STRUCTURE AND TECTONICS

Zvi Ben-Avraham

Several geophysical studies of the Dead Sea basin (Fig. 3-1) have been conducted over the past 35 years. Geophysical data for the southern part of the basin were obtained mainly by oil companies. The northern part of the basin was studied by marine geophysical techniques. The purpose of this chapter is to describe the internal structure of the Dead Sea basin, with special emphasis on the northern part of the basin in light of recent geophysical studies. In particular, the mode of deformation of the basin is discussed and compared with current models of the formation of strike-slip sedimentary basins.

BATHYMETRY

The first bathymetric survey of the Dead Sea was carried out as early as 1848 by a United States Navy expedition under the command of Lt. W. F. Lynch (1849). This survey collected much information about the shape of the Dead Sea and showed the northern part to be deep and the southern part to be shallow. The results of this survey defined the bathymetry of most of the lake for over a century. Neev and Emery (1967) were the first to conduct a modern marine geology investigation using direct sampling techniques and echo soundings but due to political borders at that time their work was confined to the southwestern quadrant of the Dead Sea.

The most detailed bathymetric survey of the northern part of the basin was carried out in 1974 by Neev and Hall (1979) as part of their geophysical study. The depth soundings were made with an Elac portable echo sounder. The records were digitized at all slope changes, and the digitized soundings were used with all previous soundings to prepare a new bathymetric map. Further depth measurements were made in 1983 and 1984 (Ben-Avraham et al., 1985; Hall and Ben-Avraham, 1985), as well as in conjunction with the gravity survey (ten Brink et al., 1993), but they were not incorporated into the existing bathymetric map of Neev and Hall (1979). A detailed description of the bathymetry of the Dead Sea basin and adjacent areas is given elsewhere in this volume (Hall, chapter 2).

The bathymetric map (Fig. 3-2) shows that nearly half of the area of the northern part of the basin is almost flat-bottomed. The maximum bottom elevation of the sea is 730 m below mean sea level. The maximum depths are found along a depression that trends north–south adjacent to the eastern margin. The bathymetric depression appears to be divided into two parts. The Jordan River delta is relatively smooth and planar, sloping 3° to an elevation of 650 m and 1° thereafter to the deep basin. Subsequent surveys have shown that the Jordan River delta is dissected by numerous channels and that the sediments are highly deformed by slumping (Niemi and Ben-Avraham, 1994; chapter 6, this volume). Overall, the bathymetry, especially of the deep northern part of the basin, is fault-controlled, as indicated by the steep slopes.

A large difference in the slopes on either side of the northern part of the basin exists, resulting in the basin's asymmetric shape. The western slopes average 7° and the eastern slopes average 30°, except in the vicinity of the Arnon River delta, where the slopes are around 8° (Neev and Hall, 1979).

CRUSTAL STRUCTURE FROM SEISMIC REFRACTION EXPERIMENTS

A deep seismic refraction experiment conducted in 1977 (Ginzburg et al., 1979a, 1979b; 1981; Perathoner et al., 1981) provided some information about the deep structure of the crust along the Dead Sea rift and adjacent areas to the west. A few years later, a seismic refraction experiment was conducted in the area east of the Dead Sea rift (El Isa et al., 1987; El Isa, 1990).

The refraction profile along the rift was about 600 km long from the Sea of Galilee to the southern part of the Gulf of Elat. Initially, the seismic data were evaluated using first-arrival information (Ginzburg et al., 1979a); then later-arrival information was used to obtain a detailed velocity structure of the crust underneath the rift and adjacent areas (Ginzburg et al., 1979b).

Information on the southern part of the basin of the Dead Sea was obtained by a reversed refraction profile and on the northern part of the basin by an unreversed refraction profile. Data for the Dead Sea region were gathered from shot points, which were located in the northern part of the basin at a 50-m water depth and in the Arava Valley south of the Dead Sea. The seismic signals were received by Mars-66 mobile seismic stations. In the Dead Sea area, the stations were located on land on the west margin of the basin. The shots in the Dead Sea provided exceptionally good energy propagation (Ginzburg et al., 1981). The structural models computed from the record sections were plotted on a longitudinal section (Fig. 3-3). The relative Bouguer anomaly is shown on the same section. The results indicate that the rift and its immediate surroundings are underlain by a thinner than usual crust, about 30 km compared with 40 km in the Negev at the southern part of Israel. The rift is characterized by the presence of a velocity transition zone just above the crust–upper mantle discontinuity (Moho). In the areas adjacent to the rift, the crust–upper mantle boundary is a sharp velocity discontinuity.

Modeling of a regional gravity profile across the Dead Sea rift just north of the Dead Sea suggests that two different crustal blocks exist on both sides of the rift (ten Brink et al., 1990), indicating an offset in the Moho. Because the seismic stations were located on the western margin of the Dead Sea basin, the depth to crystalline basement under the basin is unknown. To obtain this information, a seismic refraction experiment along the axis of the northern part of the basin of the Dead Sea was recently conducted (Ginzburg et al., 1996).

STRUCTURE OF THE DEAD SEA FROM GRAVITY ANALYSES

Marine gravity data along 300 km were collected in the northern submerged part of the Dead Sea basin in 1988 from a small utility vessel, using the Bell Aerospace BGM-3 marine gravity system (ten Brink et al., 1993). This was the first time that continuous gravity measurements were carried out in a lake with a modern marine gravimeter. The marine gravity data were integrated with land gravity data from Israel and Jordan to extend to 30 km on either side of the basin (Fig. 3-4). Free-air, Bouguer,

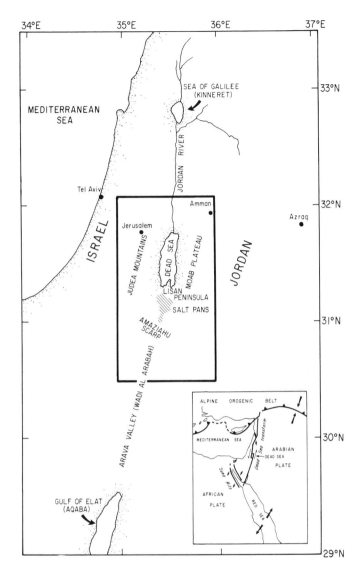

Figure 3-1 Location map of the Dead Sea basin (modified from ten-Brink et al., 1993). Rectangle in center indicates the boundaries of Figure 3-4.

blocks were interpreted by ten Brink et al., (1993) to represent a passive collapse of part of the wider transform valley into the deepening graben. The collapse was probably facilitated by movement along the normal faults that bound the transform valley. The gravity data also suggest that the Moho is not significantly elevated under the basin and that the deformation associated with the formation of the basin is confined to the crust.

MAGNETIC STUDIES

Several studies of the magnetic field of the Dead Sea basin were carried out. The results of these studies are important, especially for learning about the existence of strike-slip faulting along the basin margins, the subbottom structure, and the possible existence of subbottom and subsurface basaltic flows in the basin.

A magnetic survey over a portion of the northern part of the basin was conducted in 1974 (Neev and Hall, 1979) using a varian proton precession magnetometer. In 1983, a detailed marine magnetic study was carried out in the northern part of the basin (Frieslander and Ben-Avraham, 1989). The magnetic measurements were taken along about 900 km of track lines. The lines were spaced 1 km east–west and 1.5 km north–south. The magnetometer, a G-866 EG&G Geometrix, had an accuracy

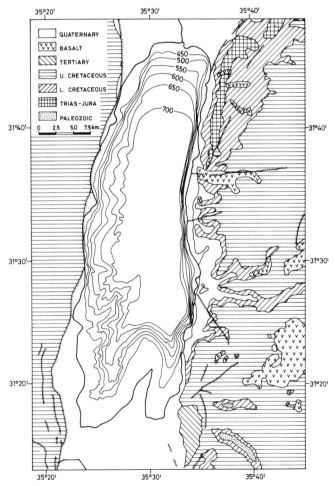

Figure 3-2 Simplified geological map of the land areas surrounding northern part of the Dead Sea (after Frieslander and Ben-Avraham, 1989). Bathymetry is shown (after Neev and Hall, 1979).

and horizontal first-derivative of the Bouguer anomaly maps of the area were prepared and analyzed. Two-dimensional gravity models along four east-west profiles across the basin (Fig. 3-5) were calculated and compared with free-air gravity profiles. The models used all the available geological and geophysical information, including bathymetry, location of faults in surface geology, seismic reflection data, and the depth to crystalline basement west and east of the basin (from seismic refraction data).

The analysis and modeling of the gravity data helped to describe the geometry of the entire Dead Sea basin (ten Brink et al., 1993). The basin is 132 km long, 7–18 km wide, and up to 10 km deep. The basin becomes narrower and shallower toward the northern and southern ends. The Bouguer anomaly along the axis of the Dead Sea basin decreases gradually from both the northern and southern ends, suggesting that the basin sags toward its deepest part in the center. The basin consists of a 7- to 10-km-wide graben, which occupies the eastern part along most of its length. Tilted fault blocks, several kilometers wide, are located along the western side of the basin. These

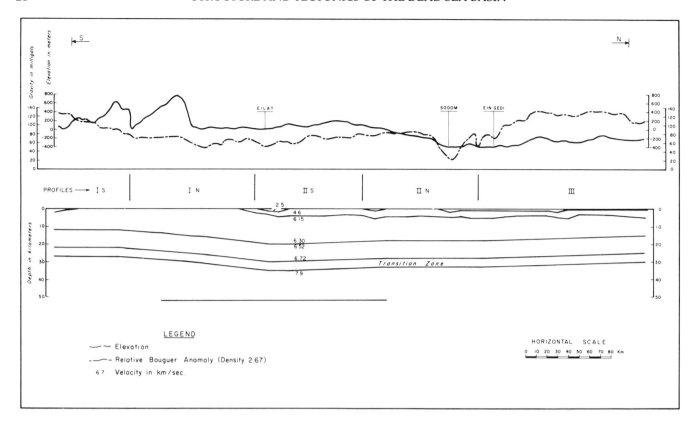

Figure 3-3 A section showing the structure of the crust, topography, and relative Bouguer gravity along the Dead Sea rift from the Sea of Galilee to the southern tip of the Gulf of Elat (after Ginzburg et al., 1981). The crustal structure was obtained from a seismic refraction profile along the rift. The location of Mount Sedom in the southern part of the Dead Sea basin and En Gedi in the northern part of the basin are marked on the topographic profile.

of 0.1 nT. An aeromagnetic survey with a dense grid was conducted in 1987 in the area west of the Lisan Peninsula between the north and south parts of the basin (Ram, 1989). The spacing between north-south lines was 100 m and between east-west lines, 1 km. Total length of the lines was 1,000 km. Measurements were taken with a Scintrex 5 Map magnetometer and logged on a Scintrex Gam 2001, both mounted on a Bell Jet Ranger helicopter. Another aeromagnetic survey was made in 1991 over the southern part of the basin by the Israel National Oil company. Earlier aeromagnetic measurements were made over the land areas west (Folkman, 1976, 1981) and east (Hatcher et al., 1981) of the Dead Sea basin. To the west, measurements were taken at an average height of 1,400 m above the Dead Sea level, which is about 300 m above the terrain. In the east, measurements were made at an average height of 1,900 m above the Dead Sea level, or about 500 m above the terrain.

The magnetic anomaly map of the northern part of the Dead Sea basin is smooth with a few isolated anomalies (Frieslander and Ben-Avraham, 1989). It can be divided into two distinct parts at 31°31'N (Fig. 3-6a). North of this latitude, the field is smooth and trends north–south and northwest (north of lat. 31°40'). South of lat. 31°31'N, the magnetic field is less smooth and very different from the field in the north. Most of the smooth north–south magnetic contours change their trend south of lat. 31°31'N to northwest, east–west, and northeast, intersecting the coast at a high angle. Generally, the magnetic contours follow the bathymetry in this area. In the central part of the basin, the magnetic contours have similar trend (Ram, 1989).

Isolated short-wavelength magnetic anomalies, some with high amplitude, are located mostly along the basin margins. These anomalies indicate the presence of small basaltic bodies buried at shallow depths. Several such anomalies also exist west of the Lisan Peninsula and in the northern portion of the southern part of the basin (Ram, 1989). The most pronounced isolated anomaly (253 nT) is a positive anomaly located on the eastern margin of the northern part of the basin about 2 km south of Wadi Zarqa Main, where young basaltic flows, 3.7 ±0.4 Ma (Barbari et al., 1979) exist. An aeromagnetic survey of Wadi Zarqa Main area (Kovach et al., 1990) indicate that the basaltic flows on land are associated with magnetic anomalies of about 3,500 nT. The anomaly at sea probably originates from the submarine continuation of these flows.

A comparison between the magnetic anomaly map of the Dead Sea and the aeromagnetic maps over the land areas to the west and east show that the magnetic anomalies extend uninterrupted from the land area on the west into the basin (Frieslander and Ben-Avraham, 1989; Ram, 1989). This configuration suggests that faulting on the western margin of the northern part of the basin has been mostly normal. In comparison, the magnetic contours are discontinuous across the eastern margin of the basin, which suggests that the eastern fault has been mainly strike-slip and caused major lithologic changes across the transform (Fig. 3-6b). Although the magnetic data suggest that a transverse fault marks the south margin of the northern part of the basin, there is no evidence of a fault on the north margin of the lake.

Modeling of the magnetic data suggests that the sedimentary fill in the northern part of the basin of the Dead Sea may reach 10.5 km (Frieslander and Ben-Avraham, 1989). The model also suggests that the basement under the northern part of the basin

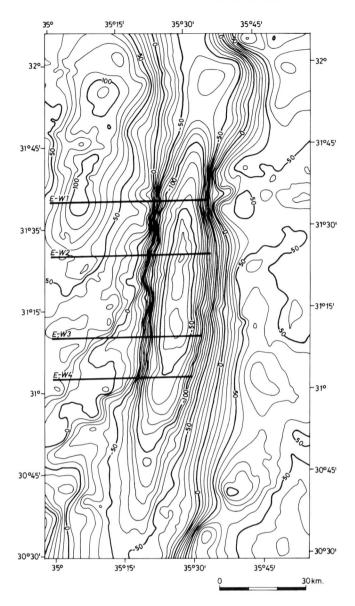

Figure 3-4 Free-air gravity anomaly map of the Dead Sea basin and its vicinity (after ten Brink et al., 1993). Contour interval is 10 mGal. Heavy lines are locations of four east-west gravity profiles (Fig. 3-5).

and brown markers, respectively. Neev and Hall (1979) suggested that the red marker represents the beginning of the present Dead Sea and is of base Holocene age. The blue and the green markers are within the Lisan Formation and Amora Formation of late to middle Pleistocene age, respectively. The deepest reflector, the brown marker, was assumed to represent the top of the Sedom Formation, which was composed mainly of rock salt of presumed Pliocene to early Pleistocene age (Fig. 3-8).

Based on the sparker profiles, Neev and Hall (1979) proposed that, prior to the beginning of the present Dead Sea, very intensive deposition of sediments took place in the central areas of the northern part of the basin. In addition, intensive erosion dissected the basin flanks. Lake Lisan, the predecessor of the present Dead Sea, was almost completely desiccated, thus marking the end of the Lisan Period. Neev and Hall (1979) suggest that because of the shrinkage of Lake Lisan, most of the dissolved salts were precipitated in the lowest part of the basin. Additional amounts of sediments were eroded at the same time and transported to this part of the lake. According to Neev and Hall (1979), the sediments of the Jordan River delta and elsewhere in the basin above the red marker were accumulated and deposited during the younger fluvial period that followed.

Diapiric structures, probably formed by the upward movement of rock salt of the Sedom Formation, were identified by Neev and Hall (1979) in the northern part of the basin. These structures were not sampled, thus their composition is not known yet. The diapirs are mainly concentrated along the major western border faults. Similar structures, although much smaller and fewer, may also exist along the eastern border fault. However, coverage is limited on the east so relative abundance is not known. Neev and Hall (1979) named two large diapiric structures that have pierced through the bottom, the En Gedi diapir and Jordan River delta diapir (Fig. 3-9). The diapirism processes have been active since the deposition of the Amora Formation and especially during the deposition of the upper Pleistocene Lisan Formation.

The sparker data provided important information about active faulting in the northern part of the basin (Fig. 3-9). On the east, a major north- to south-trending linear border fault exists. Neev and Hall (1979) interpreted secondary faults that branch off to the north-northeast. On the western margin, several border faults form a system of step faults. The trends of the individual faults are less regular. Neev and Hall (1979) also suggested that two west- to northwest-trending faults border the Lisan Peninsula on its north and south, but they could not detect a transverse west-northwest fault beneath the Jordan Delta at the northern edge of the northern part of the basin.

is probably of similar composition to that in the west, but dissimilar to that in the east (Fig. 3-7), indicating that the two sides of the Dead Sea are displaced relative to one another.

SEISMIC REFLECTION IN THE NORTHERN SUBMARINE PART OF THE BASIN

Continuous seismic reflection profiles were obtained in the northern part of the Dead Sea basin in 1974 (Neev and Hall, 1979) using a single-channel system made of a 1,000-joule EG&G sparker sound source and an eel-type 150-element hydrophone array. A few lines were also obtained in the Lynch Straits and along the northern edge of the southern part of the basin. Maximum penetration was approximately two thirds of a second, mainly in the deepest area of the northern part of the basin. Four prominent reflectors identified by Neev and Hall (1979) were named, from top to bottom, the red, blue, green,

SEISMIC REFLECTION IN THE SOUTHERN SUBAERIAL PART OF THE BASIN

Comprehensive reviews of the structure and tectonic processes of the southern part of the basin are given elsewhere in this volume (Garfunkel, chapter 4; Gardosh et al., chapter 5). They are based on commercial multichannel seismic reflection and drillhole data. Previous studies of the southern part of the basin, which were based on similar but older data, include those of Arbenz (1984), Manspeizer (1985), Kashai and Croker (1987) and ten Brink and Ben-Avraham (1989). These studies suggest various models for the structure of the Dead Sea basin.

According to ten Brink and Ben-Avraham (1989), deformation in the southern part of the basin takes place mainly on the transverse and longitudinal faults, and the intervening sediments are relatively undeformed (Figs. 3-10, 3-11). East-west

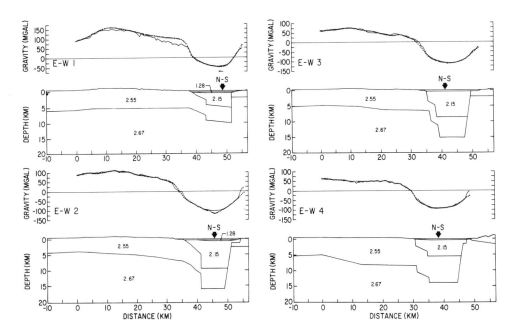

Figure 3-5 Four east-west free-air gravity profiles (dotted lines) compared with calculated gravity (solid lines) profiles from two-dimensional density-depth model (below the profiles). Density in $1 \times 10^3 \, kg/m^3$. Distance is measured east of longitude 35°00'E (after ten Brink et al., 1993). See Figure 3-4 for location of profiles.

cross sections show a full-graben geometry of flat horizontal strata of sediment infill. Whether the assumption that this is a full graben is correct depends on information, as yet unavailable, about the easternmost part of the graben. A complete cross section of the northern part of the basin suggests that the top sedimentary section has indeed a full-graben geometry. Neev and Hall (1979) and ten Brink and Ben-Avraham (1989) have shown that sedimentary strata in the northern and the southern parts of the basin are also horizontal along their long axis. There is only one exceptional area in the northern part of the basin, named the Arnon Sink, in which localized differential subsidence has occurred. These findings are in opposition to the hypothesis of Arbenz (1984), who proposed that the Dead Sea basin was formed by north–south extension on a low-angle detachment dipping to the north and leveling off at a depth of 15 km. The seismic reflection data, which show no tilting of the top sediments, do not support this model.

The seismic reflection profiles have shown that the southern part of the basin is divided into equidistant segments, 20–30 km long and 7–10 km wide, that are bounded by transverse faults (ten Brink and Ben-Avraham, 1989). This division can also be extended northward to include the Lisan segment, which as mentioned previously, is bordered by faults on both its northern and southern sides (Neev and Hall, 1979). The transverse faults, Iddan, Amaziahu, and Boqeq, are shown in the section (Fig. 3-11) along the long axis of the southern part of the basin.

Correlation of drill-hole data with seismic lines suggests that the Pliocene salt layers of the Sedom Formation underlie the Pleistocene sedimentary infill in the deep level of the southern part of the basin (Ginzburg and Kashai, 1981; Kashai and Croker, 1987; ten Brink and Ben-Avraham, 1989). According to Neev and Hall (1979), a salt layer may also underlie the entire northern part of the basin. These layers give rise to the diapirs along the margins of the entire Dead Sea basin. The lack of significant internal deformation within the sedimentary infill may be explained by the low shear strength of the salt that acts to decouple the infill from the tectonic movements of the basement and the margins (ten Brink and Ben-Avraham, 1989).

HOLOCENE SEDIMENTS AND NEOTECTONICS

High-resolution seismic reflection data of the northern part of the basin were collected during several geophysical surveys in

1983 and 1984 (Ben-Avraham et al., 1985; Hall and Ben-Avraham, 1985), using an O.R.E. subbottom profiler running at a maximum output power of 10 kW at a frequency of 3.5 kHz. Maximum penetration is about 30 ms (TWTT) in the deep part of the basin. Altogether, 1,500 km of lines were collected during these surveys. A detailed analysis of the data within the southern 20 km of the northern part of the basin between local coordinates N90 and N110 was carried out by Ben-Avraham et al., (1993). A comprehensive review of the high-resolution seismic profiles in the northern part of the basin is given elsewhere in this volume (Niemi and Ben-Avraham, chapter 6). Several hundred km of lines were also collected in the southern part of the basin, which is currently enclosed by evaporating pans with a water depth of 1 to 4 m in 1986 (Ben-Avraham et al., 1990).

The high-resolution seismic reflection profiles indicate that sediments buried at a shallow depth in the northern part of the basin are flat-lying, continuous, and unfaulted within the central basin floor (Fig. 3-12). Four prominent subsurface seismic reflectors were identified by Ben-Avraham et al., (1993) who suggested that, based on limited core data, they represent the contact between rock salt and marl layers.

A recent depocenter, the Arnon sink, was mapped in the eastern part of the central northern part of the basin (Neev and Hall, 1979). The high-resolution seismic reflection profiles (Ben-Avraham et al., 1993) suggest that both tectonism and halokinesis control the subsidence in this area. The basinal strata are deformed along recent faults that form bathymetric escarpments at the margins of the basin. Thinning of recent sediments toward the faults indicates syntectonic deposition. Neev and Hall (1979) have already pointed out that the western margin of the lake is bordered by several step faults. Ben-Avraham et al. (1993), in their detailed study of the northern part of the basin, have noted that the easternmost of these faults, which actually delimits the basin and is named the west intrabasinal fault, is segmented into three faults with different trends. The nature of the contact between the fault and the basinal sediment varies along the three segments, suggesting variations in the style of faulting along the fault.

HEAT FLOW THROUGH THE DEAD SEA BASIN

Geothermal measurements in the north of the Dead Sea basin were made in 1975 (Ben-Avraham et al., 1978). Land measure-

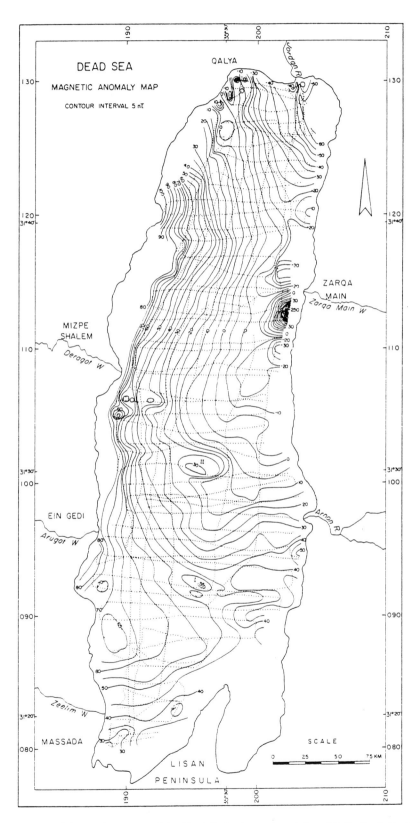

Figure 3-6a Magnetic anomaly map of the northern part of the Dead Seabasin; contour interval 5nT. Track lines are indicated by dots (after Frieslander and Ben-Avraham, 1989).

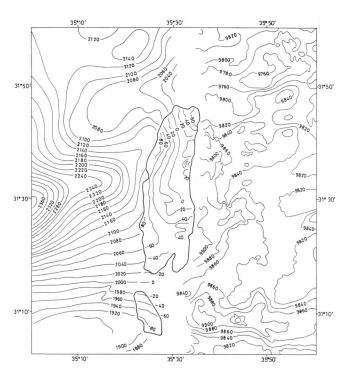

Figure 3-6b Magnetic anomaly map of the Dead Sea and its vicinity; contour interval is 20 nT. The pattern of the magnetic anomaly is simple west of the Dead Sea, and it continues uninterrupted eastward over the western part of the basin. In the east, the pattern is more complex. The map is composed of aeromagnetic maps west (Folkman, 1976) and east (Hatcher et al., 1981) of the Dead Sea and over the central part of the Dead Sea basin (Ram, 1989). The map over the northern part of the Dead Sea basin was obtained by a marine survey (Fig. 3-6A). The maps are not merged (after Frieslander and Ben-Avraham, 1989; Ram, 1989).

ments of temperature gradients in existing wells were taken west (Eckstein and Simmons, 1978) and east (Galanis et al., 1986) of the Dead Sea basin.

The measurements in the lake were taken from a small launch using a special probe (Hänel, 1968) that allows in situ measurements of temperature gradient and thermal conductivity by the same apparatus. Altogether, 19 measurements were taken in the lake. The water in the lake was found to be thermally very stable at the time of the measurements, before the first overturn in 1979 (Steinhorn et al., 1979). Measurements of the water column temperature using the heat-flow probe at the different stations in the lake, which are as far apart as 40 km, agree within less than 0.01°C for any given depth. Ben-Avraham et al., (1977) found an adiabatic temperature distribution from the bottom (335 m) to a depth of 185 m. The variations in bottom-water temperature with time were very small until the overturn. Neev and Emery (1967), who made a detailed study of the lake water and sediments, claimed that below 100 m the temperature, density, and salinity remain nearly uniform with both depth and season. This unique situation enabled scientists to take reliable heat flow measurements in the northern part of the Dead Sea basin at a water depth greater than about 100 m. The temperature gradient measurements were corrected for sedimentation and topography.

The mean value of the corrected heat-flow data for the northern part of the basin of the Dead Sea is 38 mW/m² (Fig. 3-13). This value is comparable with nearby continental values obtained by measurements in abandoned wells west of the lake (Eckstein, 1975; Eckstein and Simmons, 1978), and with heat-flow measurements at Zarqa Main and Zara, east of the lake

(Galanis et al, 1986). The average value of heat flow on land west of the Dead Sea basin, using the most reliable measurements, was calculated by Eckstein (1975) to be 42 mW/m². Thus, the heat-flow values are low throughout the entire region.

A numerical model of the heat-flow distribution from the Dead Sea basin and the surrounding land areas (Mass, 1978) indicates that the actual heat flow in this region is probably 63 mW/m². Most of the heat escapes through the diapirs, which are made of highly conductive salt. An insulating layer of low conductivity above the salt of the Sedom Formation causes the heat to flow through the margins where the diapirs are located.

Interesting results were obtained by the thermal-conductivity measurements of the bottom sediments. The values show a systematic decrease from north to south and from east to west. The highest values are 0.87 W/m °C and the lowest are 0.63 W/m °C (Ben-Avraham et al., 1978). This decrease, which correlates with the distance from the entrance of the Jordan River into the Dead Sea, may result from variations in grain size.

SEISMICITY

The Dead Sea basin is the most actively seismic region in Israel. Evidence for seismic activity in this area exists from Biblical times to the present. Several works on the seismicity of the Dead Sea basin and its surroundings have been published (Wu et al., 1973; Ben-Menahem et al., 1976, 1981; Rotstein and Arieh, 1986; Shapira and Feldman, 1987; van Eck and Hofstetter, 1989, 1990; Kovach et al., 1990). A detailed account of the seismic activity in the Dead Sea area is given in another chapter in this volume (Shapira, chapter 7).

Although poorly constrained, earthquake foci in the Dead Sea area originate probably no deeper than 12–15 km (van Eck and Hofstetter, 1989), suggesting that most of the deformation occurs within the brittle upper crust. Composite focal plane solutions of earthquake clusters indicate possible strike-slip motion along both the western and eastern longitudinal faults at the latitude of the Lisan Peninsula area (van Eck and Hofstetter, 1989), indicating some overlap between the two strike-slip fault strands. Previous research (Ben-Menahem et al., 1976) determined a left-lateral strike-slip motion for the 1927 Jericho earthquake that occurred on the western longitudinal fault north of the Dead Sea. Shapira et al. (1993) relocated the epicenter beneath the northern part of the Dead Sea basin.

The areal distribution of earthquakes within the Dead Sea basin indicate possible clustering in several distinct areas (see Fig. 7-3). One such area is the fault that separates the Lisan Peninsula from the north part of the basin. Other areas are the Arnon sink, where active subsidence is taking place, and the base of the Jordan River delta. Future earthquake activity in this area will no doubt add much to the understanding of the tectonic processes within the Dead Sea basin.

SYNTHESIS

Geophysical data from the Dead Sea basin, which include bathymetric, topographic, seismic refraction, gravity, magnetic, multichannel and single-channel seismic reflection, high-resolution seismic reflection, heat flow, and seismicity data, provide information about the geometry of the Dead Sea basin and its tectonic processes. The basin is a fault-bounded depression (Fig. 3-14). Deformation takes place mainly along the transverse and longitudinal faults, leaving the intervening sediments relatively undeformed. An exceptional area is the Arnon sink, where differential subsidence is now occurring (Neev and Hall, 1979; Ben-Avraham et al., 1993).

Earthquakes and heat-flow data suggest that most deforma-

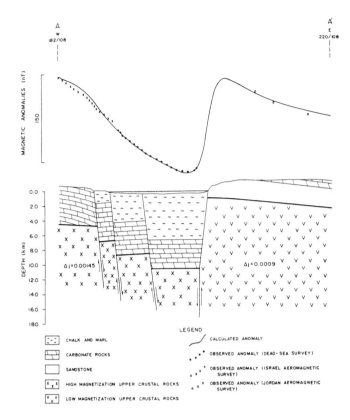

Figure 3-7 Two-dimensional model interpretation of east–west magnetic profile along lat. 31°34'N taken from the combined map (Fig. 3-6B). Aeromagnetic data west of the Dead Sea were shifted with reference to the overlapping area over the basin. The observed anomaly east of the rift was shifted so that the similarity in trend to the calculated anomaly can be seen (after Frieslander and Ben-Avraham, 1989).

tion is confined to the upper brittle crust, although basin-related deformation must include at least part of the semibrittle to ductile crust. Calculating the extension of an uncompensated Dead Sea basin (ten Brink et al., 1993) yielded extension factors that are much less than those required for decompression melting (McKenzie and Bickle, 1988). Indeed, no evidence was found for upper mantle melting or asthenospheric upwelling under the basin. The occurrence of young basaltic rocks on land east of the basin is unexplained by current models, although their origin must be linked to the formation of the Dead Sea basin.

The location and activity of the faults change with time. The longitudinal faults, which are strands of the Dead Sea transform, are segmented. Strike-slip movement takes place along several segments, with normal movement occurring on others. At present, strike-slip motion in the northern part of the basin is mainly occurring on the western intrabasinal fault (the Jordan fault); whereas in the southern part of the basin, it is occurring on the eastern border fault (the Arava fault) (Neev and Hall, 1979; ten Brink and Ben-Avraham, 1989). Some overlap between the Jordan and Arava faults may occur in the Lisan Peninsula area, as suggested by earthquake solutions (van Eck and Hofstetter, 1989), where the basin is the deepest.

The Dead Sea basin consists of at least five 25- to 30-km-long segments, (Ben-Avraham and ten Brink, 1989; ten Brink and Ben-Avraham, 1989; ten Brink et al., 1993). From south to north, these segments are the area south of the Amaziahu fault, the salt pans north of the fault, the Lisan, the area immediately north of the Lisan (the Arnon sink), and the area farther north (Fig. 3-14). The central segment is the deepest, with the maximum accumulation of sediments. Gravity models suggest that the basin may reach a depth of 10 km under the Lisan Peninsula (ten Brink et al., 1993).

The northern and southern parts of the Dead Sea basin are each divided into at least two segments. The detailed study of the Holocene sediments in the northern part of the basin (Ben-Avraham et al., 1993) indicates that the western intrabasinal

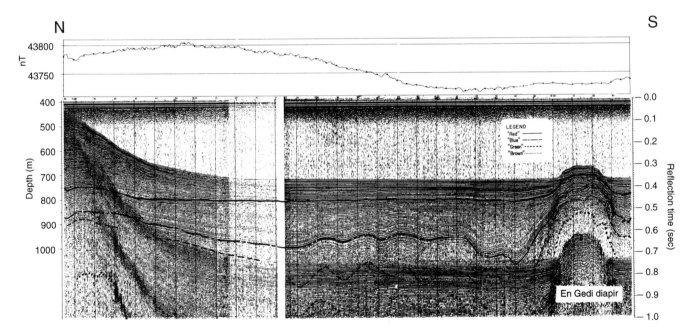

Figure 3-8 North-south longitudinal 1,000-joule sparker profile in the northern part of the Dead Sea basin. Note the En Gedi diapir on the south (at the right) and the Jordan River's submarine delta on the north (at the left) which has been accumulating since the red marker. The acoustically transparent material below the brown reflector is probably rock salt (after Neev and Hall, 1979). See Figure 3-9 for location of the profile.

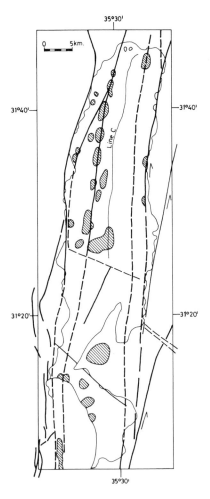

Figure 3-9 A simplified map of fault systems and salt domes in the Dead Sea region, according to Neev and Hall (1979). Thin line is location of profile C (Fig. 3-8).

fault is further divided into distinct segments on which different styles of faulting occur. This suggests that the movement on each of the strands is more complicated than previously assumed. It is possible that some of the strike-slip motion in the northern part of the basin is also taking place on the eastern longitudinal fault.

The overall asymmetry of the Dead Sea basin to the east, as expressed by the topography and bathymetry and by the extension of magnetic anomalies from the land area on the west into the basin without much offset (Frieslander and Ben-Avraham, 1989), suggests that strike-slip motion on the western intrabasinal fault in the northern part of the basin was relatively minor (Ben-Avraham, 1992; Ben-Avraham and Zoback, 1993). It is possible that the strike-slip faulting on the western margin of the northern part of the basin is relatively young and that this margin was characterized mainly by normal faulting during most of the basin's evolution. Indeed, the presence of tilted fault blocks, several kilometers wide, along the western side of the basin (Kashai and Croker, 1987; ten Brink and Ben-Avraham, 1989; ten Brink et al., 1993) suggests that normal faulting on the western margin of both the northern and southern parts of the basin was dominant.

Seismic profiles and drill-hole data suggest that rapid subsidence of the Dead Sea basin has occurred only since the Pleistocene (ten Brink and Ben-Avraham, 1989). Sedimentological evidence also supports a low topographic relief during the Pliocene and a high relief during the Pleistocene. Sands within the Sedom Formation (Pliocene) were mainly derived from the underlying Hazeva Formation (Miocene), whereas sands within the Amora Formation (Pleistocene) come from the entire Phanerozoic basement (Sa'ar, 1985). This situation indicates a shallow depression at the Dead Sea basin during the Pliocene, but the large vertical offsets are limited to the Pleistocene.

The rapid subsidence was probably associated with a slight change in the relative motion between Africa and Arabia at the beginning of the Pliocene, which introduced a component of extension across the Dead Sea (Garfunkel, 1981; Joffe and Garfunkel, 1987). It is possible that during this time a shear component was introduced into the western margin and created the west intrabasinal fault as a strike-slip fault. The *en echelon* ar-

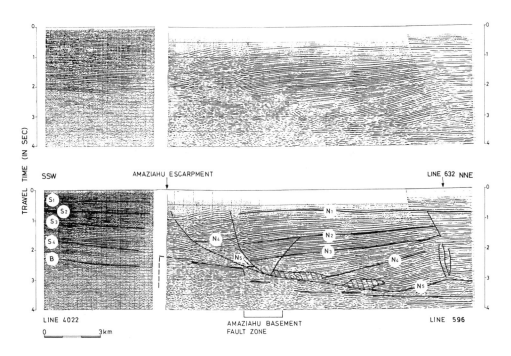

Figure 3-10 Portions of migrated seismic lines crossing the Amaziahu fault (top) and their interpretation (bottom). The 50-m high Amaziahu escarpment is the surface expression of this spectacular listric fault, which extends to a depth of 3.4 sec (about 6 km). The subhorizontal portion of the fault plane probably follows the Pliocene salt layer, and the hatched areas represent possible salt pockets. Note the tilted basement blocks. B denotes the basement reflection, and N_1 to N_5 and S_1 to S_4 denote sequence boundaries north and south of the Amaziahu fault, respectively (after ten Brink and Ben-Avraham, 1989). See Figure 3-11 for location of profile.

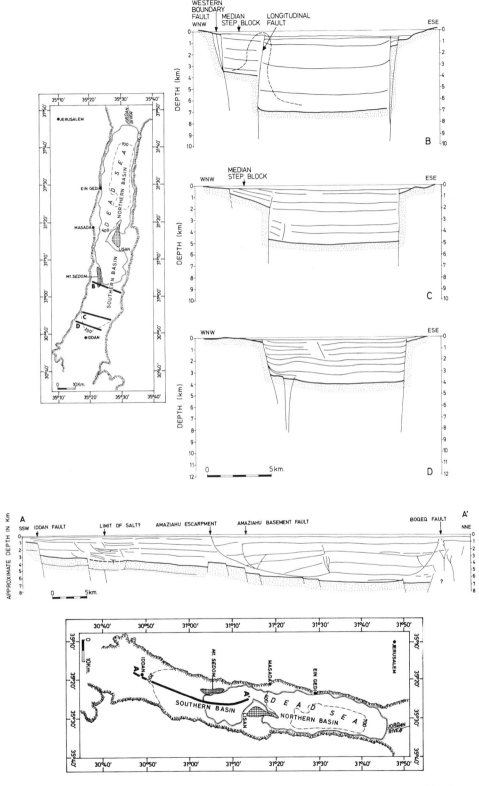

Figure 3-11 Schematic line drawings showing three east–west cross sections (top) of the southern part of the Dead Sea basin and a section along its long axis (bottom) and their location in a map view. Sections are based on the interpretation of multichannel seismic reflection profiles. The dashed line in the top cross section marks the location of Mount Sedom diapir. Despite the large vertical movement of the basement, the layers of the sedimentary infill are only mildy perturbed (after ten Brink and Ben-Avraham, 1989). The profile of Figure 3-10 is located along part of the longitudinal section.

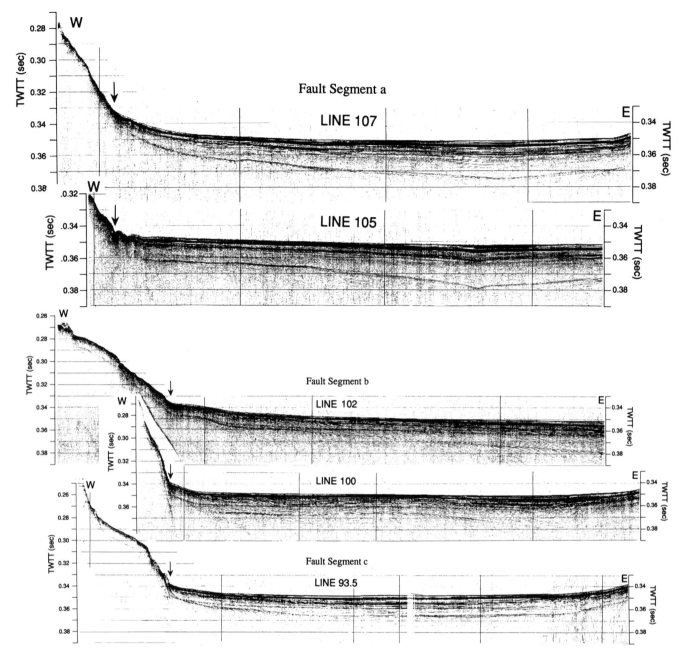

Figure 3-12 East–west 3.5-kHz seismic profiles that illustrate the various styles of deformation along the west intrabasinal fault. The line numbers correspond to the Israel local coordinates, which are marked on the map in Figure 3-6a. Arrows mark the location of the west intrabasinal fault (after Ben-Avraham et al., 1993).

rangement of the two longitudinal faults during the rapid subsidence may explain the full-graben, or close to full-graben, geometry of the basin fill. At present, most of the strike-slip motion is taking place along the western margin.

The nature of the transverse faults is unclear. They divide the Dead Sea basin into distinct segments, as previously described, above. The Amaziahu fault at the southern subbasin appears in the seismic reflection profiles as a spectacular listric normal fault (Fig. 3-10), which may be traced from the surface to a depth of almost 3.5 sec two-way travel time (Kashai and Croker, 1987; ten Brink and Ben-Avraham, 1989). On the other hand, the Boqeq fault that separates the southern part of the basin from the Lisan Peninsula, was probably characterized by normal movement during the Miocene–early Pleistocene, but thereafter, during the

late Pleistocene to the present, the movement changed to strike slip (Ben-Avraham et al., 1990). This suggests that, at least in this case, transverse faults bordering a pull-apart basin can change their nature during the evolution of a basin. Similar situations may exist in other basins along the Dead Sea rift (Ben-Avraham and ten Brink, 1989). It is possible that the change in the style of faulting along the Boqeq fault was associated with the initiation of the strike-slip movement along the western longitudinal fault in the northern part of the basin.

The geophysical data reveal the geometry of the Dead Sea basin and the nature of the faulting processes that formed this large basin. The deepest part of the basin is located at its center. It is possible that the basin grew northward with time (Zak and Freund, 1981; Garfunkel, chapter 4, this volume). Of particular interest are the changes that occurred with time in the nature and location of the main longitudinal and transverse faults that border the different parts of the basin. These faults separate the basin into distinct segments whose boundaries have changed during the evolution of the basin.

Acknowledgments

I thank Drs. Tina M. Niemi and Uri ten Brink for their helpful comments.

REFERENCES

Arbenz, J. K., 1984, Oil potential of the Dead Sea: Tel Aviv, Sismica Oil ExplorationReport 84/111, 54 p.

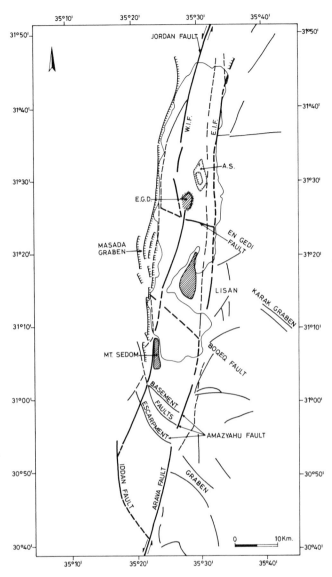

Figure 3-14 Structural interpretation map of the structural elements of the Dead Sea basin (after Neev and Hall, 1979; ten Brink and Ben-Avraham, 1989, Ben-Avraham et al., 1993). Solid lines mark faults, and dashed lines mark their extrapolations. Heavy lines are the main bounding faults of the basin. Hatched areas indicate location of Mount Sedom, Lisan, and En Gedi diapirs (E.G.D.). The area of maximum accumulation of recent sediments is the Arnon sink (A.S.)which is oriented 30° counterclockwise to the trend of the border faults ofthe basin. The west intrabasinal fault (W.I.F.) is shown to be segmented, witheach segment having a different trend. The east intrabasinal fault (E.I.F.) is probably more continuous.

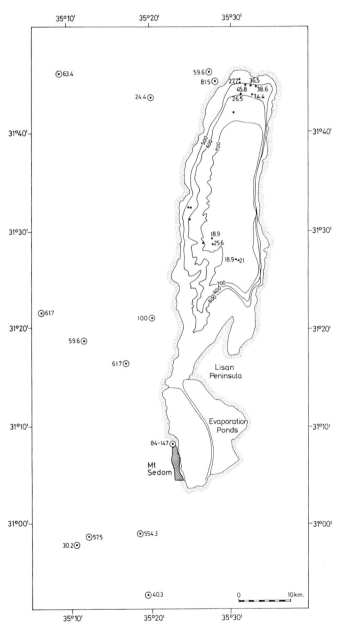

Figure 3-13 Heat-flow values in the Dead Sea and its vicinity in mW/ m². Values from the Dead Sea are from Ben-Avraham et al. (1978), whereas those from the land area to the west are from Eckstein (1975).

Barbari, F., Capaldi, G., Gasperini, P., Martinelli, G., Santracroce, R., Scandone, R., Tereil, M., and Varet, J., 1979, Recent basaltic volcanism of Jordan and its implications on the geodynamic history of the Dead Sea shear zone: Rome, International Symposium on Geodynamic evolution of the Afro-Arabian Rift system, v. 47, p. 667–683.

Ben-Avraham, Z., 1992, Development of asymmetric basins along continental transform faults: *Tectonophysics*, v. 215, p. 209–220.

Ben-Avraham, Z., Hänel, R., and Assaf, G., 1977, The Thermal structure of the Dead Sea: *Limnology and Oceanography*, v. 22,

p. 1076–1078.

Ben-Avraham, Z., Hänel, R., and Villinger, H., 1978, Heat flow through the Dead Sea rift: *Marine Geology*, v. 28, p. 253–269.

Ben-Avraham, Z., Levy, Y., Hall, J. K., and Neev, D., 1985, The structure and composition of the uppermost sedimentary layers in the Dead Sea: Tel Aviv University Report MGL TAU 1/85, 26 p.

Ben-Avraham, Z., and ten Brink, U., 1989, Transverse faults and segmentation of basins within the Dead Sea rift: *Journal of African Earth Science*, v. 8, p. 603–636.

Ben-Avraham, Z., ten Brink, U., and Charrach, J., 1990, Transverse faults at the northern end of the southern basin of the Dead Sea Graben: *Tectonophysics*, v. 180, p. 37–47.

Ben-Avraham, Z., Niemi, T. M., Neev, D., Hall, J. K., and Levy, Y., 1993, Distribution of Holocene sediments and neotectonics in the deep north basin of the Dead Sea: *Marine Geology*, v. 113, p. 219–231.

Ben-Avraham, Z., and Zoback, M. D., 1993, Transform-normal extension and asymmetric basins: An alternative to pull-apart models: *Geology*, v. 20, p. 423–426.

Ben-Menahem, A., Nur, A., and Vered, M., 1976, Tectonics, seismicity and structure of the Afro-Eurasian junction—The breaking of an incoherent plate: *Physics of the Earth and Planetary Interiors*, v. 12, p. 1–50.

Ben-Menahem, A., and Aboodi, E., 1981, Micro- and macroseismicity of the Dead Sea rift and off-coast eastern Mediterranean: *Tectonophysics*, v. 80, p. 199–233.

Eckstein, Y., 1975, The measurements and interpretation of terrestrial heat flow in Israel [Ph.D. thesis]: Jerusalem, The Hebrew University, 171 p.

Eckstein, Y., and Simmons, G., 1978, Measurements and interpretation of terrestrial heat flow in Israel: *Geothermics*, v. 6, p. 117–142.

El-Isa, Z. H., Mechie, J., and Prodehl, C., 1987, Shear velocity structure of Jordan from explosion seismic data: *Royal Astronomical Society Geophysical Journal*, v. 90, p. 265–281.

El-Isa, Z. H., 1990, Lithospheric structure of the Jordan-Dead Sea transform from earthquake data: *Tectonophysics*, v. 180, p. 29–36.

Folkman, Y., 1976, Magnetic and gravity investigations of the crustal structure in Israel, [Ph.D. thesis]: Tel Aviv, Tel Aviv University, 203 p. (in Hebrew, English abstract).

Folkman, Y., 1981, Structural features in the Dead Sea-Jordan rift zone, interpreted from a combined magnetic-gravity study: *Tectonophysics*, v. 80, p. 135–146.

Frieslander, U., and Ben-Avraham, Z., 1989, Magnetic field over the Dead Sea and vicinity: *Marine and Petroleum Geology*, v. 6, p. 148–160.

Galanis, S. P., Sass, J., Munroe, R. J., and Abu-Ajamieh, M., 1986, Heat flow at Zerqa Ma'in and Zara and a geothermal reconnaissance of Jordan: U.S. Geological Survey Open-File Report 86–631, 110 p.

Garfunkel, Z., 1981, Internal structure of the Dead Sea leaky transform (rift) in relation to plate kinematics: *Tectonophysics*, v. 80, p. 81–108.

Ginzburg, A., Makris, J., Fuchs, K., Prodehl, C., Kaminski, W., and Amitai, U., 1979a, A seismic study of the crust and upper mantle of the Jordan-Dead Sea rift and their transition toward the Mediterranean Sea: *Journal of Geophysical Research*, v. 84, p. 1,569–1,582.

Ginzburg, A., Makris, J., Fuchs, K., Perathoner, B., and Prodehl, C., 1979b, Detailed structure of the crust and upper mantle along the Jordan-Dead Sea rift: *Journal of Geophysical Research*, v. 84, p. 5,605–5,612.

Ginzburg, A., and Kashai, E., 1981, Seismic measurements in the southern Dead Sea: *Tectonophysics*, v. 80, p. 67–80.

Ginzburg, A., Makris, J., Fuchs, K., and Prodehl, C., 1981, The

structure of the crust and upper mantle in the Dead Sea rift: *Tectonophysics*, v. 80, p. 109–119.

Ginzburg, A., Ben-Avraham, Z., and Makris, J., 1996, The seismic refraction study of the north basin of the Dead Sea, Israel: *Geophysical Research Letters* (submitted).

Hall, J. K., and Ben-Avraham, Z., 1985, The Dead Sea shallow seismic survey of 5–9 June, 1982: *Geological Survey of Israel Current Research*, v. 5, p. 79–81.

Hänel, R., 1968, Untersuchungen zur Bestimmung der Terrestrischen Wärmestromdichte in Binnenseen [Ph.D. thesis]: Clausthal, Germany, Clausthal University, 121 p.

Hatcher, R. D., Zietz, I., Regan, R. D., Abu-Ajamieh, M., 1981, Sinistral strike-slip motion on the Dead Sea rift: Confirmation from new magnetic data: *Geology*, v. 9, p. 458–462.

Joffe, S., and Garfunkel, Z., 1987, The kinematics of the circum Red Sea—A re-evaluation: *Tectonophysics*, v. 141, p. 5–22.

Kashai, E. L., and Croker, P. F., 1987, Structural geometry and evolution of the Dead Sea–Jordan rift system as deduced from new subsurface data: *Tectonophysics*, v. 141, p. 33–60.

Kovach, R. L., Andreasen, G. E., Gettings, M. E., and El-Kayse, K., 1990, Geophysical investigations in Jordan: *Tectonophysics*, v. 180, p. 49–60.

Lynch, W. F., 1849, *Narrative of the United States' expedition to the River Jordan and the Dead Sea*: Philadelphia, Lea and Blanchard, 509 p.

Manspeizer, W., 1985, The Dead Sea rift: Impact of climate and tectonism on Pleistocene and Holocene sedimentation, *in* Biddle, K. T., and Christie-Blick, N., eds., *Strike-slip deformation, basin formation, and sedimentation* : Society of Economic Paleontologists and Mineralogists Special Publication 37, p. 143–158.

Mass, D., 1978, A new numerical method for the interpretation of heat flow data with application to the Dead Sea region [M.S. thesis]: Rehovot, Israel, Weizmann Institute of Science, 124 p. (in Hebrew, English summary).

McKenzie, D. P., and Bickle, M. J., 1988, The volume and composition of melt generated by extension of the lithosphere: *Journal of Petrology*, v. 29, p. 625–679.

Neev, D., and Emery, K. O., 1967, The Dead Sea: Depositional processes and environments of evaporites: Geological Survey of Israel Bulletin 41, 147 p.

Neev, D., and Hall, J. K., 1979, Geophysical investigations in the Dead Sea: *Sedimentary Geology*, v. 23, p. 209–238.

Niemi, T. M., and Ben-Avraham, Z., 1994, Evidence for Jericho earthquakes from slumped sediments of the Jordan River delta in the Dead Sea: *Geology*, v. 22, p. 395–398.

Perathoner, B., Fuchs, K., Prodehl, C., and Ginzburg, A., 1981, Seismic investigation of crust–mantle transition in continental rift systems—Jordan–Dead Sea rift and Rhinegraben: *Tectonophysics*, v. 80, p. 121–133.

Ram, E., 1989, The magnetic field over the Dead Sea-Lisan area [M.S. thesis]: Tel Aviv, Tel Aviv University, 55 p. (in Hebrew, English abstract).

Rotstein, Y., and Arieh, E., 1986, Tectonic implications of recent microearthquake data from Israel and adjacent areas: *Earth and Planetary Science Letters*, v. 78, p. 237–244.

Sa'ar, H., 1985, Origin and sedimentation of sandstones in graben fill formations of the Dead Sea rift valley: Geological Survey of Israel Report MM/3/86.

Shapira, A., Avni, R., and Nur, A., 1993, A new estimate for the epicenter of the Jericho earthquake of July 11, 1927: *Israel Journal of Earth Sciences*, v. 42, p. 93–96.

Shapira, A., and Feldman, L., 1987, Microseismicity of three locations along the Jordan rift: *Tectonophysics*, v. 141, p. 89–94.

Steinhorn, I., Assaf, G., Gat, J. R., Nishry, A., Nissenbaun, A., Stiller, M., Beyth, M., Neev, D., Garber, R., Friedman, G. M., and Weiss, W., 1979, The Dead Sea: Deepening of the mixolimnion signifies the overture to overturn of the water column: *Science*, v. 206, p. 55–57.

ten Brink, U. S., and Ben-Avraham, Z., 1989, The anatomy of a pull-apart basin: Seismic reflection observations of the Dead Sea basin: *Tectonics*, v. 8, p. 333–350.

ten Brink, U. S., Schoenberg, N., Kovach, R. L., and Ben-Avraham, Z., 1990, Uplift and possible Moho offset across the Dead Sea transform: *Tectonophysics*, v. 180, p. 71–86.

ten Brink, U. S., Ben-Avraham, Z., Bell, R. E., Hassouneh, M., Coleman, D. F., Andreasen, G., Tibor, G., and Coakley, B., 1993, Structure of the Dead Sea pull-apart basin from gravity analyses: *Journal of Geophysical Research*, v. 98, p. 21,877–21,894.

van Eck, T., and Hofstetter, R., 1989, Microearthquake activity in the Dead Sea region: *Geophysical Journal International*, v. 99, p. 605–620.

van Eck, T., and Hofstetter, R., 1990, Fault geometry and spatial clustering of micro earthquakes along the Dead Sea-Jordan rift fault zone: *Tectonophysics*, v. 180, p. 15–27.

Wu, F. T., Karcz, I., Arieh, E. J., Kafri, U., Peled, U., 1973, Microearthquakes along the Dead Sea rift: *Geology*, v. 1, p. 159–161.

Zak, I., and Freund, R., 1981, Asymmetry and basin migration in the Dead Sea rift: *Tectonophysics*, v. 80, p. 27–38.

4. THE HISTORY AND FORMATION OF THE DEAD SEA BASIN

Zvi Garfunkel

The Dead Sea basin (DSB; Fig. 4-1) is a prominent morphotectonic depression along the Dead Sea transform (also called rift). The basin is a young intracontinental plate boundary formed as a result of the late-Cenozoic breakup of the once continuous Arabo-African continent. The DSB, one of the largest of several basins along the transform, is approximately 150 km long and 15–17 km wide, and it is filled with up to 10 km of Neogene to Recent sediments. Topographically, the lowest part of the DSB is occupied by the Dead Sea with a water level over 400 m below sea level. According to Quennell (1959), the DSB and the other depressions along the transform are interpreted as rhomb-shaped pull-apart basins that formed between left-stepping fault strands.

The DSB attracted the attention of geologists in the middle of the last century (Lynch, 1849; Lartet, 1869). Since then, various aspects of the basin and its fill have been studied extensively. Syntheses and interpretations of its history, structure, tectonic setting, and geophysical characteristics were given by Neev and Emery (1967), Zak (1967), Neev and Hall (1979), Zak and Freund (1981), Garfunkel (1981), Kashai and Croker (1987), ten Brink and Ben-Avraham (1989), and ten Brink et al. (1993).

The purpose of the this chapter is to summarize the history and tectonics of the DSB and to examine two interrelated questions: How does the basin relate to the development of the Dead Sea transform, and How do we interpret its evolution and internal structure in terms of the pull-apart concept? The regional tectonic setting of the DSB and its main features are reviewed first. Then the history of the basin is analyzed based on data about its sedimentary fill. Finally, the pull-apart concept is applied to the basin. Because the DSB is one of the largest existing pull-apart basins, its features are important for the understanding of such structures in general.

TECTONIC FRAMEWORK—THE DEAD SEA TRANSFORM

Regional Setting

In the Dead Sea region, the continental crust was consolidated by the Pan-African orogeny of Late Proterozoic age. During most of the Phanerozoic, this region remained a stable platform (Garfunkel, 1988a, and references therein). Tectonic stability was interrupted, however, by Permian(?)–early Mesozoic rifting that shaped the eastern Mediterranean continental margin and produced a strongly faulted belt, 70–100 km wide, inland of the present Sinai-Israel coast. In Permian-Eocene times, the new platform edge was covered by a sediment wedge whose thickness varies from 6–8 km beneath the present Mediterranean coast to 1 km or less 250 km inland. The crust on both sides of the DSB is 30–35 km thick, but it thins markedly toward the Mediterranean continental margin (Ginzburg and Folkman, 1980; El-Isa et al., 1987), probably as a result of modification of the Pan-African crust by the early Mesozoic rifting.

In Senonian to Miocene times, the region was affected by mild compressional deformation. The resulting structures, known as the Syrian arc, include (Fig. 4-1) (1) a bundle of east-northeast- to northeast-trending folds, and (2) a group of east- to west-trending lineaments of aligned faults and folds along which some right-lateral shearing also took place. The latter form a belt that extends across Sinai and the central Negev to 200 km east of the Dead Sea; it was called the central Negev–Sinai shear belt (Bartov, 1974). The Syrian arc produced some local relief, but overall the region now crossed by the transform remained low and was covered by shallow-water marine sediments until after mid-Eocene times (ca. 40 Ma). Much of the region west of the DSB was still close to sea level in mid-Miocene times (15–12 Ma) when it was covered by a brief marine ingression.

The Cenozoic continental breakup produced several rifts and led to the separation of the Arabian plate from the major African plate. The Dead Sea transform forms a part of the new boundary of the Arabian plate, separating it from the Sinai subplate, which is an appendage of the African plate. The transform extends 1,000 km from the zone of plate divergence along the Red Sea to the Taurus-Zagros zone of plate convergence (Freund, 1965; Wilson, 1965; Fig. 4-1). The southern part of the Dead Sea transform, including the DSB, obliquely crosses the early Mesozoic strongly faulted zone, the Syrian arc, and the grain of basement exposures. The young continental breakup was accompanied by a marked change in the behavior of the region: The Syrian arc deformation virtually ceased, the region was uplifted, especially near the new plate boundaries, and widespread igneous activity took place.

Amount and history of lateral offset

Areas that are now adjacent across the transform have significantly different geological properties, but the differences disappear when a left-lateral motion of about 105 km is restored (Quennell, 1959; Freund et al., 1970; Bartov, 1974; Bandel and Khouri, 1981). This restored displacement matches all the known facies and thickness variations, as well as the erosive truncations within the Phanerozoic platformal sedimentary cover, the major structural zones, and the known features of the Pan-African basement. The magnitude of the offset is best measured by matching the lineaments of the central Negev–Sinai shear belt (Fig. 4-2). Such an offset is also supported by matching magnetic anomaly patterns across the transform (Hatcher et al., 1981).

The kinematics of the Arabian-African plate separation, deduced from the opening of the Red Sea and the Gulf of Aden (McKenzie et al., 1970) requires a left-lateral motion of 100 km nearly parallel to the Dead Sea transform. Though important as an independent line of evidence, this estimate is imprecise. Rather, the data from the Dead Sea transform allow refinement of the regional plate kinematics (Joffe and Garfunkel, 1987; LePichon and Gaulier, 1988).

The history of the motion along the Dead Sea transform is not well constrained, because young markers cannot be matched on both sides of the transform. The best constraint is provided by the regional plate kinematics. Because the transform took up most of the Arabian-African plate separation north of the Red Sea, the opening of the Red Sea must be matched by a corresponding motion along the transform. The record from the Red Sea shows that it was already a wide, evaporite-filled basin

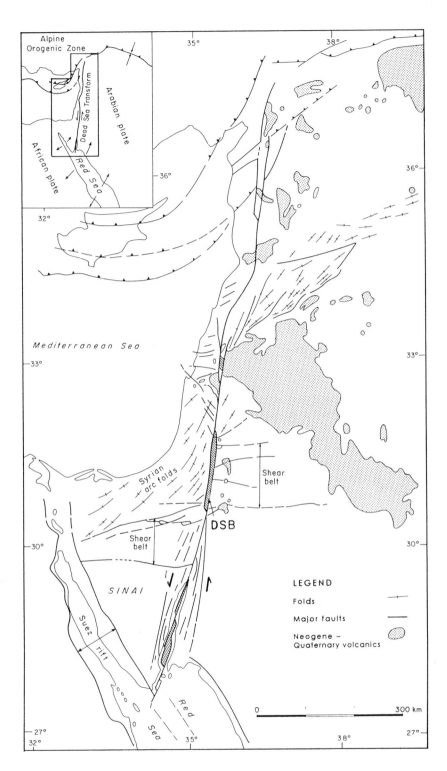

Figure 4-1 Regional setting of the Dead Sea basin. Inset shows the present plate configuration.

by 5 Ma (end of the Miocene), and that the subsequent opening was considerably less than half of the total opening (Coleman, 1974, 1984; Roeser, 1975; Izzeldin, 1987). The Suez rift took up a small part of this motion, but most of the opening of the Gulf of Suez was achieved before the Late Miocene (Garfunkel and Bartov, 1977; Moretti and Colletta, 1987; Richardson and Arthur, 1988). These observations constrain the motion along the Dead Sea transform in the last 5 Ma to about 40 km or less, which is less than half the total motion (Joffe and Garfunkel, 1987). The remaining offset must be older. At the rate of the young motion, the entire offset could have been accomplished in 13–15 Ma, that is, since the Middle Miocene. The suggestion (Schulman and Bartov, 1978; Steinitz et al., 1978) that the transform has moved only since 5 Ma conflicts with the regional plate kinematics.

The beginning of the transform motion is constrained by a system of northwest-trending dikes and small intrusions—the Red Sea dike system (Garfunkel and Bartov, 1977; Eyal et al., 1981; Steinitz et al., 1981; Coleman, 1984). These formed 25–20 Ma ago along the Red Sea–Gulf of Suez trend, but before the for-

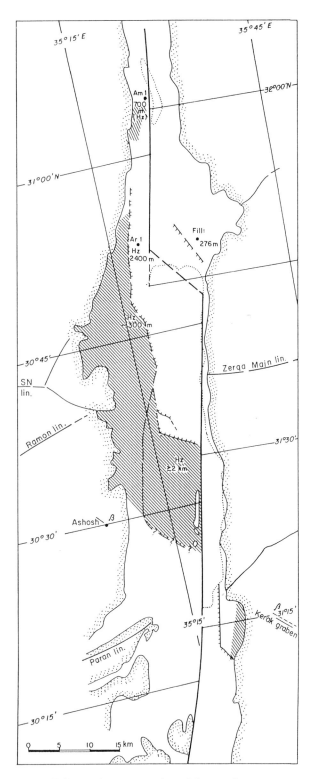

Figure 4-2 Palinspastic reconstruction of the transform segment containing the Dead Sea basin restoring the entire transform slip (compare with Fig. 4-3). Diagonal ruling: outcrops of Miocene beds within the DSB, with estimated thicknesses; dotted line: eastern shore of Dead Sea. Am 1: Amiaz-1 borehole; Ar 1: Arava-1 borehole; SN lin.: Saad-Nafha lineament.

mation of the transform because they are displaced by the entire transform offset. The most northerly bodies (Fig. 4-2) are a few plugs aligned along the northwest-trending Kerak graben east of the Dead Sea (probably underlain by a dike), about 19 Ma, and their counterparts across the transform, the 'Ashosh plug and a short associated dike about 21 Ma west of the Arava.

The transform motion may have begun before the Middle Miocene, perhaps some 18 Ma. This is suggested by the ages of the DSB and of a depression near Tiberias, which are interpreted as secondary structures produced by the transform motion (Garfunkel, 1989, and see the following). This implies an initial slower-than-average slip rate of the transform. The African-Arabian plate separation also seems to have accelerated in the Middle or Late Miocene (Izzeldin, 1987; LePichon and Gaulier, 1988). On the other hand, the earliest sedimentary-volcanic fill of the southern Red Sea shows slow rifting and crustal stretching before 25 Ma (Coleman, 1984; Brown et al., 1989). The Suez rift also nucleated before or during the emplacement of the Red Sea dikes (Garfunkel and Bartov, 1977; Richardson and Arthur, 1988). Thus, the Dead Sea transform apparently was activated after initiation of rifting along the Red Sea–Gulf of Suez trend.

Continuing young lateral motion on the transform is indicated by earthquake mechanisms along the transform (Ben-Menahem et al., 1976; van Eck and Hofstetter, 1989, 1990), as well as by the morphologic expression of young faults displacing Late Quaternary to Holocene sediments (Garfunkel et al., 1981).

Secondary and marginal structures

The slip along the Dead Sea transform produced a variety of secondary structures along bends and discontinuities of the main displacement zone. In such places, continuing lateral motion produces misfits between the bordering blocks, which leads to local deformation (Quennell, 1959; Freund et al., 1970; Garfunkel, 1981). Where the transform trace bends to the right (opposite to the sense of motion), local convergence arises, causing updoming and folding on different scales. In contrast, local divergence arises where the transform trace bends to the left, producing local depressions that tend to have rhomboidal shapes. They were called rhomb-shaped grabens or rhomb-grabens (Freund et al., 1968; Garfunkel, 1981), or pull-apart basins (Burchfiel and Stewart, 1966). Such structural relations characterize many other major continental strike-slip fault zones (e.g., Crowell, 1974; Mann et al., 1983; Christie-Blick and Biddle, 1985).

The structure of the southern half of the transform, which includes the DSB, is dominated by left-stepping longitudinal strike-slip faults and intervening rhombic pull-aparts. These structures are embedded in an almost continuous morphotectonic low—the transform valley—which is delimited by normal faults on both sides (Figs. 4-3, 4-4). Although this valley superficially looks like an extensional graben, its internal structure is very different: It contains major longitudinal strike-slip faults and pull-apart basins. Moreover, it is interrupted by a few saddles that express compressional bends of the transform trace. These features show that the structure resulted primarily from lateral motion, with transverse extension having only a secondary role. Structural analysis reveals that, along most of the southern part of the transform, the lateral motion was accompanied by a small component of transverse separation of the adjacent blocks (Garfunkel, 1981). This extensional component increased with time, indicating that the slip direction changed during the transform history. Slight changes in the direction of plate motions are to be expected, in general, but their structural consequences will be most evident along transform boundaries, where even small changes in slip direction can cause a switch among pure strike-slip, transpressional, and transtensional regimes.

Concurrent with the transform motion, deformation produced structures, mostly faults, along the margins of transform valley. Adjacent to the DSB, only mild faulting took place, but much stronger faulting occurred north and south of the basin (Fig. 4-1). The area most affected is located west of the transform and north of about lat. 32° N. Here, several dense systems of faults trending northwest and east-west formed as far as the Mediterranean Sea, more than 50 km away from the transform (Freund et al., 1970; Bartov and Arkin, 1975; Ron et al. 1984; Shaliv, 1991). The most prominent structure is the Yizreel Valley–Carmel fault system, which extends to the continental margin. This deformation is related, at least in part, to the bending to the right of the transform trace in this area. Farther north, this produced considerable faulting, doming, folding, and uplifting in the Lebanon and Hermon Mountains and farther east (Quennell, 1959; Freund et al., 1970; Garfunkel, 1981). South of the DSB, the transform margins are crossed mainly by faults subparallel to the transform. Along the southernmost transform segment, a system of anastomosing strike-slip faults was active mainly during the early stages of the transform history and displaced the previously mentioned Red Sea dikes (Eyal et al., 1981). Another system of apparently normal faults trending north-northeast affects Pliocene–Quaternary sediments in the southern Negev (Avni et al., 1993).

Uplifting and magmatism

The igneous activity accompanying the late Cenozoic breakup of the Arabo-African continent is well developed in the region crossed by the Dead Sea transform. Here, two phases of predominantly basaltic magmatic activity can be distinguished (Garfunkel, 1989). The first phase produced the latest Oligocene to Early Miocene Red Sea dikes and is probably unrelated to the transform because, as previously noted, it occurred before the transform was activated. The second, Miocene to Recent phase is broadly coeval with the transform motion.

The second phase produced mainly volcanic fields, some still active, that consist predominantly of mildly to strongly alkaline basalts (Fig. 4-1). On a regional scale, there is no obvious relation between the extent and known volume of igneous rocks and the Dead Sea transform. However, an alignment of igneous bodies is apparent along the northern two-thirds of the transform, both along its margins and along the main displacement zone (Garfunkel, 1989). In particular, igneous activity occurred intermittently since 19–17 Ma near and along the transform between southernmost Lebanon and the Beth Shean area. Drilling revealed that, in a few places, the igneous rocks beneath the transform valley are much more voluminous than could be inferred from the igneous activity visible on the surface. Farther south, basalt outcrops occur in the Grain Sabt uplift in the transform valley 30 km north of the Dead Sea (Fig. 4-3), but there was no magmatic activity on the adjacent transform margins. Another small shallow igneous body is inferred from a local magnetic anomaly under the central Dead Sea (Frieslander and Ben-Avraham, 1989). Otherwise, igneous rocks are not known from within the DSB. However, several 9–6 Ma and younger flows occur on the eastern margin of the basin along the Dead Sea and the Arava, and a 6 Ma (Steinitz

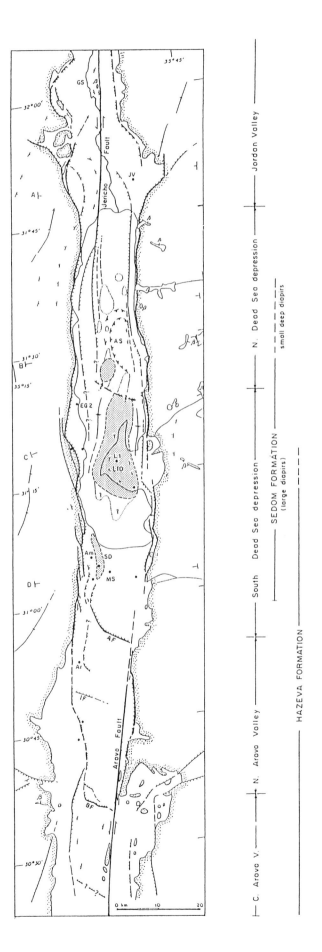

Figure 4-3 Map showing the main features of the Dead Sea basin, compiled from different sources (see text). Irregular stippling: edge of outcrops of prebasin rocks; regular stippling: diapirs expressed in the topography; dotted lines: buried diapirs not affecting the surface. ß: Neogene to Recent basalts; large dots: main boreholes. Boreholes: Am—Amiaz-1, Ar—Arava-1, EG2—En Gedi-2, MS—Melekh Sedom, JV—Jordan Valley-1, L1—Lisan-1. Other abbreviations: AF—Amazyahu fault; AS—Arnon sink; BF—Buweirida fault; GS—Grain Sabt dome; IF—Iddan fault; LiD -Lisan diapir; SD—Sedom diapir.

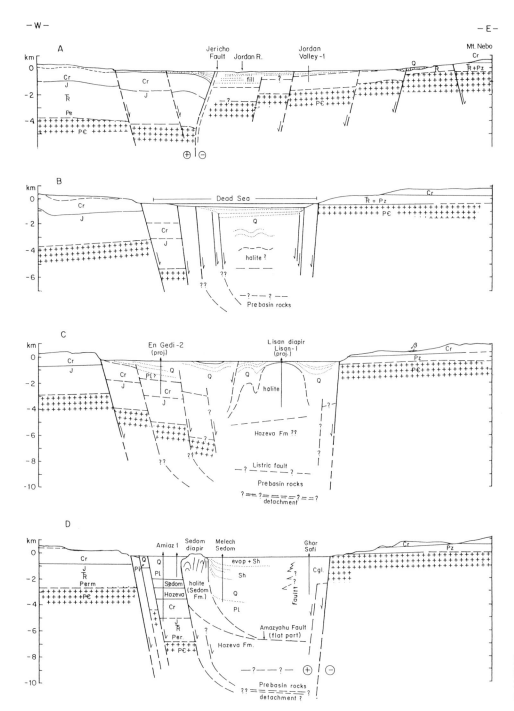

Figure 4-4 Sections across the Dead Sea basin. For location, see Figure 4-3. The deep structure is hypothetical (see text for discussion).

and Bartov, 1991; G. Steinitz, pers. com., 1995) dike and vent are present west of the central Arava (near Ein Yahav). Farther south, igneous activity is not known along the transform.

These occurrences prove that melting took place beneath or near at least part of the Dead Sea transform, including the DSB, and that the transform served as a path for magma ascent in some places. However, only small melt volumes rose to shallow levels, and they were particularly small in the DSB and along its margins. The alkaline nature of the magmas shows that the melting occurred only in the middle or lower lithosphere and involved small degrees of partial fusion. Heating at these depths is also suggested by thermometric data on ultra-

mafic xenoliths found in basalts in the Galilee and Golan areas (Stein et al., 1993).

The heat flow west of the DSB and along its western margin is low, usually 35–50 mW/m^2, but locally higher values up to 90 mW/m^2, result from hot waters that rise along faults bordering the DSB (Eckstein, 1979). In boreholes along the western margin of the Dead Sea, the temperature gradients are about 20°C/km to depths of 3–5 km (Kashai and Croker, 1987), indicating a heat flow less than 50 mW/m^2. Within the Dead Sea, the heat flow is variable: only 30 mW/m^2 in the north but probably higher in the deepest part of the lake (Ben-Avraham et al., 1978; Ben-Avraham and Ballard, 1984). Perhaps heat is conducted through the salt diapirs under the Dead Sea, reducing the heat flow elsewhere. Heat flows of 68 and 75 mW/m² were measured in Lake Kinneret and the northern Gulf of Elat, respectively (Ben-Avraham et al., 1978). East of the transform, the heat flow is high near hot springs, but generally the data are very erratic (Abu Ajamieh, et al., 1989, p. 71). Thus, the DSB and the nearby parts of the transform are not associated with any notable heat flow anomaly, except within (some?) pull-apart basins. This is compatible with the negligible stretching of the transform margins. The nature of the magmatic activity further indicates rather deep melting not very long ago, so that there was not enough time for the heating to reach the surface.

The transform margins are considerably uplifted. On both sides of the DSB, elevations reach 1 km or more above sea level, whereas farther south elevations are higher. The area west of the DSB forms a broad arch; most of its uplifting postdates marine beds 15- to 12-Ma that overlie its western flank. The eastern margin of the DSB drops gradually eastward away from the basin, indicating that it was much dissected by erosion only after the extrusion of 9- to 6-Ma basalt flows (Steinitz and Bartov, 1991). Thus, most of the uplifting of the DSB margins appears to be post-Miocene. The amount of uplifting does not correlate with visible igneous activity. Uplifting is greatest on the flanks of the Gulf of Elat (Aqaba) where magmatic activity coeval with the transform is not known.

THE DEAD SEA BASIN

The basin and its fill

The Dead Sea basin extends between two narrow portions of the transform valley near lat. 32° N and 30° N (Figs. 4-1, 4-3). Surface and subsurface data (Neev and Emery, 1967; Zak, 1967; Bender, 1974; Neev and Hall, 1979; Kashai and Croker, 1987; ten Brink and Ben-Avraham, 1989; Rotstein et al., 1991; Bartov et al., 1993) show that the basin contains a sedimentary fill of Neogene to Quaternary age with a maximum thickness of perhaps 10 km. The structurally lowest area is outlined by a sharp Bouguer gravity anomaly reaching minus 100 mgal relative to the value over the basin shoulders (ten Brink et al., 1993). The gravity low ends abruptly 10 km north of the Dead Sea near lat. 31°50'N, whereas in the south, it tapers out gradually at about lat. 30°35'N. However, the fill is quite thick—more than 2 km—some distance farther south (Bartov et al., 1993). Thus, the depression with a thick fill is 150 km long and 15–17 km in its center, but it is less than half as wide at its extremities.

The topographically lowest part of the basin, more than 350 m below sea level, forms the Dead Sea depression proper, which consists of two distinct parts (Fig. 4-3; Neev and Emery, 1967; Neev and Hall, 1979). The northern part comprises a well-defined, nearly rectangular depression 40 km long with a rather flat floor at 700–730 m below sea level that is mostly covered by water (maximum water depth of about 330 m). The southern part, about 45 km long, has a shallower floor at 350–400 m below sea level and is now mostly dry. In the south, the basin is delimited by the 50-m-high scarp of the Amazyahu fault.

The DSB is mostly covered by very young sediments, but a few exposures of older beds and subsurface data allow an outline of the stratigraphy of the basin fill, though much remains uncertain. According to Zak and Freund (1981), the basin fill is divided into three main series, as follows:

11. A fluviatile-lacustrine series of Miocene age consisting mainly of quartz-rich sandstones—the Hazeva Formation. This series is known from the southern 90 km of the Dead Sea basin, where more than 2-km-thick sections have been recorded, and thinner sections occur sporadically over a large region

outside the transform valley. The clastic material in this series was derived mostly from sources located several hundred kilometers away from the Dead Sea basin.

2. An estuarine-lagoonal series of latest Miocene (?) to Early Pliocene age consisting mainly of halite—the Sedom Formation. This series is developed mainly under the southern part of the Dead Sea, where it is 2 to 5(?) km thick and forms several diapiric structures (Figs 4-3, 4-4). This sequence was deposited during an ingression of the sea most probably from the Mediterranean through the Yizreel and Jordan Valleys into the transform valley.

3. A lacustrine-fluviatile series of Pliocene and Quaternary age that consists mainly of coarse to fine clastics derived from the areas surrounding the DSB, but also including marl, chalk, and evaporites. This series is developed under the northern two-thirds of the DSB, and its maximum thickness is 3–4 km. Most of this section is included in the Amora Formation. This and the previous series are essentially confined to the transform valley and were included in the Dead Sea Group (Zak, 1967).

Structure

The principal structural features of the Dead Sea basin are shown in Figures 4-3, 4-5, but the basin's deep structure remains hypothetical. Clearly, faults control the basin structure. Strike-slip faults—the Jericho and Arava faults—extend along the floors of the Jordan and Arava Valleys, respectively. Quennell (1959) recognized that these faults are left-stepping, which led him to interpret the Dead Sea depression between them as a pull-apart (rhomb-graben). Within the pull-apart, faults are recognizable by their vertical offsets, but some strike-slip faults that transfer the lateral motion between the Arava and Jericho faults are expected within the basin. Some earthquake mechanisms indeed record lateral motion (van Eck and Hofstetter, 1989). This motion is expected to decrease southward on the west side of the DSB and increase southward on its eastern side. Like elsewhere along the southern half of the transform, the major strike-slip faults and the pull-apart between them are embedded in the transform valley that is delimited by zones of normal faulting (Figs 4-3–4-5).

The Jericho and Arava strike-slip faults are prominent fractures that displace young sediments (Garfunkel et al., 1981). They are quite straight in map view, but minor local irregularities produce secondary structures compatible with left-lateral motion. In the Arava Valley, a few small pull-aparts and transpressional uplifts are developed. Their small size, especially of the pull-aparts, shows that they are young features resulting from recent rearrangements of the active fault strands. The Jericho fault is now transpressional; this is expressed by the deformation of the valley fill on the surface (Gardosh et al., 1990) and in the subsurface by the moderate dip of the main fault (Rotstein et al., 1991). About 30 km north of the Dead Sea, compression along the Jericho fault produced the Grain Sabt domal structure, which exposes more than 0.5 km of basin fill (Blake and Ionides, 1939; Bender, 1974). It is unlikely that this fault was always compressional, because a downfaulted valley with a thick sediment fill would not have formed next to it. Therefore, the present fault geometry is interpreted as having arisen only quite recently by shifting or bending of the main active fault strand.

The normal faults bordering the transform valley usually form prominent scarps, except where the transform margins are slightly uplifted, that is, west of the northern Arava. These faults often form several structural steps, each less than 1 km wide. The downfaulted blocks tend to be tilted and may dip 20°-30° in various directions. In places, the faults have zig-zag

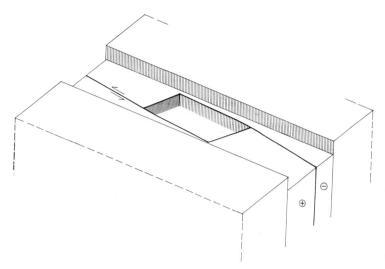

Figure 4-5 Sketch showing the structural setting of the Dead Sea basin: The pull-apart occupied by the basin and the master strike-slip faults are embedded in a valley delimited by normal faults.

shapes or are curved in plan view. On the western side of the DSB, the border faults are separated from the deepest part of the basin by a 3- to 4-km-wide step that is downthrown 1–3 km (the "median step block" of Kashai and Croker, 1987; see Gardosh, et al., chapter 5, this volume). Such a step is absent on the east. Thus, the DSB is asymmetric, with the deepest part extending along its eastern side. The Cretaceous and older sections crossed by the Amiaz-1 and En Gedi 2 boreholes (Am and EG 2 on Fig. 4-3) show that the marginal blocks were not offset laterally relative to the shoulder of the western basin. Thus, the lateral motion must have taken place within the deep part of the DSB.

The structure of the deepest part of the basin is known only in broad outline (Figs. 4-3, 4-5). It is 6–8 km wide and consists of four segments that are separated by transverse faults. The southern segment, comprising the central Arava, exposes a 2-km-thick section of the Hazeva Formation that is tilted westward (Bartov et al., 1993). Near lat. 30°35' N, the southern segment is crossed by a transverse fault downthrowing to the north—the Buweirida fault. This fault forms a well-defined scarp, attesting to continuing recent activity.

The second segment—the northern Arava—has a thicker fill than in the central Arava section. Northward thickening of the fill is recorded by the gravity anomaly that becomes increasingly negative to the north (ten Brink et al., 1993). In the Arava-1 well (Ar on Fig. 4-3), close to the basin margin, the Hazeva Formation is more than 2.5 km thick (Zak, 1967; Horowitz, 1987), so we expect a similar or thicker section in the basin center. The post-Hazeva fill may be up to several kilometers thick. In the northern part of the Arava, seismic profiles reveal a flat-lying faulted fill (Kashai and Croker, 1987; ten Brink and Ben-Avraham, 1989), but the fault offsets and pattern are not well constrained. The absence of any signs of halokinetic deformation suggests that the Sedom Formation is absent or very thin. Surface faulting is recorded by some prominent but hitherto little-studied northwest--trending scarps.

The third basin segment comprises the southern Dead Sea depression. It is separated from the northern Arava by the transverse Amazyahu fault—a listric fault that flattens at a depth of 6 km according to seismic data (Arbenz, 1984; Kashai and Croker, 1987). The sediments above the fault are of Pliocene to Quaternary age (Horowitz, 1987). In the south, sediments form a rollover structure but flatten out northward. Flat-lying sediments beneath the fault most probably belong to the Hazeva Formation, rather than to a pretransform series (see subsequent discussion). The Sedom Formation is well developed in this basin

segment and forms two large diapirs. On the west is the Sedom diapir, built of a subvertical salt wall with younger beds on the east (Zak, 1967). Seismic reflection shows that the salt rose from the level that became the flat part of the Amazyahu fault. In the north, the much larger Lisan diapir extends to a depth of 4.5–5 km, according to Abu Ajamieh et al. (1989, p. 135). It probably has a complex internal structure: Bender (1968) shows two domal highs on the Lisan Peninsula, whereas in the north, seismic data (Neev and Hall, 1979) reveal two shallow culminations. The southern border of the Lisan diapir is probably controlled by a transverse fault downthrowing to the north; this fault was already well-outlined when the Sedom evaporites accumulated (see subsequent discussion). The shallow expression of this fault is not clear, however. It may produce the deformation seen on a seismic profile southwest of the diapir (ten Brink and Ben-Avraham, 1989), but the observed features also may express halokinetic deformation.

The basin fill beneath the Lisan diapir was estimated to be some 10 km thick, based on gravity anomalies (ten Brink et al., 1993). This may well be an underestimate, because a very low density (less than pure halite) was assumed for the entire fill. Such a great thickness may imply that here the salt is much thicker than 5 km, but a more probable explanation is that a thick, prehalite series, that is, the Hazeva Formation underlies this part of the basin.

The fourth basin segment comprises the northern Dead Sea depression, which is topographically the lowest part of the DSB. In the south, it is probably delimited by a transverse fault—the En Gedi fault. An accentuated gradient of the magnetic anomaly is interpreted to record an offset of the magnetic basement, 10 km deep, by this fault (Frieslander and Ben-Avraham, 1989). The seismic survey of Neev and Hall (1979) revealed important halokinetic deformation north of this fault. In the south, diapirs deform the lake floor whereas farther north only sediments more than 200 m below the lake floor are deformed. This shows that halite of the Sedom Formation also extends under this part of the basin. An area of accentuated young subsidence—the Arnon sink—is present east of the halokinetic structures. Because these structures are not large enough to justify interpretation of the Arnon sink (AS on Fig. 4-3) as resulting from halite withdrawal, this low seems to be controlled by faults that do not reach the floor of the Dead Sea. Under the northernmost part of the lake, the upper 300 m of sediments are essentially flat, and do not show any indication of tilting of the basin floor. The steep slope at the northern end of the Dead Sea looks like it is controlled by a transverse fault, but no fracture could be identified in this area. Perhaps it is obscured by the sediments of the Jordan delta.

The crustal structure under the Dead Sea basin is poorly constrained. The presence of 10-km-thick basin fill inferred from magnetic and gravity anomalies (Frieslander and Ben-Avraham, 1989; ten Brink et al., 1993) indicate a thinned crust, but the depth of the Moho cannot be inferred with certainty.

HISTORY OF THE DEAD SEA BASIN

Reconstruction of the history of the DSB is necessary to understand its formation. In this section, the data about the basin fill are used for this purpose. It is convenient to consider three stages corresponding to the divisions of the fill.

Hazeva Formation times (Early to Late Miocene)

The Hazeva Formation, being developed both within the DSB and in the bordering regions, provides critical information about the tectonic setting in which the basin formed and about its early history. Therefore, the main features of this formation will be reevaluated—first the general setting, then the features of its occurrences in the Negev that are relevant for interpreting the record within the DSB.

The Hazeva Formation

In most outcrops, the Hazeva Formation consists of two units (Bentor and Vroman, 1957; Sneh, 1981). The lower unit (Shahak and Mashak members), a few tens of meters thick, consists of conglomerates, marls, and some limestones. The upper unit (Gidron member) consists of quartzose sandstones with some silt and shale interbeds; various pebbles, predominantly chert, form conglomerate beds or are dispersed in the sandstones. Within the DSB, this unit is more than 2.5 km thick, but in outcrops it is usually less than 200 m thick, except in a few localities (e.g., in the Karkom basin, as discussed subsequently).

The age of the Hazeva formation is constrained by several lines of evidence (Fig. 4-6). Palynological data from boreholes in the DSB (Horowitz, 1987) indicate Early to Late Miocene ages. In the northern Negev, vertebrate fossils of Burdigalian age occur in the lower part of the sandy unit (Goldsmith et al., 1988). Somewhat higher are marine-brackish beds with oyster banks that correlate with the marine Ziqlag Formation, which formed during the Langhian–early Serravallian rise of sea level (Rothman, 1967; Garfunkel, 1988a; Buchbinder et al., 1993). In the Beer Sheva valley, marine beds of Tortonian age in the Pattish Formation (Martinotti et al., 1978, Buchbinder et al., 1993) fill an erosional valley incised 200 m below the level of the Ziqlag Formation (in the north) and the Hazeva Formation (in the east) (Gvirtzman, 1970; Garfunkel, 1988a). This constrains the age of the Hazeva Formation in the northern Negev to be older than 10–8 Ma. East of the central Arava, sediments that are probably coeval with the Hazeva Formation (Bender, 1968, 1974) predate basalts dated at about 9–6 Ma (Steinitz and Bartov, 1991).

The clastics in the lower unit of the Hazeva Formation consist of carbonates and rarer cherts of local derivation, whereas the clastics in its upper part were derived from distant sources (Garfunkel and Horowitz, 1966; Garfunkel, 1978). Quartz, which forms the bulk of these sediments (carbonate and chert grains are quite uncommon), could not have been derived from the area that is still covered by the mainly carbonatic Cretaceous–Eocene beds, so it must have been derived from farther inland, that is, from southern Sinai and northwestern Arabia. These areas expose Paleozoic to Early Cretaceous mature sandstones and Late Proterozoic basement, which are appropriate sources of the Hazeva sandstones. This is confirmed by their mineralogy (Sa'ar, 1985). The pebbles dispersed among the Hazeva sand-

stones also indicate inland sources. The pebbles often consist of rock types that are unknown in the Negev, but which represent more inland Eocene–Cretaceous facies that were largely eroded. Most common are distinctive laminated cherts (so called "allochthonous" or "imported") and rarer quartzites with phosphates. Pebbles of basement rocks, mainly igneous, are sometimes also present.

In the Miocene, the source areas of the Hazeva clastics formed the eastern shoulders of the Suez rift and the Red Sea basin. These areas were significantly uplifted after the Burdigalian (Garfunkel, 1988b). This uplifting led to the erosion that supplied the Hazeva clastics and also created a regional slope of 2–5 m/km on the average, which increased with time. The clastics were transported down this slope to the DSB and west of it by a river system 300–400 km long. The Hazeva beds were deposited in response to the increased supply of clastics, which changed the balance between transportation and deposition in favor of the latter. However, even more important was the creation of an erosive and tectonic relief, which served as a sediment trap before and during the deposition of the Hazeva Formation.

The Miocene relief and structures in the Negev

In the northern Negev, west of the DSB, the Hazeva sediments filled morphotectonic basins. These basins essentially expressed the Syrian arc synclines, as well as erosional channels, up to 200 m deep, which cross some anticlinal crests (Fig. 4-7; Neev, 1960; Garfunkel and Horowitz, 1966; Harash, 1967; Shahar, 1973; Wdowinski, 1985; Zilberman, 1992). We infer that a pre-Hazeva drainage system was entrenched into the bedrock in response to regional tilting and to accentuation of the fold structures, which became directly expressed in the physiography. Initially, this led to deposition of locally derived conglomerates and of fine lake beds in some synclinal valleys. Afterward, sediments of distant origin were trapped in the relief, filled it, and even extended over the high-standing anticlinal crests. Originally, at least 200–350 m of sediments were deposited in different synclines (Fig. 4-7). Now, only 70- to 180-m-thick sections are preserved. The thickness of the beds deposited above the anticlinal crests is not known, but several points bear on this question. First, the formation of the pre-Hazeva erosive relief, and the low position of the Miocene marine-brackish beds show that the area remained somewhat uplifted until that time; there is no evidence of subsequent subsidence. Second, no feature existed that could trap much sediment after the high relief was covered. Finally, the age constraints do not leave much time—perhaps only a few Ma—for deposition after the fold relief was filled. In such a setting, it is unlikely that thick (greater than 100 m ?) sections could accumulate after the relief was filled.

Deposition ended because arching of the northern Negev (Picard, 1951; Neev 1960) blocked the sediment supply and led to erosion. As a result, the base of the Hazeva Formation in the Rotem syncline is now at an elevation of 280–320 m, whereas in the Dimona basin down the original slope, its base is higher—more than 480 m above sea level (Fig. 4-7; Arad, 1959; Harash, 1967). The structural relief of individual folds may also have been accentuated, but this is difficult to resolve.

The Hazeva Formation occurs also in structural lows along the central Negev–Sinai shear belt. The formation is tilted and faulted, which shows that these structures were active during or after its deposition. The thickest section is found within the Karkom basin, a rhombic pull-apart along the Paran lineament (Fig. 4-1; Bartov and Garfunkel, 1980, 1985). Here the lower unit is 30 m thick, whereas the overlying sandy-conglomeratic section of distant derivation is up to 1 km thick in the western part of the basin. The lower unit changes from locally derived coarse conglomerates near the basin margins to mostly fine lacustrine beds toward the basin's center, which proves that the basin was

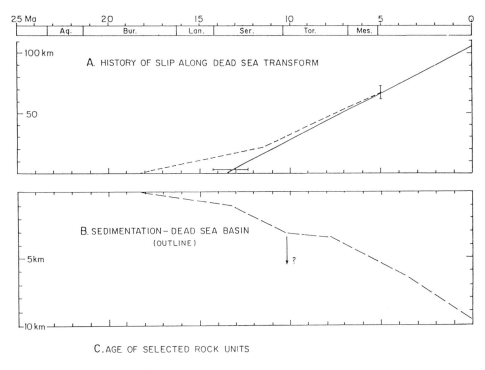

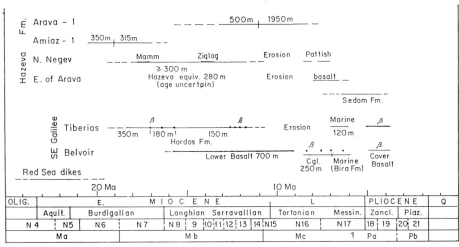

Figure 4-6 Summary of the main data relevant to the history of the Dead Sea basin. Time scale is according to Harland et al. (1989). The bottom line shows palynological zones used by Horowitz (1987).

already outlined when Hazeva sedimentation began. My observations revealed that at coord. 9701 N, 1466 E and 9702 N, 1457 E, beds about 70 m and more than 150 m above the base of the formation onlap on Eocene rocks and on the basal conglomerates, respectively, which proves continuing synsedimentary basin subsidence. Repeated occurrence of small pebbles of carbonates, which could not have traveled far, in the Hazeva sandstones shows that pre-Neogene rocks remained elevated, exposed, and were eroded outside the basin. This implies that the Hazeva Formation was not much thicker than 200–300 m, for otherwise it would have covered any relief of older rocks. To maintain such a situation, syn-Hazeva subsidence of the basin must have occurred. This was a major cause for the deposition of an anomalously thick Hazeva section within the Karkom basin.

An 80-m-thick relict of the lower unit of the Hazeva Formation is preserved approximately 5 km east of the Karkom basin (Sakal, 1967). It consists of alternating locally derived coarse clasts and lacustrine marls and limestones, a feature absent in

the Karkom basin. This indicates the existence of an eastern basin with a distinct depositional history. The eastern basin was probably separated from the Karkom basin by a topographic structural high.

Hazeva sediments are also preserved in two structural lows along the Sa'ad–Nafha lineament (Zilberman, 1981, 1992). The low position of these sediments resulted in part from tectonic deformation, but it is also due to deposition over a relief. In the Mahmal depression, they onlap an approximately 100-m-high relief of a flexure, whereas the Teref depression, they fill a somewhat larger erosive relief. The coarse nature of the locally derived basal conglomerates further indicates the presence of a nearby relief. Some 10 km south of the Teref depression, relicts of Hazeva sandstones also fill a 50-m-high erosive relief (Zilberman, 1992). These relations show that an erosive and structural relief existed in this area immediately before and during deposition of the preserved Hazeva sediments.

Sporadic sandstone occurrences in karstic pockets in Eocene beds on the high plateaus of the central Negev attest that the

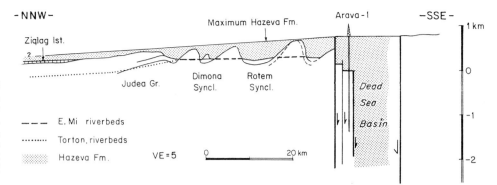

Figure 4-7 Cross section through the northern Negev (from the northern Arava to the Beer Sheva region) showing the position of the Hazeva Formation relative to the bedrock. Heavy line shows the base of pre-Hazeva valleys crossing anticlinal ridges. Heavy points show base of Late Miocene channel.

Hazeva beds extended over these areas as well (Avni, 1991; Zilberman, 1992), provided these are not relics of Late Eocene regressive sediments. However, as in the northern Negev, it is unlikely that sections much thicker than 100–200 m were deposited over the plateaus: There is no obvious feature to trap the sediments, there are no indications for regional subsidence after the formation of the erosive relief, and the available time span is quite short. Therefore, it is difficult to accept the suggestion (Zilberman, 1992) that a veneer of Hazeva sandstones many hundreds of meters thick extended over the plateaus.

Farther south, at Shekh Atia 70 km south of Elat, the Hazeva clastics, more than 200 m thick, lie on the Maastrichtian Ghareb Formation, whereas a thick Eocene section is preserved farther west. In this region, up to 400 m of high relief was eroded before the deposition of the Hazeva sediments.

West of the DSB, the Hazeva Formation fills a tectonic and erosive relief that served as sediment traps. This relief resulted partly from regional tilting, but mainly from the activity of the Syrian arc and especially of the central Negev–Sinai shear belt before and during deposition of the Hazeva sediments. Sediment thicknesses usually reached a few hundred meters, except in deep tectonic lows (e.g., Karkom basin).

The Arava and Dead Sea basin during Hazeva times

Sediments of the Hazeva Formation are known from outcrops and boreholes along a sector of the DSB more than 80 km long (Fig. 4-3). These occurrences are much thicker than the adjacent sections of the Negev, so they are thought to record deposition in a subsiding trough and are not downfaulted relics of a regional veneer of sediment. Thus the Hazeva beds in the DSB outline a prominent low along the transform that cuts across the older structural pattern of the Negev. These features show that the DSB formed early in the Miocene and that it subsided several kilometers during the deposition of the Hazeva sediments (Fig. 4-6).

In the central Arava, a section more than 2 km thick is exposed (Bartov et al., 1993), whereas the Amiaz-1 and Arava-1 boreholes (Am and Ar on Fig. 4-3) penetrated sections about 0.7 km and more than 2.4 km thick, respectively (Kashai and Croker, 1987). These sections are much thicker than those that existed in the nearby areas of the Negev (Fig. 4-7). According to palynological data (Horowitz, 1987; Fig. 4-6) the Early Miocene section in Amiaz-1 is more than five times thicker than the sections of that age in the northern Negev. The Middle and Late Miocene sections in Amiaz-1 and Arava-1 are also considerably thicker than the Hazeva Formation farther west. Because these wells are close to the basin's margin, similar or even thicker sections are expected in the basin center, for example, beneath the Amazyahu fault and farther north. This is consistent with the gravity data, suggesting that a Hazeva section a few kilometers thick may exist beneath the Lisan diapir.

The Hazeva beds of the northern Arava (Bentor and Vroman, 1957; Eidelman, 1979; Sneh, 1981) lie on Eocene, and locally older beds that form an eastward dipping flexure. On the east, this flexure is separated by a major fault from the deep part of the DSB (May, 1968; Mikhaeli and Eckstein, 1968; Bartov et al., 1993; Fig. 4-3). In the west, the basal beds usually consist of locally derived conglomerates, but eastward they pass into somewhat thicker fine lacustrine beds (Sneh, 1981). This configuration indicates that the area west of the Arava was uplifted and eroded before the arrival of distantly derived sandstones (that is, in the Early Miocene, according to the age data from farther west). The area west of the Arava is interpreted to be a part of the margin of the embryonic DSB. Farther north, sandy sections are exposed east of the Sedom diapir and near the west shore of the Dead Sea near lat. 31°15'N (Agnon, 1983), but their original setting was obliterated by faults.

These data show that the DSB began to subside relative to the neighboring areas in Early and Middle Miocene times and that its subsidence probably accelerated by the Late Miocene. The strongly subsiding depression was 70 km long and up to 8 km wide, at least in the central Arava. Because it is located along the transform, this depression is best interpreted as a pull-apart that formed during the early stages of lateral slip. This interpretation is strongly supported by the restoration of the entire transform motion (Fig. 4-2): The DSB originated where the transform trace, as constrained by the distribution of pretransform rocks, was bent to the left. According to the plate kinematic model of Garfunkel (1981), the misalignment of the transform is 6–8 km. The southernmost part of the DSB, exposing the thick Hazeva section, is located along this bend and thus may be the old part of the basin. By 10–8 Ma, the transform moved several tens of kilometers (Fig. 4-6), so the basin could have acquired its observed length if it nucleated from 40-km-long deformed zone along this bend (see subsequent discussion).

This interpretation of the DSB implies that the transform began to move in the Early Miocene, perhaps 18–16 Ma. This age is compatible with the regional plate kinematic constraints and is further supported by the history of the Tiberias region (Fig. 4-6). Here, a small basin filled with about 300 m of conglomerates and lake beds (Hordos Formation) formed next to the transform before the eruption of the approximately 17-Ma-old basalt flows. By 10–8 Ma, a depression about 30 km long and up to 10 km wide filled with 500 m of volcanics developed along the transform (Garfunkel, 1989; Shaliv, 1991). Thus, volcanism and deformation were already localized along the transform at the end of the Early Miocene, suggesting that the transform was active at that time.

Though the DSB subsided a few kilometers during the Miocene, it did not develop into a topographic low which would prevent the westward transport of the Hazeva sediments to the northern Negev. At least temporarily, sedimentation kept pace with basin subsidence, so that sediments could be transported across the basin, perhaps not always through the same outlet. Thus Hazeva deposition outside the Dead Sea basin may not have been continuous. This notion is supported by the presence of several types of sandstones (Wdowinski, 1985), which suggests several distinct depositional events.

Such early activation of the Dead Sea transform raises the following question: How did this activation relate to the central Negev–Sinai shear belt, which was also active during (and perhaps after) the deposition of the Hazeva Formation? The extensions of the Paran and Themed lineaments east of the transform, which displace probable equivalents of the Hazeva sediments (Bender, 1968), are truncated by an erosion surface on which about 9- to 6-Ma-old basalts were extruded (Steinitz and Bartov, 1991). Because it takes some time to form a flat erosion surface, the activity of these lines must have ended some 10 Ma or even earlier. Together with the other available age constraints (Fig. 4-6), this shows that the shear belt remained active when the first one-third or less of the transform motion took place, though it may also have been active before the beginning of the transform motion. In principle, the simultaneous activity of the transform and of strike-slip faults at a large angle to the transform is possible. This is occurring now in southern California, where the San Andreas fault and several strike-slip faults at an angle to it are active (Jennings, 1973). Therefore, the argument that lateral motion on the Dead Sea transform could have begun only after the central Negev–Sinai shear belt ceased to be active is not compelling. Furthermore, the structure of the lineaments in the central Negev is interpreted to show that only the Paran lineament may have moved more than 1 km during (and after?) Hazeva deposition, whereas shearing along the other lineaments was not more than several hundred meters. These lateral offsets were less than in central Sinai, where a cumulative offset of several kilometers across the shear belt is recorded by the displacement of the 25- to 20-Ma-old (latest Oligocene to earliest Miocene) dikes (Bartov, 1974; Steinitz et al., 1981). Such motions were not large enough to inhibit slip along the Dead Sea transform or to visibly bend the its trace.

From the foregoing discussion, we conclude that the Dead Sea transform formed early in the Miocene (18–16 Ma), following or during a period of regional reactivation of older folds and especially of the central Negev lineaments. Initially, all the structures acted simultaneously, but since about 10 Ma, only the transform has remained active. The DSB originated at the beginning of the transform motion. During this period, the surrounding region was somewhat uplifted and tilted, resulting in the widespread formation of an erosive relief before deposition of the Hazeva beds. The persistence of a river system transporting the Hazeva clastics shows that the transform margins were not much uplifted until 10 Ma (beginning of the Late Miocene). During this period, the subsiding DSB was a major sediment trap, but its filling usually kept pace with subsidence, so it did not develop into a deep valley that would be an obstacle for westward sediment transport. This paleogeographic setting was disrupted by uplifting of the transform margins, which must have greatly reduced the supply of clastics to the DSB. Therefore, sedimentation in the DSB could no longer keep pace with its subsidence, and the basin became a deep topographic depression.

Sedom Formation times (latest Miocene to Early Pliocene)

The deposition of thick evaporites—the Sedom Formation—within the DSB marks a new stage in its development. The Sedom diapir exposes an approximately 2-km-thick section of this unit that consists of about 80% halite, with the remainder comprising gypsum, marl, chalk, dolomite, and shale (Zak, 1967). Also present are some quartz-rich sandstones that may have been derived from the Hazeva Formation (Sa'ar, 1985). On the Amiaz block west of the diapir, the 1-km-thick Sedom Formation overlies the Hazeva Formation (Fig. 4-4; Kashai and Croker, 1987). In the Lisan diapir, more than 3.5 km of halite was drilled (Bender, 1974). Additional occurrences of salt are inferred from the distribution of halokinetic structures (Figs, 4-3, 4-4). These data show that halite underlies 70-km-long portion of the DSB. The great volume of halite and its chemistry, primarily the bromine content, show that it was deposited from a brine of marine origin, implying that the sea temporarily penetrated into the transform valley, most probably through the Jordan Valley (Zak, 1967; Zak and Bentor, 1972). It is likely that some clastics are lateral equivalents of the evaporites, but they have not been identified with certainty. Because the evaporites may have formed within a short time span (see subsequent discussion), such beds may be quite thin.

The DSB was differentiated into distinct structural units when the evaporites were deposited. Restoration of the subvertical salt wall that forms the Sedom diapir to its original horizontal position shows that originally the evaporites covered much of the basin floor (see Fig. 4-4). Thus, this area subsided enough to accommodate an approximately 2-km-thick evaporitic section. The thinner evaporitic section on the Amiaz block west of the Sedom diapir shows that a 1-km-high standing step was formed. The area farther south—both the western marginal step and the deep depression—was also higher standing, as there the halite-bearing section is thin or absent. The substantially greater volume of the Lisan diapir compared with the Sedom diapir shows that the former developed from a much thicker (by 1–2 km ?) evaporitic sequence. This implies that under the Lisan diapir, the basin subsided much more than farther south. Carnallite, which records very advanced evaporation, is present in the L-1 well in the Lisan diapir (Bender, 1974; Fig. 4-3), but it is quite rare in the Sedom section. This suggests that a depression in which terminal brines collected existed in the area of the Lisan diapir. These indications for quite different amounts of subsidence strongly suggest that a transverse fault, downthrowing to the north, separated the basin segments in which the halite-building Lisan and Sedom diapirs originated. Evaporites also were deposited north of the Lisan diapir, but the northward-decreasing relief of the halokinetic structures shows that the halite thins in this direction.

The Sedom Formation is usually considered to be of Early Pliocene age based on palynological evidence (Horowitz, 1987). Another age estimate comes from its correlation with the only marine unit that occurs in the Jordan Valley near Lake Kinneret (Tiberias), on the assumption that they both record the same marine ingression. These marine beds were considered to be of Early Pliocene age, that is, 5–4 Ma (Schulman, 1959), but radiometric dating of intercalated basalts yielded ages of 7–6 Ma (Fig. 4-6; Shaliv, 1991). These age determinations need not be in conflict, because the palynologically studied sections of the Sedom diapir and Amiaz-1 well may not be the oldest evaporites in the DSB: Still older beds may exist in the thicker section under the Lisan diapir. In fact, a thick evaporite body drilled in Zemah-1, south of Lake Kinneret, contains Late Miocene pollen (Horowitz, 1987), which shows that evaporites began to be deposited in the transform valley in the Miocene. Agnon (1983) proposed a Middle Miocene age for the Sedom evaporites, that is, that they are coeval with a part of the Hazeva Formation and that they formed when the Ziqlag ingression reached the DSB from the west. This interpretation cannot be accepted in view of the existing palynological data. It is possible, however, that sandy sediments included in the Hazeva Formation continued to accumulate in the southern DSB even after they could no longer reach the Negev and thus are lateral equivalents of the Sedom evaporites.

The great evaporite thicknesses prove that the DSB subsided considerably by the time of evaporite deposition, but understanding of the subsidence history requires knowledge of the time span represented by the Sedom Formation. Zak (1967) pointed out that evaporite deposition is potentially a very fast process: It may be five and more times faster than deposition of clastics, provided there is an adequate brine supply. At the present rate of evaporation of the Dead Sea, a 2-km halite layer can be deposited from sea water over an area of 10 km x 50 km within less than 0.2 Ma. Even if the Sedom evaporites formed 10 times more slowly, they could have formed within 2 to 3 Ma. Such a short time interval can also be deduced by assuming likely rates for the accumulation of the small amounts of clastics that are dispersed among the Sedom evaporites. If indeed the Sedom Formation formed during such a brief time span, then the basin subsided at a rate of 1 to 2 km/Ma, provided sedimentation kept pace with subsidence. However, subsidence may have been slower if the evaporites filled a preexisting, deep topographic depression. Such a depression most likely formed when the supply of Hazeva clastics was cut off and the basin became starved, so clastic sedimentation could no longer keep pace with basin subsidence.

The post evaporitic stage (Late Pliocene to the Present)

After the end of evaporite deposition, the DSB continued to subside and was filled by several kilometers of sediments. However, sedimentation did not keep pace with subsidence, so the DSB became a land-locked topographic depression into which the drainage of a large area was diverted (perhaps before deposition of the Sedom Formation). The surrounding areas continued to rise and were eroded, supplying clastics to the basin. Having no outlet, the topographically lowest parts of the DSB were occupied by lakes that necessarily became hypersaline. The lakes expanded and shrank and their water levels rose and fell in response to climatic conditions that controlled the balance between evaporation and water supply. Deposition within the DSB took place, therefore, in fluviatile environments, such as floodplains and alluvial fans, as well as in lakes. The sediments consist mainly of clastics—conglomerates, sandstones, and shales—but in the lakes, chemical sediments such as carbonates, gypsum, and halite also formed. Halite seems to have formed from dissolved older evaporites, because the chemical data do not indicate a connection with the sea (Zak, 1967). Along the basin's margins, the various facies alternate, depending on the fluctuations of the lake levels; this is seen in outcrops around the Dead Sea (Garfunkel, 1978; Gardosh et al., 1990), as well as in boreholes (Neev and Emery, 1967; Zak, 1967).

The thickness of the post-evaporitic sediments is known only from parts of the basin. Sediments are several kilometers thick east of the Sedom diapir, and they are probably not much thinner in the northern Arava. The En Gedi 2 well (EG 2 on Fig. 4-3) borehole penetrated a section of sediments more than 1350 m thick. Thicker sections are expected in the more central parts of the basin on the flanks of the Lisan diapir and under the deep northern part of the Dead Sea, though there is no direct data regarding the age of the fill in these areas. Thus, the available information shows that much of the DSB continued to subside in Pliocene–Pleistocene times and that sedimentation rates reached 1 km/Ma in some areas (Fig. 4-6). However, the sediment thicknesses do not directly reflect the tectonic subsidence in this period, because the basin was not completely filled. Moreover, salt withdrawal must be taken into consideration. Thus, diapirism resulted in the westward movement of a section of sediment about 2 km thick from below the Amazyahu fault,

thus providing space for younger sediments.

Following the terminology of Zak (1967), the bulk of post-evaporitic sediments is called the Amora Formation, but other names were applied to the exposed youngest part of these sediments. In particular, the sediments that formed during the last high stand of the lake within the basin—about 50–15 ka—are called the Lisan Formation (Neev and Emery, 1967; Begin et al., 1974). Then the water level reached 180 m below sea level, so the lake extended from Lake Kinneret to the northern Arava. Most of this area on the margin of the lake was covered by up to 40 m of sediments consisting of varves of fine clastic and chemically deposited aragonite.

The clastics in the post-evaporitic series were mostly derived from the marine Cretaceous–Eocene sediments in the areas surrounding the DSB and thus consist largely of limestone, dolomite, and chert, in grains ranging from cobble to silt size. In the finer fractions, clays are also common. Components from other sources are much less common. These include quartz grains, which were probably mostly derived from the Hazeva Formation and from Early Cretaceous and older sandstones, and also pebbles of basement rocks. The latter occur, for example, in the Amora Formation exposed on the margins of the Sedom diapir and in the northern Arava. Such clasts also occur in fluviatile sediments that were deposited by an old stream system that drained into the DSB (Garfunkel and Horowitz, 1966; Avni, 1992). Thus, some basement clasts and sand grains in the young fill of the DSB were derived from southern Sinai, but the outcrops east of the Arava were probably the main source of these clastics. Sa'ar (1985) also found heavy minerals (e.g., augite, and hornblende) derived from volcanics rocks, which are absent from older sediments. The young volcanics east of the Arava and Dead Sea are the only plausible source for these components.

Outside the DSB, some correlative sediments are present, covering an erosive relief that formed in post-Hazeva times. These include thin sheets of alluvial sediments (known as Arava and Meshar Conglomerates) that were deposited by several generations of streams in the central and southern Negev and in the Arava (Garfunkel and Horowitz, 1966). Some lake beds are also present. These include marls up to 10–20 m thick containing brackish water microfossils that occur in several localities west of the DSB, from near Jericho in the north to the margins of the Arava (Shahar et al., 1966; Eyal, 1984). These beds were called the Mazar Formation, though their contemporaneity was not established. Their position close to the present sea level shows that they formed when the base level of erosion was higher than at present, or when a high-standing water body existed in the DSB. Similar fossils were found at the base of the basin fill in the En Gedi 2 well ("Lido facies" of Neev and Emery, 1967; Zak, 1967). At present, the exact age of all these sediments is not known, so they do not provide detailed information about the history of the DSB.

The distribution of the thick post-evaporitic sediments shows that a large part of the DSB continued to subside. The distribution of known transverse faults with a record of young displacement show that basin-parallel extension and other deformation continued along much of the DSB. However, the Amazyahu fault is the only post-evaporitic structure known in some detail. Seismic reflection data (Kashai and Croker, 1987; ten Brink and Ben-Avraham, 1989) show that the Amazyahu fault is a listric fault that flattens out at a depth of 6 km and that it was active during sedimentation. The onlap of the sediments on the fault surface shows that it accommodated a longitudinal extension of several kilometers. However, the inference that a thick section of the Hazeva beds underlies this fault shows that it is a detachment within the basin fill. Moreover, its northward projection should cross the transverse fault south of the Lisan diapir well above the base of the fill. Thus, it is not clear how the motion of this fault is related to the deformation of basement beneath the fill of the DSB.

The distinctive morphology of the northern part of the Dead Sea depression suggests that it subsided faster than other parts of the basin at least in the Late Pleistocene, as there is no reason to assume that it received less sediment than the more southern parts of the basin. Its nearly rectangular shape further suggests that it was controlled by faults, though the faults controlling its northern and southern extremities are uncertain. The presence of young transverse faults farther south shows that the more southern parts of the DSB also remained active in Pliocene–Pleistocene times. If these faults have similar offsets as the Amazyahu fault, they can account for much of the transform motion in the last 3–4 Ma. It is concluded, therefore, that much of the DSB basin subsided and was longitudinally extended throughout Pliocene–Pleistocene times. The morphology of the northern basin segment thus resulted from faster subsidence than farther south, at least in the Quaternary.

The formation of diapirs was probably tectonically driven. The density of the Sedom Formation is somewhat higher than 2.2 g/cm^3, depending on the content of gypsum and clastics (Zak, 1967). The overlying, mainly clastic sediments, when water-soaked, will have a higher density if their porosity is less than 25–28%. Compaction of these sediments can produce a significant density inversion at a depth of only a couple of km. Therefore, it is most likely that diapirism was driven by the formation of structural steps, and tilting of the base of the evaporites resulting from transverse extension of the basin produced lateral pressure gradients in the evaporites that drove the diapirs. These factors can explain the elongated shapes of the Sedom and Lisan diapirs.

Summary of basin development

The foregoing account shows that the DSB developed during most, if not all, of the history of the transform (Fig. 4-6), having originated along a bend to the left (i.e., a releasing bend) of the original fracture. By the end of the Hazeva Formation (ca. 10 Ma or later), the basin was already 70 km or more in length. As the basin developed, the sites of the most important subsidence overlapped to a great extent, and tended to migrate northward (Zak and Freund, 1981). At least during the deposition of the evaporites and afterward, the basin was differentiated into parts that subsided at quite different rates. At any one time, large portions of the basin subsided simultaneously. In this respect, the present situation is quite representative: Though the northern Dead Sea depression seems to be the locus of the most accentuated subsidence, considerable subsidence also occurred farther south. Moreover, surface faulting in the southern half of the DSB shows that deformation of the basin floor is not confined to the northern depression.

The history of basin subsidence can be reconstructed in general only. During Hazeva Formation times, sediments completely filled the basin, while its shoulders were somewhat uplifted. Thus, sediment thicknesses provide a good measure of basin subsidence. The known sections show that subsidence was rather slow in the Early and Middle Miocene, but it has accelerated since the Late Miocene (Fig. 4-6). These sections do not show that subsidence was fastest during the early stages of basin development, as is common in many tectonic basins. However, since thicker Hazeva sections may exist under the deepest parts of the DSB, this point is still not clear. After the Hazeva times, sedimentation rates reached 0.5–1 km/Ma and higher. Available age determinations do not allow us to resolve spatial or temporal changes in the subsidence rates. Moreover, during this period, the basin was partially starved, so it developed into

a deep physiographic depression not only because its shoulders were uplifted, but also because sedimentation could not keep pace with basin subsidence. Therefore, sediment thickness can no longer provide a measure of subsidence.

Ongoing tectonic deformation occurs along a large part of the DSB. Most apparent is the activity of longitudinal faults. However, the presence of several transverse faults along most of the basin shows that extension parallel to the basin axis occurs simultaneously along much of the DSB. This is corroborated by the seismic activity that occurs in a broad band under much of the basin. Available seismic reflection data show that in the subsurface, like in outcrops, only a small number of transverse fractures are present. However, if the Amazyahu fault, the only one whose deep structure is known, is representative of the other faults, then they too are listric. Since they are synsedimentary fractures, together they could accommodate a total basin-parallel extension of more than 20 km during the last 2–3 Ma. It is possible that extension occurred preferentially in the north of the basin, but it clearly was not confined to the deep northern depression of the Dead Sea. This component of deformation is not very well expressed on the surface, however, because of the young age of the surficial sediments, which record only a very brief deformation period.

THE DEAD SEA BASIN AS A PULL-APART

The foregoing discussion of the geologic features of the DSB provides the basis for examining its origin, development, internal structure, and relation with the transform. This is best done within the framework of the pull-apart concept, which allows integration of the available data. Though pull-aparts are widespread structures and were much discussed (e.g., Crowell, 1974; Aydin and Nur, 1982; Mann et al., 1983; Christie-Blick and Biddle, 1985; and references therein), their internal architecture and evolution are still incompletely understood. Therefore, in the subsequent discussion, relevant principles regarding pull-aparts in general will be discussed first. These principles will then be used to analyze the DSB.

Basic geometric relations in large pull-aparts

Pull-aparts are identified on the basis of structural relations seen in map view. Because we can directly observe these relations, they provide the starting point for interpretation of the origin and architecture of pull-aparts. The basic feature of pull-aparts, including the DSB, is their position between two master strike-slip faults, which are stepping sideways in the same sense as the lateral motion. The kinematics of this fault arrangement require that divergence takes place in the area where the master faults overstep, leading to the formation of a depression (Kingma, 1958; Lensen, 1958; Quennell, 1958, 1959). As lateral motion continues, the pull-aparts grow by becoming longer in the direction of the lateral motion, whereas their width does not change as long as they are delimited by the same master faults. Conceptually, pull-aparts can be depicted as box-shaped depressions that originate by the separation of the two sides of a single fracture that linked the master faults (Fig. 4-8). However, because they are delimited by fault systems, large pull-aparts often have a more complex internal structure, including blocks of the surrounding bedrock. A more realistic model of such pull-aparts is that they originated by stretching and subsidence of fractured zones of finite length while the surrounding, little-deformed blocks moved apart.

In all cases, the pull-aparts cannot be shorter than the lateral motion that took place during their formation (length d in Fig. 4-8). This rule allows us to use the length of a pull-apart to constrain its age relative to the history of the enclosing fault zone.

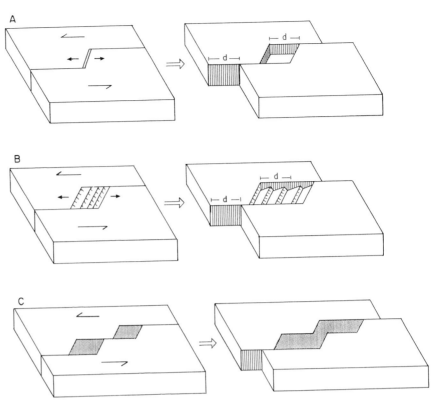

Figure 4-8 Basic relations of pull-aparts (see text for discussion). d is the amount of lateral displacement. A, pull-apart originating from a single fracture; B, pull-apart originating from a faulted zone; C, coalescing pull-aparts.

Thus, a pull-apart that is much shorter than the total offset of the enclosing fault zone must have formed late in the movement history. This implies rearrangement of the active fractures as motion progressed. As the lateral motion increases, several initially separate pull-aparts may coalesce into composite depressions (Garfunkel, 1981; Fig. 4-8). In such a case, the length of individual components, rather than of the entire depression, must be used.

Within the pull-apart, the lateral motion is transferred from one master fault to the other. The kinematics of the structure require that this took place over an area that is at least as long as the magnitude of the offset that produced the pull-apart. Therefore, when this motion is large compared with the length of the pull-apart, Ben-Avraham and Zoback (1992) propose that it cannot be confined to only one of its sides. Rather, at some period, lateral motion should have taken place (perhaps varying in magnitude) along much of the two sides of the pull-apart. At any one time, however, the lateral motion may be transferred across a single transverse structure, such as a fault. When several such transverse structures exist, lateral motion must occur on both sides of the pull-apart between these structures.

The origin of laterally stepping faults, which delimit pull-aparts, is not clear. At first sight, this fault arrangement resembles sets of *en echelon* fractures that develop during the initial stages of brittle failure induced by simple shear. Among these, fractures belonging to two sets, called Riedel and P shears, slip in the same sense as the externally applied shear (Fig. 4-9). They form angles of 10–15° with the direction of the overall shear, but they have different orientations. Such fractures were observed to form in laboratory experiments with scale models and with rock samples, and also when alluvium is deformed above active strike-slip faults (e.g., Tchalenko and Ambraseys, 1970; Wilcox et al., 1973; Freund, 1974; Bartlett et al., 1981; Christie-Blick and Biddle, 1985; Naylor et al., 1986). Similar fracture patterns also form during natural mesoscale deformation, that is, when off-

sets are less than about 1 m (e.g., Gamond, 1985).

It is doubtful, however, that such fractures are the precursors of large strike-slip faults with lateral offsets exceeding hundreds of meters. Both the Riedel and P shears are oriented at an angle to the direction of overall shearing, and therefore they are kinematically incapable of behaving as pure lateral slip faults and accommodating the overall shearing. Although opening occurs along the Riedel shears, they are not the precursors of pull-aparts. The arrangement of Riedel shears relative to the direction of overall shear is opposite to the master faults that delimit known pull-aparts, and the opening along them is almost normal to the direction of shear rather than parallel to it (Fig. 4-9). The arrangement of P shears resembles that of master faults delimiting pull-aparts, but their motion should be accompanied by compression, which is not a general feature associated with pull-aparts (see subsequent discussion). In fact, both in the laboratory and along natural faults, the initial fractures cease to be active and are disrupted as shearing continues, replaced by new through-going fractures trending parallel to the direction of the overall shear. The latter fractures, rather than the initial ones, are the most likely analogs of large natural strike-slip faults. Laterally stepping faults probably form when several large faults nucleate independently of each other along a fault zone, but it is not clear what controls this process. Therefore, kinematic considerations still provide the best basis for understanding of pull-aparts.

Very little information about the three-dimensional structure of pull-aparts is available. It is unlikely that their surficial structure extends to great depth. Pull-aparts are actually openings formed near the surface by separation of the flanking blocks, but at depth such openings are not expected to form. Moreover, at midcrustal and deeper levels, where ductile rather than brittle deformation dominates, the lateral block motion most likely occurs along displacement zones that are smoother than the fracture patterns seen on the surface. Very little data are available

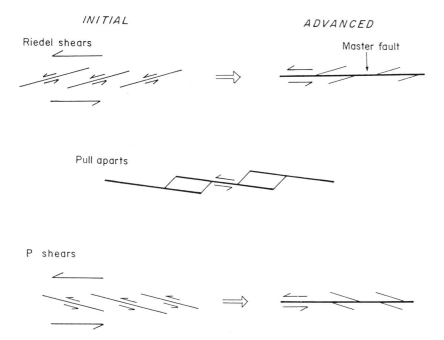

Figure 4-9 Fracture patterns formed during initial failure by shearing. Top, Riedel shear; bottom, P shears. For comparison, the fault pattern associated with pull-aparts is shown in the middle.

about such deep features, but we can use the kinematics of pull-aparts to place some constraints on their nature. For this purpose, it is important to clarify how the overall trend of a fault zone containing pull-aparts relates to the motion between the adjacent blocks.

At first sight, it might appear that the relative motion between the blocks should be along the fault zone, but this need not always be the case. In such a situation (Fig. 4-10) the block motion is at an angle to the individual laterally stepping faults (resembling the situation with P shears). With such a geometry, continuing lateral motion will cause convergence along the master faults as well as along the pull-aparts, leading to progressive narrowing of their older parts (Fig. 4-10). This will produce contractional structures, such as folds and thrusts, both within the pull-aparts and along the master faults. This will also happen when the direction of block motion is intermediate between the strike of the laterally stepping faults and the overall trend of the fault zone. However, in many cases, including the DSB, compressional structures are absent or little developed, and transverse extension is even observed. It is concluded, therefore, that in such cases the relative block motion is nearly parallel to the master faults delimiting the pull-aparts, and hence at an angle to the fault zones as a whole (it may deviate even more from the fault zone, leading to transtension). This kinematic picture is, in fact, implicitly accepted in many discussions of pull-aparts.

These considerations are especially pertinent when the fault zone separates rigid blocks or plates and when its lateral offset is large (e.g., a transform plate boundary). Then the individual laterally stepping faults are arcs centered on the corresponding Euler pole but at different distances from it, whereas the fault zone as a whole is not centered on this pole and is transtensional if it encloses only pull-aparts (Garfunkel, 1981).

These kinematic relations are compatible with two limiting possibilities of how the lateral motion is accommodated at depth (Fig. 4-11). In the first situation, the motion along the underlying fault or displacement zone is purely lateral, that is, its two sides always fit together as motion progresses. Such a deep displacement zone must be (in map view) a circular arc centered on the Euler pole describing the relative block motion. Thus, it cannot extend beneath the shallow fault zone, which is not such

an arc. Therefore, to join the laterally stepping surficial faults, the deep displacement zone must split upward into distinct segments, parts of which deviate considerably from the vertical (Fig. 4-11). In the second limiting situation, the deep displacement zone extends directly beneath the shallow fault zone and is therefore at an angle to the direction of relative block motion. Such a displacement zone is transtensional when pull-aparts are developed on the surface, and the divergence along it should match the increase in area of the pull-aparts.

These two limiting situations have very different consequences for the structure beneath pull-aparts. In the first situation (Fig. 4-11a), detachments may exist below the pull-aparts, but the blocks bordering the deep displacement zone diverge, their divergence matching the surficial block separation in the pull-aparts. Within the crust, this may be achieved (at least in part) by extension deep enough such that material from the lower crust or the mantle is expected to flow into the diverging displacement zone. In the second situation (Fig. 4-11B), the opening between the laterally slipping blocks is a rather shallow phenomenon that ends downward on a sloping fault or on a decollement or detachment; at greater depth, there is no divergence between the block. If several neighboring faults and pull-aparts are present on the surface, they all may end on a single regional detachment, together forming a thin-skinned structural complex. In all such cases, the surficial extension, expressed by the increasing area of the pull-aparts, is not matched by any localized block divergence at depth.

Application to the Dead Sea Basin

The foregoing considerations provide the framework in which to interpret the available information about the DSB. However, the existing information allows only a preliminary attempt at such an interpretation. Given its structural framework and especially its history, the DSB will be treated as a continuously evolving pull-apart. Although the possibility that the present pull-apart structure was superimposed on an older depression cannot definitely be excluded, several lines of evidence support the present approach. Thus, the stratigraphic evidence shows continuing subsidence of the DSB since early in the transform

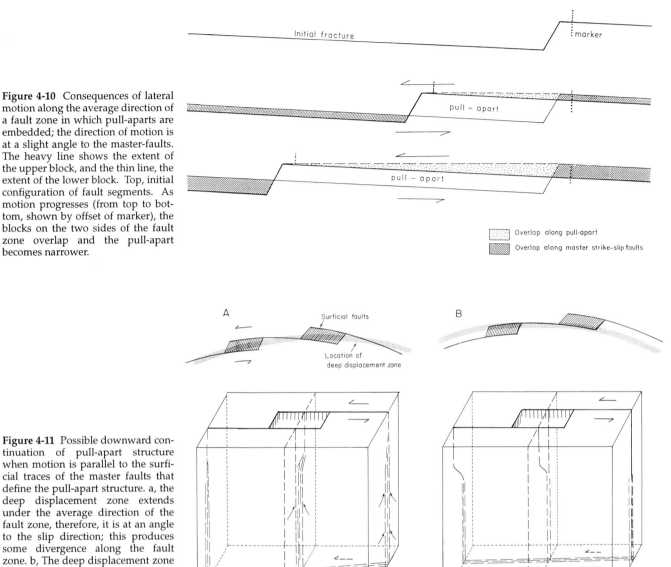

Figure 4-10 Consequences of lateral motion along the average direction of a fault zone in which pull-aparts are embedded; the direction of motion is at a slight angle to the master-faults. The heavy line shows the extent of the upper block, and the thin line, the extent of the lower block. Top, initial configuration of fault segments. As motion progresses (from top to bottom, shown by offset of marker), the blocks on the two sides of the fault zone overlap and the pull-apart becomes narrower.

Figure 4-11 Possible downward continuation of pull-apart structure when motion is parallel to the surficial traces of the master faults that define the pull-apart structure. a, the deep displacement zone extends under the average direction of the fault zone, therefore, it is at an angle to the slip direction; this produces some divergence along the fault zone. b, The deep displacement zone is parallel to the slip direction; therefore, the fractures linking it with the surficial faults are not vertical.

history (Fig. 4-6). The length of the DSB—nearly 150 km—considerably exceeds the total transform motion (105 km). The relation between the length of pull-aparts and the lateral motion that produces them, suggests that the DSB evolved during most, if not all, of the transform history. (In this respect, the DSB differs from other basins along the transform, which are much shorter than the entire offset and therefore are relatively young). The reconstruction of the entire transform offset (Fig. 4-2) places the nucleation site of the DSB along a bend in the initial transform trace, as constrained by the distribution of the pretransform rocks; the misalignment of the transform segments is some 7 km, according to the plate kinematic model of Garfunkel (1981). Together, these features are interpreted as showing that the DSB pull-apart originated when the lateral transform motion began; the pull-apart developed from a region about 40 km long along a releasing bend of the transform. The initial depression may have formed by coalescence of two overlapping lows; this is compatible with the mild bend in the western border fault

of the basin near lat. 30°46'N (Fig. 4-3; the original eastern border fault was obliterated by younger motion). However, there is no indication that the initial DSB formed by coalescence of many small pull-aparts. For comparison, the initial fill of the Karkom basin mentioned previously (Bartov and Garfunkel, 1980, 1985) proves that it originated having its full width. This may well be the usual way in which pull-aparts nucleate.

As noted previously, the absence of transverse compression along the DSB shows that the relative motion between the bordering blocks was nearly parallel to the surface traces of the master faults north and south of it (compare Fig. 4-10), which also applies to most of the southern half of the Dead Sea transform. If the entire transform motion were parallel to the overall transform direction and at an angle of merely 2–3° to the laterally stepping strike-slip faults, then the transverse shortening would have amounted to 3.5–5.2 km. This would have produced notable compressional structures in the basin, and the narrow Jordan and Arava valleys would have never developed.

The known compression in these valleys (Garfunkel et al., 1981; Rotstein et al., 1991) is too small and too young to record this amount of transverse convergence all along the DSB. The compression is interpreted, therefore, as resulting from recent local bending or rearrangement of the faults in the valley fill, rather than recording a long-term structural setting that would have prevented valley formation in the first place. The inferred kinematic picture requires, as noted previously, that the transform segment including the DSB is transtensional and that a pure strike-slip fault cannot extend directly beneath the surficial fault zone. Thus, the suggestion of Kashai and Croker (1987) that the laterally stepping surficial faults along the transform join at depth into a pure strike-slip fault cannot be accepted.

There are no direct data about the position of the main displacement zone at depth. However, gravity data (Folkman, 1981; ten Brink et al., 1993), despite their nonunique interpretation, show that the crust beneath the DSB is considerably thinned. A similar situation also exists in the Gulf of Elat (Aqaba) farther south along the transform (Ben-Avraham, 1985). Thinning of the crust is interpreted to show that the lateral motion occurs along a zone that extends from the DSB vertically downward into the middle or lower crust at least. In view of the kinematic relations discussed previously, the displacement zone at these depths is divergent, which may be accommodated by extension. If the displacement zone continues vertically to much greater depths, then upwelling or lateral flow into this zone should become important. Such a process may strongly influence magma generation beneath the transform. Such a divergent, deep displacement zone could have acted throughout the transform history, its divergence adjusting to the increasing separation along the transform recorded on the surface (Garfunkel, 1981).

The other situation compatible with the transform kinematics is that below some depth the lateral motion is taken up by a fault or displacement zone with no divergence. Such a feature should, as noted previously, be a small circle about the Euler pole of the Sinai–Arabia motion. Estimates of this motion (Garfunkel, 1981) show that if at the beginning of the transform motion this fault passed under either the Jordan Valley or the Gulf of Elat (Aqaba), then it would pass about 20 km west or east of the other part of the transform trace, respectively. An even larger deviation from the surficial transform trace corresponds to the present plate kinematics. No data are available to assess whether such a configuration exists in the mantle, but there are no indications favoring it. It is also unclear how such a deep displacement zone could adjust to changes in plate motions and to the increasing divergence along the transform.

In an analogous structural setting—the San Andreas fault—separation between plates occurs along the fault trace in the Imperial Valley and farther south, as indicated by the crustal structure (Fuis et al., 1983). In this area, the fault zone consists of laterally stepping fault strands and intervening pull-aparts. Plate separation took place along the original break in the crust. These features are best interpreted as indicating that the original displacement zone was situated practically directly beneath the surficial fault zone. This may well be the situation under all large intracontinental strike-slip faults.

The foregoing considerations reveal a few fundamental features of the DSB but do not define in detail the deep deformation within the basin. They show, however, that any model of this deformation should express the lengthening of the basin by some 105 km parallel to the transform, the transfer of lateral motion between the Jericho and the Arava faults, the subsidence of the basin, and the thinning of the underlying crust. The simplest model compatible with these requirements is that the DSB formed by uniform stretching, parallel to the transform, of the entire crust beneath an embryonic depression from which it arose. In large sedimentary basins, the resulting crustal thin-

ning is related to the amount of subsidence by the requirement of isostatic equilibrium. In the narrow DSB, local isostatic compensation is not expected. However, large deviations from equilibrium must be supported by considerable stresses; these fractures may not be sustained along the fractured transform, which may be a mechanically weak zone. Therefore, a model of local isostatic compensation may approximate the situation in the DSB, and more important, it can serve as a reference to which the available data can be compared to seek improved models.

The consequences of uniform stretching of the entire crust are shown in Fig. 4-12, with the following assumptions: The basin originated by uniform stretching of an (initially) 40-km-long area. The effects of deformation perpendicular to the transform are neglected. Thus at any time the crustal thickness is $h(t) = h_o \cdot 40/(40+l)$, where h_o is the initial crustal thickness (33 km) and l (in km) is the transform displacement at that time. The offset history is taken from curve b in Fig. 4-6A. The basin is in local isostatic equilibrium with its shoulders, and the mantle under the basin has the same temperature as under its shoulders. This assumption is adopted because the basin is so narrow. In such a case, thermal relaxation by lateral flow of heat will be important during the basin history. In the mantle, the deep divergent zone beneath the basin will be filled by lateral rather than vertical flow, which hardly disturbs the geotherm. This differs from the consequences of stretching by pure shear under wide basins, a process that results in significant upwelling and uplifting of the isotherms. At each stage, the expected sediment thickness is shown for complete basin filling, as well as for partial basin filling, which allows the formation of a valley having different depths and filled with different amounts of water. The models are compared with the known history of basin filling, taken from Fig. 4-6B.

The prediction for total sediment thickness when a valley 0.5–1 km deep forms is close to 10 km, which is about the maximum fill thickness in the center of the DSB. Moreover, this simplistic model qualitatively seems to agree with some features of the basin history. Thus, in the Late Miocene, the observed sediment thicknesses are those predicted for a starved basin that forms a deep topographic valley, perhaps filled with a lake. The great thickness of the Sedom evaporites reflects rapid, but incomplete, filling of the topographic depression. Later sedimentation, though fast, was also not enough to keep pace with subsidence. This scenario is compatible with the formation of a valley that may contain a water body. On the other hand, the model predicts thicker sections for the Hazeva times (Early to Late Miocene) than are actually documented. The discrepancy exists when other plausible histories of the early movement are examined. However, if thicker sections are indeed present beneath the deep part of the DSB, as suggested by the gravity data, the discrepancy may actually not be very large.

Thus, this simple model can serve as a starting point for understanding the subsidence of the DSB. More realistic models should take into consideration additional effects that are still difficult to quantify. Some of these effects are nonuniform stretching along the basin, transverse extension, and especially deviations from isostatic equilibrium. The thermal state beneath the basin is also important, as the assumption of complete thermal relaxation leads to an overestimate of the sediment thickness. To better evaluate the effects of stretching, additional data are needed, such as the crustal structure and especially the shape of the Moho beneath the DSB.

The actual mechanism of deformation that shaped the DSB can only be hypothesized at present. Any plausible model should account for the main structural features, such as the presence of only a few transverse faults that separate little deformed blocks, the simultaneous (though not always uniform) subsidence of large segments of the basin, and the thinning of the un-

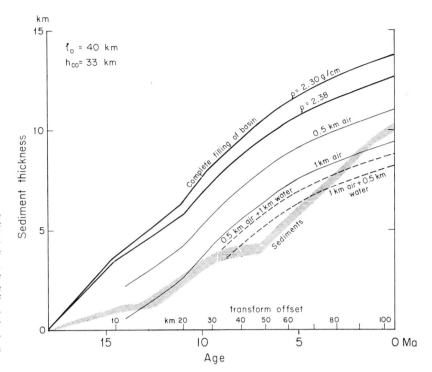

Figure 4-12 Speculative model of the subsidence of the Dead Sea basin based on the assumption of uniform stretching of the crust beneath the original depression and local isostatic compensation (see text for discussion). The graph shows the expected sediment thickness during different stages of the basin development. Upper thick curves show the expected fill thickness for two different densities of the fill. Thin curves give sediment thickness (density 2.30 g/cm^3) when the basin is incompletely filled, for different situations. Stippled area is estimated thickness of sediments in Dead Sea basin (see text for discussion).

derlying crust. The model should also describe the lengthening of the basin by 105 km and allow the transfer of lateral motion between the Arava and Jericho faults. A model that accounts for these features was proposed by Arbenz (1984), inspired by notions about listric faults and core complexes in the western United States. He assumed that the crust beneath the DSB was broken by a few transverse listric faults that had large slips parallel to the transform motion. The activity of such a fault system will lengthen the basin and will also transfer the transform motion from one master fault delimiting the pull-apart basin to the other. As the basin develops and is filled with sediments, the faults extend into the basin fill and displace the sediment. These ideas are illustrated in Figure 4-13 which combines the known shallow structure along the DSB with a hypothetical longitudinal cross section at greater depth. The deep faults are assumed to downthrow to the north because all known shallow transverse faults dip to the north, but south-dipping faults may also exist. For simplicity, the faults are also shown to sole into a single level.

This model explains many features of the DSB. The assumed fault system thins the crust and thus causes basin subsidence. The simultaneous activity of several faults explains coeval deformation and subsidence of large segments of the basin, where-

as variations in the slip rates of different faults—especially acceleration of the northern ones and perhaps also formation of new faults in the north—can explain the temporal and spatial variations in the basin development and the apparent northward shift of depocenters. This model has several potentially testable predictions (Fig. 4-13): At depths not yet imaged by seismic reflection, the basin floor should comprise a series of highs and lows, and prebasin sediments as well as the oldest units of the basin fill should underlie only the highs. On the flanks of the highs, the basin fill should have southerly dips of perhaps up to 30°, and the dips should decrease upward. Under most of the basin, the sedimentary fill is expected to be juxtaposed by flat faults directly over the crystalline basement.

This simple concept, however, is merely a first approximation. The Amazyahu fault—the only well-known fault in the basin—flattens within the basin fill, and it is not clear how it joins faults in the basement. Neither is it clear how the flat dips of the beds beneath the fault fit the model. If this fault is representative of the other transverse faults, then the actual fault pattern within the basin fill may be more complicated than the Arbenz model. This model also does not account for the asymmetry of the basin and does not incorporate extension perpendicular to the transform.

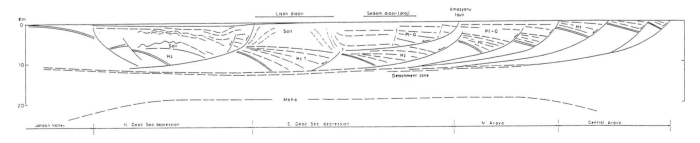

Figure 4-13 Schematic interpretation of the deep structure of the Dead Sea basin in terms of the Arbenz (1984) model. Deep structure is hypothetical.

Arbenz also assumed that the detachment coincided with the brittle-ductile transition, but later studies (summarized by Lister and Davis, 1989) showed that this simple concept should be abandoned. Rather, core complexes were formed by the activity of complicated systems of flat, midcrustal extensional faults. Moreover, sloping faults penetrating deep into the crust may also exist. Thus, there is no preferred a priori geometry of midcrustal deformation beneath the Dead Sea. Clearly, we need more data on the deep structure of the DSB before the mechanisms that produced it can be understood.

Examination of the available knowledge about the DSB within the framework of the pull-apart concept provides important insights and constraints about its formation, development, and deep structure. The discussion also highlights the question of which type of data are needed.

CONCLUDING REMARKS

The foregoing discussion led to a reevaluation of the existing knowledge about the DSB, especially about its fill and development, and revealed several new features about the basin history. The data are best interpreted as showing that the basin formed in the Early Miocene. Several kilometers of Miocene sediments underlie a substantial part of the basin. Large portions of the basin were deformed and subsided simultaneously throughout its history, though the area with the most subsidence shifted northward. Thus, the DSB seems to have evolved continuously without any major structural reorganization, but this point needs further confirmation. The basin developed during most, if not all, of the history of the transform.

Given the structural setting of the DSB—along a transform between two laterally stepping master strike-slip faults—it is interpreted as a pull-apart. This provides a unifying conceptual framework in which to integrate the basin's features and to evaluate the significance of particular observations. In particular, this concept highlights the relations between the deep structure of the basin and its history and shallow structure. It also allows us to use the kinematics of the surficial structures to constrain processes at depth.

Following this approach, some important problems regarding the DSB were examined. Because some important pieces of information are lacking, only a preliminary discussion is possible by making several assumptions, which, though plausible, still need confirmation. This allows integration of the available data, as well as a crude quantitative discussion of the basin's subsidence. The Arbenz (1984) model leads to a viable interpretation of the basin's deep structure, but it still needs considerable elaboration and verification.

Though many features of the DSB are still incompletely understood, it seems that more is known about its deep structure than about any other large pull-apart basin. Therefore, the considerations regarding this basin and the questions that it raises may apply to other large pull-apart basins as well.

Acknowledgments

I am greatly indebted to Zvi Ben-Avraham for prompting me to write this chapter. Without his continuing encouragement, this work would not have been completed. A very helpful review by T. Niemi is greatly appreciated.

REFERENCES

Abu Ajamieh, M. M., Bender, F. K., Eicher, R. N., El Kaysi, K. K., Nimri, F., Qudah, B. H., and Sheyyab, K. H., 1989, *Natural Resources in Jordan—Inventory, evaluation, development program*: Amman, Jordan, Natural Resources Authority, 224 p.

Agnon, A., 1983, Development of sedimentary basins and morphotectonics in the south of the western fault scarp of the Dead Sea [M.S. thesis]: Jerusalem, Israel, The Hebrew University, 70 p. (in Hebrew).

Arad, A., 1959, Interim report on the geology of Dimona area: Jerusalem, Geological Survey of Israel, unpublished report, 15 p. (in Hebrew).

Arbenz, J. K., 1984, Oil potential of the Dead Sea area: Sismica Oil Exploration Ltd., Tel Aviv, Report 84/111, 54 p.

Avni, Y., 1991, The geology, paleogeography and landscape evolution in the central Negev highlands and the western Ramon structure: Jerusalem, Geological Survey of Israel Report GSI/6/91, 170 p. (in Hebrew).

Avni, Y., Garfunkel, Z., Bartov, Y., and Ginat, H., 1993, The influence of the Plio–Pleistocene fault system on the tectonic and geomorphological structure in the margin of the Arava Valley [abs.]: Israel Geological Society, Annual Meeting, p. 7.

Aydin, A., and Nur, A., 1982, Evolution of pull-apart basins and their scale independence: *Tectonics*, v. 1, p. 91–105.

Bandel, K., and Khouri, H., 1981, Lithostratigraphy of the Triassic of Jordan: *Facies*, v. 4, p. 1–23.

Bartlett, W. L., Friedman, M., and Logan, M. J., 1981, Experimental folding and faulting of rocks under confining pressure. Part IX. Wrench faults in limestone layers: *Tectonophysics*, v. 79, p. 255–277.

Bartov, Y., 1974, A structural and paleogeographic study of the central Sinai faults and domes [Ph.D. thesis]: Jerusalem, Israel, The Hebrew University, 143 p. (in Hebrew, English abs.).

Bartov, Y., and Arkin, Y., 1975, Photogeological map of Israel, scale 1:500 000: Tel Aviv, The Survey of Israel.

Bartov, Y., Frieslander, U., and Rotstein, Y., 1993, New observations on the structure and evolution of the Arava rift valley [abs.]: Israel Geological Society, Annual Meeting, p. 11.

Bartov, Y., and Garfunkel, Z., 1980, Relations between clastic sediments and tectonic history of the Karkom graben [abs.]: Israel Geological Society, Annual Meeting, p. 4–5.

Bartov, Y., and Garfunkel, Z., 1985, Field trip to the Karkom graben [abs.]: Israel Geological Society, Annual Meeting, p. 152–153.

Begin, B. Z., Ehrlich, A., and Nathan, Y., 1974, Lake Lisan—The Pleistocene precursor of the Dead Sea: Jerusalem, Geological Survey of Israel Bulletin 63, 30 p.

Ben-Avraham, Z., 1985, Structural framework of the Gulf of Elat (Aqaba), northern Red Sea: *Journal of Geophysical Research*, v. 90, p 703–726.

Ben-Avraham Z., and Ballard, R. D., 1984, Near-bottom temperature anomalies in the Dead Sea. *Earth and Planetary Science Letters*, v.71, p. 356–360.

Ben-Avraham, Z., Hanel, R., and Vilinger, H., 1978, Heat flow through the Dead Sea rift: *Marine Geology*, v. 28, p. 253–269.

Ben-Avraham, Z., and Zoback, M. D. 1992, Transform-normal extension and asymmetric basins: An alternative to pull-apart models: *Geology*, v. 20, p. 423–426.

Bender, F., ed. 1968, The Geological map of Jordan, scale 1:250 000, 5 sheets: Geological Survey of the Federal Republic of Germany, Hannover.

Bender, F., 1974, *Geology of Jordan* (supplementary edition of vol. 7): Berlin-Stuttgart, Germany, Gebrüder Borntraeger, 196 p.

Ben-Menahem, A., Nur, A., and Vered, M., 1976, Tectonics, seismicity and structure of the Afro-Eurasian junction—the breaking of an incoherent plate: *Physics of the Earth and Planetary Interiors*, v. 12, p. 1–50.

Bentor, Y. K., and Vroman, A., 1957, The geological map of the Negev, scale 1:100,000, Sheet 19: Arava Valley, with explanatory notes: Geological Survey of Israel, 66 p.

Blake, G. S., and Ionides, M. G., 1939, Report on the water resources of Transjordan and their development, incorporating a report on the geology, soils and minerals, and hydrogeology: London, Crown Agency for Colonies, p. 43–127.

Brown, G. F., Schmidt, D. L., and Huffman, A. C., Jr., 1989, Geology of the Arabian Peninsula: Shield area of western Saudi Arabia: U.S. Geological Survey Professional Paper 560–A, 188 p.

Buchbinder, B., Martinotti, G. M., Siman-Tov, R., and Zilberman, E., 1993, Temporal and spatial relationships in Miocene reef carbonates in Israel: *Paleogeograpy, Paleoclimatology, Paleoecology*, v. 101, p. 97–116.

Burchfiel, B. C., and Stewart, J. H., 1966, 'Pull-apart' origin of the central segment of Death Valley, California: *Geological Society of America Bulletin*, v. 77, p. 439–442.

Christie-Blick, N., and Biddle, K. T., 1985, Deformation and basin formations along strike-slip faults, *in* Biddle, K.T, and Christie-Blick, N., eds, *Strike-slip deformation, basin formation, and sedimentation*: Society of Economic Paleontologists and Mineralogists Special Publication 37, p. 1–34.

Coleman, R. G., 1974, Geological background of the Red Sea: *Initial Reports DSDP*, v. 23, p. 813–820.

Coleman, R. G., 1984, The Red Sea: A small ocean basin formed by continental extension and sea floor spreading: International Geological Congress 27th, v. 23, p. 93–121.

Crowell, J. C., 1974, Origin of late Cenozoic basins in southern California, *in* Dickinson, W. R. ed., *Tectonics and Sedimentation*: Society of Economic Paleontologists and Mineralogists, Special Publication 22, p. 190–204.

Eckstein, Y., 1979, Review of the heat flow data from the Eastern Mediterranean region: *Pure and Applied Geophysics*, v. 117, p. 150–179.

Eidelman, A., 1979, The geology of the Arava margins in the En Yahav Ramat Zofar area: Jerusalem, Geological Survey of Israel Report MM/10/79, 66 p. (in Hebrew, English abs.).

El-Isa, Z., Metchie, J., Prodehl, C., Makris, J., and Rihm, R., 1987, A crustal structure study of Jordan derived from seismic refraction data: *Tectonophysics*, v. 138, p. 235–253.

Eyal, A., 1984, The geology of the northern Arava and its western margins in the Ein Yahav—Hazeva region [M.Sc. thesis]: Jerusalem, Israel, The Hebrew University, 66 p. (in Hebrew, English abs.).

Eyal, M., Eyal, Y., Bartov, Y., and Steinitz, G., 1981, The tectonic development of the western margin of the Gulf of Elat (Aqaba) rift: *Tectonophysics*, v. 80, p. 39–66.

Folkman, Y., 1981, Structural features in the Dead Sea–Jordan rift zone, interpreted from a combined magnetic–gravity study: *Tectonophysics*, v. 80, p. 135–146.

Freund, R., 1965, A model for the development of Israel and adjacent areas since the Upper Cretaceous times: *Geological Magazine*, v. 102, p. 189–205.

Freund, R., 1974, Kinematics of transform and transcurrent faults: *Tectonophysics*, v. 21, p. 93–134.

Freund, R., Garfunkel, Z., Zak, I., Goldberg, M., Weissbrod, T., and Derin, B., 1970, The shear along the Dead Sea rift: *Philosophical Transactions of the Royal Society of London*, Ser. A, v. 267, p. 107–130.

Freund, R., Zak, I., and Garfunkel, Z., 1968, Age and rate of movement along the Dead Sea rift: *Nature*, v. 220, p. 253–255.

Frieslander, U., and Ben-Avraham, Z., 1989, The magnetic field over the Dead Sea and its vicinity. *Marine and Petroleum Geology*, v. 6, p. 148–160.

Fuis, G. S., Mooney, W. D., Healy, J. H., McMechan, G. A., and Lutter, W. J., 1983, A seismic refraction survey of the Imperial Valley region, California: *Journal of Geophysical Research*, v. 89, p. 1165–1189.

Gamond, J. F., 1985, Bridge structures as sense of displacement criteria in brittle fault zones: *Journal of Structural Geology*, v. 9, p. 609–620.

Gardosh, M., Reches, Z., and Garfunkel, Z., 1990, Holocene tectonic deformation along the western margins of the Dead Sea: *Tectonophysics*, v. 180, p. 123–137.

Garfunkel, Z., 1978, The Negev—Regional synthesis of sedimentary basins: Proceedings of 10th Congress of the International Association of Sedimentologists, Guidebook to Precongress Excursions, p. 35–110.

Garfunkel, Z., 1981, Internal structure of the Dead Sea leaky transform (rift) in relation to plate kinematics: *Tectonophysics*, v. 80, p. 81–108.

Garfunkel, Z., 1988a, The pre-Quaternary geology of Israel, *in* Yom-Tov, Y., and Tchernov, E., eds., *The Zoogeography of Israel*: Dordrecht, Holland, W. Junk. p. 7–34.

Garfunkel, Z., 1988b, Relations between continental rifting and uplifting: Evidence from the Suez rift and northern Red Sea: *Tectonophysics*, v. 150, p. 33–49.

Garfunkel, Z., 1989, Tectonic setting of Phanerozoic magmatism in Israel: *Israel Journal of Earth Sciences*, v. 38, p. 51–74.

Garfunkel, Z., and Bartov, Y., 1977, The tectonics of the Suez rift: Jerusalem, Geological Survey of Israel Bulletin 71, 44 p.

Garfunkel, Z., and Horowitz, A., 1966, The Upper Tertiary and Quaternary morphology of the Negev, Israel: *Israel Journal of Earth Sciences*, v. 15, p. 101–117.

Garfunkel, Z., Zak, I., and Freund, R., 1981, Active faulting in the Dead Sea rift: *Tectonophysics*, v. 80, p. 1–26.

Ginzburg, A., and Folkman, Y., 1980, The crustal structure between the Dead Sea rift and the Mediterranean Sea: *Earth Planetary Science Letters*, v. 51, p. 181–188.

Goldsmith, N. F., Hirsch, F., Friedman, G. M., Tchernov, E., Derin, B., Gerry, E., Horowitz, A., and Weinberger, G., 1988, Rotem mammals and Yeroham Crassostreids: Stratigraphy of the Hazeva Formation (Israel) and the paleogeography of Miocene Africa: *Newsletter Stratigraphy*, v. 20, p. 73–90.

Gvirtzman, G., 1970, The Saqiye (Late Eocene to Early Pleistocene) in the coastal plain and Hashephela regions, Israel: Jerusalem, Geological Survey of Israel Report OD/5/67, 170 p. (in Hebrew, English abstract).

Harash, A., 1967, The geology of the Yeruham–Dimona plain, [M.S. thesis]: Jerusalem, Israel, The Hebrew University, 71 p. (in Hebrew).

Harland, W. B., Armstrong, R. B., Cox, A. V., Craig, L. E., Smith, A. G., and Smith D. G., 1989, *A geologic time scale*: Cambridge, England, Cambridge University Press, 263 p.

Hatcher, R. D., Zeiz, I., Reagan, R. D., and Abu-Ajameh, M., 1981, Sinistral strike-slip motion on the Dead Sea rift: confirmation from new magnetic data: *Geology*, v. 9, p. 458–462.

Horowitz, A., 1987, Palynological evidence for the age and rate of sedimentation along the Dead Sea Rift, and structural implications: *Tectonophysics*, v. 141, p. 107–115.

Izzeldin, A. Y., 1987, Seismic, gravity and magnetic surveys in the central part of the Red Sea: Their interpretation and implications for the structure and evolution of the Red Sea: *Tectonophysics*, v. 143, p. 269–306.

Jennings, C. W., compiler, 1973, State of California, preliminary fault and geologic map, scale 1:750,000: California Division of Mines and Geology.

Joffe, S., and Garfunkel, Z., 1987, The plate kinematics of the circum Red Sea—A reevaluation: *Tectonophysics*, v. 141, p. 5–22.

Kashai, E. L., and Croker, P. F., 1987, Structural geometry and evolution of the Dead Sea–Jordan rift system as deduced from new subsurface data: *Tectonophysics*, v. 141, p. 33–60.

Kingma, J. T., 1958, Possible origin of piercement structures, local unconformities and secondary basins in the Eastern Geosyncline, New Zealand. *New Zealand Journal of Geology and Geophysics*, v. 1, p. 269–274.

Lartet, L., 1869, *Essai sur la géologie de la Palestine et des contrees avoisinantes, telles que l'Egypte et l'Arabia*: Paris, France, Masson, 292 p.

Lensen, G. J., 1958, A method of graben and horst formation: *Journal of Geology*, v. 66, p. 579–587.

LePichon, X., and Gaulier, J. M., 1988, The rotation of Arabia and the Levant fault system: *Tectonophysics*, v. 153, p. 271–294.

Lister, G. S., and Davis, G. A., 1989, The origin of metamorphic core complexes and detachment faults formed during Tertiary continental extension in the northern Colorado River region , U.S.A.: *Journal of Structural Geology*, v. 11, p. 65–94.

Lynch, W. F., 1849, *Narrative of the United States Expedition to the River Jordan and the Dead Sea*: Philadelphia, Pa, U.S.A., Lea & Blanchard, 508 p.

Mann, P., Hempton, M. R., Bradley, D. C., and Burke, K., 1983, Development of pull-apart basins: *Journal of Geology*, v. 91, p. 529–554.

Martinotti, G. M., Gvirtzman, G., and Buchbinder, B., 1978, The Late Miocene transgression in the Beer Sheva area: *Israel Journal of Earth Sciences*, v. 27, p. 72–82.

May, P. R., 1968, Gravimetric estimation of the depth of aquifers in the Hazeva area, Arava valley: *Israel Journal of Earth Sciences*, v. 17, p. 30–43.

McKenzie, D. P., Davies, D., Molnar, P., 1970, Plate tectonics of the Red Sea and East Africa: *Nature*, v. 224, p. 125–133.

Mikhaeli, A., and Eckstein, Y., 1968, Hydrogeology of the Ein Yahav–Hazeva area: Tahal, report PM/591, 43 p. (in Hebrew).

Moretti, I., and Colletta, B., 1987, Spatial and temporal evolution of the Suez rift subsidence: *Journal of Geodynamics*, v. 7, p. 151–168.

Naylor, M. A., Mandl, G., and Sijpenstejn, C. H. K., 1986, Fault geometries in basement-induced wrench faulting under different initial stress states: *Journal of Structural Geology*, v. 8, p. 737–752.

Neev, D., 1960, *A pre-Neogene erosion channel in the southern coastal plain of Israel*: Jerusalem, Geological Survey of Israel Bulletin 25, 20 p.

Neev, D., and Emery, K. O., 1967, The Dead Sea: Depositional processes and environments of evaporites: Jerusalem, Geological Survey of Israel Bulletin 41, 147 p.

Neev, D., and Hall, J. K., 1979, Geophysical investigations in the Dead Sea: *Sedimentary Geology*, v. 23, p. 209–238.

Picard, L., 1951, Geomorphogeny of Israel—Part 1: The Negev: *Research Council of Israel Bulletin*, v. 8G, p. 1–30.

Quennell, A. M., 1958, The structural and geomorphic evolution of the Dead Sea rift: *Quarterly Journal of the Geological Society of London*, v. 114, p. 2–24.

Quennell, A. M., 1959: Tectonics of the Dead Sea rift: International Geological Congress 20th, Association of African Geological Surveys, p. 385–405.

Richardson, M., and Arthur, M. A., 1988: The Gulf of Suez–northern Red Sea Neogene rift: A quantitative basin analysis: *Marine Petroleum Geology*, v. 5, p. 247–270.

Roeser, H. A., 1975, A detailed magnetic survey of the southern Red Sea: *Geologisches Jahrbuch*, v. D13, p. 131–153.

Ron, H., Freund, R., Garfunkel, Z., and Nur, A., 1984, Block rotation by strike slip faulting: Structural and paleomagnetic evidence: *Journal of Geophysical Research*, v. 89, p. 6256–6270.

Rothman, S., 1967, Miocene mollusca of the Lakhish area, central Israel: *Israel Journal of Earth Sciences*, v. 16, p. 140–164.

Rotstein, Y., Bartov, Y., and Hofstetter, A., 1991, Active compressional tectonics in the Jericho area, Dead sea rift: *Tectonophysics*, v. 198, p. 239–259.

Sa'ar, H., 1985, The provenance and deposition of sandstones in the sedimentary fill of the Dead Sea rift [M.S. thesis]: Jerusalem, Israel, The Hebrew University, 92 p. (in Hebrew).

Sakal, E., 1967, The geology of Reches Menuha [M.S. thesis]: Jerusalem, Israel, The Hebrew University, 96 p. (in Hebrew).

Schulman, N., 1962, The geology of the central Jordan Valley, [Ph.D. thesis]: Jerusalem, Israel, The Hebrew University Jerusalem, 103 p. (in Hebrew, English abs.).

Schulman, N., and Bartov, Y., 1978, Tectonics and sedimentation

along the rift valley: Proceedings of 10th Congress of the International Association of Sedimentologists, Guidebook II, p. 37–96.

Shahar, Y., 1973, The Hazeva Formation in the Oron-Ef'e syncline: *Israel Journal of Earth Sciences*, v. 22, p. 31–50.

Shahar, Y., Reiss, Z., and Gerry, E., 1966, A new outcrop of marine Neogene in the Negev: *Israel Journal of Earth Sciences*, v. 15, p. 82–84.

Shaliv, G., 1991, Stages in the tectonic and volcanic history of the Neogene basin in the Lower Galilee and the Valleys: Geological Survey of Israel Report GSI/11/91, 100 p. (in Hebrew, English abs.).

Sneh, A., 1981, The Hazeva Formation in the Arava Valley: *Israel Journal of Earth Sciences*, v. 30, p. 81–91.

Stein, M., Garfunkel, Z., and Yagoutz, E., 1993, Chronothermometry of peridotitic and pyroxenitic xenoliths: Implications for the thermal evolution of the Arabian lithosphere: *Geochimica et Cosmochimica Acta*, v. 57, p. 1325–1337.

Steinitz, G., and Bartov, Y., 1991, The Miocene–Pliocene history of the Dead Sea segment of the Rift in light of K–Ar ages of basalts: *Israel Journal of Earth Sciences*, v. 40, p. 199–208.

Steinitz, G., Bartov, Y., Eyal, M., and Eyal, Y., 1981, K-Ar age determination of Tertiary magmatism along the western margin of the Gulf of Elat: *Geological Survey of Israel Current Res. 1980*, p. 27–29.

Steinitz, G., Bartov, Y., and Hunziker, J. C., 1978, K-Ar age determinations of some Miocene–Pliocene basalts in Israel: Their significance to the tectonics of the Rift Valley: *Geological Magazine*, v. 115, p. 329–340.

ten Brink, U. S., and Ben-Avraham, Z., 1989, The anatomy of a pull-apart basin: Seismic reflection observations of the Dead Sea basin: *Tectonics*, v. 8, p. 333–350.

ten Brink, U. S., Ben-Avraham, Z., Bell, R.E., Hassouneh, M., Coleman, D. F., Andreasen, G., Tibor, G., and Coakley, B., 1993, Structure of the Dead Sea pull-apart basin from gravity analysis: *Journal of Geophysical Research*, v. 98, p. 21,877–21,894.

Tchalenko, J. S., and Ambraseys, N. N., 1970, Structural analysis of the Dasht-e Bayaz (Iran) earthquake fractures: *Geological Society of America Bulletin*, v. 81, p. 41–60.

van Eck, T., and Hofstetter, A., 1989, Microearthquake activity in the Dead Sea: *Geophysical Journal International*, v. 99, p. 605–620.

van Eck, T., and Hofstetter, A., 1990, Fault geometry and spatial clustering of microearthquakes along the Dead Sea–Jordan rift fault zones: *Tectonophysics*, v. 180, p. 15–27.

Wdowinski, S., 1985, The geology of southern Hebron Mountains [M.S. thesis]: Jerusalem, Israel, The Hebrew University, 74 p. (in Hebrew, English abs.).

Wilson, J. T., 1965, A new class of faults and their bearing on continental drift: *Nature*, v. 207, p. 343–347.

Wilcox, R. E., Harding, T. P., and Seely, D. R., 1973, Basin wrench tectonics: *American Association of Petroleum Geologists Bulletin*, v. 57, p. 74–96.

Zak, I., 1967, The geology of Mount Sedom [Ph.D. thesis]: Jerusalem, Israel, The Hebrew University, 208 p. (in Hebrew, English abs.).

Zak, I., and Bentor, Y. K., 1972, Some new data on the salt deposits of the Dead Sea area, Israel: Proceedings of the Symposium on Geology of Saline Deposits, p. 137–146.

Zak, I., and Freund, R., 1981, Asymmetry and basin migration in the Dead Sea rift: *Tectonophysics*, v. 80, p. 27–38.

Zilberman, E., 1981, The geology of the central Negev–Sinai shear zone, Part A—The Sa'ad Nafha lineament: Jerusalem, Geological Survey of Israel Report Hydro/1/81, 59 p.

Zilberman, E., 1992, Remnants of Miocene landscape in the central and northern Negev and their paleogeographical implications: Geological Survey of Israel Bulletin 83, 54 p.

5. HYDROCARBON EXPLORATION IN THE SOUTHERN DEAD SEA AREA

Michael Gardosh, Eliezer Kashai, Shalom Salhov, Haim Shulman, and Eli Tannenbaum

The widespread occurrence of surface asphalt shows, both on land and floating on the lake, raised the interest in the petroleum potential of the Dead Sea area as early as 1912 (Nissenbaum, 1991). On October 29, 1953, the first exploration well drilled in the State of Israel, the Mazal-1, was spud 16km southwest of the Dead Sea shore. By 1993, a total of 20 exploration wells had been drilled. Although most of the wells encountered various types of hydrocarbon shows, no commercial discovery has been made. The lack of commercial discovery does not reflect the hydrocarbon potential of the basin because most of the wells were drilled during an early exploration phase, at times when seismic lines were not available and the internal structure of the basin was insufficiently understood.

The Israel National Oil Company is presently conducting a multistage exploration project in the southern Dead Sea. Of the three wells drilled between 1990 and 1993, Sedom Deep-1 (total depth, T.D. 6,445 m) is the deepest well in the basin. Initial results from the wells confirm that the basin is one of the most important targets for exploration in Israel. In this chapter, we review the structure, stratigraphy, and hydrocarbon potential of the southwestern Dead Sea (SDS) in light of this new exploration data.

TECTONIC AND GEOLOGIC SETTING

The Dead Sea lake (Fig. 5-1) occupies part of a long and narrow continental trough—the Dead Sea basin—located in the Dead Sea rift. Tectonically, the Dead Sea rift is a transform fault system that separates the Arabian plate on the east from the African plate on the west, connecting the spreading zone of the Red Sea in the south to the Taurus collision zone in the north (Quennell, 1958; Garfunkel, 1981; Fig. 5-1).

The formation of the Dead Sea rift is generally attributed to a left-lateral shear. A cumulative lateral displacement of 105 km was estimated by Quennell (1958) and Freund et al. (1970) since the time of the rift initiation near the Oligocene-Miocene boundary (Cochran, 1983). A sinistral strike-slip motion takes place on several faults 100-400 km long found within the rift (Garfunkel et al., 1981; Kashai and Croker, 1987); two of these are exposed at the eastern and western margins of the Dead Sea basin (Fig. 5-1).

Uplifted rift shoulders bordering deep depressions filled with young sediments are common along the rift. The Dead Sea basin is one of the best examples of such a depression.

On the highlands west of the basin floor (500–1,000 m above the level of the Dead Sea lake), Middle Cretaceous carbonates and Upper Cretaceous clays, marls, chalks, and cherts are exposed (Fig. 5-2). Among the latter are Senonian bituminous shales containing up to 20% organic matter. These shales are considered to be the main source rock of hydrocarbons in the Dead Sea basin (Tannenbaum, 1983). The eastern highlands are composed mostly of Paleozoic to Lower Cretaceous Nubian sandstone (Fig. 5-2). The different rock units on the two opposite sides of the basin reflect both relative lateral motion and differential uplift.

The Mesozoic strata exposed on the basin rim are buried in the basin under a thick sedimentary section termed the Dead Sea Group (Zak, 1967; Zak and Freund, 1981). The basal unit of

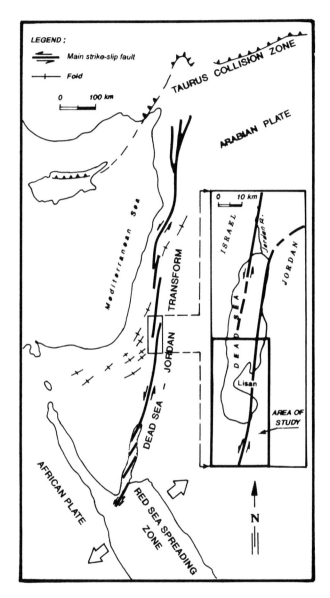

Figure 5-1 Outline of the Dead Sea rift, also termed the Dead Sea–Jordan Transform, and the main strike-slip faults that connect a spreading zone in the south with a collision zone in the north. Folds of the Syrian Arc fold system extend in a southwest–northeast direction across the rift zone. The area of study, in the inset, is the southern part of the Dead Sea basin, one of several deep troughs found along the rift. The shorelines of the Dead Sea lake is at the 1967 water level. Today, the area south of the Lisan Peninsula is occupied by shallow industrial salt ponds (modified after Kashai, 1989).

the section is the Hazeva Formation (Bentor and Vroman, 1957). Relics of this continental clastic unit can be found along the edge of the basin (Fig. 5-2) and in wells drilled into it (Kashai and Croker, 1987). Overlaying the Hazeva Formation are massive evaporite beds named the Sedom Formation (Zak, 1967). These beds are exposed in the Mt. Sedom salt diapir (Fig. 5-2). The upper part of the Dead Sea Group is composed of the Amora, Samra, and Lisan Formations (Zak, 1967), a sequence of marl, clay, sand, gravel, and some evaporites, the top of which is exposed along most of the basin floor (Fig. 5-2).

It is difficult to date the Dead Sea Group. Ages are mostly derived from correlation to units found outside the SDS basin. A chronology commonly used in the Dead Sea literature is Miocene Age for the Hazeva Formation, late Miocene to Pliocene Age for the Sedom Formation, and Pleistocene Age for the Amora, Samra, and Lisan Formations (Zak, 1967; Zak and Freund, 1981; Kashai, 1989). Although some modifications were recently suggested (Shaliv, 1989; Agnon, 1993), we refer to the more commonly used ages.

EXPLORATION HISTORY

Surface hydrocarbon shows, such as floating asphalt blocks on the Dead Sea lake, asphalt seeps, and impregnated asphaltic gravels and sands in the surrounding area, have been well known since Biblical times (Nissenbaum, 1978). This "tar belt" is particularly developed on the western side of the SDS, where most of the exploration wells are located (Fig. 5-2).

The history of hydrocarbon exploration in the area can be divided into three phases. During the first phase, between 1953 and 1980, 14 wells were drilled (Table 5-1). Because information on the subsurface of the SDS at that time was very limited, many of the wells were located on surface highs and in the vicinity of asphalt shows (Table 5-1). Some were located on local gravity anomalies, but none were based on seismic data, which were not yet available. The main result of this early phase is one significant light oil show (30–35° API) in Masada-1 (Table 5-1, Fig. 5-2) that was tested and found to be noncommercial.

During this phase, gas was found at the Zohar structure, 10 km west of the Dead Sea. It was followed by the discovery and development of three small gas fields at Zohar, Kidod, and Hakanaim (Fig. 5-2; Coats et al., 1963).

During the second phase of exploration between 1981 and 1985, significant subsurface data were collected. Several seismic programs were completed by companies such as Oil Exploration (Investment) Ltd. and Sismica Oil Exploration. Most of the SDS area, west of the international border and east of the western highlands, was covered by a grid of two-dimensional seismic lines. The result of this work was the drilling of three wells on subsurface structures identified in different parts of the basin: Amazyahu-1, Har Sedom-1, and Zuk Tamrur-1 (Table 5-1, Fig. 5-2). In the Zuk-Tamrur-1 well, light oil (28° API) was discovered. However, commercial production was not established because of a high gas/oil ratio and a low production rate (10–50 barrels per day).

The third phase of exploration was initiated by the Israel National Oil Co. (INOC) in 1989. New seismic lines were shot to improve the seismic coverage, and many old lines were reprocessed. New gravity and aeromagnetic data were acquired and interpreted (Edcon Inc., 1991), and basin modeling techniques were applied (Tannenbaum, 1990, 1991). Three wells were drilled by the INOC: Admon-1, Amiaz East-1, and Sedom Deep-1 (Table 5-1, Fig. 5-2). Sedom Deep-1 is the deepest well in the basin (T.D. 6,445 m), and the first one to penetrate the salt beds of the Sedom Formation in the deep parts of the basin. The most significant oil show found in these wells is a thick layer (ca. 30 m) of asphalt-bearing sands in the Amiaz East-1 well.

To date, no exploration well has been drilled in the northern part of the Dead Sea north of the En Gedi-2 well (Fig. 5-2). Most of this area is covered by the hypersaline, approximately

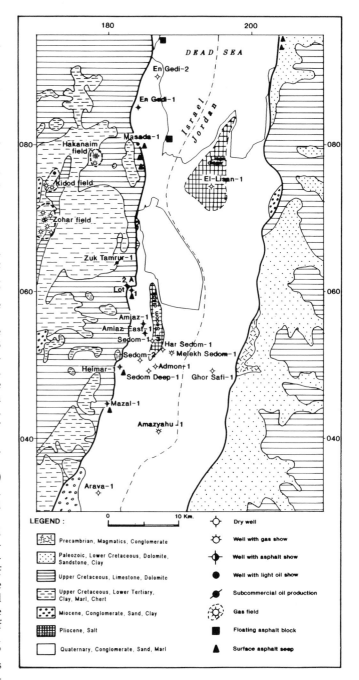

Figure 5-2 Schematic geologic map of the study area. The Paleozoic and Mesozoic rocks east and west of the basin are elevated 100–1,000 m above the basin floor and are delimited by a pronounced topographic escarpment. Note the wide distribution of the Upper Cretaceous to Lower Tertiary clay marl and chert (Mt. Scopus Group) west of the basin. Bituminous-rich rocks of this unit buried in the basin are the source rocks for the many hydrocarbon shows, both on the surface and in wells drilled in the SDS. Two large salt diapirs are shown. Mt. Sedom is exposed on the surface, whereas the top of the Lisan diapir was reached by shallow wells at a depth of several hundred meters. The map also shows the 20 exploration wells drilled in the last 40 years and the three small gas fields discovered on the highlands west of the basin. (The geology is simplified after the Geological Map of Israel, 1:250,000 and 1:500,000).

300-m-deep Dead Sea lake, and its hydrocarbon potential is poorly known.

Modern exploration of the SDS area is largely based on a seis-

Table 5-1 Summary of information on exploration wells drilled in the SDS basin.

Well	Spud Year	Total Depth (m)	Age at T.D.[a]	Structural Step	Exploration Model	Reservoirs Penetrated[a]	Hydrocarbons
Mazal-1	1953	597	E	Rim block	Tar sands at the surface; Assume tilted block	D, E	Asphalt shows
En Gedi-1	1955	758	A	Intermediate block	Unknown	A	Asphalt shows
Sedom-1	1956	2,751	C	Intermediate block	Assumed subsurface high based on gravity anomaly	—	—
Sedom-2	1957	1,744	A	Intermediate block	Assumed subsurface high based on gravity anomaly	A	—
En Gedi-2	1957	2,751	E	Intermediate block	Assumed subsurface high based on gravity anomaly	A, D, E	—
Arava-1	1959	2,738	A	Intermediate block	Stratigraphic well	A	—
El Lisan-1[b]	1966	3,672	B	Deep block	Assumed horst underneath salt diapir	—	Oil shows?
Heimar-1	1966	2,437	H	Rim block	Asphalt seep at the surface; Assumed tilted block	D, E, F, G, H	Asphalt shows
Lot-1	1967	2,190	F	Rim block	Asphalt seep at the surface; Assumed tilted block	D, E, F, G, H	Asphalt shows
Melekh Sedom-1	1968	3,480	A	Deep block	Assumed subsurface high	A	Gas shows
Ghor Safi-1[b]	1971	2,783	A	Deep block	Unknown	A	—
Lot-2	1975	1,223	F	Rim block	Asphalt seep at the surface; A faulted anticline mapped at the surface	D, E, F	Asphalt shows
Amiaz-1	1976	4,605	F	Intermediate block	A tilted block based on gravity anomaly and surface mapping	A, C, D, E, F,	Asphalt shows
Masada-1 After deepening	1954 1980	2,743	G	Rim block	A horst block at the surface	D, E, F, G,	Light oil show 30–35° A.P.I.
Amazyahu-1	1981	2,500	A	Intermediate block	A rollover anticline on a listric fault mapped in the subsurface	A	Gas shows
Zuk Tamrur-1	1981	2,747	H	Rim block	A tilted block mapped at the surface and the subsurface	D, E, F, G, H	Light oil 30° A.P.I.
Har Sedom-1	1985	1,818	B	Deep block	Upturned beds on a salt diapir flank mapped at the subsurface	A	Subcommercial
Admon-1	1990	2,179	A	Deep block	Upturned beds on a salt diapir flank mapped at the subsurface; Seismic "bright spots"	A	—
Amiaz East-1	1990	3,644	C	Intermediate block	A tilted antithetic block, mapped in the subsurface	A, C	Asphalt shows
Sedom Deep-1	1992	6,448	C	Deep block	A fault block underneath salt diapir, mapped at the subsurface	A, C	—

Note: Wells drilled west of Zur-Tamrur-1 are not considered to be within the basin area.

[a]A, Pleistocene; B, Pliocene; C, Miocene; D, Upper Cretaceous; E, Lower Cretaceous; F, Jurassic; G, Triassic; H, Permian.

[b]Only partial information was found on this well in Jordan.

mic data set composed of about 100 multichannel seismic profiles. As the result of varied and occasionally difficult surface conditions, the energy sources were either vibroseis, dynamite, or detonating cord, with the vibroseis lines being generally of superior quality. The older lines are recorded on 24 and 48 channels. The introduction of 96- and 120-channel recording systems in the early 1980s significantly improved the quality of the deep data.

Data processing was done in several stages, with some of the lines reprocessed up to five times, in 10 different centers. Despite the great variation in acquisition and processing parameters, the data quality, especially in the upper 2–3 seconds, is fair to good.

The direction and length of the lines are greatly limited by the obstacles created by the basin's narrow geometry, international border, badland topography, and industrial ponds. Most of the lines are, therefore, only 5 to 10 km long. Very few lines were shot across the western escarpment and the elevated shoulder of the basin, an area intensively dissected by deep gorges. At this stage, the necessary improvement of data quality may be achieved only through advanced acquisition and processing techniques or by shooting longer lines across the entire width of the basin.

STRUCTURE

Main Structural Steps

Three structural steps were previously identified in the basin (Kashai et al., 1973; Kashai and Croker, 1987): the rim block, the intermediate block, and the deep-sunken block (Fig. 5-3). Recent seismic and well data (e.g., Sedom Deep-1) confirm the presence of these steps.

The rim block is a belt of intensely faulted, exposed Mesozoic rocks topographically elevated 100–300 m from the Dead Sea lake (Figs. 5-2 and 5-3). Faults strike approximately north-south and dip 70–80° predominantly to the east but occasionally to the west, creating horst and graben structures and rotated fault blocks (Fig. 5-3). The sense of motion is apparently vertical.

The rim block is delineated on the east by several discontinuous, subparallel, normal faults with large vertical throw, commonly known as the "border faults" (Neev and Emery, 1967). On seismic lines, these faults separate coherent reflections of young sedimentary fill on the east from less coherent reflections of Mesozoic and Paleozoic rocks on the west (Fig. 5-4). The border faults are the main tectonic features along which the enormous subsidence of the Dead Sea basin took place. A thick stratigraphic section of the basin fill is found on the downthrown side and is lacking on the upthrown side, indicating that the border faults have been intermittently active from the early phase of basin formation through recent time (Kashai and Croker, 1987).

The Iddan fault (ten Brink and Ben-Avraham, 1989) at the southern edge of the basin (Fig. 5-3) is considered to be a border fault, although its structural configuration is different. Unlike the other border faults, it crosses the basin in a northwest-southeast direction. On seismic line 2 (Fig. 5-5), it separates about 2 seconds of coherent, parallel, and high- and low-amplitude reflections on its downthrown side from an area that lacks coherent reflections on its upthrown side.

On the basis of its seismic character, the downthrown side was identified as part of the Plio-Pleistocene sedimentary fill (Kashai and Croker, 1987; ten Brink and Ben-Avraham, 1989; Grossowicz and Gardosh, 1992; Gardosh and Grossowich, 1993). The top of the upthrown side is composed of the Hazeva Formation, exposed in the area west of the fault (west of Arava-1 in Fig. 5-2). These observations suggest that the Iddan Fault has been the southern border fault of the basin since Plio-Pleistocene times. As the fault trace does not reach the surface (Figs.

5-3 and 5-5), the fault has not been active in the recent past.

The intermediate structural step extends along the entire western part of the SDS (Fig. 5-3). West of Mt. Sedom, at the Amiaz plain, the step is 4 km wide. It narrows to about 1-2 km toward the north and widens again in the Masada plain (Fig. 5-3). Like the rim block, the intermediate block is internally deformed by sets of steep normal faults, in both synthetic and antithetic directions, down-to-the-east step blocks and rotated fault blocks (Figs. 5-4 and 5-5).

The intermediate block is regionally tilted to the south. The depth of Upper Cretaceous rocks underneath the basin fill is about 1,400 m below the Dead Sea level in the Masada plain (En Gedi-2), about 3,300 m below the Dead Sea level in the Amiaz plain (Amiaz-1), and it is postulated to be about 4–5 km deep northeast of the Iddan fault (Fig. 5-5).

In the area between the Masada plain and the Amiaz plain, the intermediate block is bordered on the east by the Sedom fault (Fig. 5-3 and 5-6). This is a subsurface normal fault originally identified on seismic profiles underneath the Mt. Sedom diapir (Kashai and Croker, 1987; ten Brink and Ben-Avraham, 1989). It was interpreted on line 3 (Fig. 5-6), where it separates parallel continuous reflections on the downthrown side from the chaotic reflections on the upthrown sides between 2 and 3 seconds. Subsequent drilling of Sedom Deep-1 established the location of the Sedom fault when it was penetrated at about 5,500 m.

South of Mt. Sedom, the Sedom fault curves to the southeast in a direction similar to the Iddan fault. Two other subsurface normal faults cross the basin in a northwest-southeast direction in this area (Fig. 5-3). These four sets of faults form two shallower intermediate blocks separated by a deeper trough (Fig. 5-3). The northern intermediate block is seen in the center of line 4 (Fig. 5-7). It was originally noted by Arbenz (1984) and was later interpreted by Shulman and Kashai (1991), who named it the Neot Hakikar ridge.

Two deep-sunken blocks are found east of the Sedom fault and south of the Neot Hakikar ridge. The southern one is a small trough limited to the southern edge of the basin, whereas the extensive, northern deep block extends from the Neot Hakikar ridge northward into the area of the Lisan diapir where it probably deepens (Fig. 5-3 and 5-7).

The large vertical throw of the Sedom fault, which forms the deep block, is visible on line 3 (Fig. 5-6), where the coherent high- and low-amplitude reflections of the Pleistocene sedimentary fill east of the fault extend to about 3 seconds. Sedom Deep-1, located in the deep block east of the fault, penetrated the entire Pleistocene fill, as well as the Sedom and Hazeva Formations underneath it. The sediments at the bottom of the well at 6,445 m are estimated to be of the lowermost part of the Hazeva Formation. Hence, the depth to the base of the fill in the western part of the deep block is 6–7 km below the Dead Sea level. Farther east, the base of the fill is deeper and is estimated to be at 7–8 km below the Dead Sea level.

The considerable increase in thickness of the Hazeva Formation across the Sedom fault, from 650 m in Amiaz-1 to about 2,000 m in Sedom Deep-1, indicates syntectonic deposition and dates the initial activity of the fault to the Miocene. The Sedom Formation is also offset by the fault (Figs. 5-4 and 5-6) indicating that the Sedom fault was active through Lower Pleistocene time.

The areas delineated by the three structural steps outlined previously are also three different exploration provinces. In the rim block, the Paleozoic and Mesozoic sedimentary section can be penetrated by relatively shallow wells, 2–3 km deep; in the intermediate block, the Paleozoic and Mesozoic reservoir rocks are 3–5 km deep; and in the deep block, most of the Mesozoic section is deeper than 6–7 km and beyond the depth of commercial drilling.

Tectonically, these structural steps seem to be the result of

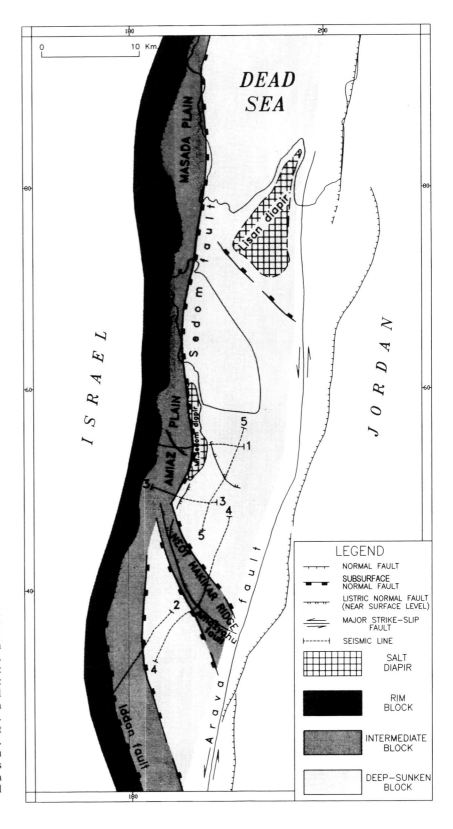

Figure 5-3 Tectonic map showing the three main structural steps of the basin: the rim, intermediate, and deep-sunken blocks. The normal faults that separate the intermediate and deep blocks offset the lower part of the basin's sedimentary fill and are mapped at near base-fill level. Their extension east of the Israel-Jordan border is postulated. Many small listric faults, mostly related to salt flow, offset the upper part of the sedimentary fill. Only some of them, including the large Amazyahu fault, are shown here. Also shown are the locations of the seismic lines in Figures 5-4 to 5-8. (Based on seismic information, Kashai and Croker, 1987 and Geological Map of Israel 1:250,000).

rifting and extension in east-west and northeast-southwest directions. No simple relation was found to the much-discussed sinistral shear in the Dead Sea rift and its associated rhomb-shaped or pull-apart basins (Garfunkel, 1981; Aydin and Nur, 1982; Ben-Avraham, 1985; Kashai and Croker, 1987; ten Brink and Ben-Avraham, 1989).

Several recent works (Frieslander and Ben-Avraham, 1989; Ben-Avraham and Zoback, 1992) suggest that the strike-slip motion in the Dead Sea area takes place on the eastern side of the basin, whereas on the western side most of the deformation is vertical and normal. Ten Brink et al. (1993) proposed that the Dead Sea basin is a full graben that developed by crustal

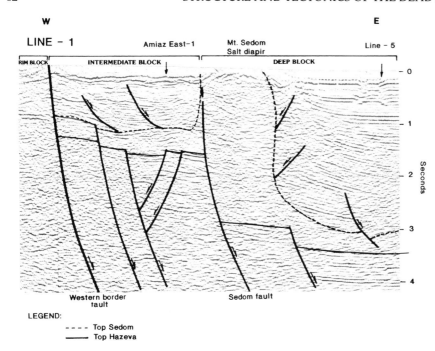

Figure 5-4 Seismic time line 1, showing the western border fault and the two structural steps, the intermediate and deep blocks, separated by the Sedom fault. Note the deformed sediments at the eastern flank of the Mt. Sedom diapir. (For location, see Fig. 5-3).

stretching along its long axis, and sagging of the northern and southern ends towards the center. The fault pattern described here on the western side of the SDS basin is in agreement with both models. A more detailed discussion on the tectonic framework of the basin is beyond the scope of this work.

Halokinesis

Halokinesis, or salt tectonics, at the SDS results from the subsurface flow of the massive salt beds of the Sedom Formation. At the Mt. Sedom and Lisan Peninsula areas, this salt emerges to the surface in large piercing diapirs (Fig. 5-3).

Thick, subhorizontal salt beds were encountered by Amiaz-1 (1,500–2,700 m) and Amiaz East-1 (1,700–3,000 m) in the intermediate block, and by Sedom Deep-1 (3,800–4,700 m) in the deep central block (Figs. 5-4 and 5-6). Chaotic areas underneath the coherent, parallel reflections of the Pleistocene sedimentary fill, occasionally in mounded shapes, were interpreted as salt beds or salt bodies on many of the SDS seismic lines (Fig. 5-8). These two observations suggest that the Sedom salt layer extends in the subsurface throughout most of the SDS area (as previously suspected by Kashai and Croker, 1987).

Small listric normal faults within the Pleistocene fill are found in all parts of the basin. In places such as those on line 5

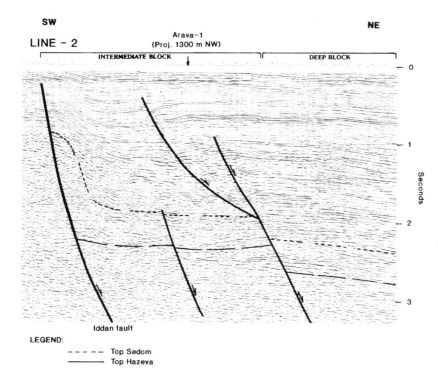

Figure 5-5 Seismic time line 2, showing the Iddan fault and the two structural steps northeast of it. Sediments of the Hazeva Formation exposed on the surface southwest of the Iddan fault are assumed to be below 2 seconds at the downthrown side. The Iddan fault is therefore considered to have been the southern border fault of the basin since Plio-Pleistocene times. The identification of the Top Sedom and Top Hazeva horizons is postulated by the seismic character of the overlying sediments and by the indication for salt-related deformation northeast of the Iddan fault. (For location see Fig. 5-3).

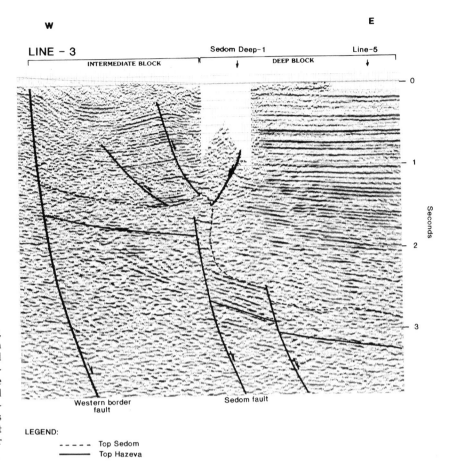

W   **E**

LINE - 3   Sedom Deep-1   Line-5

INTERMEDIATE BLOCK   DEEP BLOCK

Western border fault   Sedom fault

LEGEND:

- - - - - Top Sedom

———— Top Hazeva

Figure 5-6 Seismic time line 3. The Sedom fault, separating the intermediate block on the west from the deep-sunken block on the east, was penetrated by the Sedom Deep-1 well. The well also penetrated the Sedom and Hazeva Formations on the downthrown side of the fault. A small salt swell on top of the fault is the subsurface southern extension of the Mt. Sedom diapir. Small rim synclines found in the overlaying sediments west of the salt swell were formed by salt flow eastward. (For location see Fig. 5-3).

(Fig. 5-8), salt swells and pillows are located underneath the faults. In these cases, the development of the faults can be directly related to salt withdrawal and gliding of the overlying sediments into the area of salt evacuation (Jenyon, 1986).

Line 5 (Fig. 5-8) also shows a two-layer tectonic style typical of the SDS basin. Whereas the lower part of the section is offset by steep normal faults, the upper part is affected by shallow listric faults. The development of salt swells and the initiation of salt flow is most likely related to small movements on the underlying faults. Hence, the location of salt-related structures and diapirs is controlled by the deep-seated normal faults.

The most spectacular listric fault in the basin is the

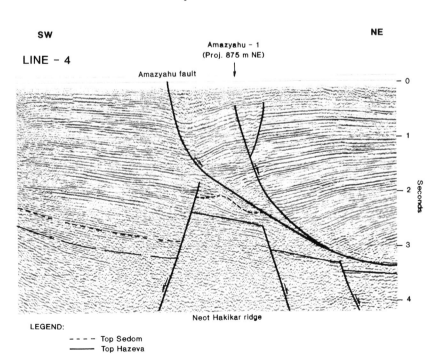

SW   **NE**

LINE - 4   Amazyahu - 1
                                      (Proj. 875 m NE)

Amazyahu fault

Neot Hakikar ridge

LEGEND:

- - - - - Top Sedom

———— Top Hazeva

Figure 5-7 Seismic time line 4. The Neot Hakikar Ridge, in the center of the section, is an intermediate block separating two deep blocks. Note the similar seismic character of the young sediments on top of the Sedom salt northeast and southwest of the ridge. The Amazyahu listric growth fault is a detachment plane on top of the Sedom salt; it developed as a result of salt flow southward toward the Neot Hakikar ridge. (For location, see Fig. 5-3).

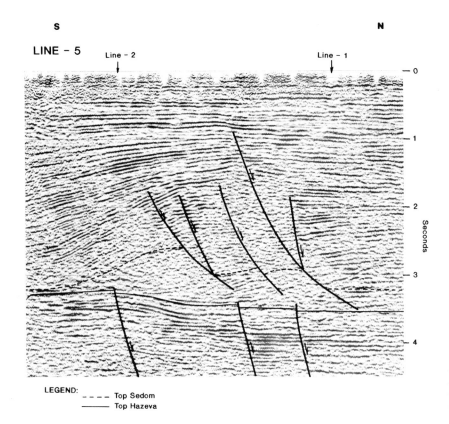

S

LINE - 5

Line - 2

Line - 1

N

LEGEND: - - - - Top Sedom
———— Top Hazeva

Figure 5-8 Seismic time line 5. The flow of Sedom salt into small pillows and swells results in the formation of listric faults in the overlying sediments, shown here at the central part of the basin. (For location, see Fig. 3).

Amazyahu fault (Figs. 5-3 and 5-7). This listric normal fault is one of several detachment planes developed on top of the Sedom salt northeast of the Neot Hakikar ridge. An area of chaotic reflections underneath the fault on line 4 (Fig. 5-7) is interpreted as a salt layer flowing from the deeper part of the basin toward the Neot Hakikar ridge. As much as a 50% increase in the thickness of most of the fill layers on the downthrown side of the fault (Fig. 5-7) indicates continuous salt-flow and fault activity from the Early Pleistocene to the Holocene.

Ten Brink and Ben-Avraham (1989) calculated the volume created by the extension of the sedimentary fill on the Amazyahu fault to be roughly similar to the volume of the Mt. Sedom diapir; they suggested a genetic relationship between the two phenomena. This suggestion is supported by recent seismic interpretation of Grossowicz and Gardosh (1992), who noted that, in the subsurface, the Amazyahu fault curves northward toward the southern edge of Mt. Sedom (Fig. 5-3) and the salt layer underneath the fault actually connects to the salt body of the diapir. Both Mt. Sedom and the Amazyahu fault, therefore, result from the same salt flow but in different directions and intensities. (See Jenyon, 1986, figure 5.36, for a similar situation.)

The Sedom diapir exposed east of the Amiaz Plain extends northward and southward in the subsurface as a nonpiercing diapir along the edge of the intermediate block. The initiation of this complex salt body is related to faulting, increased overburden, and subsequent upward flow along the Sedom fault (Zak, 1967; ten Brink and Ben-Avraham 1989).

In line 1 (Fig. 5-4), almost the entire Pleistocene fill from about 0.5 to 3 seconds is pushed upward on the eastern flank of the Sedom diapir, suggesting that in this area salt flowed continuously westward and upward from the deep-sunken block, throughout the Pleistocene. In line 3 (Fig. 5-6), however, most of the deformation of the young sedimentary fill takes place in

the intermediate block, where shallow-rim synclines (between 0.2 and 1 seconds) were formed as a result of salt evacuation and eastward flow. These indications for two directions of flow suggest that Mt. Sedom developed like a mushroom diapir (Weinberger and Agnon, 1993) rather than a simple salt wall (Zak, 1967).

Intense folding of the young sedimentary fill on the downthrown side of the Iddan fault, interpreted on line 2 (Fig. 5-5), may be the result of a basin-edge diapirism and salt flow along the Iddan fault.

STRATIGRAPHY

The stratigraphy of the SDS area is divided into two parts: (1) prebasin rocks and (2) basin-fill rocks that accumulated syntectonically within the Dead Sea depression.

Prebasin section

The stratigraphy of the Paleozoic, Mesozoic, and Early Tertiary prebasin rocks found in the SDS wells and in outcrops on its rim is similar to that found in the rest of eastern Israel.

The oldest rocks penetrated by wells are Permian sandstones with some carbonates and shales (Fig. 5-9, Table 5-1). The Triassic and Jurassic sequence, overlying the Permian, is composed mostly of carbonates and shales, occasionally rich in organic matter, and some sands (Fig. 5-9). Throughout the Triassic and Jurassic, the Dead Sea area was part of the eastern Mediterranean platform, hence depositional environments were mostly epicontinental shallow marine to fluvial (Druckman, 1974; Goldberg and Friedman, 1974). Considerable thinning of the Jurassic section of the SDS, which is only 500–1,000 m in Masada-1, Heimar-1, and Lot-1 compared to more than 2,000 m in wells in the Mediterranean coastal area (Helez Deep-1), indicates a structurally elevated position in Jurassic times.

The Cretaceous rocks are divided into three distinct units (Fig. 5-9): (1) the Lower Cretaceous continental sandstone of the Kurnub Group, (2) the Cenomanian-Turonian shelf limestone and dolomite of the Judea Group, and (3) the Upper Cretaceous to Lower Tertiary sequence of chalk marl and clay with phosphatic and chert intercalations, deposited in generally open marine environment, named the Mt. Scopus Group.

Thick sections of the Mt. Scopus Group, containing bituminous marls and clays (10–20%) mostly of Campanian and Maastrichtian age, are found within Syrian Arc synclines (Fig. 5-1) throughout southern Israel. In the SDS, 200- to 300-m-thick sections were found in outcrops south of the Masada plain (described by Bentor and Vroman, 1960, as the "En Boqeq oil shale") and in wells (Amiaz-1, En Gedi-2). In the literature, the unit is commonly termed the "Senonian bituminous rocks" (SBR).

Eocene and Oligocene limestone and chalk, exposed in the Negev area west of the Dead Sea, were not identified in SDS wells and were probably eroded by the Hazeva fluvial system in Miocene times. Remnants of rocks of this age have been identified in wadi incisions on the western rim blocks (Agnon, 1993).

Basin-fill section

The Hazeva Formation

The Hazeva Formation (Bentor and Vroman, 1957), consisting mostly of coarse- to fine-grained clastic sediments of fluvial and lacustrine origin, is widely exposed in southern Israel. Its thickness in outcrops is commonly 50–150 m. However, within wells drilled in the SDS, it is significantly thicker (Figs. 5-9 and 5-10) and is therefore considered as the oldest basin-fill unit.

Located on an intermediate block, Amiaz-1 and Amiaz East-1 penetrated 630 m and 600 m, respectively, of the Hazeva Formation (Fig. 5-10). A considerably greater thickness was encountered in the Sedom Deep-1 well located on a deep-sunken block. Underneath the Sedom salt, the well penetrated 1,750 m (from 4,700 to 6,445 m) of predominantly clastic Hazeva sediments with some carbonate (Fig. 5-10). Microfauna and pollen found at the base of the section suggest that the lowermost 50 meters may be of pre-Hazeva Oligocene age. This seems to be the thickest continuous Hazeva section ever found, both in outcrops and wells, in southern Israel.

The lower part of the Hazeva Formation in Sedom Deep-1 (5,350–6,445 m) consists of interlayered conglomerates, sands, marls, and red shales similar to the Hazeva rocks in other wells and outcrops. The upper 650 m (4,700–5,350 m) contain thick carbonate beds and some evaporites, which are not common in other Hazeva sections (Fig. 5-10). The upper Hazeva in the Sedom Deep-1 is probably a locally developed unit of late Miocene age that may correspond to the Mc palynozone of Horowitz (1987) found beneath the Sedom Formation in the bottom of the Sedom-1 well. The presence of limestones and some evaporites in this unit may indicate a brackish depositional environment, possibly that of a terminal lake.

The great thickness of the unit in Sedom Deep-1 indicates the existence in Miocene times of a Hazeva depocenter in the SDS area. Large amounts of quartzose sands were transported intermittently across and into the Dead Sea rift from as far away as the Arabo-Nubian Massif south and east of the Dead

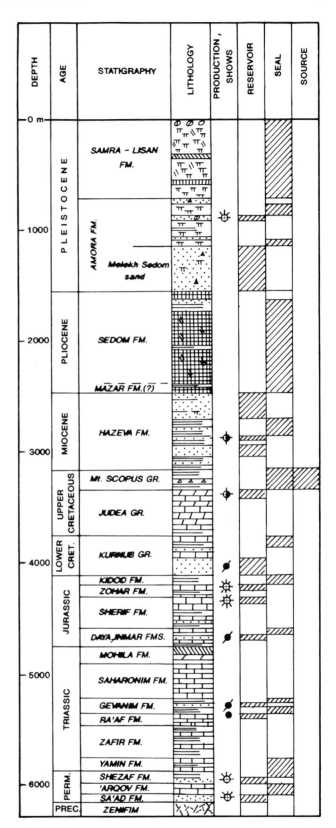

LEGEND :

-◈- Asphalt show   ☼ Gas show

● Light oil show   ☼ Gas production

◪ Subcommercial oil production

Figure 5-9 Generalized columnar section showing the stratigraphy and the main reservoir, seal, and source rocks of the SDS area. Hydrocarbon shows in the reservoir rocks are taken from wells drilled within the basin and on the highlands to the west of it. (For lithology, see legend in Fig. 5-12).

AMIAZ EAST - 1   **SEDOM DEEP - 1**

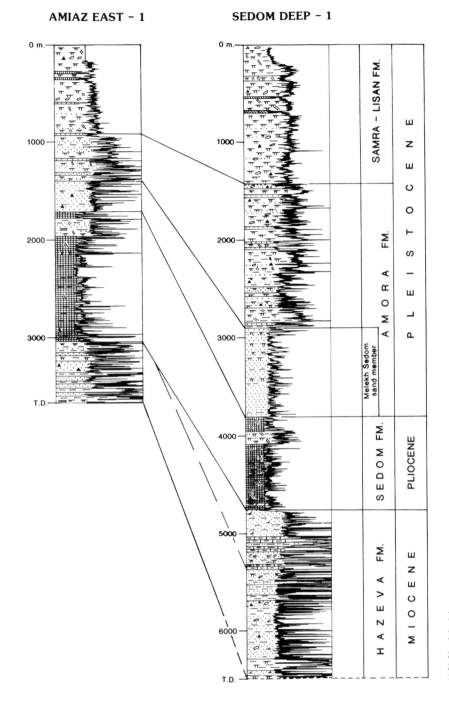

Figure 5-10 A stratigraphic cross section between Amiaz East-1 and Sedom Deep-1 (for location of wells, see Fig. 5-2). The correlations of the stratigraphic units are based on the lithology and the gamma ray log character. (For lithology see legend in Fig. 5-12).

Sea (Sa'ar, 1985). The Hazeva drainage system probably flowed toward the Mediterranean Sea across the rift zone in earlier Miocene times and into the subsiding rift zone in later Miocene times. The northern limit of the basin may have been at the northern part of the SDS area, as indicated by the lack of Hazeva sediments in En Gedi-2 (Neev and Emery, 1967).

The detailed shape and depth of the basin during Hazeva times remains uncertain because of insufficient well and seismic data. However, thick outcrops of Hazeva sediments (up to 2,000 m composite thickness) recently found in the En Yahav area south of the Dead Sea (Y. Bartov, pers. com., 1993) may suggest that the basin extended along the rift from the Lisan area in the north to about 100 km south of the present Dead Sea.

The Sedom Formation

The Sedom Formation (Zak, 1967) consists mainly of halite, 20–30% anhydrite, potash salts, organic-rich shales, and dolomites. The unit was deposited on top of the Hazeva Formation (Fig. 5-10) within a sea tongue that invaded the Dead Sea rift from late Miocene to Pliocene times (Schulman, 1962; Zak, 1967; Shaliv, 1989). The Mazar Formation, a stratigraphic equivalent to the lower part of the Sedom Formation (Fig. 5-9) is exposed in various locations southwest of the SDS, where it consists of interlayered gray marls, sands, and conglomerates (Shahar et al., 1966; Elron, 1980). Within the Dead Sea basin, the Mazar Formation was recognized only in En Gedi-2 (depths of 1,378–

1,420 m) (Zak, 1967; Kashai and Croker, 1987), where it overlies prebasin Senonian rocks.

The original thickness of the Sedom Formation cannot be well established because of later salt flow. However, the 900 and 1,300 m found in the layered salt beds in the Sedom Deep-1 and Amiaz East-1, respectively, (Fig. 5-10), may be taken as minimal values. Seismic reflection information indicates an extensive distribution of the Sedom salt in the SDS. The southern limit of the basin during Sedom times was probably the Iddan fault. The Sedom basin probably extended northward over the entire area presently occupied by the Dead Sea lake, as suggested by small subsurface salt diapirs identified by Neev and Hall (1976) north of the Lisan Peninsula and south of the Jordan River delta (Fig. 5-1).

The Amora and Samra-Lisan Formations

The upper part of the basin fill is the sedimentary section that overlies the Sedom Formation (Figs. 5-9 and 5-10). Although it is the thickest series of sediment in the SDS (maximal thickness over 4,000 m), its stratigraphy is not adequately defined. The names Amora, Samra, and Lisan Formations were given to parts of this section. However, the 400-m-thick Amora sediments of Zak (1967) exposed on the eastern flank of the Mt. Sedom diapir and the 50- to 150-m-thick Lisan and Samra sediments (Begin et al., 1974) found on the margins of the basin, are not well defined in the subsurface.

Based on gamma ray log character and lithology in Amiaz East-1 and Sedom Deep-1, we propose the following definitions (Fig. 5-10):

1) The Amora Formation is predominantly a coarse-grained clastic unit overlying the uppermost massive salt beds of the Sedom Formation. Its thickness ranges from about 1,000 m in the intermediate block to more than 2,500 m in the deep block. The lower part of the Amora Formation is composed of a 300- to 900-m-thick layer of fine- to coarse-grained quartzose sand, with a minor amount of shale. This sand layer is the Melekh Sedom Sand Member, previously identified in the Melekh Sedom-1 well by Horowitz (1987). The upper part of Amora Formation is composed of interlayered quartzose sands, marls, shales, and some conglomerate.

2) The Samra-Lisan Formation is a 1,000- to 1,500-m-thick unit overlaying the Amora Formation. A pronounced change in gamma ray log character and lithology (Fig. 5-10) marks the stratigraphic contact between the units. The Samra-Lisan Formation is composed predominantly of gray marl with varying amounts of anhydrite, gypsum, and salt, as well as a minor amount of sand and conglomerate. The evaporites appear either as layers 5–20 m thick or as very fine laminae within the marl. The stratigraphic contact between the older Samra Formation and the younger Lisan Formation is well defined in outcrops on the basin margins (Picard, 1943; Begin et al., 1974; Gardosh et al., 1992). It is not identified in the subsurface at the Mt. Sedom area. However, it is assumed that the Lisan Formation in the strict sense composes the upper 100 to 300 m of the unit (Fig. 5-10).

As a whole, the Amora and Samra-Lisan Formations display a gradual transition from predominantly coarse-grained clastic sediments in the lower part to very fine-grained clastic to chemical sediments in the upper part.

The Amora Formation, as defined here, is roughly equivalent to the QI and QII palynozones of Horowitz (1987), which were deposited in the preglacial period during lower to middle Pleistocene times. Its depositional environment is fluvial to lacustrine and possibly eolian (Weinberger, 1993). The Samra-Lisan Formation, as defined here, is equivalent to the QIII to QIX palynozones of Horowitz (1987), which were accumulated in the glacial period during middle to upper Pleistocene times.

The formation was deposited in large brackish to saline lakes that occupied the Dead Sea area during the pluvials (the local equivalent of glacials). From Amora to Lisan times, the Dead Sea basin extended from the Iddan fault area in the south to the Jordan River delta in the north (Figs. 5-1 and 5-3).

HYDROCARBON POTENTIAL

Source Rocks

A variety of noncommercial hydrocarbon shows are known in the Dead Sea area (Fig 5-2). Light and heavy oils are found in the subsurface in many of the exploration wells. Asphalt, which appears as huge blocks floating on the Dead Sea and as seepages along basin faults, has been known since historical times (Langotzky, 1963; Nissenbaum, 1978). Commercial gas was found and exploited in the Zohar-Kidod-Hakanaim fields. These shows can be viewed as a "tar belt" of a potential oil province.

Organic geochemical studies show that all the oils and asphalts in the area belong to one geochemical family and hence are derived from a common source rock (Amit and Bein, 1979; Nissenbaum and Goldberg, 1980; Tannenbaum, 1983; Spiro et al., 1983). However, the subject of the source rock for the Dead Sea oils has been debated at length. Goldberg and Starinsky (1972) suggested that these oils were flushed from traps in the northern Negev to their present location by brines. Fleischer et al. (1977) preferred the theory that oils migrated from the coastal plain. Kashai et al. (1973), Breger (1978), and Kashai (1980) suggested that Senonian bituminous rocks (SBR) buried in the basin could be the source for the oils in the region. Nissenbaum and Goldberg (1980) showed nonconclusive geochemical correlation between the SBR and oils, whereas Amit and Bein (1979) concluded that the association of the SBR with the hydrocarbons is coincidental, and the source of the latter is unknown.

Tannenbaum (1983) and Tannenbaum and Aizenshtat (1985), using several geochemical parameters, showed a good multiparameter correlation between the organic constituents of the oils and asphalts and those of the SBR. In particular, a detailed correlation (based on carbon isotopic composition and distribution of biomarkers) was shown between the Amiaz-1 asphaltic oil and the associated SBR. Additional studies (Spiro et al., 1983; Tannenbaum, 1991) support these findings.

The Senonian bituminous rocks are organic rich carbonates, cherts, and phosphorites of the Mt. Scopus Group (Fig. 5-9) found in Israel and in neighboring countries. The SBR were deposited in the highly productive environment of small morphotectonic basins with dysoxic to anoxic bottom water (Spiro, 1980). As much as 20% of the SBR is organic, mostly sapropelic Type II kerogen. The source of the organic matter is mainly marine algae, with some terrestrial contribution. Reworking of the deposited organic matter by bacteria gave rise to its present properties and its prominent enrichment by heteroatoms. Because of its high sulfur content (up to 12%), the kerogen is better described as Type II-S, which thermally degrades at lower temperatures compared to other oil-prone kerogens (Hunt et al., 1991).

Maturation

The SBR were not buried to a significant depth until the formation of the Dead Sea basin in the Miocene. Since then, the SBR has been buried in different basin blocks to various depths to 8 km and has been exposed to temperatures up to 200°C—temperatures that prevail in the "oil window" zone in many sedimentary basins. The differences in the thermal histories of the

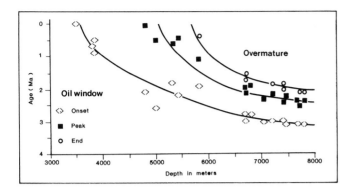

Figure 5-11 The oil window for the SBR in the Dead Sea basin. The window is shown at the current depth of source rocks versus time in millions of years of oil generation. The curves are calculated based on the maturation model described in the text. According to the diagram, most of the SBR buried below approximately 6,000 m are expected to be overmature.

SBR in the basin blocks generated oils at different maturation levels.

Two distinct levels of oil maturity are known in the Dead Sea area: (1) the immature asphaltic oils found in Amiaz-1 and as floating blocks on the Dead Sea, and (2) the light, mature oils found in the margins of the basin (e.g., Masada-1 and Zuk Tamrur-1 drill holes) (Fig. 5-2).

The asphaltic oil in Amiaz-1 flowed into the borehole directly from the source (SBR) at a depth of 3,450 m (90°C). On the basis of its geochemical characteristics, this oil is considered an early product of the oil-generation process (Nissenbaum and Goldberg, 1980; Tannenbaum, 1983). Despite the uncertainty of the exact ages of the basin-fill formations, a reasonable reconstruction of the burial and thermal histories has been carried out. The calculated time-temperature index (TTI) (Waples, 1980) for the SBR in Amiaz-1 is 0.6 (Tannenbaum, 1991).

The TTI is widely used for maturation assessments of petroleum source rocks. The onset of oil generation is at TTI = 15, the peak is at TTI = 75, and the end of the oil window is at TTI = 160. However, these figures are based on kinetic considerations and empirical evaluations related to vitrinite maturation. The kinetic parameters of maturation reactions of the SBR are considerably different. Based on our results, the SBR tend to thermally degrade at lower TTI values than vitrinite, mainly because of its high sulfur content (Tannenbaum and Aizenshtat, 1984, 1985). The early degradation of high-sulfur kerogens in different basins is now well documented (Lewan, 1985; Orr, 1985; Petersen and Hickey, 1987; Hunt et al., 1991). Some of these studies indicate that oil generation in the Monterey Formation of California starts at TTI values of less than 1.0 (Petersen and Hickey, 1987; Hunt et al., 1991).

To quantify the maturation level of the SBR buried to different depths in the Dead Sea basin, Tannenbaum (1991) suggested the following "oil windows":
   Onset of oil generation, TTI = 0.6
   Peak of oil generation, TTI = 3.0 (light oil generation)
   End of oil generation, TTI = 7.0
These values are different from the oil-window values of Waples (1980) and Hunt et al., (1991). They were suggested after the recognition that the oil window is not universal (e.g., Wood, 1988; Hunt et al., 1991) and that its limits are dependent on the type of organic matter in the source rock and the rates of heating during burial. In other words, every source rock can have its own specific "window."

The TTI value for the onset of oil generation in the SBR (Fig. 5-11) is determined empirically based on the findings in Amiaz-1. The other values for the oil window were chosen based on the assumption that kinetics similar to those applied in the Lopatin/Waples model (Arrhenius equation) can be applied here. In any case, these values seem conservative when compared to those suggested by Hunt et al. (1991) for the SBR.

In the deep-sunken basin blocks, the maximum burial of the SBR is expected to be greater than 7 km (see Sedom Deep-1, Figs. 5-6 and 5-10). The corresponding temperatures are higher than 160°C, and the calculated TTI values are greater than 50. These values indicate that the oil-generation process in these SBR has been completed (Fig. 5-11). The light oil found in Masada-1 and Zuk Tamrur-1 is direct evidence for the advanced maturation stages of the SBR in the basin.

Migration

The immature, viscous asphaltic oil generated in the SBR in the intermediate block may stay in its source rock (as was the case in Amiaz-1 drill hole) or migrate to the overlying Hazeva Formation (as in Amiaz East-1 drill hole).

Light oils generated in the deep basin blocks initially migrate into the overlying Hazeva Formation or downward into the Turonion dolomites. Secondary migration depends on the local hydrodynamic conditions. In general, most of the petroleum is expected to migrate upward, northward, westward, and southward through faults and carrier beds. The conditions for eastward migration are unknown.

Oil maturation

In the course of low-temperature oil-alteration processes (mainly biodegradation and water washing), hydrocarbons of low to medium molecular weight are removed from the oils. Consequently, heavy oils and asphalts, which are difficult to exploit, are formed. The most significant alteration process is biodegradation, which prevails in reservoirs where the oils are in contact with meteoric waters.

Along the western margins of the Dead Sea basin, oil shows are found throughout the sedimentary section (Figs. 5-2 and 5-9). At depths of more than 2 km, light oils are encountered in Triassic formations. At shallower depths, the section is flushed with meteoric waters (Starinsky, 1974; Fleischer et al., 1977) and the oils are heavy. At the surface, only heavy asphaltic oils and asphalts are encountered.

Using geochemical characterization and classification, Tannenbaum et al. (1987) and Tannenbaum (1990) showed a very good correlation between the quality of the oils in the basin rim and the extent of meteoric water invasion into the reservoirs. According to these studies, the surface asphalts represent a residue of 10–20% of the original light oils.

SUMMARY

The current understanding of the petroleum-generation process in the Dead Sea area is as follows: The SBR has been buried in the basin since the Miocene to different depths, resulting in the generation of oils with varying degrees of thermal maturation. The time of oil generation is about 3 million years ago to the present (Fig. 5-11). Some of these oils were trapped within the basin, and some migrated to the basin rims. At shallow depths, these oils were subjected to alteration processes, causing generation of heavy oils and asphalts. A calculation of the oil-generation capacity of the SBR, based on a conservative estimation of its distribution in the deep-sunken blocks, indicate a potential of about 8–10 billion barrels of oil (Tannenbaum, 1991).

Classification of hydrocarbon shows

In the Dead Sea area, hydrocarbons are classified as follows:

Immature asphaltic oils generated at an early stage of thermal maturation:

(1) Asphalt blocks floating on the salt waters of the Dead Sea, indicating submarine seepages along basin faults.

(2) Asphaltic oil encountered in the Amiaz-1 drill hole at 3,450 m. This oil flowed into the borehole from its source rock (the SBR).

(3) Asphaltic oil in the Hazeva Formation in Amiaz East-1 at 3,240 m. This oil most probably migrated from a nearby SBR.

Light oils in the Triassic formation in Masada-1 and in Zuk Tamrur-1 (basin rim). These are mature oils that were generated in the deep basin and migrated upward and westward.

Heavy oils in various formations in most of the wells east of the Zohar gas field (basin rim). These oils represent different degrees of light-oil alteration by biodegradation and water washing.

Asphalt seepages along the western border faults. These asphalts are the end products of the alteration processes of light and heavy oils.

Natural gas (commercial) in the Zohar-Kidod field. The gas, which migrated upward and westward from the deep basin, is another product of the oil-generation processes.

TRAPPING REGIMES

Reservoirs and Seals

The geologic conditions that prevailed in the Dead Sea area during much of its history resulted in the deposition of abundant high-quality reservoir rocks and seal rocks, often in close association (Fig. 5-9).

Many of the various sandstone and carbonate reservoirs either are commercially producing or contain subcommercial amounts of hydrocarbons in the area west of the SDS or within the basin itself.

The oldest reservoir rocks in the basin are the Permian Shezaf Arqov and Sa'ad sandstones and limestones (Fig. 5-9), which are the stratigraphic equivalents of the producing Khuff reservoirs of the Persian Gulf.

The Triassic Ra'af limestone (Fig. 5-9), which has good reservoir properties where secondary porosity is developed, was tested at Masada-1 and several liters of light oil were recovered. Another Triassic unit, the Gevanim Formation, produced light oil from a thin sand layer in Zuk Tamrur-1.

The Jurassic Inmar sandstones, capped by the tight Daya limestone (Fig. 5-9) produced approximately 3,000 bbl of heavy oil in the Gurim structure (10 km west of the SDS). The Jurassic Zohar and Sherif fractured limestones, overlain by the Kidod Shale, are the main reservoirs in the Zohar gas field.

The Lower Cretaceous Kurnub sandstones (Fig. 5-9) display excellent reservoir properties in several Dead Sea wells. The porosity of these sandstones is 10–30%, the permeability is high, and secondary cementation is minor. This unit produced approximately 9,000 bbl of heavy oil in the Gurim structure.

The Upper Cretaceous Judea limestones and dolomites (Fig. 5-9) are the main water aquifer of central and southern Israel. Their porosity is related to secondary solution and karsting. They are overlain by the SBR, which are the primary source rocks of the basin, as well as an effective seal.

The Miocene Hazeva sands (Fig. 5-9) display 20–30% porosity and high permeability in the intermediate block wells of Amiaz-1 and Amiaz East-1. In the latter, a 30-m interval of asphalt-bearing sand was found. However, reservoir properties in the Hazeva sands seem to deteriorate with depth as a result of secondary cementation, as was found in Sedom Deep-1. The Hazeva contains abundant shale and marl interbeds and is capped by the massive salt beds of the Sedom Formation. This combination of porous sands overlain by salt is an ideal situation for oil trapping.

The Pleistocene Melekh Sedom sands (Fig. 5-9), as well as many smaller sandy intervals in the Amora Formation, are poorly consolidated and highly porous and permeable. These are overlain by various thin seals within the formation, as well as by the thick impermeable marls of the Samra-Lisan Formation (Fig. 5-9). Gas shows were found in these sediments at the Amazyahu-1 and Melekh Sedom-1.

Traps

Reservoir rocks at all stratigraphic levels may be found in a variety of structural and stratigraphic trapping conditions at the rim, intermediate, or deep-sunken blocks of the SDS. At least five different potential trapping styles can be identified (Fig. 5-12): rotated fault blocks, folds, listric faults, salt diapirs, and stratigraphic traps.

Rotated fault blocks All three provinces (rim, intermediate and deep) contain down-faulted, rotated blocks in either synthetic or antithetic position, bounded by normal faults (Fig. 5-12). This trapping style, which is well known in oil provinces associated with rifts, was the target of many SDS wells (Table 5-1). Indeed, two of them (Zuk Tamrur-1 and Masada-1) contained oil. The main uncertainty regarding these traps is fault closure and the sealing capacity of the fault planes. Although tectonic activity may cause occasional opening of fault planes, we assume that they are usually sealed for vertical migration, particularly at depth, because of clay smearing, grain crushing, or filling with asphaltic oil.

Folds The Syrian Arc folding phase which affected the region from Late Cretaceous to Early Tertiary times, produced anticlines and monoclines of various size, which are well exposed in the Negev area (Fig. 5-1). After the opening of the Dead Sea rift, these folds were downfaulted and subsided into the basin. We assume that folds of the Syrian Arc are present in the prebasin rocks of the SDS (Fig. 5-12). As oil traps, these folds have two major advantages: (1) closure may not be totally controlled by faults, thereby diminishing the problem of hydrocarbon leakage along fault planes, and (2) the formation of traps predate oil maturation. The identification of folds in the subsurface is however difficult mainly because of (1) the intense faulting in the basin that has masked older structures, and (2) the reduced quality of seismic data under the salt layer of the Sedom Formation.

Listric faults Listric growth faults often associated with salt movement within the young sedimentary fill of the SDS basin offer another trapping mechanism in the intermediate and deep blocks. Pleistocene reservoirs, such as the Melekh Sedom sands, may be trapped in rollover anticlines or tilted beds at the downthrown side of these faults (Fig. 5-12). The Amazyahu-1 Well (Table 5-1), drilled into the northern downthrown side of the Amazyahu fault (Fig. 5-7), tested this type of structure. Trapping conditions in these young structures are ideal. However, to date, no significant oil shows and only minor gas shows were found in the young sedimentary fill of the basin. It is possible that the Sedom salt layer, which occupies large parts of the SDS basin, effectively isolates Pleistocene reservoirs from the Senonian source rocks underneath.

Salt Diapirs Large piercing diapirs (Mt. Sedom and Lisan) and abundant smaller nonpiercing salt swells and pillows are found in the intermediate and deep blocks of the SDS basin (Figs. 5-4 and 5-8). Various types of traps on the flanks, tops, and bottoms of salt structures are well known in many oil provinces, such as the North Sea and Gulf of Mexico (e.g., see

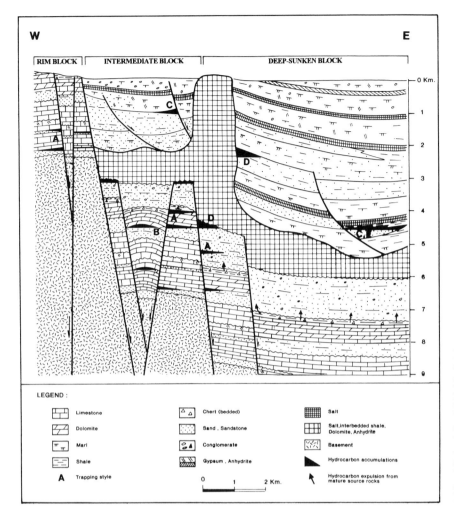

Figure 5-12 Schematic east–west geologic cross section approximately parallel to seismic line 1, showing the three structural steps of the basin. Mature source rocks in the deep blocks expel hydrocarbons into the overlying Hazeva Formation. These light hydrocarbons migrate both laterally along porous beds and vertically along faults towards the basin margins. Hydrocarbons may accumulate in several types of potential traps shown in the section: A = rotated blocks bounded by normal faults, B = buried folds, C = tilted beds on listric faults, D = salt diapir flanks and bottoms, E = stratigraphic traps and pinchouts.

Jenyon, 1986). The potential for this kind of trapping mechanism also exists in the SDS (Fig. 5-12). Two wells, Har Sedom-1 and Admon-1, were drilled east of Mt. Sedom into an uplifted Pleistocene reservoir on the eastern flank of the diapir. Har Sedom-1 was abandoned because of technical problems, whereas Admon-1, drilled by the INOC in 1991, was a dry hole. The latter well was intended to test high-amplitude reflections on several seismic lines suspected to be direct hydrocarbon indications; instead, thin salt beds in the Samra-Lisan Formation were found.

Stratigraphic traps Several types of stratigraphic traps may be found in both prebasin and basin-fill sediments: unconformities, pinchouts of sand, channel fill, and other buried topography (Fig. 5-12). No exploration well has yet been designed to test stratigraphic traps in the prebasin strata, primarily because of the insufficient quality of seismic data.

EXPLORATION STRATEGY

The capacity of the SDS basin to produce hydrocarbons from the Senonian bituminous rocks is the most significant factor in the Dead Sea exploration. The SBR were deposited in Syrian Arc–type synclines (Fig. 5-1) and were later buried in the Dead Sea rift. Although the unit is widely exposed west of the SDS area (Fig. 5-2), its distribution within the basin is not well known because of insufficient geologic and seismic data.

A conservative estimate of the extent of the SBR in areas that either passed or are currently in the oil window led Tannen-baum (1991) to calculate the hydrocarbon-generation capacity in the SDS basin to be 8–10 billion barrels of oil. We assume that only a minor amount of these hydrocarbons has been lost through alteration processes in the basin's margin, or was already found in the small gas fields west of the basin (about 60 billion cubic feet in the Zohar, Kidod, and Hakanaim fields). The trapping efficiency of small and confined basins like the SDS is generally considered to be high. Hence it is assumed that a significant percentage of the hydrocarbons generated in the basin could have been trapped in commercial accumulations.

At least 11 stratigraphic intervals (Fig. 5-9) contain reservoir rocks. The abundant shows on both the surface and subsurface suggest that the conditions for migration and accumulation of hydrocarbon exist. Lateral migration through the many reservoirs in the lower part of the sedimentary section is probably more common than vertical migration along faults. At least one regional seal, the Sedom salt, as well as several other potentially extensive seals, such as the SBR and Kidod shales (Fig. 5-9), are found within the basin.

Abundant potential traps, particularly downfaulted and rotated blocks, are found in all three structural provinces. The main difficulty with past exploration in the basin was the identification and definition of these traps. The location of 14 wells drilled in the early phase of exploration was based only on surface asphalt seeps, surface mapping, or interpreted gravity anomalies (Table 5-1). The validity of gravity anomalies is questionable because the huge gradient generated by the boundary faults masks local features. Drilling targets and depth ranges were often rather inadequately defined. As a result, several wells did not penetrate reservoir rocks or were abandoned at shallow depths (Table 5-1).

In view of the basin's structural complexity on one hand and its potential on the other hand, exploration efforts should continue in all three structural provinces. Drilling in the deep blocks has the major advantage of proximity to the area of oil generation. Most of the Mesozoic reservoirs are, however, beyond the depth of commercial drilling. The Hazeva Formation, ideally located above the Senonian source rocks, is within reach. Only one out of the six wells drilled in a deep block (Sedom Deep-1) tested the Hazeva Formation, as originally proposed by Kashai (1980). More wells in better structural position should be drilled into the unit at the deep blocks.

The Paleozoic and Mesozoic reservoirs are considerably shallower in the intermediate block. Only two wells (Amiaz-1 and En Gedi-2) (Table 5-1) penetrated some of these reservoirs in the intermediate block. If indeed lateral migration in the lower parts of the section is taking place, these deep reservoirs may have great potential in the intermediate blocks.

The structurally highest block is the rim. The only shows of light oil were found there (Masada-1, Zuk Tamrur-1) (Table 5-1). The quality of seismic lines in the rim block is generally poor because of the topography of the area. However, in view of the technological progress in data acquisition, more effort should be made to define traps seismically and to drill into the Mesozoic and Paleozoic rocks in the rim block.

In summary, although the SDS basin has been explored for 40 years, it is still in the relatively early stages of exploration, and its potential as a significant hydrocarbon province should continue to be tested.

Acknowledgments

We are indebted to the Israel National Oil Company and its managing director, Y. Ran, for supporting this project. We thank the INOC-Dead Sea Limited Partnership and its partners Naphta Ltd., Naphta Exploration-Limited Partnership, Delek Oil Exploration Ltd., M.G.N. Oil and Gas Resources, Mandel Resources Corp., and Ashdod Refinery Ltd. for permission to use the geological and geophysical data. The critical reading and helpful comments of S. Baker of the INOC and T. Niemi are greatly appreciated. We thank M. Bouni for the excellent drafting and production of the figures, and Y. Rinkoff and D. Kedar for typing the manuscript.

REFERENCES

Agnon, A., 1993, Field trip No. B-1, part a.: Israel Geological Society, Annual Meeting, March 15–18, p. 81–97 (in Hebrew).

Amit, O., and Bein, A., 1979, The genesis of the asphalt in the Dead Sea area: *Journal of Geochemical Exploration*, v. 11, p. 211–215.

Arbenz, J. K., 1984, Oil potential of the Dead Sea area: Tel Aviv, Sismica Oil Exploration Ltd., Internal Report 84/111, 54 p.

Aydin, A., and Nur, A. 1982, Evolution of pull-apart basins and their scale independence: *Tectonics*, v. 1, p. 91–105.

Begin, Z. B., Ehrlich, A. and Nathan, Y., 1974, Lake Lisan: The Pleistocene precursor of the Dead Sea: Jerusalem, Geological Survey of Israel Bulletin 63, 30 p.

Ben-Avraham, Z., 1985, Structural framework of the Gulf of Elat (Aqaba), Northern Red Sea: *Journal of Geophysical Research*, v. 90, p. 703–726.

Ben-Avraham, Z., and Zoback, M. D., 1992, Transform-normal extension and asymmetric basins: An alternative to the pull-apart models: *Geology*, v. 20, p. 423–426.

Bentor, Y. K., and Vroman, A., 1957, The Geological Map of Israel, 1:100,000 Scale, Series A, Sheet 19, Arava Valley, (with explanatory text): Jerusalem, Geological Survey of Israel, 66 p.

Bentor, Y. K., and Vroman, A., 1960, The Geological Map of Israel, 1:100.000 Scale, Series A, Sheet 16, Mt. Sedom (with explanatory text): Jerusalem, Geological Survey of Israel, 77 p.

Breger, I. A., 1978, Fossil fuels of Israel and bitumens of the Green River Formation of Utah: Relationships and analogies [abs.], *in* Friedman, G. M., ed., 10th International Congress on Sedimentology, v. 1, p. 84.

Coates, J., Gottesman, E., Jacobs, M. and Rosenberg, E., 1963, Gas discoveries in the western Dead Sea region: 6th World Petroleum Congress, Proceedings, Frankfurt/Main, Section 1, paper 26, 15 p.

Cochran, J. R., 1983, A model for the development of the Red Sea: *American Association of Petroleum Geologists Bulletin*, v. 67, p. 41-69.

Druckman. Y., 1974, The stratigraphy of the Triassic sequence in southern Israel: Jerusalem, Geological Survey of Israel Bulletin 64, p. 1–92.

Edcon Inc., 1991, Processing and interpretation, gravity and aeromagnetic survey, southern Dead Sea area: Tel Aviv, I.N.O.C. Ltd., Internal Report, 48 p.

Elron, E., 1980, The geology of the Nahal Zin area (Northern Hazeva Sheet): Jerusalem, Geological Survey of Israel Report MM/6/79, (in Hebrew with English abstract), 54 p.

Fleischer, E., Goldberg, M., Gat, J. R., and Margaritz, M., 1977, Isotopic composition of formation water from deep drillings in southern Israel: *Geochimica et Cosmochimica Acta*, v. 41, p. 511–525.

Freund, R., Garfunkel, Z., Zak, I., Goldberg, M., Weissbrod, T. and Derin, B., 1970, The shear along the Dead Sea rift: *Philosophical Transactions of the Royal Society London*, Ser. A, v. 267, p. 107–130.

Frieslander, U., and Ben-Avraham, Z., 1989, The magnetic field over the Dead Sea and its vicinity: *Marine and Petroleum Geology*, v. 6, p. 148–160.

Gardosh, M., Kaufman, A., and Yechieli, Y., 1991, A reevaluation of the lake sediment chronology in the Dead Sea basin based on new 230 TH/U dates [abs.]: Israel Geological Society, Annual Meeting, p. 42–43.

Gardosh, M. and Grossowich, Y., 1993, Mapping of Seismic Units in the Plio–Pleistocene Sedimentary Fill of the Southern Dead Sea Basin [abs.]: Israel Geological Society, Annual Meeting, March 15–18, p. 39.

Garfunkel, Z., 1981, Internal structure of the Dead Sea Leaky Transform (rift) in relation to plate kinematics: *Tectonophysics*, v. 80, p. 81–108.

Garfunkel, Z., Zak, I., and Freund, R., 1981, Active faulting in the Dead Sea Rift: *Tectonophysics*, v.80, p. 1–26.

Goldberg, M., and Starinsky, A., 1972, Hydrocarbon migration to the Dead Sea Area: Oil Exploration in Israel—A Symposium, Geological Survey of Israel, 22 p.

Goldberg, M., and Friedman, G. M., 1974, Paleoenvironment and paleogeographic evolution of the Jurassic system in southern Israel: Jerusalem, Geological Survey of Israel Bulletin, 44 p.

Grossowicz, Y., and Gardosh, M., 1992, Seismic interpretation of Pleistocene formations in the southern Dead Sea area: Tel Aviv, I.N.O.C. Ltd., Internal Report, 31 p.

Horowitz, A., 1987, Palynological evidence for the age and rate of sedimentation along the Dead Sea rift, and structural implications: *Tectonophysics*, v. 141, p. 107–116.

Hunt, J. M., Lewan, M. K., and Hennet R., 1991, Modeling oil generation with time–temperature index graphs based on the Arrhenius equation: *American Association of Petroleum Geologists Bulletin*, v. 75, p. 795–807.

Jenyon, M. K., 1986, *Salt tectonics*: New York, Elsevier, N.Y., 191 p.

Kashai, E. L., 1980, Sedom-3, recommendation for drilling: Tel Aviv, Oil Exploration (Investment) Ltd., Internal Report 80/24, 24 p.

Kashai, E. L., 1989, A review of the relation between the tectonic, sedimentation and petroleum occurrences of the Dead Sea–Jordan Rift system, *in* Manspeizer, W., ed., *Triassic–Jurassic Rifting*: Amsterdam, Elsevier, p. 883–909.

Kashai, E. L., Fleischer, L. and Schlein, N., 1973, The petroleum potential of the Dead Sea basin [abs.]: Oil Exploration in Israel in the Seventies, Symposium, Jerusalem, p. 20–21.

Kashai, E. L., and Croker, P. F., 1987, Structural geometry and evolution of the Dead Sea–Jordan Rift system as deduced from new subsurface data: *Tectonophysics*, v. 141, p. 36–60.

Langotzky, Y., 1963, Asphalt in the Dead Sea region: Jerusalem, Geological Survey of Israel Report 129/62, 28 p. (in Hebrew).

Lewan, M. D., 1985, Evaluation of petroleum generation by hydrous pyrolysis experimentation: *Philosophical Transactions of the royal Society London*, London, v. 315, p. 123–134.

Neev, D., and Emery, K., O., 1967, The Dead Sea depositional processes and environments of evaporites: Jerusalem, Geological Survey of Israel Bulletin, 147 p.

Neev, D., and Hall, J. K., 1976, The Dead Sea Geophysical Survey—Final Report No. 2: Jerusalem, Geological Survey of Israel Report MG/6/76, 21 p.

Nissenbaum, A., 1978, The Dead Sea asphalts—Historical aspects: *American Association of Petroleum Geologists Bulletin*, v. 62, p. 837–844.

Nissenbaum, A., 1991, Oil exploration in the Holy Land, 1884–1955: *Israel Journal of Earth Sciences*, v. 40, p. 245–250.

Nissenbaum, A., and Goldberg, M., 1980, Asphalts, heavy oils, ozocerite and gases in the Dead Sea basin: *Organic Geochemistry*, v. 2, p. 167–180.

Orr, W. L., 1985, Kerogen/asphaltene/sulfur relationship in sulfur-rich Monterey oils: *Organic Geochemistry*, v. 10, p. 499–516.

Petersen, N. F., and Hickey, P. J., 1987, California Plio-Miocene Oils: Evidence of early generation, *in*, Meyer, R. F., ed., *Exploration for heavy crude oil and natural bitumen*: American Association of Petroleum Geologists Studies in Geology #25, p. 351–359.

Picard, L., 1943, Structure and evolution of Palestine with comparative notes on neighbouring countries: Jerusalem, The Hebrew University Geological Department Bulletin, v. 4, no. 2–4, 187 p.

Quennell, A. M., 1958, The structure and geomorphic evolution of the Dead Sea rift: *The Quarterly Journal of the Geological Society of London*, v. 64, p. 1–24.

Sa'ar, H., 1985, Origin and sedimentation of sandstones in basin fill formations of the Dead Sea rift valley: Jerusalem, Geological Survey of Israel Report MM/3/86, 80 p. (in Hebrew with English Abstract).

Schulman, N., 1962, The geology of the central Jordan valley [Ph.D. thesis]: Jerusalem, The Hebrew University, 103 p. (in Hebrew with English abstract).

Shahar, Y., Reiss, Z., and Gerry, E., 1966, A new outcrop of marine Neogene in the Negev: *Israel Journal of Earth Science*, v. 15, p. 82–84.

Shaliv, G., 1989, Stages in the tectonic and volcanic history of Neogene continental basins in northern Israel [Ph.D. thesis]: Jerusalem, The Hebrew University, 102 p. (in Hebrew with English abstract).

Shulman, H., and Kashai, E., 1991, Tectonic review of the southern Dead Sea basin [abs.]: Israel Geological Society, Annual Meeting, April 22–25, p. 100.

Spiro, B., 1980, Geochemistry and mineralogy of bituminous rocks in Israel [Ph.D. thesis]: Jerusalem, The Hebrew University, 152 p. (in Hebrew with English abstract).

Spiro, B., Welte, D. H., Rullkoetter, J., and Schaefer, R. G., 1983, Asphalts, oils and bituminous rocks from the Dead Sea Area—A geochemical correlation study: *American Association of Petroleum Geologists Bulletin*, v. 67, p. 1,163–1,175.

Starinsky, A., 1974, Relationship between Ca-chloride brines and sedimentary rocks in Israel [Ph.D. Thesis]: Jerusalem, The Hebrew University, 176 p. (in Hebrew with English abstract).

Tannenbaum, E., 1983, Researches in the geochemistry of oils and asphalts in the Dead Sea area, Israel [Ph.D. Thesis]: Jerusalem, The Hebrew University, 117 p. (in Hebrew with English abstract).

Tannenbaum, E., 1990, Basin modeling of the southern Dead Sea basin—Progress report: Tel Aviv, I.N.O.C. Ltd., Internal Report, 18 p.

Tannenbaum, E., 1991, Basin modeling of the southern Dead Sea basin, with an emphasis on oil generation migration and entrapment: Tel Aviv, I.N.O.C. Ltd. Internal Report, 18 p.

Tannenbaum, E., and Aizenshtat, Z., 1984, Formation of immature asphalt from organic-rich carbonate rocks, correlation of maturation indicators: *Organic Geochemistry*, v. 6, p. 503–511.

Tannenbaum, E., and Aizenshtat, Z., 1985, Formation of immature asphalt from organic-rich carbonate rocks, geochemical correlation: *Organic Geochemistry*, v. 8, p. 181–192.

Tannenbaum E., Starinsky, A., and Aizenshtat, Z., 1987, Light oils transformation to heavy oils and asphalts—An Assessment of the amounts of hydrocarbon removed and the hydrological-geological control of the process, *in* Meyer, R. F., ed., *Exploration for heavy crude oil and natural bitumen*: American Association of Petroleum Geologists Studies in Geology #25, p. 221–231.

ten Brink, U., and Ben-Avraham, Z., 1989, The anatomy of a pull-apart basin-Seismic reflection observation of the Dead Sea basin: *Tectonics*, v. 8, p. 333–350.

ten Brink, M. S., Ben-Avraham, Z., Bell, R. E., Hassouneh, M., Coleman, D. F., Andreasen, G., Tibor, G., and Coakley, B., 1993, Structure of the Dead Sea pull-apart basin from gravity analyses: *Journal of Geophysical Research*, v. 98, p. 21,877–21,894.

Waples, D. W., 1980, Time and temperature in petroleum formation—Application of Lopatin's method to petroleum exploration: *American Association of Petroleum Geologists Bulletin*, v. 64, p. 916–926.

Weinberger, G., 1993, A sedimentological study of the pre-glacial Pleistocene QI palynozone of Sedom-2 borehole, southern Dead Sea, Israel: *Journal of African Earth Sciences*, v. 17, p. 249–255.

Weinberger, R., and Agnon, A., 1993, Field Trip No. B-1, part b: Israel Geological Society, Annual Meeting, March 15–18, p. 98–111 (in Hebrew).

Wood, D. A., 1988, Relationships between thermal maturity indices calculated using Arrhenius equation and Lopatin method—Implication for petroleum exploration: *American Association of Petroleum Geologists Bulletin*, v. 72, p. 115–134.

Zak, I., 1967, The geology of Mt. Sedom [Ph.D. thesis]: Jerusalem, The Hebrew University, 208 p., (in Hebrew with English abstract).

Zak, I., and Freund, R., 1981, Asymmetry and basin migration in the Dead Sea rift: *Tectonophysics*, v. 80, p. 27–38.

6. ACTIVE TECTONICS IN THE DEAD SEA BASIN

TINA M. NIEMI AND ZVI BEN-AVRAHAM

The Dead Sea, more than 400 m below sea level, is an actively subsiding basin formed along the Dead Sea–Jordan transform plate boundary. In general, models for the development of the Dead Sea are based on evidence from land geology that indicates strike slip entering the Dead Sea basin along the eastern margin on the Arava fault in the south and stepping over to the western margin to the Jordan fault north of the Dead Sea (Fig. 6-1a). The basin, therefore, is thought to be a depression formed by a north–south extension between the *en echelon* faults—a classic example of a pull-apart basin (left-stepping fault traces in a left-lateral, strike-slip fault system).

Whereas early models of the tectonic development of the Dead Sea and cumulative displacement of about 105 km across the Dead Sea–Jordan transform were based solely on field map compilations (e.g., Quennell, 1959; Freund et al., 1970), geophysical data are now available to interpret tectonic processes in the Dead Sea (see Ben-Avraham, chapter 3, this volume). These data are much more extensive in the southern Dead Sea basin because of active hydrocarbon exploration (e.g. Arbenz, 1984; Kashai and Croker, 1987; ten Brink and Ben-Avraham, 1989; Gardosh et al., chapter 5, this volume). There remain many questions on the structure of the northern, deep-water portion of the Dead Sea basin.

Specific data on the geometry of faults submerged in the northern Dead Sea basin were first collected in a geophysical survey by Neev and Hall (1976, 1979), who presented a sketch map of tectonic elements that is based on interpretation of the seismic reflection data and satellite imagery (Fig. 6-1b). In this structure map, strike-slip faults are confined to the eastern margin. They map three *en echelon*, northeast-trending, strike-slip faults, of which only the northernmost fault bounds the offshore portion of the Dead Sea between its north coast and the Arnon (Wadi Mujib) delta. The fault system along the western

margin of the basin was interpreted as a series of subparallel, down-to-the-east, normal faults. Cross faults are inferred along the flanks of a horst block of the Lisan Peninsula.

Additional data on the sense of motion of the submarine faults in the northern Dead Sea basin was interpreted by Frieslander and Ben-Avraham (1989) from an extensive grid of marine magnetic data. These authors concluded, on the basis of the continuity of regional magnetic trends on the west and mismatches to the east, that most of the western margin of the basin is characterized by normal faults, and the eastern margin, by strike-slip faults (Fig. 6-1c). The structure of the Dead Sea basin was also inferred from marine and land gravity data (ten Brink et al., 1993). The western margin faults were interpreted as "passive collapse" of walls of the basin along normal faults.

The distribution of earthquake epicenters along the plate boundary suggests that the Dead Sea is one of the most seismically active sections of the Dead Sea transform north of the Gulf of Elat (Aqaba) (van Eck and Hofstetter, 1990). Because of the poor positioning of seismometers, the spatial resolution does not permit assigning individual earthquakes to specific faults. Portable seismic networks allow for increased accuracy when plotting epicentral locations and enable the solution of earthquake focal plane mechanisms. Along the Lisan Peninsula region of the southern Dead Sea basin, the complexity of the area is seen by fault-plane solutions for left-lateral, strike-slip motion on both sides of the basin, in addition to right-lateral, normal, reverse, and oblique slip (van Eck and Hofstetter, 1989).

These previous studies show that faulting in the Dead Sea cannot be explained by a simple overlap of two *en echelon* strike-slip faults (e.g., Wilcox et al., 1973; Crowell, 1974). The active faults within the Dead Sea basin, their history, and their neotectonic behavior are not fully understood. High-resolution

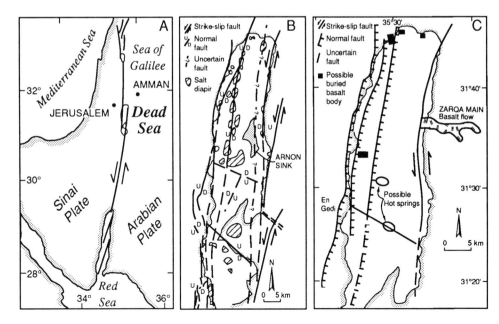

Figure 6-1 a, Simplified map of the Dead Sea–Jordan transform; b, Tectonic map of the Dead Sea based on interpretation of seismic reflection and magnetic data by Neev and Hall (1979); c, Fault map of the Dead Sea based on magnetic data (Frieslander and Ben-Avraham, 1989).

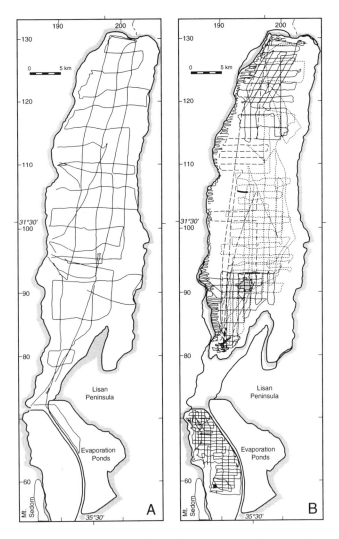

Figure 6-2 Ship tracks of high-resolution geophysical surveys in the northern Dead Sea basin. a, Map showing the location of single-channel Sparker seismic reflection lines collected in the 1974 survey (Hall and Neev, 1975; Neev and Hall, 1976, 1979); b, 3.5-kHz seismic reflection data collected in the Dead Sea.

seismic reflection data were collected in the Dead Sea basin to map the seafloor geomorphology and subbottom structure. In this chapter, we review the methods of data acquisition and the arguments for the age of the sediment, and we discuss interpretations of basin subsidence, active faulting, and paleoseismic data for the northern Dead Sea basin.

HIGH-RESOLUTION SEISMIC REFLECTION DATA

Approximately 2,000 kilometers of high-resolution seismic reflection data have been collected during seven geophysical surveys in the northern basin of the Dead Sea (Fig. 6-2). Geophysical images of the faults in the northern Dead Sea basin were first collected in 1974 during a single-channel, Sparker seismic reflection and magnetic survey (Hall and Neev, 1975; Neev and Hall, 1976, 1979). These data were reanalyzed for this study. The Sparker system with an energy output of 1,000 joules has a deeper penetration (750 msec) and lower resolution

than the 3.5-kHz system. Details of the layers directly beneath the seafloor are obscured on the Sparker data. A combination of both types of seismic reflection acquisition systems (Sparker and 3.5 kHz) provides the best means for interpreting neotectonic and sedimentary processes.

The 3.5-kHz seismic reflection data used in this study were collected using an Ocean Research Equipment (O.R.E.) subbottom profiler. The acoustic system contains an underwater transducer mounted in a towfish that emits and receives continuous wave pulses. The O.R.E. subbottom profiler was operated at a maximum output power of 10 kW at a frequency 3.5 kHz. The analog data were recorded on an EPC 19-in. dry paper graphic recorder using a 1/4-second sweep. Subbottom penetration ranged from 10 to 40 msec. Assuming a 2,000 m/sec velocity for marl sediments, the depth of penetration is approximately 10 –40 m with a bed thickness resolution on the order of tens of centimeters. Most of the surveys used a Motorola Miniranger navigation system that provided continuous (every 2.5 sec) positioning through ship-to-shore range measurement. The shore-based transponders were located at En Gedi and Mizpe Shalem for the southern surveys, and Mizpe Shalem and Kalia for the northern surveys. Only in the 1993 survey was a differential global positioning satellite (GPS) system used for navigation.

The first shallow seismic survey in the Dead Sea in June 1983 collected 650 km of 3.5-kHz seismic reflection data. In November 1983, a two-day survey was conducted along two major canyons that feed into the basin: from the north, the Jordan River, and from the south, the Lynch channel. Approximately 45 km of side-scan sonar images were collected along the Jordan River delta and canyon, whereas 20 km of 3.5-kHz seismic reflection data were recorded in the Lynch channel.

Two seismic reflection and bathymetric surveys were conducted in 1984. In November 1984, a survey collected 145 km of data from the shallow water region along the western margin of the Dead Sea. Seismic profiles were oriented along east-west lines, spaced every 250 m, that extended from the shoreline to a 50-m water depth. For the first time, bathymetric data were collected in digital form. In December 1984, approximately 380 km of 3.5-kHz seismic reflection data and digital bathymetric data were recorded.

An intensive survey in an approximately 2 km^2 area along the southern margin fault of the Dead Sea's northern basin was conducted in June 1986. The location was selected on the basis of previously recorded, near-bottom water temperature anomalies (Ben-Avraham and Ballard, 1984). The survey was unique because it combined 3.5-kHz seismic reflection profiling, remotely-controlled bottom photography, grab and gravity core sampling of bottom sediments, and water column sampling. Biological, geochemical, and geophysical analyses of the data were conducted at various research institutes. Preliminary results of these data are reported in Ben-Avraham et al. (1987).

And finally, a total of 200 km of new, analog high-resolution seismic reflection data and digital bathymetric data were collected during a survey in October 1993. Analyses of the data from all seven marine geophysical surveys (Fig. 6-2) are the basis for the interpretations we present herein.

AGE OF SEDIMENTS

There are few direct age determinations for the sediments in the northern Dead Sea basin. Five radiocarbon ages have been determined for organic carbon extracted from sediments cored in the northern basin (Stiller et al. 1988). Three cores taken in shallow water along the western shore near En Gedi yielded the following radiocarbon ages: 2,160 ±250 yr B.P. (core M2 at 100–110 cm), 1,360 ±170 (core DSEG2 at 105–115 cm), 5,100 ±180

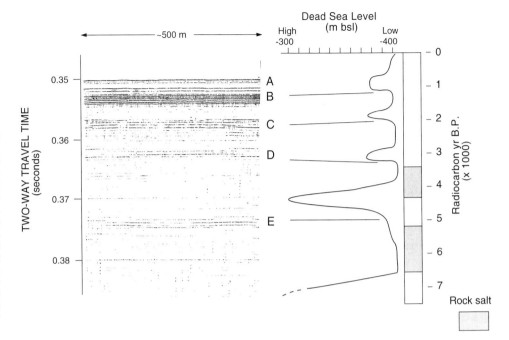

Figure 6-3 Comparison of seismic reflectors A–E with levels of the Dead Sea. Five prominent seismic reflectors (A-E) were identified on the 3.5-kHz seismic reflection data. Limited core data suggest that these reflectors represent the top of rock-salt layers. Based on the Dead Sea level curve presented in Frumkin et al. (1991), we correlate the seismic reflectors with five major lowstands of the Dead Sea during the past 5,000 years. Asterisks mark uncorrected radiocarbon dates on disseminated organic carbon in marl layers below halite beds as reported in Neev and Emery (1967).

yr B.P. and 7,700 ±570 yr B.P. (core DSEG1 at 415–435 cm and 540–550 cm, respectively). One 1.8-m core, recovered from a water depth of 310 m, had a radiocarbon age of 650 ±180 yr B.P. (core B1 at 165–175 cm). These data indicate sediment accumulation rates in shallow water of 0.5–0.9 mm/yr and 3 mm/yr in deep water. Increased sediment accumulation rates in the deep basin are most likely a result of the greater influx of both chemical and detrital sediments from the Jordan delta to the deep basin.

Other lines of evidence also suggest that the sediment accumulation rates within the deep basin of the Dead Sea basin are three to ten times greater than within the shallow portions of the basin. Neev and Emery (1967) report a radiocarbon age of 930 ±165 yr B.P. on plant material taken 20-40 cm below the seafloor within a submarine canyon at water depths of 32-40 m in the area connecting the northern and southern basins (Lynch Straits). A sediment accumulation rate of 1 mm/yr in the littoral zone of the Dead Sea was calculated on the basis of an 8- to 10-cm-thick evaporatic sequence (aragonite, gypsum, and marl) deposited on a tree root that was submerged in 1887 and reemerged in 1960 (Neev and Emery, 1967). These rates are compatible with radiocarbon determination on organic carbon of 4,410 ±320 and 9,580 ±150 yr B.P. from subbottom depths of 3.5-4.0 m and 10.2-10.8 m, respectively, from a borehole in the southern Dead Sea basin (Bentor, 1961).

Late Pleistocene to Holocene sediment accumulation rates reported by Neev and Hall (1979) were calculated from the thickness of the Jordan River delta wedge and the assumption that most of the delta was formed only after the final recession of the late Pleistocene Lake Lisan, which filled the Dead Sea region (Begin et al., 1974). Neev and Hall (1979) traced the reflector at the base of the delta from its northern end, where it is 330-msec-thick, into the central portion of the basin. The maximum sediment thickness of the age-equivalent sediments in the central deep basin of the Dead Sea is 130 msec. If we sssume a seismic velocity of 2,000 m/s (1 msec = 1 m) for marl sediments and the radiometric age (^{14}C and U-series) of the top of the Lisan Formation of 13-17 ka (Kaufman, 1971; Vogel and Waterbolk, 1972; Kaufman et al., 1992), the Holocene sediment

accumulation rate of the deep basin of the Dead Sea is approximately 8–10 mm/yr, whereas it is about 20–25 mm/yr in the delta region. However, considering the large-scale fluctuation of Lake Lisan with lowstands (minus 370 m msl) that occurred at 22–24 ka and around 30 ka (Begin et al., 1985), the assumption that the delta began to form only after the final retreat of Lake Lisan may be in error. If we assume that the delta wedge began forming during the lowstands of Lake Lisan between 22 and 30 ka, the rate of sediment accumulation ranges from 3 to 5 mm/yr for the deep basin and 11 to 15 mm/yr in the delta region.

If we use the late Quaternary sediment accumulation rate of Neev and Hall (1979) of 8–10 mm/yr, the 25-msec section of the 3.5-kHz profiles may represent approximately 2,500 years of sediment accumulation. However, if we use a sediment accumulation of 3–5 mm/yr derived from both radiocarbon analyses (Stiller et al., 1988) and the sedimentation rate since the incipient development of the Jordan delta wedge during lowstands of Lake Lisan (Begin et al., 1985), the 25-msec record may represent deposition over the past 5,000-8,000 years. We favor the latter rate.

Four prominent seismic reflectors (B–E) were identified in the deep-water basin that, on the basis of lithology of sediments in limited short cores (Elazari-Volcani, 1943; Neev and Emery, 1967; Garber, 1980; Stiller et al., 1983; Levy, 1984, 1988; Stiller et al., 1988), represent the top of rock-salt layers. By using an actualistic model provided by a recent change in Dead Sea sedimentation from marl to halite (see Gavrieli, chapter 14, this volume), we interpret the strong seismic reflections as halite deposited during Dead Sea lowstands. We correlate the four subbottom reflectors to the major periods of halite accumulation and lowstands of the Dead Sea over the past 5,000 years (Fig. 6-3; see Frumkin, chapter 22, this volume).

BASIN SUBSIDENCE

Sediment accumulation patterns within the central portion of the northern Dead Sea basin were presented in Ben Avraham et al. (1993). The data revealed that sediments buried at a shallow

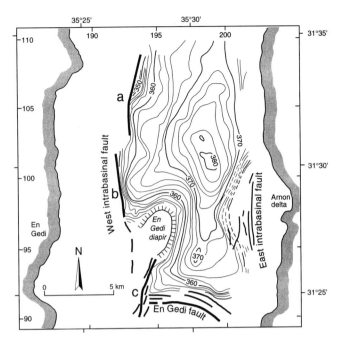

Figure 6-4 Structure map to the top of reflector E. Contour interval is 2 milliseconds. (See Fig. 6-3 for seismic stratigraphy.)

depth are flat-lying, continuous, and unfaulted within the central basin floor.

Isopach and structural maps of recent sediments indicate an asymmetry of active subsidence. Maximum sediment thickness is reached within a rhomb-shaped depression centered closer to the eastern margin of the basin (Fig. 6-4). The long axis of the main, rhomb-shaped depocenter of recent sediments trends approximately N20°W and is oriented about 20–30° counter-clockwise to the regional trend of the faults bordering the basin. The trend and shape of the recent depocenter suggests tectonically controlled subsidence, perhaps as a pull-apart.

Comparison of the 3.5-kHz seismic profiles with the Sparker seismic profiles of Neev and Hall (1976) indicates recent subsidence has shifted only slightly throughout the recent past. The main locus of recent deposition overlies a trough, referred to by Neev and Hall (1979) as the Arnon sink (Fig. 6-1b). The coincidence of the major accumulation of recent sediments directly overlying the Arnon sink and the proximity of the depocenter to the En Gedi and Lisan Peninsula diapirs and associated rim depression suggests that local subbasinal subsidence is related to the lateral movement of rock salt at depth. Therefore, both tectonism and halokinesis apparently control subsidence within a recent depocenter and it is not possible to distinguish between these two controlling factors in the shallow seismic stratigraphic record.

ACTIVE FAULTING

Analyses of the 3.5-kHz data provide a new interpretation of the fault pattern and style of deformation in the northern Dead Sea basin. A preliminary map of the location of active faults within the northern basin of the Dead Sea is given in Figure 6-5. Faults were identified based on several criteria that affect the surface and subbotton reflectors: (1) truncation, (2) offset, (3) folding, (4) sediment thinning, (5) bathymetric escarpments, and (6) change in reflector or surface echo type. We interpret these data to indicate that there are three fault systems: north-

to northwest-trending faults, northeast-trending faults, and east- to west-trending faults.

North- to northwest-trending faults

The west intrabasinal fault system marks the major bathymetric escarpment within the central portion of the basin and probably correlates to the eastern margin of the median block (Arbenz, 1984; Kashai and Croker, 1987; ten Brink and Ben-Avraham, 1989). In map view, the west intrabasinal fault appears to be segmented (Figs. 6-4, 6-5). In cross section, sediment thickness thins toward the escarpment (Line 104, Fig. 6-6), suggesting a period of quiescence on the fault. Small faults are also located up to 500 m basinward of the main fault escarpment and have apparent normal offset. The fault escarpment and small basinward faults suggest that the sense of motion on the faults that trend north to northwest is normal.

Faults with a similar trend are present along the eastern margin of the Dead Sea. Truncation of recent sediments along this trend which is seen on the Sparker data set (Neev and Hall, 1976; 1979), especially north of local coordinate 120, suggests strike-slip deformation at this location.

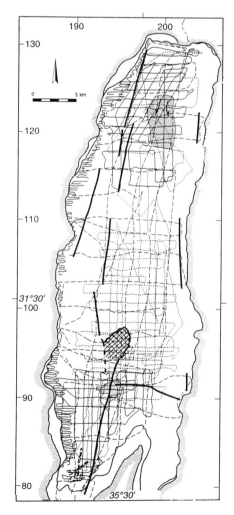

Figure 6-5 Map of active faults within the northern basin of the Dead Sea.

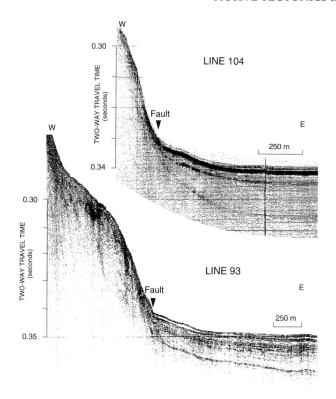

Figure 6-6 High resolution seismic reflection profiles of faults in the Dead Sea. a, Line 104 showing details of recent stratigraphy, sediment accumulation along N-trending intrabasinal fault; b, Line 93 showing truncation of basinal sediments along northeast-trending fault. (See Fig. 6-2 for location of lines.)

Northeast-trending faults

Two faults strike along a northeast trend: the submarine extension of the Jordan (Jericho) fault in the north and a fault along the west flank of the Lisan Peninsula (Segment c on Fig. 6-4 and Fig. 6-5). In cross section, the southern fault sharply truncates the recent sediment in places, suggesting it is a strike-slip fault (Line 93, Fig. 6-6). In the north, the trend of the fault is the submarine extension of the strike-slip Jordan fault. A focal plane mechanism for the July 11, 1927, M_L 6.2 earthquake indicates strike-slip motion (Ben-Menahem et al., 1976). The epicenter of this earthquake has been relocated to the Mizpe Shalem area (Shapira et al., 1993) and probably is associated with the northeast-trending fault extending from the shallow water to deeper water, as mapped from the 3.5-kHz data.

East- to west-trending faults

An east- to west-trending, down-to-the-north fault was mapped by Neev and Hall (1976, 1979). This fault was later called the En Gedi fault (Ben-Avraham and ten Brink, 1989; ten Brink and Ben-Avraham, 1989). Analyses of the 3.5-kHz seismic reflection data along this trend indicates a complex region of faults, channels, and escarpments of uncertain origin. Very detailed mapping at the fault intersection of the east- to west-trending En Gedi fault and a longitudinal, northeast-trending fault was based on the interpretation of the 1986 survey (Fig. 6-7). These data indicate a cross-cutting relationship between the fault system, with the northeast-trending faults apparently

truncating the En Gedi fault (Fig. 6-5). A fault trendng east-west and located within the Lynch Straits may correlate with the En Gedi fault. If this is so, the western portion of the En Gedi fault may have been translated left-laterally a few kilometers to the southwest. The elongated Mt. Sedom diapir may have been offset from the Lisan Peninsula diapir by 15-20 km. Further investigation into these hypotheses is necessary.

A very interesting feature at the fault intersection of the En Gedi fault and the northeast-trending intrabasinal fault is an area of acoustic blackout (Fig. 6-7) which may be the site of hydrothermal activity.

Flanking the submerged extension of the Lisan Peninsula are a series of escarpments that are characterized by several distinct terraces, changes in slope, and pinnacles. The surface reflection is dominated by seismic diffractions, indicating a rough surface. These features may represent upturned sedimentary beds deformed by the rising Lisan Peninsula diapir and similar to strata that flank the Mt. Sedom diapir (Zak, 1967).

Fault geometry

From analyses of the high-resolution seismic reflection record, we mapped the pattern of active faults within the northern

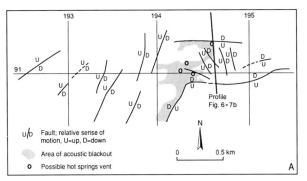

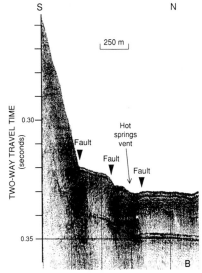

Figure 6-7 Possible hot springs on the floor of the Dead Sea. A, Detailed mapping of the faults and areas of acoustic blackout (shaded area). Circles mark the possible location of hydrothermal vents or hot springs; B, Profile across an area of acoustic blackout located at the intersection of a longitudinal, northeast-trending fault and the east- to west-trending En Gedi fault. These areas may be the sites of hydrothermal activity.

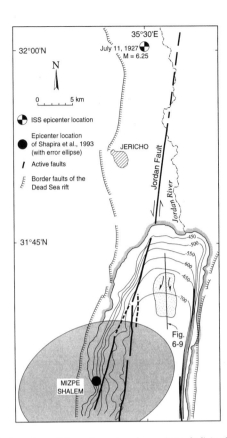

Figure 6-8 Location of the submarine slump (stippled) in the northern Dead Sea. Active faults that deform subbottom strata in the northern basin of the Dead Sea are mapped based on analyses of 3.5-kHz and Sparker high-resolution seismic reflection data (Neev and Hall, 1976). The map also shows the International Seismological Summary catalogue epicenter of the July 11, 1927, Jericho earthquake and fault plane solution of Ben-Menahem et al. (1977). The location of the epicenter beneath Dead Sea may fit the seismological data better (Shapira et al., 1993). Bathymetric contours are in meters below sea level.

Dead Sea basin. The location of the basinal faults, as well as the nature and magnitude of fault movements contribute to the asymmetry of the basin. The basinal strata are clearly deformed along the intrabasinal faults, which form bathymetric escarpments at the margins of the northern basin of the Dead Sea and indicate their recent activity. Active deformation in the Dead Sea is also manifested as tectonic subsidence.

Three fault trends are recognized. The style of deformation along faults of different trends suggests that they may represent several types of faulting, predominately oblique-slip and normal. Variation in the recency of faulting is also indicated by changes in the style of syntectonic deposition and by deformation of subbottom reflectors along submarine escarpments.

Although all faults mapped in the northern Dead Sea basin are active, northeast-trending faults apparently cut faults that trend north-south and east-west. A similar cross-cutting fault relationship is seen on the Landsat satellite image of the transform north and south of the Dead Sea. Recently active strike-slip faults in the Jordan and Arava valleys trend at a clockwise angle to normal faults that flank the fault valley. North of the Dead Sea, compressional features across the active fault (Gardosh et al., 1990; Rotstein and Bartov, 1992) indicate a slight reorganization of plate motion in the Quaternary.

PALEOSEISMOLOGY

Earthquakes caused by fault motion on the Dead Sea–Jordan transform fault and associated structures have been felt throughout history in the Near East. A long earthquake record has been compiled in several catalogs (e.g., Willis, 1928; Kallner-Amiran, 1950; Ambraseys, 1962; Ben-Menahem, 1979; Poirier and Taher, 1980; Turcotte and Arieh, 1988; Ben-Menahem, 1991; Al-Tarazi, 1992). A survey of ancient, earthquake-damaged structures that have been excavated at archaeological sites shows a concentration along the transform north of the Dead Sea (Karcz et al., 1977; Russell, 1980, 1985). Despite such catalogs and surveys, few attempts have been made to assign rupture segments to the historic earthquakes because of uncertainty and exaggerations in felt reports (e.g., Rotstein, 1987; Turcotte and Arieh, 1988). Reches and Hoexter (1981) reported the only direct physical evidence for the timing of historic ground rupture along the plate boundary fault.

The July 11, 1927, Jericho earthquake (M_L 6-2) is the largest earthquake to have been instrumentally recorded on the Dead Sea–Jordan transform and thus is a crucial seismic event for seismotectonic studies. Ben-Menahem et al. (1976) suggested that the 1927 earthquake was generated by left-lateral slip of an estimated 40 ±10 cm on a 40-km fault segment, extending from within the Dead Sea northward. No evidence of ground rupture of the earthquake was detected in trenches along the fault 7 km north of the Dead Sea (Reches and Hoexter, 1981) nor has it been adequately documented elsewhere along the fault trace. The location of the epicenter beneath the Dead Sea may fit the seismological data better (Shapira et al., 1993). From the analyses of the high-resolution seismic reflection data, we suggest a means for locating prehistoric rupture through the study of slumped sediments at the bottom of the Dead Sea (Niemi and Ben-Avraham, 1994).

Seismically triggered slump

A large, submarine slump (2×10^8 m³) mapped from 3.5-kHz seismic reflection data, lies adjacent to a segment boundary of the main fault flanking the margins of the deep northern basin of the Dead Sea (Fig. 6-8). The headwall scarp is located about 5.5 km south of the mouth of the river in the stable bottomset region of the Jordan River delta on a slope of less than 1° at a water depth of 250 m. Given the properties of the extremely stable Dead Sea sediments—evaporite lithology (Elazari-Volcani, 1943; Neev and Emery, 1967; Garber, 1980; Levy, 1984, 1988), low porosity (Stiller et al., 1983), high effective angle of internal friction (Almagor, 1990)—and that gravity-induced sliding on slopes of less than 3° occurs only where silt and clay are overpressured and underconsolidated (Crans et al., 1980), we conclude that this large submarine slump was triggered by an earthquake.

The young appearance of the slump, which had only minor, thin sediment drapes covering the slide blocks (Fig. 6-9), suggests that the slump was triggered by the 1927 Jericho earthquake. The simultaneity of the earthquake and submarine slumping is corroborated by eyewitness reports that described a wave that rose suddenly in the middle of the Dead Sea and piled water 1 m high along its north shore (Abel, 1927; Blanckenhorn, 1927; Shalem, 1927; Alt, 1928; Brawer, 1928). The displaced bottom mass generated by the slumping would have provided the impulse for the observed wave. An eyewitness described the wave as moving northward from a point within the middle of the northern Dead Sea basin (R. Avni, pers. com., 1993).

Although other small, seismically triggered slumps, probably produced by distant seismic sources, have been recognized in

Figure 6-9 North-south 3.5-kHz seismic reflection profile showing cross sectional detail of submarine slump and correlation of seismic reflectors with major, radiocarbon-dated lowstands of Dead Sea (Frumkin et al., 1991); m bsl is meters below sea level. Headwall scarp of slump is marked by an arrow. Two seismically triggered slump events are labeled 1 and 2. The basal glide plane of the upper slump is within a thick seismically transparent interval between reflectors D and E. Evidence of an earlier seismically triggered slump is found in a lower sequence that was apparently deformed sometime after the deposition of reflector E and before deposition of the parallel reflectors A–C involved in the 1927 slump.

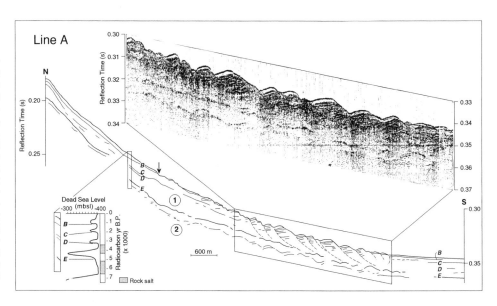

the shallow-water, steeper-sloped region of the Jordan River delta (Almagor, 1990), none have the same appearance or areal extent of the 1927 submarine slump. The kilometer-wide scale of the seismically triggered slump suggests that it formed from intense ground shaking in close proximity to slip on an adjacent fault segment in the Dead Sea. If slip in the 1927 earthquake was restricted to the Dead Sea basin, more strain has accumulated on the Jordan fault north of the Dead Sea than had been previously suspected.

Recurrence Interval

If we assume that all slumps in the location of the 1927 slump are seismically triggered and that each earthquake on the Dead Sea fault segment produces an accompanying submarine slump, then a record of buried slumps in the same location would yield a repeat time of earthquakes on that fault segment. The seismic reflection profiles provide a record of buried seismically triggered slumps. On the 3.5-kHz profiles (Fig. 6-9), only the upper reflectors are involved in the 1927 slumping. The lower reflectors have a different deformation pattern and apparently slumped in an earthquake prior to 1927. So, at least one earthquake occurred after the deposition of the lower sequence and before deposition of the parallel reflectors involved in the 1927 slumping. Therefore, at least two earthquakes occurred during an interval of a few thousand years. This is corroborated by older buried slumps visible on Sparker seismic reflection data (Niemi and Ben-Avraham, 1994).

A long recurrence interval for maximal magnitude earthquakes has previously been inferred for the transform fault north of the Dead Sea, on the basis of historical earthquakes (Ben-Menahem, 1981). Reches and Hoexter (1981) found convincing evidence in trench exposures for one or two major earthquakes in the past 2 ka along the Jordan (Jericho) fault; this evidence also supports a long interval of earthquake recurrence. A long recurrence interval of strike-slip earthquakes may be attributed to slow, long-term slip along the Dead Sea transform. Although a minimum left-lateral slip rate of 0.7 mm/yr for the past 3–4 ka was calculated from compressional features along

the Jordan fault north of the Dead Sea (Gardosh et al., 1990), estimates of slip rate for the Holocene remain insufficient. Given an estimated geologic slip rate of 6–10 mm/yr (Zak and Freund, 1966; Freund et al., 1970; Garfunkel et al., 1981; Joffe and Garfunkel, 1987), the long recurrence interval of major earthquakes suggested by buried seismically triggered slumps in the Dead Sea implies that a large percentage of strain is released aseismically (Ben-Menahem, 1981), that the present-day slip rate is slower (Garfunkel et al., 1981), or that strain is partitioned onto transform-fault normal extensional and compressional structures (Ben-Avraham and Zoback, 1993).

In contrast, seismologists have calculated repeat times for earthquakes with M greater than or equal to 6.1 ranging from 140 to 500 yr. These short intervals are based on various frequency-magnitude relations of instrumental earthquakes and of presumed plate-rupturing historical earthquakes that are thought to have occurred somewhere along the Dead Sea-Jordan transform (Ben-Menahem et al., 1977; Ben-Menahem and Aboodi, 1981; Rotstein, 1987).

Clearly, additional physical evidence gleaned from trenching investigations, fault segmentation models from field mapping, and more historical data would help resolve some of the problems of defining the ages and locations of earthquakes along the Dead Sea–Jordan transform.

SUMMARY

To assess the geologic hazard of the faults in the Dead Sea, 2,000 km of high-resolution seismic reflection data were collected. These data were used to construct isopach, structural, and fault maps and to interpret the pattern of sediment accumulation and active faulting. Recent faulting is indicated by syntectonic deposition, tectonic subsidence, and deformation of subbottom reflectors along submarine escarpments. The style of deformation along faults of different trends suggests that they may represent several types of faulting, predominately oblique-slip and normal. A large submarine slump seen in the seismic data, as well as historical accounts of a Dead Sea seismic sea wave, suggest that the slump formed from intense ground shaking in close proximity to slip on an adjacent submarine fault segment in the July 11, 1927, Jericho earthquake. The estimated number and age range of buried slumps suggest a long earthquake recurrence time.

Acknowledgments

We thank Captain Gonen, G. Amit, and I. Liebermann for assistance in geophysical data acquisition, D. Neev and J. K. Hall for the generous use of their data; G. Almagor, A. Shapira, and R. Avni for helpful discussions, and the Israel Ministry of Energy for support.

REFERENCES

Abel, F.-M., 1927, Le récent tremblement de terre en Palestine: *Revue Biblique*, v. 36, p. 571–578.

Al-Tarazi, E., 1992, Investigation and assessment of seismic hazard in Jordan and its vicinity [Ph.D. thesis]: Ruhr-Universität Bochum, Germany, 194 p.

Almagor, G., 1990, The physical properties, consolidation and shear strength of sediments from the Jordan delta, Dead Sea: Geological Survey of Israel Report GSI/2/90, 36 p. (in Hebrew).

Alt, D. A., 1928, Erbeben in Palästina: *Palästinajahrbuch*, v. 24, p. 5–10.

Ambraseys, N. N., 1962, A note on the chronology of Willis's list of earthquakes in Palestine and Syria: *Bulletin of the Seismological Society of America*, v. 52, p. 77–80.

Arbenz, J. K., 1984, Oil potential of the Dead Sea area: Tel Aviv, Sismica Oil Exploration Ltd., Report 84/111, 54 p.

Begin, Z. B., Ehrlich, A., and Nathan, Y., 1974, Lake Lisan, the Pleistocene precursor of the Dead Sea: Jerusalem, Geological Society of Israel Bulletin 63, 30 p.

Begin, Z. B., Broecker, W., Buchbinder, B., Druckman, Y., Kaufman, A., Magaritz, M., and Neev, D., 1985, Dead Sea and Lake Lisan levels in the last 30,000 years: Geological Survey of Israel Report GSI/29/85, 18 p.

Ben-Avraham, Z., and Ballard, T. D., 1984, Near bottom temperature anomalies in the Dead Sea: *Earth and Planetary Sience Letters*, v. 71, p. 356–360.

Ben-Avraham, Z., Oren, A., Ayalon, A., Gavrieli, I., Ganor, Y., Tibor, G., Katz, A., and Kempler, D., 1987, Acoustic and photographic mapping of fault zone hotsprings on the deep seafloor of the Dead Sea: Tel Aviv, Tel Aviv University Report MGL 10/87, 28 p. (in Hebrew).

Ben–Avraham, Z., ten Brink, U., and Charrach, J., 1990, Transverse faults at the northern end of the southern basin of the Dead Sea graben: *Tectonophysics*, v. 180, p. 37–47.

Ben-Avraham, Z., and Zoback, M. D., 1993, Transform-normal extension and asymmetric basins: An alternative to pull-apart models: *Geology*, v. 20, p. 423–426.

Ben-Avraham, Z., Niemi, T.M., Neev, D., Hall, J.K., and Levy, Y., 1993, Distribution of Holocene sediments and neotectonics in the deep north basin of the Dead Sea: *Marine Geology*, v. 113, p. 219–231.

Ben-Menahem, A., 1979, Earthquake catalogue for the Middle East (92 B.C.–1980 A.D.): *Bollettino di geofisica teorica ed applicata*, v. 21, p. 245–310.

Ben-Menahem, A., 1981, Variation of slip and creep along the Levant rift over the past 4500 years: *Tectonophysics*, v. 80, p. 183–197.

Ben-Menahem, A., 1991, Four thousand years of seismicity along the Dead Sea rift: *Journal of Geophysical Research*, v. 96, p. 20,195–20,216.

Ben-Menahem, A., Nur, A., and Vered, M., 1976, Tectonics, seismicity and structure of the Afro-Eurasian junction—The

breaking of an incoherent plate: *Physics of the Earth and Planetary Interiors*, v. 12, p. 1–50.

Ben-Menahem, A., Aboodi, E., Vered, M., and Kovach, K.L., 1977, Rate of seismicity of the Dead Sea region over the past 4000 years: *Physics of the Earth and Planetary Interiors*, v. 14, p. P17–P27.

Ben-Menahem, A., and Aboodi, E., 1981, Micro– and macroseismicity of the Dead Sea rift and off-coast eastern Mediterranean: *Tectonophysics*, v. 80, p. 199–233.

Bentor, Y. K., 1961, Some geochemical aspects of the Dead Sea and the question of its age: *Geochimica et Cosmochimica Acta*, v. 25, p. 239-260.

Blanckenhorn, M., 1927, Das Erdbeben im Juli 1927 in Palästina: *Zeitschriften des Deutches Palästinische Vereinung*, v. 50, p. 288–296.

Brawer, A.J., 1928, Earthquakes in Palestine from July 1927, to August 1928: Jerusalem, *Jewish Palestine Exploration Society*, p. 316–325 (in Hebrew).

Crans, W., Mandl, G., and Haremboure, J., 1980, On the theory of growth fault: A geomechanical delta model based on gravity sliding: *Journal of Petroleum Geology*, v. 2/3, p. 265–307.

Crowell, J.C., 1974, Origin of Late Cenozoic basins in southern California: *S.E.P.M. Special Publication* 22, p. 190–204.

Elazari-Volcani, B., 1943, Bacteria in the bottom sediments of the Dead Sea: *Nature*, v. 152, p. 274–275.

Freund, R., Garfunkel, Z., Zak, I., Goldberg, M., Weissbrod, T., and Derin, B., 1970, The shear along the Dead Sea rift: *Royal Society of London Philosophical Transactions*, v. A267, p. 107–130.

Frieslander, U., and Ben-Avraham, Z., 1989, Magnetic field over the Dead Sea and vicinity: *Marine and Petroleum Geology*, v. 6, p. 148–160.

Frumkin, A., Magaritz, M., Carmi, I., and Zak, I., 1991, The Holocene climatic record of the salt caves of Mount Sedom, Israel: *The Holocene*, v. 1, p. 191–200.

Garber, R. A., 1980, The sedimentology of the Dead Sea [Ph.D. thesis]: Troy, New York, Rensselaer Polytechnical Institute, 169 p.

Gardosh, M., Reches, Z., and Garfunkel, Z., 1990, Holocene tectonic deformation along the western margins of the Dead Sea: *Tectonophysics*, v. 180, p. 123–137.

Garfunkel, Z., Zak, I., and Freund, R., 1981, Active faulting in the Dead Sea rift: *Tectonophysics*, v. 80, p. 1–26.

Hall, J. K., and Neev, D., 1975, The Dead Sea Geophysical Survey, 19 July– 1 August, 1974—Final Report No.1: Jerusalem, Geological Survey of Israel Report 2/75, 27 p.

Joffe, S., and Garfunkel, Z., 1987, Plate kinematics of the circum Red Sea—A reevaluation: *Tectonophysics*, v. 141, p. 5–22.

Kallner-Amiran, K. H., 1950, A revised earthquake-catalogue of Palestine: *Israel Exploration Journal*, v.1, p. 223 –246.

Karcz, I., Kafri, U., and Meshel, Z., 1977, Archaeological evidence for subrecent seismic activity along the Dead Sea–Jordan rift: *Nature*, v. 269, p. 234–235.

Kashai, E. L., and Croker, P. F., 1987, Structural geometry and evolution of the Dead Sea–Jordan rift system as deduced from new subsurface data: *Tectonophysics*, v. 141, p. 33–60.

Kaufman, A., 1971, U-series dating of Dead Sea basin carbonates: *Geochimica et Cosmochimica Acta*, v. 35, p. 1269–1281.

Kaufman, A., Yechieli, Y., and Gardosh, M., 1992, Reevaluation of the lake-sediment chronology in the Dead Sea basin, Israel, based on new ^{230}Th/U dates: *Quaternary Research*, v. 38, p. 292–304.

Levy, L., 1984, Halite from the bottom of the Dead Sea: Geological Survey of Israel Report MG/48/84, 16 p.

Levy, L., 1988, Recent depositional environments in the Dead Sea: Geological Survey of Israel Report GSI/42/88, 21 p.

Neev, D., and Emery, K. O., 1967, The Dead Sea: Depositional processes and environments of evaporites: Jerusalem, Geological Survey of Israel Bulletin 41, 147 p.

Neev, D., and Hall, J. K., 1976, The Dead Sea geophysical survey, 19 July–1 August 1974—Final Report 2: Geological Survey of Israel, Marine Geology Division, Report No. 6/76, 21 p.

Neev, D., and Hall, J.K., 1979, Geophysical investigations in the Dead Sea: *Sedimentary Geology*, v. 23, p. 209–238.

Niemi, T. M., and Ben-Avraham, Z., 1994, Evidence of Jericho earthquakes from slumped sediments of the Jordan River delta in the Dead Sea: *Geology*, v. 22, p. 395–398.

Poirier, J. P. and Taher, M. A. 1980, Historical seismicity in the Near and Middle East, North Africa, and Spain from Arabic documents (VIIth–XVIIIth Century): *Bulletin of the Seismological Society of America*, v. 70, 2,185–2,201.

Quennell, A. M., 1959, Tectonics of the Dead Sea rift: 20th International Geological Congress, Mexico, 1956, p. 385–405.

Reches, Z., and Hoexter, D. F., 1981, Holocene seismic and tectonic activity in the Dead Sea area: *Tectonophysics*, v. 80, p. 235–254.

Rotstein, Y., 1987, Gaussian probability estimates for large earthquake occurrence in the Jordan Valley, Dead Sea rift: *Tectonophysics*, v. 141, p. 95–105.

Rotstein, Y., Bartov, Y., and Hofstetter, A., 1991, Active compressional tectonics in the Jericho area, Dead Sea rift: *Tectonophysics*, v. 1983, p. 239–259.

Russell, K. W., 1980, The earthquake of May 19, A.D. 363: *American School of Oriental Research Bulletin*, v. 238, p. 47–64.

Russell, K. W., 1985, The earthquake chronology of Palestine and northwest Arabia from the 2nd through the mid-8th Century A.D.: *American School of Oriental Research Bulletin*, v. 260, p. 37–59.

Shalem, N., 1927, Il recente terremoto in Palestina: *Societ o(`,a) Sismologica Italiana Bollettino*, v. 27, p. 3–17.

Shapira, A., Avni, R., and Nur, A., 1993, A new estimate for the epicenter of the Jericho earthquake of July 11, 1927: *Israel Journal of Earth Sciences*, v. 42, p. 93–96.

Stiller, M., Kaushansky, P., and Carmi, I., 1983, Recent climatic changes recorded by salinity of pore waters in the Dead Sea sediments: *Hydrobiologia*, v. 103, p. 75–79.

Stiller, M., Carmi, I., and Kaufman, A., 1988, Organic and inorganic ^{14}C concentrations in the sediments of Lake Kinneret and the Dead Sea (Israel) and the factors which control them: *Chemical Geology*, v. 73, p. 68–78.

ten Brink, U. S., and Ben-Avraham, Z., 1989, The anatomy of a pull-apart basin: Seismic reflection observations of the Dead Sea basin: *Tectonics*, v. 8, p. 333–350.

ten-Brink, U., Ben-Avraham, Z., Bell, R. E., Hassouneh, M., Coleman, D. F., Andreasen, G., Tibor G., and Coakley, B., 1993, Structure of the Dead Sea pull-apart basin from gravity analyses: *Journal of Geophysical Research*, v. 98, p. 21,877–21,894.

Turcotte, T., and Arieh, E., 1988, Catalogue of earthquakes in and around Israel, *in* Preliminary Safety Analysis Report: Tel Aviv, Israel Electric Corporation, Ltd., Appendix 2.5A, p.1—18, 6 tables.

van Eck, T., and Hofstetter, A., 1989, Microearthquake activity in the Dead Sea region: *International Geophysical Journal*, v. 99, p. 605–620.

van Eck, T., and Hofstetter, A., 1990, Fault geometry and spatial clustering of microearthquakes along the Dead Sea–Jordan rift fault zone: *Tectonophysics*, v. 180, p. 15–27.

Wilcox, R.E., Harding, T.P., and Seely, D.R., 1973, Basic Wrench Tectonics: *American Association of Petroleum Geologists Bulletin*, v. 57, p.74–96.

Willis, B., 1928, Earthquakes in the Holy land: *Bulletin of the Seismological Society of America*, v.18, 73–103.

Vogel, J. C., and Waterbolk, H. T., 1972, Groningen radiocarbon dates: Geological samples Dead Sea Series (Lisan): *Radiocarbon*, v. 14, p. 46-47.

Zak, I., 1967, The geology of Mount Sedom [Ph.D. thesis]: Jerusalem, The Hebrew University, 208 p. (in Hebrew, English summary).

Zak, I., and Freund, R., 1966, Recent strike-slip movements along the Dead Sea rift: *Israel Journal of Earth-Sciences*, v. 15, p. 33–37.

7. ON THE SEISMICITY OF THE DEAD SEA BASIN

AVI SHAPIRA

The Dead Sea basin appears to be the most seismically active section along the Dead Sea–Jordan transform system and, as such, has attracted the attention of seismologists working in this region. This chapter provides a short review of the seismicity of the Dead Sea basin as reflected in recent seismological studies of the area.

The mapped geological faults in the Dead Sea basin are seismically active. The seismicity parameters of this region are updated and reviewed herein. This review is based on the seismological work of a number of researchers, mainly from the former Seismological Laboratory of the Geological Survey of Israel, the Department of Applied Mathematics at the Weizmann Institute of Science, and the Seismological Division of the Institute for Petroleum Research and Geophysics (IPRG). These seismological findings are already used for earthquake hazard assessment, yet far more seismic data are required to reach definitive conclusions with regard to many of the seismotectonic features of the Dead Sea basin.

REGIONAL SETTING

The Dead Sea basin is part of the Dead Sea rift system, which is considered a transform boundary between the African and Arabian plates. Many investigators, including Neev and Emery (1967), Freund et al. (1968), Steinitz et al. (1978), Bartov et al. (1980), Garfunkel et al. (1981), Joffe and Garfunkel (1987), Kashai and Croker (1987), Rotstein et al. (1991), Garfunkel (chapter 4, this volume), among others, describe the geotectonics of the Dead Sea rift system and suggest that this transform is characterized by a left-lateral, strike-slip motion. The average geological slip rate is about 5–10 mm/year.

The geometry of the transform generally consists of left-stepping, *en echelon* faults. In the overlap areas (e.g., Rotstein et al., 1990), a series of depressions, frequently referred to as rhomb-shaped grabens or pull-apart basins, appears. The Dead Sea lake is located in one of these basins.

LOCATION OF EPICENTERS

Documented evidence of strong earthquakes along the Jordan rift system dates to Biblical times. Catalogs of earthquake occurrences in the last 4,000 years have been published in numerous papers, including Ambraseys and Melville (1985), Amiran (1951), Arieh (1967), Ben-Menahem (1979), Ben-Menahem et al. (1976), Shapira (1979), Ben-Menahem and Aboodi (1981), Turcotte and Arieh (1988), and Ben-Menahem (1991). On the basis of macroseismic observations, it is difficult to determine accurately the location of these historical earthquakes. However, the events presented in Table 7-1 seem to have occurred either in or in close proximity to the Dead Sea basin during the last 1,000 years (E. Arieh, pers. com., 1992). These earthquakes were probably the strongest and most destructive to occur during that period of local history.

The epicenter location of the July 11, 1927, earthquake has been corrected and is now placed in the Dead Sea basin (Shapira et al., 1993). Until recently, in accordance with the determination of the Ksara station in Lebanon (as quoted in the ISS or International Seismological Summary, 1927), this earthquake was located at 32.0°N, 35.5°E. The location of this earthquake was never computed, despite its unique importance as the strongest event recorded by seismometers. Shapira et al. (1993) determined its location, for a fixed depth of 10 km, to be in the northern basin of the Dead Sea, a few tens of kilometers south of previous estimates.

During the period 1903 to 1981, information regarding the seismic activity in the region was obtained through macroseismic observations and instrumental data from a few regional stations, namely, Ksara (Lebanon), Helwan (Egypt), Jerusalem (Israel), and Eilat (Israel). Stronger earthquakes were also occasionally reported by some European stations (such as the earthquake of 1927). Since 1982, the seismic data for the region, including the Dead Sea basin, have been obtained by the Israel Seismic Network (ISN; Fig. 7-1; Shapira, 1982). Since 1986, new seismic stations in Jordan have provided additional informa-

Table 7-1 Strongest earthquakes in the Dead Sea basin in the last 1,000 years

| Date A.D. | After Turcotte and Arieh (1988) | | | After Arieh (1992) |
	Maximum Intensity*	Approximate Location	Maximum Estimated Magnitude	Magnitude
1060	VIII	Dead Sea	6.1	5.9 ±0.2
1160	VIII	Dead Sea	6.1	5.8 ±0.2
1293	IX	Dead Sea	—	6.2 ±0.3
1458–1459	VIII	Dead Sea	—	5.9 ±0.3
May 23, 1834	X	East of Lisan Peninsula	6.3	5.8 ±0.2
July 11, 1927	IX	Dead Sea	6.2	6.2 ±0.1

*Modified Mercalli scale

tion on the seismic activity in the Dead Sea and thus improving location accuracy.

Figures 7-2 and 7-3 show the epicenter locations in the Dead Sea for the periods 1923–1981 and 1982–1993, respectively. The threshold magnitude level for the two maps is obviously different. As demonstrated by Shapira et al. (1986), despite the high concentration of seismic stations in and around the Dead Sea basin, focal-depth determinations are not sufficiently accurate. These lead to mislocations of the epicenters and thus, it is rather difficult to associate the epicenters with the mapped faults in the area. Apparently, all mapped faults can be associated with located earthquakes.

FREQUENCY-MAGNITUDE RELATIONSHIP

Ben-Menahem (1979), Shapira and Feldman (1987), and Arieh and Rabinowitz (1989) analyzed the rate of seismicity in the Dead Sea basin. The investigators agree with regard to the b-value in the frequency-magnitude relationship Log $(N) = a - b$ M, where $b = 0.80$–0.86. The values corresponding to the seismicity level (a) vary with the definition of the area constituting the Dead Sea basin. When we delineate the Dead Sea basin to be within lat. 31.0°N and 31.8°N and long. 35.3°E and 35.7°E, the frequency-magnitude relationship for the Dead Sea basin yields the following equation:

$$N = (9.5 \pm 1.6) \frac{e^{-\beta(M_L - M_\phi)} - e^{-\beta(M_M - M_\phi)}}{1 - e^{-\beta(M_M - M_\phi)}} \text{ for } M_M > M_L > M_\phi (1)$$

where $M_M = 7.5$ and $M_\phi = 2.2$, N is the annual number of earthquakes of magnitude M_L and greater, and $\beta = 2.03 \pm 0.02$ (corresponding to a b-value of 0.88 ± 0.01). This relationship is plotted in Figure 7-4. The evaluation of Equation 7-1 is based on the following assumptions:

1. Since the year 1000 A.D., the seismicity data are complete for all earthquakes of magnitude 6 and greater. For the period 1900–1992, the data are complete for earthquakes of magnitude $M_L > 5.5$. For magnitudes 4.0 and above, the data can be considered complete from 1964. Since 1983, all earthquakes of magnitude 2.0 and greater have been detected and located by the ISN.

2. The maximum magnitude for an earthquake in the Dead Sea is 7.5. Actually, the highest magnitude reported for an earthquake in the Dead Sea is only 6.5 (see Table 7-1). The Dead Sea, however, is part of the Dead Sea–Jordan transform system, where earthquakes of magnitude 7.5 are believed possible.

EARTHQUAKE SOURCE MECHANISM

Several publications describe the source mechanism of earthquakes in the Dead Sea. Ben-Menahem et al. (1976) suggest that the earthquakes of July 11, 1927 (M = 6.2) and October 8, 1970 (M = 4.6) are the result of left-lateral, strike-slip motion. A similar solution was obtained by Arieh et al., (1982) for the M = 5.1 earthquake in the Lisan Peninsula area on April 23, 1979. Van Eck and Hofstetter (1989) obtained fault plane solutions of microearthquakes and a few low-magnitude earthquakes in the southern Dead Sea basin. Their composite solution shows left-lateral, strike-slip motions; the individual solutions, however, show dip slips of both normal and reverse faulting. The source mechanisms of these earthquakes are presented in Figure 7-3. Two recent earthquakes of magnitude 4.1 that occurred on the same day (August 2, 1993) show both types of mechanism (see Fig. 7-3). The northern event, which occurred about 10 km east of the western shores of the lake opposite Mizpe Shalem, is

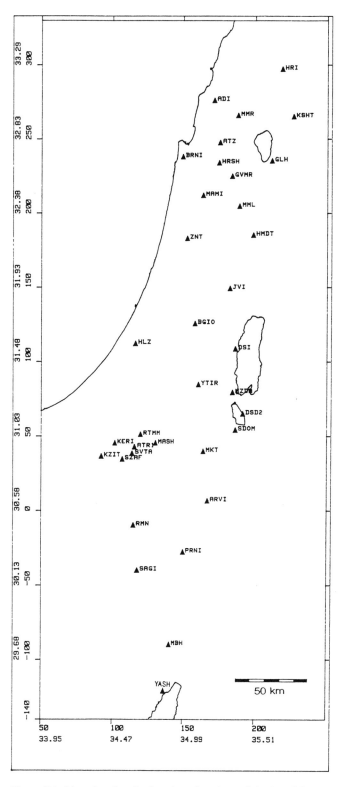

Figure 7-1 Map showing the location of stations of the Israel Seismic Network.

associated with a northwest- to southeast-striking fault of left-lateral motion with a component of normal slip. The second earthquake, which occurred on the fault along the western shores of the lake near Masada, is associated with a classic left-

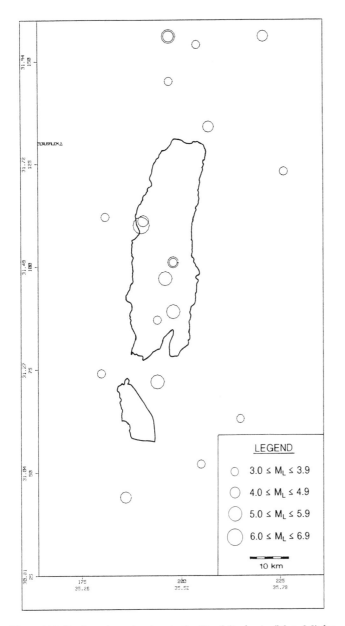

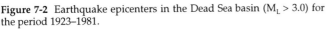

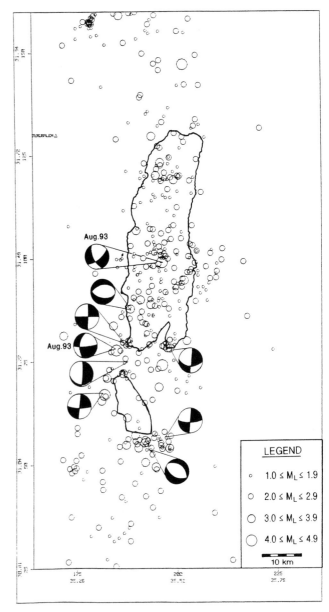

Figure 7-2 Earthquake epicenters in the Dead Sea basin ($M_L > 3.0$) for the period 1923–1981.

Figure 7-3 Earthquake epicenters in the Dead Sea basin ($M_L > 1.0$) for the period 1982–1993. Also shown are fault plane solutions for earthquakes in the Dead Sea basin (after van Eck and Hofstetter, 1989) and the $M_L = 4.1$ earthquakes which occurred on August 2, 1993, in the Dead Sea.

lateral slip striking due north. These solutions demonstrate the complexity of the seismotectonics of this region, yet the available source mechanism solutions generally agree with the seismotectonic model of the Dead Sea basin.

SCALING LAWS FOR EARTHQUAKES IN THE DEAD SEA

Much of the work related to the study of scaling relationships for various parameters in the time and frequency domains for the Dead Sea area was carried out by van Eck (1988) and van Eck and Hofstetter (1989). Their work is based on the analysis of microearthquakes that have occurred in the Dead Sea basin since 1981. The main results are as follows:

1) The correlation between the local magnitude, M_L, and the seismic moment, M_0, agrees with the correlation obtained in California (Bakun, 1984): $\log M_0 = 1.2\, M_L + 17.0$.

2) For earthquakes with seismic moments in the range 1.2×10^{19} to 2.3×10^{23} dyne cm, the stress drop increases from 0.6 to 92 bars.

3) The observed f_{max} for S-waves in the Dead Sea area is, on average, 7.6 Hz.

4) The measured corner frequency, f_c, is influenced by the propagation path. The empirical correlation between the "true" corner frequency, f_o, and the observed f_c is: $\log f_o = \log f_c + 0.0019R$, where R is the epicentral distance in km.

5) A study of the anelastic attenuation coefficient, $V = \pi f / Q$

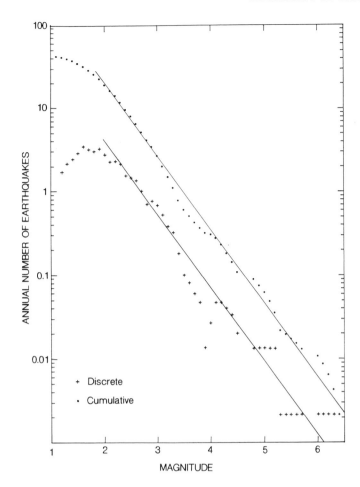

Figure 7-4 Frequency-magnitude relationship for the Dead Sea basin.

v, for Coda waves suggests that the quality factor, Q, is frequency-dependent (v is the propagation velocity of S-waves). The empirical relation is: $Q = Q_0 f^n$ where Q_0 is 65 (average) and $n = 1.05$. When $n \approx 1$, the coefficient, V, becomes independent of the frequency and, consequently, the attenuation of the seismic waves (geometric and anelastic) does not depend on the frequency of the propagating waves.

These observations are currently valid for short epicentral distances only. The Q_0 value seems to vary significantly with distance and is site dependent.

Acknowledgments

I am grateful to Dr. G. Shamir for providing the fault plane solutions of the August 2, 1993, earthquakes and for critical reading of the manuscript. Thanks are due to all colleagues who made their results available to me. Special thanks are due to the technical staff of the IPRG Seismological Division who maintained the seismic stations in good working order and to the team of analysts who provided the basic source parameters of the earthquakes. I am grateful to Miss I. Chelinsky for drawing the figures.

Much of the work summarized in this article has been sponsored by the Earth Sciences Research Administration of the Ministry of Energy and Infrastructure.

REFERENCES

Ambraseys, N. N., and Melville, C. P., 1985, An analysis of eastern Mediterranean earthquake of 20 May 1202, *in* Lee, W. H. K., ed., Proceedings, IASPEI/UNESCO working group on historical earthquakes: Tokyo.

Amiran, D. H. K., 1951, A revised earthquake-catalogue of Palestine: *Israel Exploration Journal*, v. 1, p. 223-246.

Arieh, E., 1967, Seismicity of Israel and adjacent areas: Jerusalem, Geological Survey of Israel Bulletin 43, p. 1-14.

Arieh, E., Rotstein, Y., and Peled, U., 1982, The Dead Sea earthquake of April 23, 1979: *Bulletin of the Seismological Society of America*, v. 72, p. 1621-1634.

Arieh, E., and Rabinowitz, N., 1989, Probabilistic assessment of earthquake hazard in Israel: *Tectonophysics*, v. 167, p. 223-233.

Bakun, W. H., 1984, Seismic moments, local magnitudes, and coda duration magnitudes for earthquakes in Central California: *Bulletin of the Seismological Society of America*, v. 74, p. 439-458.

Bartov, Y., Steinitz, G., Eyal, M., and Eyal, Y., 1980, Sinistral movement along the Gulf of Aqaba—Its age and relation to the opening of the Red Sea: *Nature*, v. 285, p. 220-221.

Ben-Menahem, A., 1979, Earthquake catalog for the Middle East (92 B.C.–1980 A.D.): *Bolletino Geofisica Teoretica e Applica*, v. 21, p. 245-313.

Ben-Menahem, A., 1991, Four thousand years of seismicity along the Dead Sea rift: *Journal of Geophysical Research*, v. 96, p. 20,195-20,216.

Ben-Menahem, A., Nur, A., and Vered, M., 1976, Tectonics, seismicity and structure of the Afro–Eurasian junction—The breaking of an incoherent plate: *Physics of the Earth Planetary Interior*, v. 12, p. 1–50.

Ben-Menahem, A., and Aboodi, E., 1981, Micro- and macroseismicity of the Dead Sea rift and offshore coast eastern Mediterranean: *Tectonophysics*, v. 80, p. 199–233.

Freund, R., Zak, I., and Garfunkel, Z., 1968, Age and rate of the sinistral movement along the Dead Sea rift: *Nature*, v. 220, p. 253– 255.

Garfunkel, Z., Zak, I., and Freund, R., 1981, Active faulting in the Dead Sea rift: *Tectonophysics*, v. 80, p. 1–26.

Joffe, S., and Garfunkel, Z., 1987, Plate kinematics of the circum Red Sea—A re-evaluation: *Tectonophysics*, v. 141, p. 5–22.

Kashai, E. L., and Croker, P. F., 1987, Structural geometry and evaluation of the Dead Sea–Jordan rift system as deduced from new subsurface data: *Tectonophysics*, v. 141, p. 33–60.

Neev, D., and Emery, K. O., 1967, The Dead Sea: Depositional processes and environments of evaporites: Jerusalem, Geological Survey of Israel Bulletin, v. 41, 147 p.

Rotstein, Y., Bartov, Y., and Frieslander, U., 1990, Evidence for shifting of the Dead Sea transform and changes in the local structural setting in the rift: IPRG Report No. 874/260b/90, 11 p.

Rotstein, Y., Bartov, Y., and Hofstetter, A., 1991, Active compressional tectonics in the Jericho area, Dead Sea rift: *Tectonophysics*, v. 1983, p. 239–259.

Shapira, A., 1979, Redetermined magnitudes of earthquakes in the Afro–Eurasian junction: *Israel Journal of Earth Sciences*, v. 28, p. 107–109.

Shapira, A., 1982, Detectability of regional seismic networks: analysis of the Israel seismic networks: *Israel Journal of Earth Sciences*, v. 41, p. 21–25.

Shapira, A., Manor, O., Oman, S., and Sheshinsky, R., 1986, The empirical relation of earthquake epicenters to mapped faults in Israel: *Israel Journal of Earth Sciences*, v. 35, p. 149–157.

Shapira, A., and Feldman, L., 1987, Microseismicity of three locations along the Jordan rift: *Tectonophysics*, v. 141, p. 89–94.

Shapira, A., Avni, R., and Nur, A., 1993, A note: New estimate of the Jericho earthquake epicenter of July 11, 1927: *Israel Journal of Earth Sciences*, v. 42, p. 93–96.

Steinitz, G., Bartov, Y., and Hunziker, J. C., 1978, K–Ar age definition of some Miocene–Pliocene basalts in Israel—Their significance to the tectonics of the rift valley: *Geology Magazine*,

v. 115, p. 329–240.

Turcotte, T., and Arieh, E., 1988, Catalog of earthquakes in and around Israel, *in* Preliminary safety analysis report: Tel Aviv, Israel Electric Corporation Ltd., Appendix 2.5A, p. 1–18, 6 tables.

van Eck, T., 1988, Attenuation of coda waves in the Dead Sea region: *Bulletin of the Seismological Society of America*, v. 78, p. 770-779.

van Eck, T., and Hofstetter, A., 1989, Microearthquake activity in the Dead Sea region: *International Journal of Geophysics*, v. 99, p. 605–620.

van Eck, T., and Hofstetter, A., 1990, Fault geometry and spatial clustering of microearthquakes along the Dead Sea–Jordan rift fault zone: *Tectonophysics*, v. 180, p. 15–27.

II. PHYSICAL, CHEMICAL, AND BIOLOGICAL ASPECTS OF THE DEAD SEA

8. THE HYDROGRAPHY OF A HYPERSALINE LAKE

David A. Anati

The Dead Sea water column is composed, to a first approximation, of two water bodies: the deep water, constituting most of the lake's volume, and a shallow upper layer a few meters thick. The temperature and salinity profiles can both be either stabilizing or destabilizing, depending on the regime and the season, and examples of the various combinations are documented. If salinity is destabilizing, conditions for double-diffusive mixing may be attained, and its effect is then reflected in the water parcel's trajectory in the temperature-salinity (T–S) space. The trajectories of the Dead Sea brines since 1977 belong to one of three different categories: upper layer under a meromictic regime, upper layer under a holomictic regime, and lower layer under a holomictic regime. The lower layer during the meromictic regime of 1979–1982 remained constant in its properties, and its trajectory is thus represented by a single point. The overall stability of the lake and its seasonal variation for a period covering a meromictic regime, as well as a holomictic regime, is presented, and the typical differences in its magnitude are shown to give a good forecasting criterion for the expected state. The local stability in the water column is examined for several situations and values up to $N^2 = 0.3$ s^{-2} are found—two orders of magnitude greater than stabilities of other natural water bodies under similar climatic conditions. The stability of the Dead Sea water column can offer an interpretation to some microbiological findings and their changes with time.

BACKGROUND

The Dead Sea (Fig. 8-1) is a hypersaline terminal lake with a surface area of 760 km^2 and a maximal depth of 324 m, located in the deepest terrestrial spot on earth. The 1993 surface level was about 410 m below mean sea level (MSL), and the lowest level ever recorded is 411 m below MSL in December 1991. The Dead Sea brines, being saturated with halite, aragonite, and gypsum (Gavrieli et al., 1989), precipitate salts mostly at the periphery but also in the interior of the lake (Anati, 1993). Every winter (the rainy season), the lake receives a freshwater inflow from floods and from the Jordan River runoff, but the year-to-year variability in the amount of inflowing winter freshwater is rather high: In a rainy year, it can be more than twice as much as the average, and in a droughty year, it can be negligible.

Prior to 1979, the Dead Sea was meromictic, with a permanent pycnocline at a depth of about 40 m (Neev and Emery, 1967). Seasonal and year-to-year variations were limited to the upper layers, whereas the deep layer was referred to as "the fossil water body" (Steinhorn and Gat, 1983). How long the Dead Sea had been meromictic is under dispute; the two most popular theories are 1,400 years and 350 years, respectively, with evidence in favor of the latter being somewhat more convincing. The droughty climate of the second half of the 20th century, combined with the 1964 diversion of Jordan River freshwater for irrigation and domestic use, caused the gradual weakening of the permanent pycnocline, which terminated in the 1979 historic overturn of the Dead Sea water column (Steinhorn, 1985), practically creating a new lake with a new set of hydrographic regimes.

In some rare cases, the inflowing freshwater mixes to great depths, diluting the whole volume of the lake, but mostly the freshwater mixes only with the uppermost few meters of the lake's water. The mixture then remains atop as a buoyant upper layer floating on the denser Dead Sea brine, thus creating a stable halocline (the base of the buoyant upper layer) at depths between 5 and 15 m, depending on the year. This strong halocline acts as a "virtual bottom", obstructing the penetration of turbulence through it, and all the vertical turbulent fluxes (of salt, heat, chemical concentrations, microorganisms, etc.) are therefore weakened, reducing the actual effective fluxes to minimal values. As the spring heating begins, consequently, all the heat remains trapped above this virtual bottom (penetrating radiation is negligible below 4 m) and reinforces the stability of the water column to values considerably above those observed elsewhere in natural basins. In the Dead Sea in 1980 and again in 1992, for example, the vertical density difference across a transition layer 1 to 2-m-thick was more than three times that of the whole North Atlantic Ocean over its 4 km depth.

A convenient measure of the overall stability of a lake is the energy per unit area, W, required to mix the whole water column against its stable stratification. This quantity W is computed from the observed potential density profile $\rho(z)$, of mean density ρ_o, taking into account the horizontal area $A(z)$ at every given depth z of the lake:

$$W = \frac{g}{A(0)} \int (\rho - \rho_0) A(z) z \, dz \qquad (1)$$

The vertical direction is taken as *depth*, i.e., positive downward.

The overall stability W of the Dead Sea changes along the seasonal cycle. Its minimal value, always greater than or equal to zero (no Rayleigh instabilities have been observed since the Dead Sea hydrographic survey began in 1977), is reached every year around mid-December. If this December minimum is of zero stability, it means that the water column is totally mixed and will stay mixed as long as the buoyancy flux across the surface remains negative. A negative buoyancy flux, derived either by cooling or by an evaporation-induced salinity-rise, does increase the density ρ; but if the water column is mixed, then the density is homogeneous over depth, $\rho = \rho_o$, and W as defined by equation 1 remains zero regardless of the temporal changes in ρ. This mixed state with $W = 0$ normally persists for about 2 months, until late February, when the buoyancy flux becomes positive and the new seasonal pycnocline begins to form (see Fig. 8-2a, December–February of all the years from 1982 on represented in the figure). The regime described, called *holomictic*, is typical of droughty periods, such as 1982–1991.

Occasionally, a rainy winter occurs, during which the Dead Sea receives freshwater inflows exceeding the amount that will evaporate by next winter's first flood. This happened in 1980 with a surface rise of about 1.5 m, and again in 1992 with a surface rise of about 2 m versus an average annual surface drop of 0.8 m (Anati and Shasha, 1989a). Under suitable weather conditions, mainly weak winds, the December minimum of stability may then be greater than zero, in which case the water column will remain further stratified through two or more seasonal cycles. The onset of this regime, called *meromictic*, is typi-

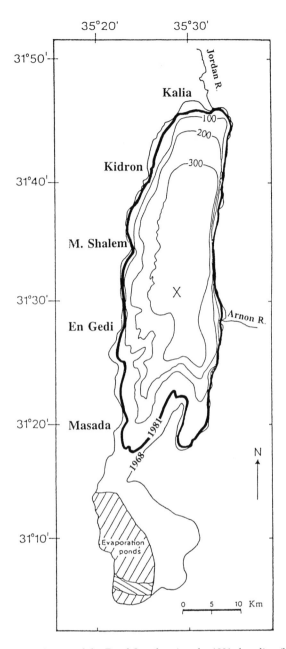

Figure 8-1 A map of the Dead Sea showing the 1981 shoreline (heavy line) and bathymetry (thin lines). The map approximately describes subsequent (post-1981) Dead Sea shorelines and bathymetry, but not previous ones; the 1968 shoreline is added for comparison. X marks the main standard site of the hydrographic stations.

the water column in different years may react differently with respect to stability, to similar amounts of winter freshwater inflow, depending on weather conditions. It will be shown subsequently, however, that during the 15-year Dead Sea hydrographic survey (1977–1993), statistically, the W decreased in the autumn at a rather regular rate ($\approx 1 \ kJ/m^2/d$). Therefore, only in a limiting situation can a single storm at the critical time change the whole regime form meromictic to holomictic.

If W is a measure of the overall stability of the water column, the vertical gradient of the buoyancy $b = g \ (\rho\text{-}\rho_o)/\rho_o$, better known as the square of the Brunt-Väisälä frequency N^2, is a measure of the local stability at any given depth z of the water column. Its precise expression can be approximated to a high degree of accuracy as:

$$N^2 = \frac{g}{\rho_0}\frac{\partial \rho}{\partial z} \qquad (2)$$

where $\rho = \rho(z)$ is the density, ρ_o is a reference density (for instance, the average density of the layer considered), and g is the acceleration of gravity. $N^2(z)$ is a powerful tool: It indicates at what depth internal seiches may be expected, the natural frequency of oscillations of the pycnocline, and to what extent the pycnocline may act as a virtual bottom, and, as will be shown, it describes the vertical structure of the column's stability.

THE CONCEPT OF SALINITY IN A HYPERSALINE LAKE

The absolute salinity S of a brine is defined as the mass of dissolved matter per mass of brine (Forschhammer, 1865). Therefore, this is a dimensionless number; because of its smallness, it is usually expressed in parts per thousand and designated either by g/kg or by the symbol ‰. Other ways to define or express salinity are also used, but not all of them are recommended; for example, the unit TDS (total dissolved mass of salts per unit volume of brine) is temperature dependent, and therefore is not a valid unit unless the temperature at the time of determination is specified.

Although salinity is a very common parameter in oceanography and limnology, direct measurements of salinity are very rare. Instead, some other variable is usually measured: chlorinity, electrical conductivity, optical refraction, density, sound velocity, etc., from which the salinity can be inferred by calibration. Substitute parameters of physical quantities in all domains are quite acceptable, one of the most common among them being the difference in thermal expansion between glass and mercury as a measure of temperature.

Some of the standard salinity definitions and almost all standard methods of salinity determination accepted in oceanography are based on the assumption that all interconnected seas and oceans have the same relative ionic composition (Marcet, 1819). These methods are not applicable, of course, to stable pools (such as those found in the bottom of the Red Sea and eastern Mediterranean) and, in our context, to continental lakes and terminal lakes, such as the Dead Sea. Consequently, no universal oceanic formula can be applicable to the Dead Sea brines. Some alternative ways to solve the problem of salinity determination in the Dead Sea have been considered in the past, and because of the cardinal importance of this parameter, they deserve further explanation.

The optical refraction index of an undersaturated Dead Sea water sample at a fixed temperature and pressure can be calibrated to salinity. Levi et al. (1984) showed that for the ionic composition of the 1981–1982 Dead Sea brines, the refraction index is a monotonic function of salinity to high concentrations. Applied research was abandoned because some practical requirements, such as an extremely clean optical fiber, could not be met, and the best expected accuracy was unacceptable.

The electrical conductivity at a fixed temperature and pres-

cal of rainy winters, such as the winter of 1979–1980 or that of 1991–1992.

The amount of freshwater inflowing into the Dead Sea in a given winter, although a very important factor in itself, is not the only operator affecting the stability W. Another factor that affects the stability of the pycnocline after its formation is wind-induced stirring, which is proportional to the third power of the wind's velocity (see, e.g., Kraus, 1972). This means, for instance, that a 1-day storm with winds 10 times faster than the average wind is equivalent (in terms of energy input) to 3 years of average wind. This illustrates the high sensitivity of the system to unpredictable weather events. Conceivably, therefore,

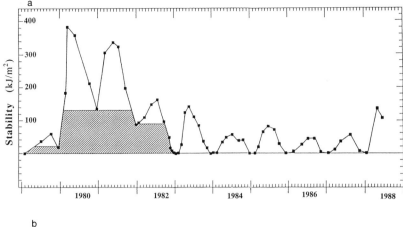

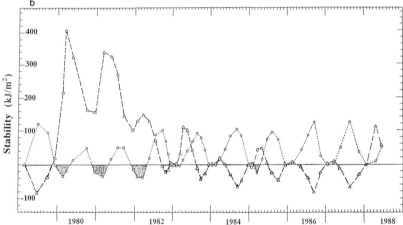

Figure 8-2 The stability of the water column of the Dead Sea between the 1979 historic overturn and 1988. a, the total stability W as defined by equation 1 in the text. The hatched area represents the meromictic component; b, the separate contributions of the thermal structure (dotted lines) and the haline structure (dashed lines). Dotted areas denote a destabilizing thermocline.

sure could, in principle, be calibrated to salinity through the conversion formula for the particular ionic composition on hand. However, the conductivity of Dead Sea brines at high concentrations ceases to be a monotonically increasing function of salt concentration (Lazar, 1982) and reaches its maximal value around the present day density ($\approx 1,236$ kg/m^3). This method, consequently, is not applicable to the Dead Sea brines.

Dilution of the brine could adapt a certain hypersaline sample to electrical conductivity measurements, but the procedure would then include two weighings and one conductivity measurement. Each one of the three would then contribute its own share to the final inaccuracy. Consequently, this method cannot compete with one single weighing to determine density as a measure of salinity.

The use of density at precisely monitored temperature and pressure has been specifically recommended as a measure of salinity (UNESCO, 1976). The density of an undersaturated Dead Sea brine increases monotonically with salinity, and it is a valid salinity indicator with a one-to-one conversion to the Forschhammer unit (mass/mass). This is the routine method chosen as the most meaningful to evaluate the salinities of the Dead Sea brines, with density measured by pycnometry (weighing a known volume), or by hydrometry (using a calibrated floating body), or by other means, such as the gradient column method (introducing the brine into a column of varying density) first advanced by Kaushansky and Starobinets (1982) and further used by Stiller et al. (1983). The salinity value itself is normally expressed in sigma-units (σu), i.e., density in the MKS system (kg/m^3) excess to the standard reference density of 1,000 kg/m^3. Its notation is by the Greek letter σ with the fixed temperature in degrees centigrade as a subscript (σ_{20}, σ_{25}, etc.), with atmospheric pressure being implied. Designating

the salinity directly in density units circumvents the undesirable use of conversion formulae, corrections, or other assumptions, and is no less legitimate than designating pressure, for example, in units of mm of mercury.

The equation of state of the Dead Sea brines may change through the years after changes in their relative ionic composition (see Gavrieli, chapter 14, this volume). The latest checks were performed on a December 1990 brine (partially reported by Anati, 1993), and yielded the following relations:

Thermal expansion coefficient:

$$\frac{\partial \rho}{\partial T} = 0.4309 \pm 0.0005 \ \sigma u / K \tag{3a}$$

Total-salts expansion coefficient:

$$\frac{\partial \rho}{\partial S} = 0.936 \pm 0.002 \ \sigma u / \text{‰} \tag{3b}$$

Halite expansion coefficient:

$$\frac{\partial \rho}{\partial [\text{NaCl}]} = 0.805 \pm 0.004 \ \sigma u / \text{‰} \tag{3c}$$

The reference point in use, adapted from Steinhorn (1981) who summarized previous works,

$$\sigma_{20} = 235.00 \leftrightarrow S = 277 \pm 2 \frac{kg}{g} \tag{3d}$$

is not known as accurately as desirable and consequently, the absolute accuracy of salinity estimates, S (mass/mass), is considerably lower than their relative precision.

Special problems, not encountered in undersaturated brines,

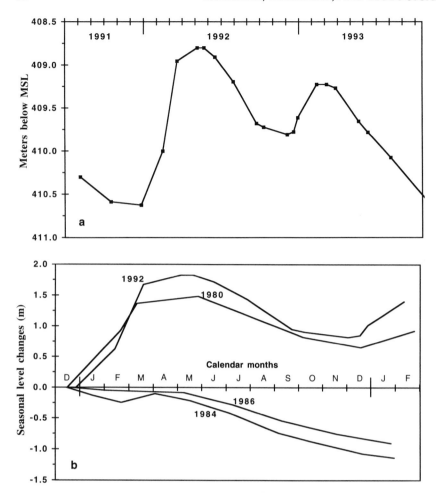

Figure 8-3 Surface level changes in the Dead Sea. a, in the last few years, reference to the Mediterranean MSL; b, during the two most rainy years (1980 and 1992) and the two most droughty years (1984 and 1986) since the 1979 historic overturn of the Dead Sea water column, reference to the minimal level observed in the month of December preceding each calendar year.

emerge when a brine is saturated with respect to a certain salt. In our context, the Dead Sea has been saturated with respect to NaCl at least between 1983 (Gavrieli et al., 1989; Anati and Stiller, 1991) and 1991 (Anati, 1993). Samples of Dead Sea water collected in a certain cast, therefore, may have contained salt crystals in suspension while still in situ. It is difficult to detect their presence immediately because NaCl crystals are not visible to the naked eye if smaller than about 4 μ. The first problem is: What would one like to measure—the density (salinity) of the liquid phase only (that is, as if it had no crystals at all in suspension) or that of the mixture (brine + crystals)? The density of the mixture (brine + crystals) is of course the desired one because, when the crystals in situ are not accelerated (that is, they have reached their final Stokes velocity), the density with a physical meaning for hydrostatic pressure-gradients determinations, stability evaluations, and dynamic computations is that of the mixture. This is rather fortunate, because there would be no known standard technology to safely strip the sample of the salt crystals in suspension and remain sure that the liquid phase was left absolutely unchanged in the process, if this were desirable.

By the time the sample is brought to the laboratory for an evaluation of its density, the amount of salt crystals in suspension may have changed, perhaps by changes in ambient temperature or transitions to steady-state, and the sample may well constitute a liquid quite different from the one intended to be examined. The second question posed by the situation is therefore: What can be measured with good repeatability? Restoring the precise amount of in situ suspended salt crystals must be ruled out, at least on the ground that the precise amount is not known. The customary procedure is to heat the sample to

ensure that any possible salt crystals would have been dissolved, and only then measure their density. The repeatability of the measure is good, and the salinity S (mass/mass) is unaltered in the process. But the measured density results are slightly lower than the real in situ density of the original mixture (expansion coefficients (eq. 3b and eq. 3c) are different from each other), and since density is the parameter actually measured, this difference is conceptually relevant. It has been shown, however, (Anati, 1993) that, by today's accuracy standards, the maximal possible deviation, even under extreme conditions, would be negligible, and the procedure has therefore been considered acceptable.

In this chapter, salinities will be given in units of σ_{20}.

DATA ACQUISITION

Hydrographic measurements in the Dead Sea before 1977 were scanty and sporadic. During the last century, one sample of Dead Sea water was brought to Britain and analyzed and one temperature profile was taken, and in the 1930s more hydrography was done by D. Ashbel and by The Dead Sea Works, Ltd. The first systematic survey was undertaken in 1959–1960 (Neev and Emery, 1967), and measured chemistry, bathythermograms, reversing thermometers temperatures, currents (by drogues and current meters), and water transparency.

The present Dead Sea hydrographic survey began in 1977, two years before the historic overturn of the water column, with a Weizmann Institute 2-year project (Steinhorn, 1985), which was then continued for 6 more years. Solmat Systems Ltd. joined the survey between 1980 and 1985, concentrating on the detailed structure of the top 50 m of the lake (reported in

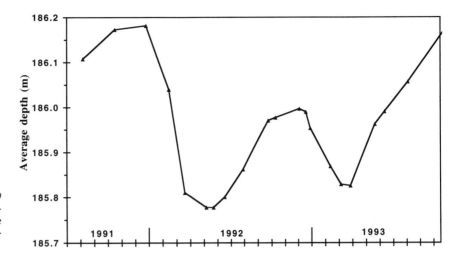

Figure 8-4 Changes in the average depth D of the Dead Sea. For the present configuration of the Dead Sea shores, the average depth D diminishes with a rise in the surface level (compare with Fig. 8-3a).

Anati et al., 1987). The project continues to date with both methods (deep casts and detailed shallow profiles) conducted by the Hebrew University in a similar monitoring pattern. The standard parameters measured are temperature and salinity (density). Profiles of penetrating solar radiation at noon are occasionally measured, whenever sea state and cloudiness permit.

Roughly 200 one-day-cruises have taken place since 1977, about half of them during 5 years of particularly intensive research, 1980–1985. Most of the data were taken from the deepest site in the lake (x in Fig. 8-1), other sites were occasionally visited for a study of the three-dimensional thermohaline structure of the lake or by necessity because of conditions, such as daylight shortage, sea state, or shallow-anchorage.

In deep casts, temperature was measured by reversing thermometers at vertical spacings ranging from 5 m (upper portion) to 50 m (lower portion). Water samples for salinity determination were taken by means of standard sampling bottles at the same vertical spacings.

In shallow profiles, temperature was measured by means of one single thermistor lowered to the desired levels, at vertical spacings as close as 10 cm at sharp "elbows," and every meter by default. The use of the same probe at all depths circumvents problems of intercalibration and does not require any a priori commitment as to the levels to be measured. It comes, however, at the expense of simultaneity, whenever time scales of 10–15 min are relevant. Water samples for salinity determination in shallow profiles were pumped through a hose, the end of which was lowered to the desired depth. Vertical spacings were as close as 25 cm at sharp elbows, and every 2 m by default.

The relative precision of the thermistors used is 0.02°C, and their absolute accuracy 0.15°C. The relative precision of pycnometric salinity determinations is 0.02 sigma-units (σu), with an absolute accuracy of 0.1 σu, whereas the hydrometric salinity determinations have an accuracy and a precision of 0.2 σu. The conversion formula to salinity S in units of mass of dissolved matter per mass of brine (eq. 3d) leaves the relative precision at 0.02 σu <=> 0.02 g/kg, but lowers the absolute accuracy of the salinity S by a factor of 20, to 2 g/kg.

CHANGES OF SURFACE LEVEL

The Dead Sea, being a terminal lake, cannot keep a constant surface level or a constant surface area (unless, theoretically, the water flowing in were to balance evaporation exactly at any moment). Figure 8-1 illustrates the particularly marked

changes in surface area that occurred during the period 1968–1981, when the shallow bottom of the southern basin became exposed.

The average trend since 1918 has been droughty. The average in surface level drop between 1974 and 1980, for example, was 0.6 m/yr, and the net level drop in every single year during the period 1981–1988 falls in the range of 0.8 ±0.1 m/yr (Anati and Shasha, 1989a). The Dead Sea level rises in winter in unpredictable and short events (runoff and sudden floods), and drops at a fairly constant rate the rest (most) of the year. In a droughty year, the total level rise in winter is smaller than the yearly drop, and in a rainy year it is larger. An exactly balanced year, the common situation in most of the world's lakes (which are nonterminal), has never been recorded in the Dead Sea.

The surface level of the lake is measured continuously by the Dead Sea Works Ltd. and recorded on the day of each cruise of the hydrographic survey. Figure 8-3a presents the surface level changes in the last few years, and Figure 8-3b shows the changes during the two most rainy years and the two most droughty years since the 1979 historic overturn of the secular stratification (Steinhorn, 1985; Anati et al., 1987). Because the difference in surface level changes among the years can be rather large, there cannot be a typical curve of the Dead Sea seasonal level changes.

As displayed in Figure 8-3b, the two rainy years chosen, 1980 and 1992, received fairly similar amounts of water and, as will be show subsequently, were rather similar in overall stability W. Nevertheless, they developed considerable differences in the details of their local stability structures $N^2(z)$, which is attributable to casual differences in weather conditions. The average depth of the lake is about 186 m (Fig. 8-4) and varies following changes in surface levels. Average depth is an important parameter, useful in balances and bulk calculations. For the particular configuration of the present Dead Sea shores, a rise in the surface level diminishes the average depth D (compare Fig. 8-3a with Fig. 8-4).

A critical topographic slope at the level of the lake's shore is given by the relation

$$\left(\frac{\partial A}{\partial z}\right)_{\text{crit}} = \frac{A(0)}{D} \tag{4}$$

in which the average depth D is unaffected by level changes. If the topography is steeper than critical, a level rise will increase the average depth D, whereas if the topography is less steep than critical, a level rise will flood additional shallow areas at the periphery of the lake, diminishing the average depth D. As seen by comparing Figure 8-3a with Figure 8-4, the topographic

Salinity and Density (σu)

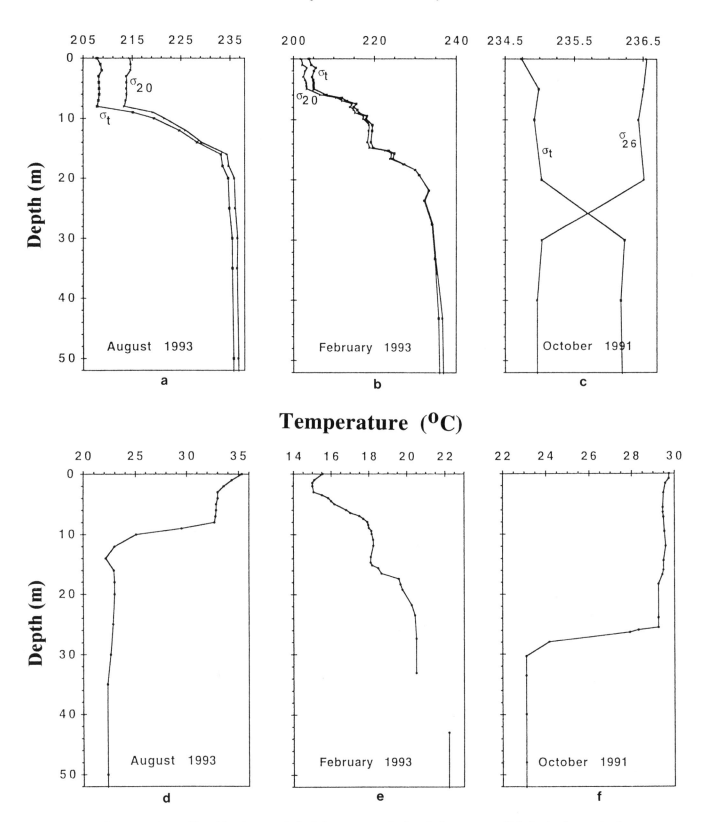

Temperature (°C)

Figure 8-5 Representative profiles of density, σ$_t$, and of salinity, σ$_{20}$ or σ$_{14}$, (a-c), and of temperature (d-f), for the three main classes of stratification; August 1993, temperature and salinity are both stabilizing; February 1993, temperature is destabilizing; October 1991, salinity is destabilizing.

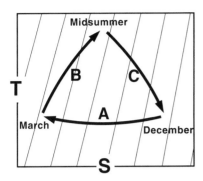

Figure 8-6 A schematic representation of the trajectory of the upper layer Dead Sea brine in the T-S space in a hypothetical hydrographically balanced year. Between December and March, floods and Jordan River runoff dilute the brine (leg A), between March and August, the heat flux is positive and evaporation increases the salinity (leg B), and between August and December, the heat flux is negative and evaporation increases the salinity (leg C). Diagonal thin lines represent isopycnals.

slope at the level of the Dead Sea shores is less steep than critical.

TYPICAL PROFILES

The vertical thermohaline structure of the Dead Sea water column can be neutral or stable; as mentioned previously, Rayleigh instabilities have never been reported. A neutral stability, i.e., a totally mixed column with $W = 0$, occurs with every overturn and normally lasts for about 2 months (Fig. 8-2a).

Stable Dead Sea thermohaline structures may be classified in three main classes, as follows:

Temperature and salinity are both stabilizing. This structure is found invariably in spring and summer under a meromictic regime, and occasionally toward the end of the rainy season or just after it under a holomictic regime. Strong stratifications with this structure are not uncommon. A recent example, from August 2, 1993, is shown in Figure 8-5a and 8-5d. On that date, the density step was 26 σu. Most of it, 22 σu, was due to the salinity stratification, and only 4 σu was due to the temperature stratification.

Temperature is destabilizing, salinity is stabilizing. This structure is found in winter and spring, sometimes continuing through early summer, under a meromictic regime, but it has rarely been observed in holomictic years; compare, for instance, March 1985 in Figure 8-2b with the winter conditions during the meromictic period of 1979–1982 shown in the same figure. In the February 1993 example, (Fig. 8-5b, e), a stabilizing salinity difference of about 30 σu between surface water and deep water was weakened by 3 σu by the temperature's destabilizing 7°C difference. Double-diffusive "layering" (as distinguished from double-diffusive "fingering") is theoretically possible with this kind of stratification, but in the Dead Sea this feature has never been observed nor has it been suggested by indirect evidence.

Temperature is stabilizing, salinity is destabilizing. This structure occurs only during holomictic period. It begins as early as February in a very droughty year (1979) and as late as September in less droughty years (1982–1983). Once started, this structure is bound to maintain itself until the overturn. The present (1994) regime, being meromictic since the end of 1991, has not witnessed this structure nor any destabilizing halocline at all; the last observed example of this structure was on October 16, 1991 (Fig. 8-5c, f). As a rule, the stability accom-

panying this stratification is much weaker than that of other structures. In the examples shown in Figure 8-5, the October 1991 stratification is almost 20 times weaker than the other two. A necessary condition for double-diffusive fingering is satisfied with this kind of stratification, and indeed the feature has been documented in the Dead Sea on several occasions, albeit indirectly, through the energetics of the rate of deepening of the pycnocline, as well as through the trajectory of the upper water parcel in the T-S space, maintaining a constant density ration of $R_\rho = 1.7$ (see Fig. 3 in Anati and Stiller, 1991).

TRAJECTORIES IN THE T-S SPACE

As mentioned previously, an exactly hydrographically balanced year has never been recorded in the Dead Sea. Nevertheless, it is instructive to consider what would be the expected trajectory of a parcel of Dead Sea brine in the temperature-salinity (T-S) space if a balanced year were to occur.

There are no floods or rains in the Dead Sea except during a short rainy season, typically between December and April. The summer evaporation by far exceeds the meager contribution of the few springs, minor rivers, and the (summer) Jordan River inflow. Therefore, the salinity increases continuously outside of the rainy season and decreases (by the same amount, in a balanced year) during the short rainy season. Concurrently, the temperature follows the seasonal pattern with a warm August and a cold February. The path in the T-S space is therefore composed of three separate legs that form a triangular trajectory, as depicted schematically in Figure 8-6.

Increases in density occur mainly after midsummer (leg C in Fig. 8-6) and decrease in density occur in winter (leg A). During the spring and early summer (leg B), heating and evaporation compensate each other to a large extent, and the path is more isopycnal than during the rest of the seasonal cycle.

To compare the observed trajectories of the Dead Sea brines to the schematic trajectory of a hypothetically balanced year (Fig. 8-6), we will distinguish between regimes (meromictic vs. holomictic), as well as between the upper layer, only a few meters thick but in contact with the atmosphere, and the deep-water body, constituting the bulk of the Dead Sea brines but more isolated.

Upper layer, meromictic

The actual trajectory of the upper-layer water type in the T-S space observed during the meromictic stage of 1979–1983 is shown in Figure 8-7. During this period, evaporation exceeded precipitation and runoff; therefore, by contrast to the balanced year, the curve is not closed; a gradual increase in salinity is observed, so that the multiyear trajectory is looping. The seasonal dilution (corresponding to leg A in Fig. 8-6) varies considerably from year to year, whereas seasonal heating (leg B) and cooling (leg C) remain around 13.5°C with only minor variations.

The point for December 1982 in Figure 8-7 also represents the deep-layer water type during the whole meromictic period. This date represents (by definition) the occurrence of the 1982 overturn of the whole water column, that is, the beginning of the second (holomictic) stage reported here.

Upper layer, holomictic

The actual trajectory of the upper-layer water type in the T-S space observed during the holomictic stage of 1982–1991 is shown in Figure 8-8. During this period, the upper layer was consistently thicker by a few meters than during the meromictic

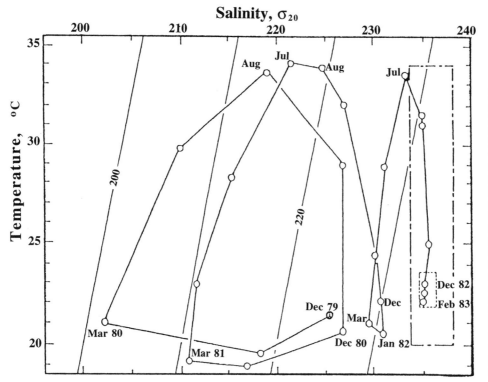

Figure 8-7 A T-S diagram showing the evolution of the Dead Sea surface water during the meromictic period of 1979 to 1983. The year-to-year variability in the winter legs (corresponding to leg A in Fig. 8-6) is appreciable compared to the regularity of the spring-summer legs (B) and the summer-autumn legs (C). Thin diagonal lines denote isopycnals (potential density) in units of σ_{20}. Points of measurements (open circles) are averaged over the diurnal cycle. Error bars are not indicated in the figure because they are smaller than the symbols (see Fig. 8-9). The dash-dot box corresponds to the whole frame of Figure 8-8, and the dotted box to the whole frame of Figure 8-9.

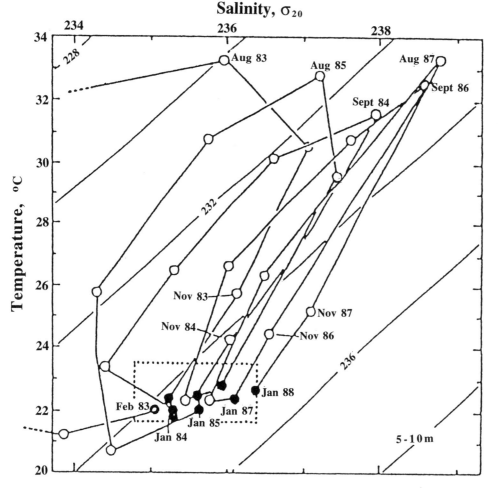

Figure 8-8 A T-S diagram showing the evolution of the upper water mass (5–10 m) during most of the holomictic stage of 1983–1991. The figure's frame is an enlargement of the dash-dot box shown in Figure 8-7. The diagonal lines denote isopycnals (potential density) in units of σ_{20}. Open circles denote stable (stratified) states, and full dots denote neutral (mixed) states. The first loop (March–July 1983) includes salinities below $\sigma_{20} = 234$ and was omitted to allow expansion of the main features. Compare the salinity decrease in the cooling period (August–January) with the salinity increase in the same season during the meromictic stage (Fig. 8-7). Error bars are not indicated on the figure because they are smaller than the symbols (see Fig. 8-9). The dotted box corresponds to the whole frame of Figure 8-9.

stage (Anati et al., 1987). At the same time, the temperature excursion was consistently smaller than in the meromictic stage (10°C vs. 13.5°C). These two differences compensated each other so that the total heat stored in the upper layer remained roughly the same. Apparently, a negative feedback mechanism keeps the total stored enthalpy within boundaries.

During the period of strong thermal stratification build-up (March to midsummer, leg B), the upper layer underwent a salinity increase (as expected according to Fig. 8-6), but at a much slower rate than in the meromictic stage. In the spring of 1980, for example, the salinity increased about five times faster than in the spring of 1985 (compare Fig. 8-7 with Fig. 8-8). This difference occurred despite a similar evaporation rate and a similar surface salinity in the two stages, and it occurred during a season with no inflowing freshwater, which might have explained a partial dilution. The difference can be attributed to salt extraction: The (supersaturated) Dead Sea brines precipitated salts during this period, and the ensuing effect, as evaluated from shallow cores and sediment traps, was shown to fit this observed reduction in the rate of salinity increase (Anati and Stiller, 1991).

The most conspicuous difference between these holomictic upper-layer trajectories (Fig. 8-8) and those found during the meromictic stage (Fig. 8-7), is the salinity trend after the midsummer point. A clear decrease is indicated in leg C during the holomictic stage rather than the expected increase (compare with Figs. 8-6 and 8-7). This occurred although evaporation is estimated as maximal in this season (Stanhill, 1985). The observed decrease is explained by a double-diffusive mixing between upper and lower layers, as corroborated by the "density ratio" R_ρ, as well as by the energy changes during the same period (Anati and Stiller, 1991).

Lower layer, meromictic

In a meromictic regime, the deep waters of the Dead Sea are practically isolated from any outside influence during more than a complete seasonal cycle. In a holomictic regime, the deep waters are isolated during most of one seasonal cycle. It might be expected, therefore, that these deep waters would

remain constant in all their properties in a meromictic regime, changing only in the short, mixed winter season (typically December to February) in a holomictic regime. During the 1979–1982 meromictic regime, the deep water's properties indeed remained constant, as expected, with a temperature of T = 23.15 ±0.03°C and a salinity of σ_{20} = 235.05 ±0.10. In the beginning of the present (1994) meromictic regime, on the other hand, a slight decrease in salinity was found (see subsequent discussion).

Lower layer, holomictic

During the holomictic regime of 1982–1991, the situation was rather complex; the climatic signature of an overturn depends on whether the upper water trajectory hits the deep water isopycnal at a higher temperature and higher salinity, or at a lower temperature and lower salinity than the deep-water point (Anati et al, 1987). The former option brings about a rise in the deep-water temperature and salinity, whereas the latter causes a drop in the deep-water temperature and salinity. The probability of a rise in temperature and salinity is considerably higher than that of a drop in these properties because it occurs in an arid period, when overturns are common occurrences. The opposite (a drop of temperature and salinity of the deep waters) occurs in rainy years, when the stable stratification is more likely to prevent overturns. We may, therefore, anticipate that the multiyear trend will show a general secular rise in temperature and salinity, superimposed, of course, on the seasonal signal.

The actual trajectory of the deep Dead Sea waters during the holomictic years 1982–1988 is shown in Fig. 8-9. The secular trend described previously results in an increase of roughly 0.16°C/yr in temperature and 0.26 σu/yr in salinity. The same trend can be found also in the upper-layer brines just before overturns (November lines in Fig. 8-8) and just after overturns (January). If the results of our 15-year survey are typical of the geological evolution of the Dead Sea, therefore, the bulk of the Dead Sea brines does not reflect past average climatic conditions, but mainly the climatic conditions prevailing during particularly droughty periods.

The bulk of Dead Sea brines is thus irreversibly bound to

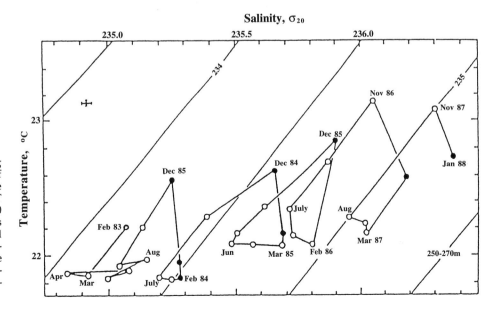

Figure 8-9 A T-S diagram showing the evolution of the deep waters of the Dead Sea during most of the holomictic stage of 1982–1989. The T and S values are taken from 250–270 m below the surface. The figure's frame is an enlargement of the dotted boxes in Figures 8-7 and 8-8. Diagonal lines and symbols are as in Figure 8-8. Error bar of single measurements is shown in the upper left corner.

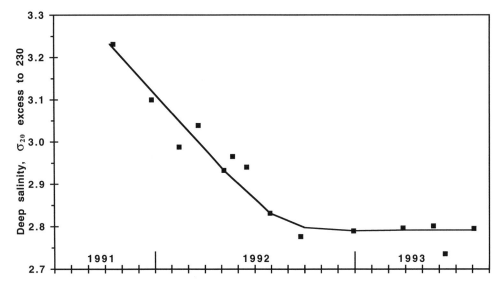

Figure 8-10 Changes in the salinity of the deep Dead Sea water, 1991–1993. Averages of all measurements between 100 and 310 m depth. The deep layer became isolated by strong, stable stratification toward the end of 1991. One point (Sept. 1992) was shifted by 0.04 σu because of a defective salinity determination.

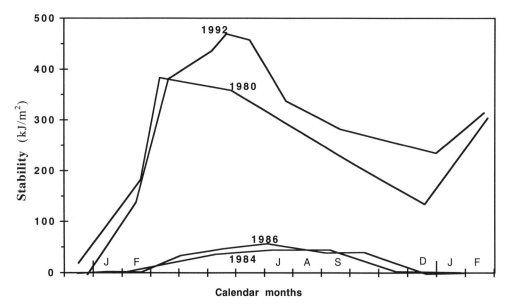

Figure 8-11 Changes in the overall stability W of the Dead Sea during the two most rainy years (1980 and 1992) and the two most droughty years (1984 and 1986) since the 1979 historic overturn of the Dead Sea secular stratification.

increase in salt concentration and in the relative amount of the more soluble salts during holomictic periods, and to preserve its concentration and relative ionic composition during meromictic periods, but hardly ever to be diluted with freshwater or enriched with less-soluble salts.

The details shown in Figure 8-9 indicate that changes in the deep-water properties occur not only in the short mixed periods (full dots) as might have been expected, but during the rest of the year as well (open circles), even though the deep water is then sealed by the stable stratification. The trajectory in the T-S space is again a loop, as it is for the upper layer of water, but this time it is out of phase with the schematic trajectory depicted in Figure 8-6 or those shown in Figure 8-7. The deep water attains its highest temperature in December, and its lowest salinity in June, July, or August. The processes causing this looping trajectory have been analyzed in detail by Anati and Stiller (1991), and may be summarized as follows for legs A–C

of Figure 8-6:

Leg A. With the beginning of the spring stratification (around March), the deep water becomes isolated and the winter cooling stops, while halite precipitation continues to lower the total salt concentration. An examination of the salinity changes throughout the water column (see Fig. 7 in Anati and Stiller, 1991) confirms this effect at all depths below the pycnocline.

Leg B. With the beginning of double-diffusive mixing (June–August), the deep water mixes with the warmer, more saline upper-layer water, more or less along their common isopycnal.

Leg C. With the occurrence of the overturn (December), the deep water becomes part of the whole water column and follows it in all its properties: Winter cooling, changes in salinity (in both directions), and chemistry.

The two types of trajectories documented for the upper layer in recent history (Figs. 8-7 and 8-8), and the third type docu-

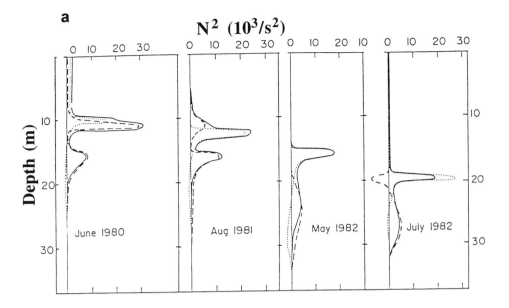

a   **N² (10³/s²)**

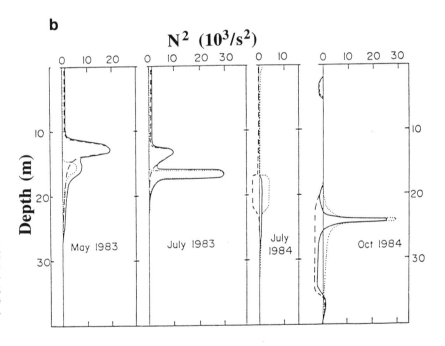

b   **N² (10³/s²)**

Figure 8-12 Examples of Dead Sea local stability N^2 (solid lines) during the period 1980–1984. a, under a meromictic regime; b, under a holomictic regime. The dotted lines mark the contribution of the thermocline and the dashed lines mark the contribution of the halocline.

mented for the deep water (Fig. 8-9) do not necessarily cover all possibilities. Other regimes may eventually occur, and different types of trajectories cannot be ruled out.

The time scale of salt-crystal formations and precipitation from a supersaturated brine is several days, at most, in the laboratory (E. Sass, pers. com., 1992). Nevertheless, salt depletion persisted in the supersaturated Dead Sea deep-layer brines through the relatively long time spans of stable stratification (March–June, Fig. 8-9) during which the layer remained isolated. How long would it have taken to reach its steady state? Deep-water salinities measured in the new meromictic period, which began toward the end of 1991, suggest that the time scale in question could be about 1 year (Fig. 8-10). As of early 1994, more research on the kinetics of this feature is necessary before this suggestion can be accepted.

THE OVERALL STABILITY W

During a meromictic period, the deep-water body is isolated and, as mentioned previously, any interaction with the atmosphere (by means of heat flux, evaporation, wind stress, breaking of surface waves, etc.) is suppressed. The lake's deep waters remain almost unchanged. Any minute alteration observable in their properties (less than 0.5 σu in 1 year, in the case shown in Fig. 8-10) is scrutinized for a possible explanation in terms of some internal process. In addition, from the point of view of the air-sea interaction, in a meromictic period, the lake behaves as a shallow lake in all respects, including seasonal heat-storage and microorganism habitat. It is, therefore, not surprising that there is great interest in forecasting any possible overturn of the water column. The standard indicator for such

Table 8-1 Typical stabilities

	N. Atlantic Ocean	Straits of Tiran	Levantine Basin	Lake Kinneret	Dead Sea 1980	Dead Sea 1992
$\partial\rho/\partial z[\sigma u/m]$	1.1/75	1.2/53	1.4/24	0.94/3	4/1	17.5/0.5
$N^2[10^3/s^2]$	0.14	0.22	0.56	3.1	32	304
τ	8.8 min	7.1 min	4.4 min	1.9 min	35 s	11 s

a forecast is the amount of work the wind must put into the lake (accumulated) to mix it thoroughly; in other words, the W, as defined by equation 1.

Figure 8-2 shows the overall stability W for the period 1979–1988. Figure 8-11 displays the seasonal changes of W for the same 4 years (two rainy and two droughty) represented in Figure 8-3b. The similarity of these curves to those displayed in Figure 8-3b is not casual. Rather, it indicates that a main factor for the realization of a stable stratification is indeed the amount of freshwater flowing into the lake during the rainy season.

After a rainy winter, W reaches its peak early in the season (March–May) when the seasonal thermal structure is not yet maximal but the effect of floods is still fresh. After a droughty winter, the main contribution to W comes from the thermal structure, which reaches its peak in July and August. The minimal W value always occurs in December, when the rainy season, with its accompanying freshwater inflow, normally begins.

A statistical study of the 10-year period 1979-1988 (Anati and Shasha, 1989b) suggests that the rate of decrease of W after the summer peak is fairly steady through the years,

$$\frac{\partial W}{\partial t} = -28 \pm 5 \frac{kJ}{m^2 mo} \tag{5}$$

and that the critical summer peak W value for a meromictic year (i.e., when no overturn of the whole water column will take place) is around

$$W_c = 250 \frac{kJ}{m^2} \tag{6}$$

A summer peak larger than equation 6 would indicate that a meromictic year is to be expected, and a smaller one, that an overturn is to be expected. According to equation 6, therefore, the meromictic regime begun in the winter of 1991-1992 is likely to last at least 2 years, that is until December 1994, or longer if more rainy winters occur by then.

From the curves displayed in Figure 8-11, we can see that the difference in W between rainy years and droughty years (1980 and 1992 vs. 1984 and 1986 in our example) is appreciable. The long-term average W of the Dead Sea overall stability, therefore, is not a good first approximation for any specific year, and consequently cannot constitute a good forecasting guess either. On the other hand, the two rainy years taken together and the two droughty years taken together do not differ dramatically from each other in this parameter. It will be presently shown that, in the local stability N^2, there can be marked differences even among years, such as 1980 and 1992, having similar freshwater inflows (Fig. 8-3b), as well as similar overall stabilities (Fig. 8-11).

THE LOCAL STABILITY N^2

The Brunt-Väisälä frequency N, as defined by equation 2, can be interpreted as the frequency of oscillation of a water parcel displaced vertically within the pycnocline and then released free to move frictionless under its own buoyancy as a restoring force. Its square N^2, is a scaled expression of the density gradient $\partial\rho/\partial z$ (see eq. 2) and therefore a measure of the local (gravitational) stability.

To put the local stability N^2 of the Dead Sea pycnoclines in the right perspective, it is instructive to look first at some typical N^2 values found in other water basins, particularly those located at about the same latitude with roughly the same climate (insolation, air temperature, wind regime, etc.) as the Dead Sea. No formal documentation and references are given, as these are merely examples.

The North Atlantic Ocean has a pycnocline with one of its strongest values just south of Bermuda, with $N^2 \approx 0.14 \times 10^{-3}/s^2$.

The Strait of Tiran has a characteristic two-layer flow with a rather constant pycnocline thickness. The maximal density difference is found around the month of September, when $N^2 \approx 0.22 \times 10^{-3}/s^2$.

The Levantine Basin receives waters of North Atlantic origin and modifies them into the warmer and more saline "Levantine Water." The highest local value of stability there, $N^2 \approx 0.56 \times 10^{-3}/s^2$, is found south of Cyprus in the summer.

Lake Kinneret is a holomictic freshwater basin with its strongest overall stability W in August but its maximal local stability of $N^2 \approx 3.1 \times 10^{-3}/s^2$ in November, at a depth of about 25 m.

Examples of the Dead Sea stability structure $N^2(z)$ in the years 1980–1984 are presented in Figure 8-12, and it is seen that the stabilities are between 10 and 200 times larger than the previous examples. The stabilities presented in Figure 8-12a are similar in magnitude, vertical scale, and structure to those presented in Figure 8-12b, meaning that the local stability N^2 in those years was not much affected by the regime (meromictic or holomictic) or by the total stability W (compare with Fig. 8-2), or by the yearly amounts of freshwater entering the Dead Sea (compare with Fig. 8-3). Note the occasional occurrences of two pycnoclines in the same profile (e.g., June 1980, August 1981, July 1983), representing three-layer structures.

Representative examples of $N^2(z)$ for the 1992 Dead Sea are shown in Figure 8-13, where their maximal values are roughly one order of magnitude greater than those shown in Figure 8-12. The points of measurement are purposely marked to show that often they are not dense enough to ensure the resolution of details, and, since we deal with the vertical gradient (eq. 2), they may also fail to represent the maximal values attained.

The Dead Sea stability, therefore, reaches very high values compared to those found elsewhere in natural basins. In the African lake of Sonachi, values approaching those of the 1980 Dead Sea are reported (MacIntyre and Melack, 1982). In artificially controlled solar ponds, values of almost 200 ($10^{-3}/s^2$) are maintained by human intervention (Tabor and Weinberger, 1981), but the record in a natural basin of this size is probably held by the 1992 Dead Sea, as shown in Figure 8-13.

The examples of this section are summarized in Table 8-1 where τ denotes the natural period of oscillation.

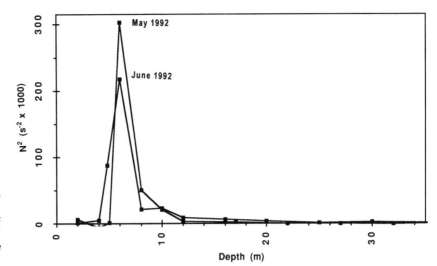

Figure 8-13 Examples of Dead Sea local stability N^2 after the 1992 rainy winter. The contribution of the thermocline is negligible, and the contribution of the halocline alone is nearly the total local stability N^2. The scale of N^2 is one order of magnitude greater than that of Figure 8-12.

DEAD SEA HYDROGRAPHIC CURIOSITIES

The Dead Sea is unique in many respects, so it generally cannot be treated by standard methods. The high stability reported in the previous section is just one example of this. The chemistry of its brines, the short response time scales to meteorological events, and the inadequacy of the Boussinesq approximation all indicate that we should not approach the Dead Sea in routine ways or with preconceptions. After years of acquaintance with the Dead Sea whims, we find there are always new ways by which the Dead Sea is capable of surprise. Two examples will be presented to illustrate how unpredictable a Dead Sea pycnocline can be.

A liquid core

The water sampling in our hydrographic work is performed with a standard sampling bottle, basically a vertical cylinder 40 cm high, open at both ends (upper and lower), with an appropriate mechanism for closing tightly on command at both ends to trap the water inside once the bottle has been lowered to the desired depth. This device has been serving the Dead Sea hydrographic survey for several years and there have been no complaints about its performance.

The standard procedure for measuring the salinity of the sample is to fill two or three 150-cc glass bottles from the sampling bottle and, later in the laboratory, measure the density of their content by pycnometry at a precisely monitored temperature. This procedure also has been used within the framework of the Dead Sea hydrographic survey for several years, with no complaints and good repeatability with those two or three glass bottles.

In 1992, for the first time, the laboratory was unable to reproduce the salinity (density) values of several samples within the standard claimed precision (± 0.02 σu). Repeated pycnometric measurements revealed, however, that all the measured salinities of a certain sample fell in two or three groups, depending on the number of glass bottles filled form the same sampling bottle. The riddle was solved: With a stratification as strong as that of the 1992 Dead Sea, the water inside the cylinder of the sampling bottle retained the ambient stratification. The first glass bottle to be filled, taking the lowest water in the sampling bottle, gave then the highest salinity value. In fact, every subsequent glass bottle gave a lower salinity value than the previous ones filled from the very same sampling bottle.

Those geoscientists who are used to dealing with cores taken from the (solid) bottom of the ocean may think that this liquid core is a normal phenomenon. On the contrary, in standard oceanographic and limnological research in the world's lakes, seas, and oceans, it has not yet been reported to my knowledge, that a 40 cm sampling bottle was too tall for proper resolution of the vertical structure.

An unusual tracer

The turbulence in an open water basin originates mainly at the surface, with the effect of winds and waves, and the turbulent field propagates downward by entraining waters from below into the turbulent field. A virtual bottom in this respect is a barrier; the propagation of turbulence can go as far down as the virtual bottom but no further.

During the spring and early summer of 1980, the Dead Sea developed a three-layer structure, that is, a structure with two pycnoclines (see Fig. 8-12), in which case only the uppermost layer (the one in contact with winds and waves whence turbulence originates) is expected to be turbulent and mixed, whereas the intermediate and deep layers are expected to remain quiescent and therefore in a quasi-laminar state.

During the subsequent months, with the characteristic negative buoyancy flux of later summer and autumn, the density of the uppermost layer was observed to increase gradually toward that of the intermediate layer below, whereas the densities of the intermediate and deep layers, being isolated from surface sources, remained unchanged in the process. The upper pycnocline therefore weakened gradually, and by early September 1980, it ceased to constitute a virtual bottom. The upper and intermediate layers then united, in a so-called minioverturn, creating from both of them a new mixed upper layer of mean temperature, mean salinity, with a combined thickness of both original layers.

This scheme and the concept of virtual bottom found use in the interpretation of some microbiological data. All the technical details and other information about the microbiological nature of the events reported here can be found in Oren and Shilo (1982) and in Oren (chapter 19, this volume). The information relevant to the present section follows.

The green planktonic alga *Dunaliella parva* occasionally populates the upper waters of the Dead Sea. Temperatures and salt concentrations in the 1980 Dead Sea were both rather high but within the *Dunaliella* range of tolerance, and thus, from this aspect, *Dunaliella* cells could have survived at almost any depth. One limiting factor was radiation: Near the Dead Sea surface, the solar radiation at noon is strong enough to kill *Dunaliella* cells exposed to it, and below a depth of about 5 m, the daily mean radiation is too weak for cell growth. One might expect, therefore, the *Dunaliella* cells would be found only in a rather thin euphotic sublayer, say at a depth between 2 and 4 m.

However, observations showed something quite different. In the spring and early summer of 1980, there was a sharp boundary at about a 10-m depth, above which those cells were evenly distributed and below which they were absent. Since this depth coincided with the uppermost thermohalocline of that period (the uppermost pycnocline shown in Fig. 8-12), a tempting explanation could have been environmental, namely, that thermohaline conditions above the critical depth were more favorable than below it. This argument would also explain the sharpness of the transition at a 10-m depth and, to some extent, also the evenness of the *Dunaliella* distribution above it, as temperatures and salinities did not vary much within the upper layer.

With the disappearance of the upper pycnocline (at the minioverturn of September 1980), the *Dunaliella* population remained evenly distributed above a certain depth and absent below it. However, this depth shifted abruptly to a few meters deeper, coinciding with the depth of what had been the lower pycnocline of the spring structure.

If we examine the whole range of temperatures and salinities where *Dunaliella* cells have been found and compare this to the whole range of temperatures and salinities where *Dunaliella* cells have not been found, we soon see that there was enough overlap to rule out environmental explanations. If, on the other hand, we accept that the Dead Sea pycnoclines constitute virtual bottoms and that the particular pycnocline that is closest to the surface obstructs the penetration of turbulence to lower layers but leaves the upper layer in a turbulent state, we can propose a different scheme. Perhaps the *Dunaliella* cells were mixed by turbulence throughout the surface mixed layer (which includes the thin euphotic sublayer) but not permitted to cross its bottom, thus producing an even distribution of the cells above it and their absence below it. This scenario is similar to the previously described observations, whether before or after the minioverturn at which time the surface mixed layer changed (increased) its vertical dimension.

Regarding *Dunaliella* as a live tracer in the so-called Dead Sea, we may learn something about the nature of turbulent mixing. For example, we have learned that in the first (three-layer) period every *Dunaliella* cell must have spent, on the average, at least 20% of its time in the thin euphotic sublayer, with gaps no longer than 1 week.

In the second period (after the minioverturn and the disappearance of the upper pycnocline), the *Dunaliella* conditions were harsher because the added depths were practically dark and vertical distances became greater; the statistical chance of every single cell staying long enough within the thin euphotic sublayer to live, therefore, became slimmer. Indeed, countings show that during this period the *Dunaliella* population density declined exponentially at a rate of 35% per month.

A third period began with the arrival of winter floods. The temperatures and salt concentrations (previously higher than optimal) were both lowered and thus became more favorable to the *Dunaliella* cells. Based on environmental considerations alone, *Dunaliella* would be expected to thrive anew and perhaps start a new, massive bloom. Observation showed quite the opposite: With the arrival of the winter floods, a sharp decline in the *Dunaliella* concentrations was observed, and by early

spring no *Dunaliella* cells could be found.

The turbulent mixing–virtual bottom approach allows us to conclude that, as the new floods created a new shallow pycnocline, the majority of the *Dunaliella* cells remained trapped below the thin euphotic sublayer, causing their extinction by light starvation.

Acknowledgments

Part of the field work, data analyses, and interpretations presented in the present work was accomplished while I was affiliated with Solmat Systems Ltd., and with the Weizmann Institute of Science. Thanks to M. Stiller, A. Oren, and A. Lerman for useful comments and remarks on the first draft of the manuscript. I express my appreciation to skipper Moti Gonen and his skillful crew; the 15-year-old Dead Sea hydrographic survey could not have been as fruitful without their dedication and determination. This work was supported in part by the Israeli Ministry of Energy and Infrastructure, and by the Israel Council for Higher Education.

REFERENCES

Anati, D. A., 1993, How much salt precipitates from the brines of a hypersaline lake? The Dead Sea as a case study: *Geochimica et Cosmochimica Acta*, v. 57, p. 2,191–2,196.

Anati, D. A., Stiller, M., Shasha, S., and Gat, J. R., 1987, Changes in the thermohaline structure of the Dead Sea: *Earth and Planetary Science Letters*, v. 84, p. 109–121.

Anati, D. A., Shasha, S., 1989a, Dead Sea surface-level changes: *Israel Journal of Earth Science*, v. 38, p. 29–32.

Anati, D. A., and Shasha, S., 1989b, The stability of the Dead Sea stratification: *Israel Journal of Earth Science*, v. 38, p. 33–35.

Anati, D. A., and Stiller, M., 1991, The post-1979 thermohaline structure of the Dead Sea and the role of double-diffusive mixing: *Limnology and Oceanography*, v. 36, p. 342–354.

Forschhammer, G., 1865, On the composition of sea-water in different parts of the ocean: *Philosophical Transactions of the Royal Society of London*, v. 155, p. 203–262.

Gavrieli, I., Starinsky, A., and Bein, A., 1989, The solubility of halite as a function of temperature in the highly saline Dead Sea brine system: *Limnology and Oceanography*, v. 34, p. 1,224–1,234.

Kaushansky, P., and Starobinets, S., 1982, A temperature gradient column for the rapid measurement of the density of brines: *Marine Chemistry*, v. 11, p. 289–292.

Kraus, E. B., 1972, *Atmosphere-ocean interaction*: New York, Oxford University Press, 275 p.

Lazar, B., 1982, Geochemical studies of heavy metals in natural waters [Ph.D. thesis]: Jerusalem, The Hebrew University, 174, 166 p.

Levi, Y., Bodenheimmer, Y., and Varshawski, Y., 1984, Optical measurements in the Dead Sea: Jerusalem, College of Technology, 8 p.

MacIntyre, S., and Melack, J. M., 1982, Meromixis in an equatorial African soda lake: *Limnology and Oceanography*, v. 27, p. 595–609.

Marcet, A., 1819, On the specific gravity, and temperature, in different parts of the ocean, and particular seas; with some account on their saline content: *Philosophical Transactions of the Royal Society of London*, v. 109, p. 161-208.

Neev, D., and Emery, K. O., 1967, The Dead Sea: Depositional processes and environments of evaporites: Geological Survey of Israel Bulletin 41, 147 p.

Oren, A., and Shilo, M., 1982, Population dynamics of *Dunaliella parva* in the Dead Sea: *Limnology and Oceanography*, v. 27, p.

201 –211.

Stanhill, G., 1985, An updated energy balance estimate of evaoporation from the Dead Sea: Israel Meteorological Research Papers, v. 4, p. 98-116.

Steinhorn, I., 1981, A hydrographical and physical study of the Dead Sea during the destruction of its long-term meromictic stratification [Ph.D. thesis]: Rehovot, Weizmann Institute, 323 p.

Steinhorn, I., 1985, The disappearance of the long-term meromictic stratification of the Dead Sea: *Limnology and Oceanogra-phy*, v. 30, p. 451 –462.

Steinhorn, I., and Gat, J. R., 1983, The Dead Sea: *Scientific American*, v. 249, p. 102 –109.

Stiller, M., Kaushansky, P., and Carmi, I., 1983, Recent climatic changes recorded by the salinity of pore waters in the Dead Sea sediments: *Hydrobiologia*, v. 103, p. 75 –79.

Tabor, Z. H., and Weinberger, Z., 1981, Non-convective solar ponds, *in*, *Solar energy Handbook*, Kreider, J. F., and Kreith, F., ed.,: Chap. 10: New York, McGraw-Hill, p. 10.1-10.29

UNESCO, 1976, Technical papers in marine sciences, 24.

9. SURFACE CURRENTS AND SEICHES IN THE DEAD SEA

Ziv Sirkes, Florian Schirmer, Heinz-Herman Essen,
and Klaus-Werner Gurgel

SURFACE CURRENTS

Lake currents are mainly driven by wind. A steady wind will generate, after a certain time, a circulation pattern that is unique to a particular lake. Knowing the typical circulation patterns of a lake is important for many different problems such as dispersion of pollutants, erosion and sedimentation at beaches, and drifting of objects and people.

Two approaches are generally used to investigate and to understand the circulation pattern of a lake. In most cases, the research combines both of them. One approach is to use wind measurements, which are representative of the winds above the lake, as input for a numerical circulation model. Results of the model for different wind regimes will give typical circulation patterns of the lake. However, each model uses several parameters that must be calibrated to give satisfactory results. The calibration is usually done by comparing the results of the model with results of measurements, and this is the second approach to understanding lake circulation.

In all lake circulation studies until now only current meters were used to measure current velocities. An intensive measurement campaign would include current measurements at various stations and depths. Such measurements were conducted over a period of 3 years by researchers from the Israel Oceanographic and Limnological Research (Hecht and Ezer, 1982; Hecht and Ezer, 1983; Ezer, 1984; Hecht et al., 1984).

A new remote sensing technique for measuring surface currents has been developed. Called CODAR (COastal raDAR), the technique measures the average current in the upper 0.5 m of a water body on a two-dimensional grid of 1.5 km by 1.5 km over 18-minute time intervals. CODAR measurements have many advantages over current meter measurements. The spatial density of data is much greater and very similar to the results of numerical models, which also give currents on a two-dimensional grid, thus making it easier to compare results. CODAR makes use of the ground wave, which is attached to the sea surface. A ground wave develops as the superposition of the incident and reflected wave, which cancel each other in the case of nonconducting water. Thus, this technique can be applied only to water bodies that are electrically conducting, that is, salt water. Only a few lakes around the world are appropriate for using CODAR. One of these lakes is the Dead Sea.

We took CODAR measurements at the Dead Sea during a 2-week period (November-December 1984). The experiment was carried out continuously for 9 days. These data were complemented by the following: (1) conventional current meter measurements at one station and two depths; (2) wave height measurements taken by a wave-rider buoy; (3) wind measurements at a number of locations on the western shore of the Dead Sea and at one station in the sea; and (4) a profile of salinity and temperature at one station.

Measurement Techniques

Hydrographic Measurements

On November 27, 1984, between 10:30 and 10:44 Israel Standard Time (= UTC + 2 hr), a temperature profile was taken near En Gedi in the central lake (Fig. 9-1; HS), and water samples from different depths were collected for salinity analysis (Anati, 1984). From these two types of measurements, it was possible to calculate the density structure of the Dead Sea at that time (Steinhorn, 1980).

Wind Measurements

Winds are measured routinely by different institutions at several locations on the western shore of the Dead Sea and at one station in the sea, as follows:

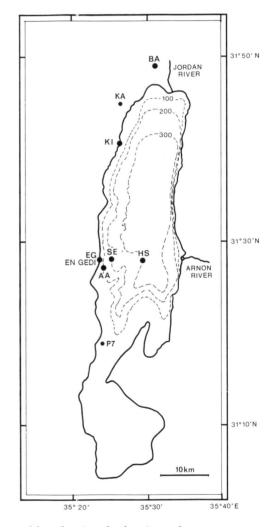

Figure 9-1 Map showing the locations of measurements and the bathymetry of the Dead Sea (after Hall, 1978).BA: Bet Ha'arava—Wind measurements; KA: Kalia—Wind measurements; KI: Kidron—CODAR station and wind measurements; EG: En Gedi—CODAR station; SE: Floating station—Wind measurements; P7: P7-Masada—Lake level measurements; AA: Aanderaa mooring; HS: Hydrographic station. The depth contours in meters, refer to the lake surface.

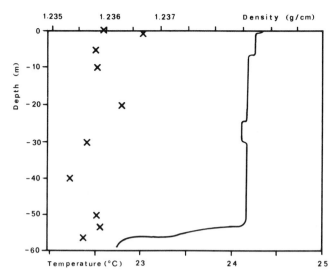

Figure 9-2 Profile of temperature (solid line) and density (crosses) at the hydrographic station (HS) in the central Dead Sea on November 27, 1984 (see Fig. 9-1, HS).

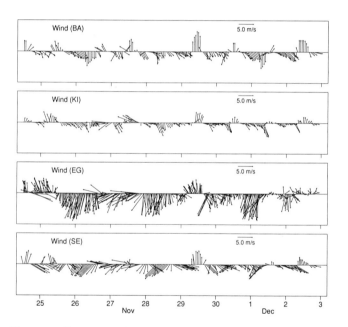

Figure 9-3 Time series of wind vectors measured from November 25 to December 2, 1984. The measuring positions are given in Figure 9-1.

1) The Israel Meteorological Service maintained a wind station at Bet Ha'arava (Fig. 9-1; BA). Wind speed and direction were measured with a Woelfle anemometer. This instrument records analog data on a continuous strip of paper at a rate of 1 cm/hr. The paper strip is analyzed manually, resulting in a time series of hourly averages of wind speeds and directions.
2) Prof. G. Stanhill from the Volcani Center maintained an automatic weather station at Kidron (Fig. 9-1; KI), approximately 400 m from the northern CODAR station. Wind speeds and direction were measured and hourly averages recorded on magnetic tape.
3) The Israel Navy maintained a Woelfle anemometer station at En Gedi (Fig. 9-1; EG), not far from the southern CODAR station.

(4) A research group from the Meteorology Department of the Hebrew University, Jerusalem, maintained a floating automatic weather station off En Gedi at a water depth of 200 m (Fig. 9-1; SE). Half-hourly averages of wind speed and direction are recorded on magnetic tape.

All measurements were taken 3 m above the ground except the measurements at sea, which were taken 7 m above the sea surface.

Wave-rider Buoy

The wave-rider buoy measured wave heights every 2 hours (Fig. 9-1, HS). The data were transmitted to the station at Kidron, where the respective spectra were computed and the output was printed on paper.

Aanderaa Current Meters

Three Aanderaa current meters at depths of 5 m, 10 m, and 40 m were moored off En Gedi at a water depth of 100 m (Fig. 9-1; AA). They were deployed on November 25 and recovered on December 2, 1984. During recovery, we discovered that the lowest current meter was entangled in the ropes and had not worked at all. The rotor of the current meter from a 10-m depth was covered with salt and stuck. Data analysis showed later that it had worked properly until shortly before its recovery. The rotor of the uppermost current meter was also partly covered with salt, but it still rotated freely.

CODAR Measurements

CODAR was developed for measuring surface currents in coastal areas. The system is based on the physics of Bragg scattering. An electromagnetic wave is backscattered from surface waves of half its wavelength running toward or away from the radar site. The Doppler shift of the received signal is determined by the phase velocity of the scattering surface wave, which is known from the dispersion relation. Deviations from this theoretical value are assumed to be due to underlying currents. A detailed discussion of the theoretical background and the realization of the measuring system of CODAR is given by Barrick et al. (1977).

CODAR has been operated successfully in several experiments (Essen et al., 1984; Essen et al., 1989) conducted at different locations by researchers from the University of Hamburg. CODAR uses a frequency of 29.85 MHz, which corresponds to a wavelength of about 10 m. One CODAR site measures radial current velocities with respect to its position. Two distinct CODAR sites are necessary to obtain two-dimensional currents. In general, the range of each site is between 30 and 50 km, depending on sea state. Although the surface of the Dead Sea sometimes looked very calm, we always got return signals. But in a few cases the range was only 20 km. The two CODAR stations were located at Kidron (Fig. 9-1; KI) and En Gedi (Fig. 9-1; EG). To obtain a high spatial resolution (1.2 km), 8-μs transmit pulses were used instead of 16-μs pulses, as in the previously cited experiments.

Data Analysis and Results

Hydrographic Measurements

Figure 9-2 displays a profile of temperature and density in the upper 60 m of the central Dead Sea on November 27, 1984 (see Fig. 9-1; HS). There is no stratification, so the Dead Sea was homogeneous, at least in the upper 60 m, during the 2-week experiment period.

Wind Measurements

A comparison of the time series at Bet Ha'arava, Kidron, En Gedi, and the floating station off En Gedi (see Fig. 9-1, BA, KI,

EG, SE, respectively) is shown in Figure 9-3, which reveals the following findings:

(1) The wind patterns in BA and KI are very similar in wind direction, which is almost the same for both most of the time.

(2) Winds in the north (BA, KI) are not in the same directions as winds in the south (EG, SE). This is true especially for the east-west component of the wind. There is a moderate correlation between the winds in the north and in the south for the north-south component of the wind.

(3) There is often a discrepancy in wind direction between EG and SE, even though these stations are less than 4 km apart. The wind at the sea station is somewhat higher than that over land; this may partly be due to the different heights at which the data were collected.

Wave Measurements

The wave-rider measurements were collected for the purpose of obtaining a qualitative idea of the wave and wind conditions at the Dead Sea. The wave-rider buoy was calibrated for Atlantic Ocean conditions and therefore, did not give good quantitative results in the Dead Sea; however, we believe that the relative amplitudes are correct.

Figure 9-4 displays the time series obtained from data analysis of the wave-rider buoy measurements. The time series show five occurrences of relatively high waves during our experiment. The times of occurrence of higher waves coincide with relatively higher wind speed, as measured at the floating station. For a more in depth discussion on wave measurements, see discussion by Hecht et al. (chapter 10, this volume).

Currents Measured by Aanderaa Current Meters

Figure 9-5 displays time series of wind at the floating station (see Fig. 9-1; SE) and current velocities as measured by CODAR and Aanderaa current meters at 5-m and 10-m depths (see Fig. 9-1; AA). The CODAR measurements will be discussed subsequently. With respect to the subsurface currents, some interesting features can be seen immediately from this figure.

1) There is a high correlation between the current speed at the two depths, as well as a good correlation between the currents and the wind velocities.

2) Except for a very short period on November 26, the velocity at the 5-m depth is always equal to or greater than the velocity at the 10-m depth. This finding is in accordance with the prediction of the simple Ekman theory.

3) The current meters record a velocity of 1.5 cm/sec even when no current exists. During the measurement period, there were three occurrences of very low wind speeds. On November 27, this occurrence was quite brief. The current speed of both current meters dropped considerably, but because it was still much higher than the threshold of the current meters, the rotors continued to turn. On November 30, the wind ceased for a longer period, thus allowing the current meter at 10 m to come to a standstill. However, it is evident from Figure 9-5, that the rotor turned again after the wind started to increase. The shape of the time series is highly correlated to the shape of the time series of the other current meter, which had not come to a standstill, and to the shape of the time series of the wind. Only on December 1 did the rotor of this current meter come to a final standstill, from which it did not recover.

The high correlations between the subsurface speeds at 5 m and 10 m and the wind (r = 0.79 and 0.71, respectively), as well as between themselves (r = 0.78), is evident in Figure 9-6. There is a better correlation between wind speed and current at a depth of 5 m than at 10 m. The dependence for both is nearly linear. These high correlations should be expected, since we are observing a wind-induced circulation, in general, and comparing the measurements of wind and current from two very close

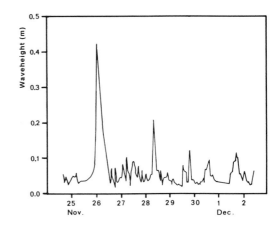

Figure 9-4 Time series of maximal wave heights measured from November 25 to December 2, 1984.

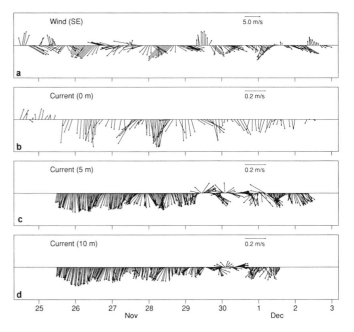

Figure 9-5 Time series of wind and current vectors. A, wind at the floating station (see Fig. 9-1, SE); B, current velocities measured by CODAR at Aanderaa mooring; C, current velocities measured by Aanderaa at a 5 m depth and; d, current velocities measured by Aanderaa at a 10 m depth (see Fig. 9-1, AA).

locations. It is also well known that current speed decreases with depth in wind-driven circulation.

Regarding north and east components separately, we see a distinct behavior in the time series for both components. From the beginning of the Aanderaa records until November 29, the wind was blowing from the north most of the time. The east component is very irregular and seems to be distributed randomly around zero. There is a reasonable correlation between the north components of subsurface currents at 5 m and 10 m and wind (r = 0.47 and 0.46, respectively). This is not true for the east components, where the currents are often in the opposite direction from wind (r = 0.27 and 0.19, respectively). Both north and east components of subsurface currents at 5 m are highly correlated to those at 10 m (r = 0.86 and 0.62, respectively).

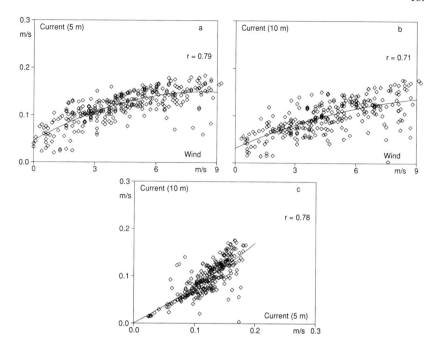

Figure 9-6 Correlation between speeds of Aanderaa current meters and wind. A, Aanderaa 5 m versus wind; B, Aanderaa 10 m versus wind; C, Aanderaa 10 m versus Aanderaa 5 m.

At approximately midday on November 29, the features of the time series changed suddenly (Fig. 9-5); accordingly, the east components of wind and subsurface currents are highly correlated. This change lasted until the end of November 30, followed by a period of a few hours in which the reading of the direction in both current meters was undetermined. Current directions were more random for the measurement at the 10 m depth than for that at 5 m. We do not have any explanation for this peculiar behavior. The measurements thereafter showed that the currents followed the wind with high correlation in the north components.

One explanation of the generally higher correlation between wind and current in the north component compared to the east component is the location of the mooring, close to a north-south running coast. Winds blowing from west have only a short fetch, and westward-directed currents are restrained by the coast.

Surface Currents Measured by CODAR

The CODAR measurements were taken over 9 days, a minimum of 12 times each day. The current velocities are mean values, averaged over 18 minutes in time (measuring time), vertically over about 0.5 m from the surface (as deep as the scattering waves sense currents), and horizontally within a circle of radius 2 km (depending on pulse length and processing algorithms). Each CODAR measurement yields a map of two-dimensional currents.

From the whole set of measurements, time series may be derived for each point in the coverage area. One example is presented in Figure 9-5, which shows the CODAR-measured surface currents within the averaging circle around the current-meter mooring. The correlations of the corresponding CODAR-measured speeds to the wind and subsurface speeds measured by the Aanderaa current meters are given in Figure 9-7. The correlation between CODAR-measured current and wind speeds ($r = 0.44$) is lower than those of subsurface currents and wind (see Fig. 9-6), and the correlation between surface and subsurface speeds is relatively low ($r = 0.40$, and $r = 0.30$, respectively). But, considering the components separately, the north component of surface currents (CODAR) is more highly correlated to wind ($r = 0.76$) than that of subsurface currents at 5 m and 10 m ($r = 0.47$, and $r = 0.46$, respectively). The

correlation of surface and subsurface currents is on the order of 0.5. The respective values of the east component are considerably lower. Care should be taken in comparing CODAR and current meter data, because the first represent horizontal averages and the second, point measurements, which may be different, especially close to the coast.

We present current maps, concentrating on a few interesting "episodes" of circulation patterns in chronological order. Because the grid distance is 1.5 km, adjacent current vectors are not independent of each other, as there is some overlap of the averaging areas.

The sequence in Figure 9-8 shows the typical circulation pattern in the Dead Sea during the first 3 days of our measurement campaign (November 25–27). Two different flow regions exist in the Dead Sea. The south-southwesterly directed currents, as visible in the lower maps of Figure 9-8, are present also for the following 2 days. The currents in the northern part change continuously in a clockwise direction. During the early morning (4 UTC = 6 IST), the current is directed southwestward. At noon (10 UTC), there is a frontal structure with main current directed eastward. Six hours later (16 UTC), the current changes to the south, and then during the night (22 UTC), again to the southwest. On the night of November 25/26, we recorded the strongest winds and highest waves during our measurements (see Figs. 9-3, 9-4), and the current velocities are correspondingly high.

Current maps from November 26 and 27 are shown in Figure 9-9. The two maps of each day are separated by only 1 hour and show the variability during this short time. The circulation pattern from November 26 (Fig. 9-9a, b) occurred very often during our campaign. We would like to call it the most frequent circulation pattern in the Dead Sea. Peculiarly, the observed wind measurement that accompanies this picture is always the same, at least in direction. However, the wind directions at Bet Ha'arava and Kidron (see Fig.9-3; BA, KI) are, although consistent between themselves, perpendicular to the currents in the northern part of the Dead Sea. In the southern part, there is good agreement between the current pattern measured with CODAR and the results from the Aanderaa current meter at 5 m (see Fig. 9-5). The direction of both current measurements is nearly parallel to the wind direction.

Another episode of circulation patterns is shown in Figure 9-

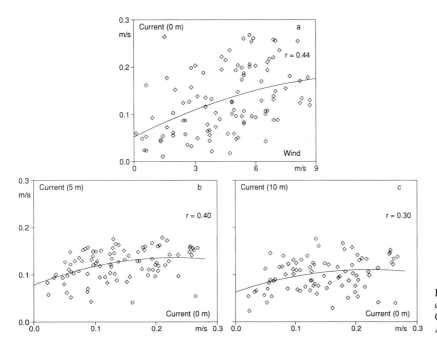

Figure 9-7 Correlation between speeds of Aanderaa current meters and wind. A, CODAR versus wind; B, CODAR versus Aanderaa 5 m; C, CODAR versus Aanderaa 10 m.

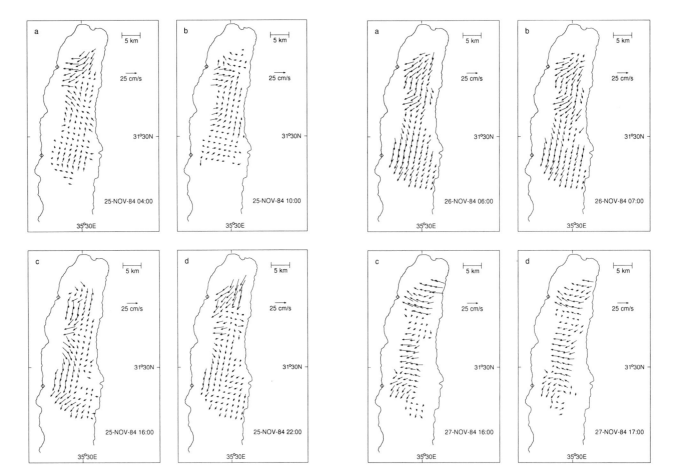

Figure 9-8   Maps of currents measured by CODAR. Six-hour maps from November 25, 1984. Times are in UTC.

Figure 9-9   Maps of currents measured by CODAR.   One-hour maps from November 26 and 27, 1984. Times are in UTC.

9 c, d. The similarity in wind direction at Kidron, En Gedi, and the floating stations (see Fig. 9-3; KI, EG, SE) suggests an easterly wind over the whole Dead Sea. This wind generates westward flow over major parts of the Dead Sea with an anticyclonic gyre in its northern part. This gyre remains in place for many hours until the wind changes direction (at 22 UTC).

The most interesting sequence during our whole measurement period is depicted in Figure 9-10, which consists of current maps from November 29. In the morning (at 8 UTC), no predominant flow pattern could be detected (Fig. 9-10), only gyres on the longitudinal axis of the lake are visible. Two hours later (Fig. 9-10b), there suddenly is a very distinct flow pattern over the whole lake. In the north, currents are northward (parallel to the wind at Kidron). In the southern part, the currents are mainly to the east and turn to the left on the eastern boundary. At the floating station (see Fig. 9-1; SE), the surface currents are nearly in the opposite direction to the wind and to the Aanderaa measured currents (see Fig. 9-5). The flow pattern continues to develop with time. For the next 6 hours, winds are northward at all stations, and the current pattern in the south remains remarkably unchanged while the circulation in the northern part turns clockwise (not shown).

Around 16 UTC, the wind changes to the northwest, causing the whole flow to follow (Fig. 9-10c). The northeastern edge of the Dead Sea displays a southwesterly flow, which, in our opinion, is a real phenomenon because it continues on the next map. At 18 UTC, the wind direction remains unchanged in the northern part of the lake, while it has turned clockwise in the south. The currents have also changed direction (Fig. 9-10d). Nearly all over the lake, the currents are now to the south; the southwesterly flow in the northeastern edge has advanced downstream. Finally, at 20 UTC, this flow covers a significant portion of the northern part of the Dead Sea. The flow in the other regions is to the south.

During weak winds, the flow pattern in the Dead Sea is dominated by gyres, which indicate relaxation phenomena of the basin (Fig. 9-11).

SEICHES

Internal oscillations (seiches) are a well-known phenomenon in freshwater lakes during summer stratification. Their periods and amplitudes are usually measured by thermistor chains, which are lowered to the depth of the thermocline (Mortimer, 1974; Kanari, 1975). Current measurements are also often used to verify calculated periods of internal seiches (Schwab, 1977; Oman, 1982; Hutter et al., 1983).

The amplitude of internal seiches in a two-layered lake has its maximum at the density interface and decreases exponentially away from it (Krauss, 1966). Its manifestation at the surface is related to the amplitude at the pycnocline and is proportional to the density difference across the pycnocline.

The density difference across the pycnocline (thermocline) in freshwater lakes is of the order of 10^{-3} g/l, so that even though the amplitude of internal seiches can reach more than 10 m, its manifestation at the surface barely exceeds 1 cm. Since the amplitude of surface seiches is in most cases a few centimeters, it is very hard to detect internal oscillations by means of lake level measurements.

The density of water often reaches high values in saline lakes. Therefore, in a highly saline, stratified lake the potential for larger density differences between the layers is significantly greater, especially when the lake inflow is composed primarily of freshwater. As a consequence, the amplitudes of internal oscillations at the surface can reach higher values in saline lakes

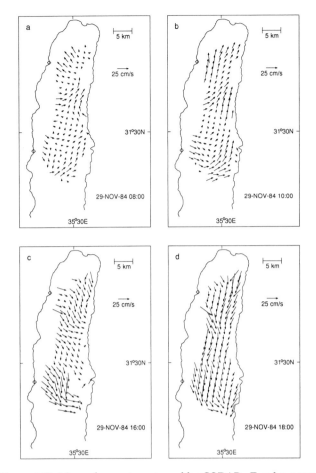

Figure 9-10 Maps of currents measured by CODAR. Two-hour maps from November 29, 1984. Times are in UTC.

than in freshwater lakes for a given amplitude at the pycnocline. The existence of strong density stratification during some time spans and the lack of it during the remaining time makes the Dead Sea an optimal site for the study of surface manifestations of internal seiches.

Measurements of seiches

Wind directions and speeds were measured at the northern end of the Dead Sea by the Israel Meteorological Service with a Woelfle anemometer at a height of 3 m above the water surface. The instrument recorded wind directions and speeds on a paper strip at a rate of 10 mm/hr. The station was at Kalia (Fig. 9-1; KA) until November 1982, after which it was moved to Bet Ha'arava (Fig. 9-1; BA).

Lake level was measured at the southern end of the northern basin opposite Masada (Fig. 9-1; P7), using a strip-chart recorder developed by the Israel Hydrological Service.

Since most lakes are freshwater where temperature alone contributes to density changes, the oscillations of the pycnocline (thermocline) are usually measured with thermistor chains. One of the thermistors is generally located in the thermocline zone and directly monitors the thermocline oscillations. Temperature was measured at En Gedi 100 for the time period July 5–9, 1981 with a thermistor chain (Ben-Yaakov, 1983). These measurements were taken only 2 weeks before the data reported here and were conducted under similar hydrographic and wind conditions.

Data analysis and results

The wind regime during the summer of 1981 was very regular (Fig. 9-12), mainly as a result of the daily Mediterranean Sea breeze. Time series of hourly mean values of the north and east component of wind velocity during 60 days (July 21–September 19, 1981) were used for spectral analysis. The Fourier spectral density estimate from these time series shows major peaks at periods of 24 and 12 hours, accompanied by minor peaks at higher harmonics (8, 6, and 4.8 hours).

The wind regime during the autumn of 1984 was less regular than during the summer. The Mediterranean Sea breeze almost completely disappeared, and the winds were influenced by local weather. The spectral density estimate of the east-west component contained only a diurnal component, which is less dominant than in summer, whereas the north-south component contained a diurnal and a semidiurnal period; but both are less distinct than during the summer of 1981.

Figure 9-13a displays a typical 3-day limnogram for the summer of 1981. Phenomena with periods longer than 3 hours have larger amplitudes than oscillations of short periods (surface seiches and gravity waves). The most striking feature in this time series is the rapid drop of lake level with time (Fig. 9-14). The average rate of retreat, caused by an excess of evaporation over inflow, was 5.4 mm/day. The detrended time series (not shown) shows a daily cycle whose amplitude is modulated (beats)—an indication of the existence of two oscillations with close periods. The smoothed spectral density estimate (Fig. 9-15) reveals a dominant peak of diurnal period and several other peaks of shorter period.

Figure 9-13b shows a typical 3-day limnogram from the recording of October 22–December 12, 1984. The most significant oscillations visible have a period of 34 min, which is the period of the first mode of surface seiches; no long-period oscillations can be seen (see Fig. 9-13a). During this time, the mean rate of drop in lake level was 2.3 mm/day, which is about half the value of the summer 1981. A time series of 15 days representative of the autumn of 1984 (November 18, 1984) was detrended and submitted to spectral analysis. Its spectrum revealed a diurnal cycle and periods of surface seiches. Although the time series was long enough to separate a 24-hr period from one of 21 hours (Jenkins and Watts, 1968), the latter is not apparent in the spectrum.

Figure 9-16 shows the time series from three thermistors. The thermistor at a depth of 10.75 m was obviously located in the thermocline; it reveals peak-to-peak variations of up to 13°C. Usually, time series of isotherm depth are used for spectral analysis instead of the original time series of temperature at constant depth. Figure 9-17 gives the spectrum of the time series of isotherm depth, obtained by Maximum Entropy Spectral Analysis (MESA), with an indication of the "worst-case" shift from the correct frequency (Marmorino and Mortimer, 1978). Phenomena with periods around 80, 20.5, and 11.4 hours seem to be significant at the 95% level; shorter periods are slightly below the 95% confidence level (peaks at 25.4 and 13.2 hours are not significant).

A numerical model of internal seiches

A numerical model for predicting periods and relative amplitudes in the Dead Sea (Rao and Schwab, 1976; Schwab, 1977) was used to simulate a two-layered lake of constant depth taking into account the rotation of the earth. Periods of internal seiches, calculated by this model, are almost identical to those obtained by spectral analysis of the measurements of lake levels and from the thermistor chain measurements (Table 9-1).

INTERPRETATION AND DISCUSSION

The goal of research in Dead Sea circulation is to understand the circulation pattern of the lake as a function of the wind regimes. A comparison of the measured winds on the western shore with the measured current patterns indicates that quite often the currents are perpendicular to the measured winds (on the shore) in the northern part of the Dead Sea.

Wind directions nearly perpendicular to surface currents suggests that the three wind stations on the western shore of the Dead Sea do not always represent the entire wind regime over the Dead Sea. We therefore consulted meteorologists from different institutions in Israel; they confirmed our suspicions and offered explanations to the aforementioned problems.

We conclude that the winds in the Dead Sea are influenced by four mechanisms, as follows:

1) the large-scale synoptic conditions over the region,
2) the Mediterranean Sea breeze, which is caused by differential heating and cooling of the Mediterranean Sea compared to the land,

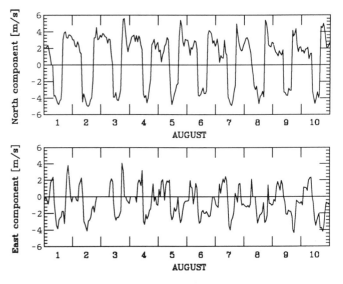

Figure 9-11 Maps of currents measured by CODAR. Two-hour maps from December 1, 1984. Times are in UTC.

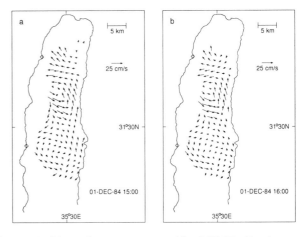

Figure 9-12 North and east components of wind at Kalia, August 1–10, 1981.

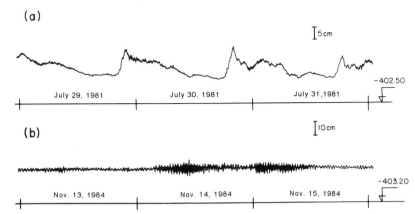

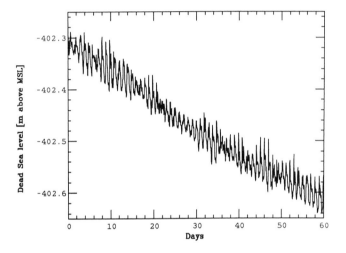

Figure 9-13 Dead Sea limnograms.

Figure 9-14 Time series of lake level, July 21–September 19, 1981.

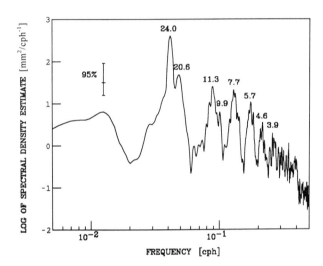

Figure 9-15 Spectral density estimate of fluctuations in lake level, summer 1981.

3) the local sea breeze of the Dead Sea, whose mechanism is similar to that of the Mediterranean Sea breeze, and

4) the catabatic winds at the Dead Sea, which develop when the slopes of the hills surrounding the Dead Sea cool down after sunset. A layer of a few ten meters of air cools, gets denser (and therefore heavier), and starts to slide downhill. The speed of these winds is dependent on the steepness and length of the slope. The hills on the northeastern part of the Dead Sea are higher and farther away from the Dead Sea than the hills on the west. This is why the catabatic winds from the northeast are much stronger than the winds from the northwest. Measurements on the west shore of the Dead Sea show northwesterly winds, whereas most of the northern part of the sea is influenced by the catabatic winds from the northeast. Anabatic winds occur when the hill slopes are heated during the day and result in winds diverging from the Dead Sea. Anabatic and catabatic winds can also be used to explain the discrepancy in wind direction between En Gedi and the floating station off En Gedi, since En Gedi is much more influenced by the topography of its surroundings.

It is clear from this discussion that the three wind stations at the shore of the Dead Sea often do not give a reliable picture of the wind regime over the entire Dead Sea. Therefore, these wind regimes cannot be compared with the current maps, and they cannot serve as input for a numerical model to simulate the measured currents throughout the Dead Sea. The wind

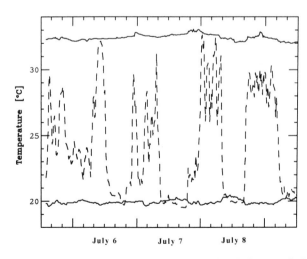

Figure 9-16 Time series of three thermistors at En Gedi 100, July 5–9, 1981. Thermistors were at depths of 5.25, 10.75, and 16.75 m.

Table 9-1 Periods (in hours) of internal seiches in the Dead Sea

Mode	Spectral analysis of lake levels	Thermistor chains	Two-dimensional numerical model of seiches
1	20.6	20.5	21.3
2	11.3	11.4	11.5
3	9.9	8.8	9.9
4	7.7	7.3	8.0
5	—	6.3	6.3
6	5.7	5.4	5.7

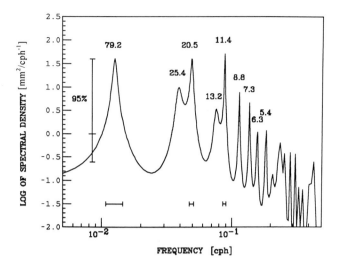

Figure 9-17 MESA spectrum of thermocline oscillations.

measurements at the floating station are more representative, however, but only for the southern part of the Dead Sea.

The irrelevance of the wind data also makes the interpretation of the current maps more difficult. Nevertheless, some important features have been learned about the circulation pattern in the Dead Sea:

(1) Different flow patterns exist in the northern and southern part of the northern basin of the Dead Sea. In the south, the flow is frequently to the south, whereas in the north, it tends to rotate clockwise with time.

(2) There exists a rather typical night circulation, with currents in the south flowing to the south and currents in the north flowing to the southwest.

(3) After the wind stops, the currents do not disappear. Instead, gyres emerge as a result of the relaxation processes in the lake.

(4) A maximum velocity of 30 cm/sec was measured in the central part of the Dead Sea.

Manifestations of internal seiches at the surface of the Dead Sea during stratification have been confirmed by comparing lake level data from the summer of 1981 with thermistor chain measurements and with a numerical model of internal seiches. Furthermore, these periods did not exist in the wind data and therefore were not caused by direct wind forcing. The absence of periods of internal seiches in lake level data during a vertically homogeneous regime in the Dead Sea further confirms that the periods in lake level data seen in the summer of 1981 were caused by internal seiches.

Acknowledgments

Our special thanks are extended to Prof. J. R. Gat for his help and advice throughout this research. We thank Prof. J. Stanhill, Prof. A. Cohen, the Israel Meteorological Service, and the Israeli Navy for the wind data, and Dr. D. A. Anati for hydrological data. This work was supported by the Israel Ministry of Energy and Infrastructure under Contract 1828, by a grant of the Mediterranean-Dead Sea Company and by the German Science Foundation (DFG, Sonderforschungsbereich 94).

REFERENCES

Anati, D. A., 1984, The top 50 metres of the Dead Sea: Solmat Publication No. 28, 11 p.

Barrick, D. E., Evans, M. W., and Weber, B. L., 1977, Ocean surface currents mapped by radar: *Science*, v. 198, p. 138–144.

Ben-Yaakov, S., 1983, Oxygen, temperature and light intensity in the Dead Sea: Internal Report to the Ministry for Energy and Infrastructure, 8 p. (in Hebrew).

Essen, H.-H., Freygang, T., Gurgel, K.-W., and Schirmer, F.,

1984, Oberflächenströmungen vor Sylt—Radarmessungen im Herbst 1983: *Deutsche Hydrographische Zeitschrift*, v. 37, p. 201–215.

Essen, H.-H., Gurgel, K.-W., and Schirmer, F., 1989, Surface currents in the Norwegian Channel measured by radar in March 1985: *Tellus*, v. 41A, p. 162–174.

Ezer, T., 1984, Simulation of Drift Currents in the Dead Sea [M.S. thesis]: Jerusalem, The Hebrew University, 97 p. (in Hebrew with English abstract).

Hall, J. K., 1978, Dead Sea Geophysical Survey, Bathymetric Chart; Marine Geology Division, Geological Survey of Israel.

Hecht, A., and Ezer, T., 1982, Measurements of currents and waves in the Dead Sea: Israel Oceanographic and Limnological Research Reports, Series 1, H 7/82, 13 p. (in Hebrew).

Hecht, A., and Ezer, T., 1983, Measurements of currents, temperatures and electrical conductivity in the Dead Sea: Israel Oceanographic and Limnological Research Reports, Series 1, H 2/83, 5 p. (in Hebrew).

Hecht, A., Ezer, T., and Mandelzweig, R., 1984, Currents, temperatures and meteorology in the Dead Sea—Summary report for the third year of research; Israel Oceanographic and Limnological Research Reports, Series 1, H 6/84, 9 p. (in Hebrew).

Hutter, K., Salvade, G., and Schwab, D. J., 1983, On internal wave dynamics in the northern basin of the Lake of Lugano: *Geophysical and Astrophysical Fluid Dynamics*, v. 27, p. 299–336.

Jenkins, G. M., and Watts, D. G., 1968, *Spectral analysis and its applications*: San Francisco, U.S.A., Holden-Day, 525 p.

Kanari, S., 1975, The long period internal waves in Lake Biwa: *Limnology and Oceanography*, v. 20, p. 544–553.

Krauss, W,. 1966, *Methoden und Ergebnisse der Theoretischen Ozeanographie, Interne Wellen*: Berlin, Germany, Gebrüder Borntraeger, 248 p.

Marmorino, G. O., and Mortimer, C. H., 1978, Internal waves observed in Lake Ontario during the International Field Year for the Great Lakes (IFYGL) 1972: 2. Spectral analysis and modal decomposition: Center for Great Lakes Studies, University of Wisconsin, Milwaukee, Special Report 33.

Mortimer, C. H., 1974, Lake hydrodynamics: *Communications of the International Association for Theoretical and Applied Limnology*, v. 20, p. 124–197.

Oman, G., 1982, Das Verhalten des geschichteten Zürichsees

unter aeusseren Windlasten: ETH Zürich, Mitteilungen der Versuchsanstalt für Wasserbau, *Hydrologie und Glaziologie*, v. 60.

Rao, D. B., and Schwab, D. J., 1976, Two-dimensional normal modes in arbitrary enclosed basins on a rotating earth. Applications to Lakes Ontario and Superior: *Philosophical Transac-tions of the Royal Society of London*, Series A, v. 218 p. 63–96.

Schwab, D. J., 1977, Internal free oscillations in Lake Ontario: *Limnology and Oceanography*, v. 22, p. 700–708.

Steinhorn, I., 1980, The density of the Dead Sea water column as a function of temperature and salt concentration: *Israel Journal of Earth Sciences*, v. 29, p. 191–196.

10. WIND WAVES ON THE DEAD SEA

ARTUR HECHT, TAL EZER, AVRAHAM HUSS, AND AVIV SHAPIRA

There is very little documented information on wind waves on the Dead Sea. The information we have was contributed by Neev and Emery (1967), who based their report solely on visual observations. In general, they report a diurnal cycle during which the sea is calm and mirror-like until about 1000 hours. This is followed by a gradual intensification of the wave field, which reaches its peak at about 1500 hours and then dies out to become calm again at about 2000 hours. According to their report, northerly storms in the autumn, winter, and spring produced higher waves, which reached a maximum height of 1.1 m, a length of 10 m, and a period of a few seconds. They compare the wave field on the Dead Sea to that on Lake Kinneret, where, for similar winds but much shorter fetches, waves of 2 or even 3 m high were reported. Neev and Emery (1967) attribute this paradox to the higher density of the Dead Sea.

Direct wave measurements on the Dead Sea were taken, for the first time from October 1982 to January 1983. These measurements were taken by the Israel Oceanographic and Limnological Research Ltd. (IOLR) and are the subject of the present investigation. Subsequent wave, wind and current data are presented in Sirkes et al. (chapter 9, this volume).

DATA ACQUISITION AND ANALYSIS

Wave heights were measured with an Environmental Devices Corporation (ENDECO) buoy deployed in 45 m of water, about 3.5 km east of the mouth of the Mishmar River (Fig. 10-1). The buoy measured vertical accelerations induced by the waves, integrated the measurements twice, and transmitted the resulting wave heights ashore, where they were recorded. The buoy did not measure wave directions. The accuracy and characteristics of the ENDECO wave-measuring buoy are described in Middleton et al. (1976) and Brainard (1980). The resolution and accuracy of the measurements were ±6 cm. The buoy did not respond significantly to waves of less than 10 cm in height; therefore, our definition of calm seas included waves of up to 10 cm high. The response of the buoy to waves with periods of 3.3 to 14.3 s did not affect the measurement of the wave height. For waves with periods of less than 3.3 s, wave amplitudes were attenuated. The attenuation was as much as 7 dB (that is, 50%) for 1.5-second waves, and as much as 12 dB (that is, 75%) for 1-second waves. The attenuation was corrected according to tables provided by the manufacturer. However, our estimates indicate that the contribution of waves shorter than 2 s was insignificant. The buoy response also attenuates the amplitudes of waves with periods larger than 14 s. However, this is irrelevant because, as we will show, such waves could not possibly develop on the Dead Sea.

We carried out two series of wave measurements. During the first series, October 19, 1982, to January 4, 1983, the analog data from the ENDECO buoy were recorded for 5 minutes every 4 hours. From February 1983 to November 1983, the data from the same buoy were recorded digitally for 10 minutes every hour. The length of the record and the intervals between records were a compromise determined by the minimum length and time interval required for a reasonable analysis ver-

sus the capability of the available recording instrument (the analog paper recorder being far more limited than the digital magnetic tape).

As far as the analog data are concerned, the diurnal distribution of the measurements, the length of each record, and the length of the entire series are not suitable for sophisticated mathematical analysis and for a full description of the wave climate in the Dead Sea. Moreover, there are gaps in the recordings resulting from power failures on shore at the recording station. Far more data were collected during the series of digital measurements; however, we were still plagued by power failures and transmission errors. Unfortunately, some of the power failures occurred during storms, when wave development was particularly interesting.

Some of the transmission errors resulted in wave records that were obviously impossible, such as wave heights larger than 10 m. It was easy to weed out and ignore such erroneous records. However, some marginal occurrences were more difficult to decide on. For instance, although everyone would agree that a 10-m wave could not develop on the Dead Sea, what about a 4-m wave? We felt it was necessary to formulate a criterion to eliminate records showing waves that were not obviously wrong but were practically impossible. We decided to base this criterion on the wind climatology of the Dead Sea and the wave prediction method of the Coastal Engineering Research Center Shore Protection Manual, CERC-SPM (Coastal Engineering Research Center, 1984).

According to Figure 20 in Neev and Emery (1967), the strongest winds, 5.5 m/s, were observed during the evening hours of June, July, and August, whereas toward the end of the year and the beginning of winter, the wind maxima diminished to about 2.5 m/s. These results were based on winds measured inland at the northern and southern ends of the Dead Sea, which may be affected by local topography and may not be representative of the wind field at sea. The only wind measurements concurrent with wave measurements were acquired on shore at En Gedi. However, these measurements, in addition to being ashore, were sporadic and sometimes inaccurate. These measurements indicated maximum wind speeds of at most 8 m/s.

Between January 1984 and November 1988, the IOLR measured meteorological parameters at sea. These measurements were acquired by Aanderaa transducers, from a buoy located about 3.5 km east of En Gedi (Fig. 10-1). The meteorological data were averaged over 30-minute intervals and organized into a continuous time series. Technical difficulties caused many interruptions in the continuity of these measurements and, at present, not all the data are available. However, in spite of its limitations, this is the longest and the most detailed set of "at sea" meteorological measurements presently available. Some of the data were interpreted and reported by Weiss et al. (1987), as well as Weiss and Cohen (1988). In the present investigation, we will use some of this meteorological data (i.e., the winds at two levels, at 3.5 m and 6 m about the sea surface, air temperatures, and sea surface temperatures) for climatological information relevant to the analysis of the waves measured on

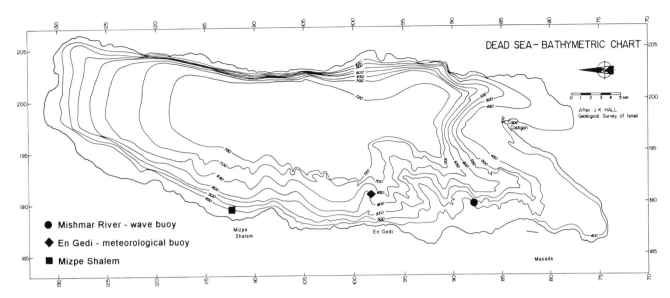

Figure 10-1 Location of the wave buoy off the Mishmar River bed and the meteorological station off En Gedi.

the Dead Sea. The information on the mast height enabled us to compute wind factors compensated for a standard height. The simultaneous air and sea temperature measurements enabled us to compute the air instability factor as required for the wave prediction nomogram of the CERC-SPM (Coastal Engineering Research Center, 1984).

These meteorological data show that, occasionally, the winds reached velocities of 16 m/s at the lower wind gauge and 17 m/s at the higher one. These particular peaks were westerly winds associated with depressions that crossed Israel in the winter and early spring. However, any analysis of the relation between the wave pattern and the coincident weather pattern has to take into consideration that the buoy was located closer to the western and southern shores of the Dead Sea than to its northern and eastern shores (Fig. 10-1). Therefore, with respect to the buoy, winds from the northern quarter (i.e., 0° ±30°) have the largest fetches (about 35 km) and could produce far larger waves than winds of equal speeds from other directions. Thus, intensive western winds do not produce high waves, at least not at the site of the buoy. Very strong northerly winds occurred as well during the winter and early spring. These winds were of the order of 11 m/s (once, even as high as 12.2 m/s). At the time, the air to sea temperature difference was about - 2°C. From the Coastal Engineering Research Center (1984), we find that given these winds, a fetch of 35 km, and a duration of 4 hours, the significant waves and the significant periods are expected to be 1.7 m and 5.5 s (a significant length of 47 m), respectively.

We must stress that the correction factors and the nomogram in the manual are intended for the open ocean, which is certainly not the case in the Dead Sea. Moreover, the empirical evaluations of the manual do not take into consideration such local factors as the particularly high density of the Dead Sea and the vertical component of the winds that result from the geographical position of this sea (i.e., 400 m below sea level with some very steep orographic features surrounding it). At present, we cannot estimate the magnitude of these factors, but we expect that the vertical component of the winds would enhance the wave field. Therefore, we decided to reject a record (the entire record) that showed significant waves heights

larger than 3 m. Obviously, to some degree, this is an arbitrary decision.

Subject to this criterion, out of a total of 251 analog records, 12 had to be rejected. Out of a total of 2,058 digital records, 45 had to be rejected. Of the rejected records, only five indicated significant wave heights between 3 and 4 m, and another six indicated significant wave heights between 5 and 6 m. The rest of the rejected records showed wave height that exceeded 10 m and thus were certainly wrong. Moreover, because we managed to acquire only some sporadic measurements since mid-May 1983, we limited our investigation to the data obtained from February to May 1983, that is a total of 1,890 digital records.

Data were analyzed according to Tucker (1963; see also Draper, 1966, 1976). The following parameters were determined for every record: t, the exact duration of the record in seconds; N_C the number of crests; N_Z, the number of upward zero crossings; H_1, the sum of the highest crest and the lowest trough; H_2, the sum of the second highest crest and the second lowest trough; and $T_Z = N_Z/t$, the mean zero crossing period; and $T_C = N_C/t$, the mean crest period. If we assume that the Dead Sea wave heights, like open sea wave heights, are Rayleigh distributed, the measured parameters can provide us with some of the statistical characteristics of the wave field (Longuet-Higgins, 1952) and, in particular, permit the computation of the significant wave height, H_S (or $H_{1/3}$ as it is named sometimes), for each of the wave records. The significant wave height and the significant wave period are defined as "the average height and period of the one-third highest waves" (Tucker, 1963, p. 306).

Another computed parameter is the spectral width, $\varepsilon^2 = 1 - (N_Z/N_C)^2$. This is a measure of the range of frequencies present relative to the mean wave frequency, and it controls the shape of the probability distribution of the heights of the wave crests (Tucker, 1963). Draper (1963, p. 300) describes the significance of this parameter as follows:

If the wave components cover a wide range of frequencies, the long waves will carry short waves on top of them and there will be many more crests than zero crossings, so that T_C will be much smaller than T_Z and ε will be nearly one. If, on the other hand, there is a simple swell which contains only a narrow range of frequencies, each crest will be associated with a zero crossing, so that T_C will approximately equal T_Z and ε will be nearly zero.

Finally, since the wind measured at En Gedi coincided only sporadically with the wave measurements, there were no data for the computation of the correlation between the winds and the waves.

RESULTS

Analog Measurements

The significant wave heights, H_S, are given in Table 10-1, where the gaps indicate missing data and 0 indicates calm seas,

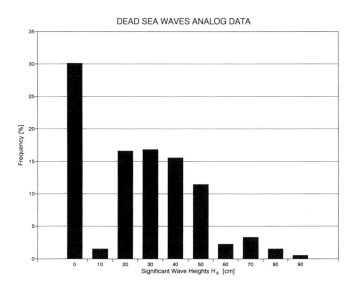

Figure 10-2 Analog data measurements; histogram of significant wave heights (H_S)

Table 10-1 Dead Sea wave measurements: Analog Data (significant wave heights in cm)

	October 1982 Time (Hours)						November 1982 Time (Hours)						December 1982 Time (Hours)						January 1983 Time (Hours)					
Day	00	04	08	12	16	20	00	04	08	12	16	20	00	04	08	12	16	20	00	04	08	12	16	20
01								34	44	40	21								20	28	41			34
02									27			33							34	32	32	26	00	00
03							67	55	46	22	00	00							13	15	28	16	00	00
04							41	35	50	29	00	00							00	16	30	31		
05							50	40		16	00	00												
06							10	14	13	00	00	00					30	00						
07							00	41	44	62	28		00	19	00	28	11							
08							00	42	22	18	23	00	00	09	23	00	00	00						
09							33	25	00	80	45	30	00	30	40	33	21	22						
10							23	24	18	26	18	23	41	46	52	38	34	37						
11							00	11	00	00	00	00	61	52	64	34	47	25						
12							00	00	11	18	00	00	20	47	43	41	16	00						
13							24	41	42	37	15	21	22	15	15	16		18						
14							40	36	38	23	25	33	13	00	28	15		00						
15							38	35	43	15	00	31	00	00	16	18	00	00						
16							36	68	67	86			30	60	48	41	32	24						
17								29	33	20	00	00	32	38	19	41	49	45						
18							13	16	16	12	00	00	66	61	47	39	37	25						
19				00	00	34	00	00	00	00	00	00	24	41	40	27	09	00						
20	32	16	00	00	00	20	00	00	12	00	23	19	00	00	00	09	00	00						
21	28	18	17	00	27	39	00	00	19	00	00	13	00	00	24	33	18	00						
22	43	31	42	40	00	00	31	60	60	29	00	17	12	24	40	58	44	39						
23	25	25	22	41	41		24	00	11	14	14	28	13	33	53	61	73	49						
24				00	00	00	34	23	00	47	00	00	31	36	71	38	40	44						
25	37	43	20	00	00	32	20	30	42	37	22	00	38	38	45	36	09	00						
26	35	43	28	33	00	29	00	11	00	00	00	14	00	09	15	00	00	00						
27	37	14	25	20	00	00	11	19	32	19	00	27	00	00	00	00	00	21						
28	59	40	52	24	00	00	23	50	71	34	12	20	17	58	43	48	19	19						
29	76	87	75	31			62	66					29	10	00	18	00	16						
30		29	27			00							00	00	17	00	00	00						
31	24	11				00							00	23	33	43	22	13						

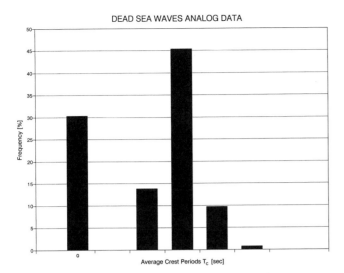

Figure 10-3 Analog data measurements; histogram of average crest periods (T_C). To complete the histogram to 100%, the value for calm seas was included.

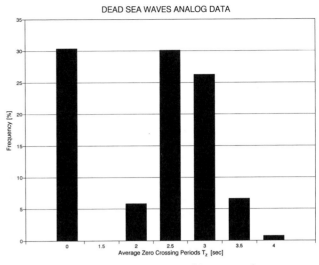

Figure 10-4 Analog data measurements; histogram of average zero crossing periods (T_Z). To complete the histogram to 100%, the value for calm seas was included.

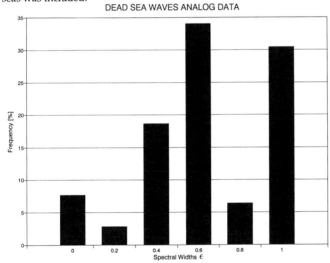

Figure 10-5 Analog data measurements; histogram of spectral width (e). To complete the histogram to 100%, the value for calm seas was included.

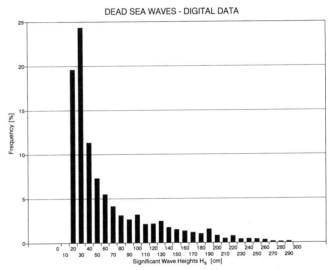

Figure 10-6 Digital data measurements; histogram of significant wave heights (H_S).

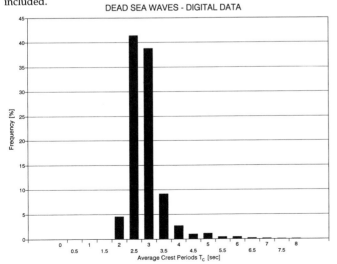

Figure 10-7 Digital data measurements; histogram of average crest periods (T_C).

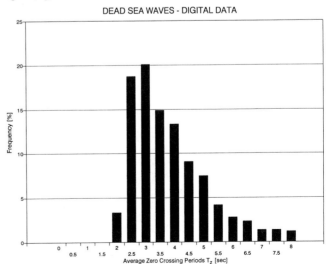

Figure 10-8 Digital data measurements; histogram of average zero crossing periods (T_Z).

defined as waves lower than 10 cm. During this particular period, the autumn of 1982, our measurements showed that 30% of the waves had significant heights lower than 10 cm and that 5% of the waves had significant wave heights larger than 70 cm (Fig. 10-2). About 30% of the time, the sea was calm, and 5% of the time the sea was rough. The waves were generally short, that is, they had average crest periods of 2.5 seconds and average zero crossing periods of 2.5 and 3 seconds (wave lengths of 10–14 m, as determined from the deep-water wave dispersion equation $L = (g/2\pi)T^2$, Figs. 10-3 and 10-4). For most of the measurements, the spectral width parameter is about 0.5, indicating a balanced combination of short and long waves (Fig. 10-5).

In general, calm seas lasted for long periods of time (e.g., Table 10-1, November 17–21), as opposed to rough seas, which lasted for short periods of time, that is, 8 to 12 hours of high waves (e.g., Table 10-1 on October 29, November 9, 16, 28, and December 24). The change from relatively calm seas to rough seas could be rather abrupt. For instance, the sea was completely calm at 1600 and at 2000 hours of October 28, 1982. Four hours later, by midnight of the same day, the significant wave heights measured 76 cm, and by 0400 hours, the significant wave heights reached 87 cm, with individual waves as high as 140 cm (from $H_{max}=1.6 \cdot H_S$). Again, on November 9, we measured calm seas at 0800 hours and 80-cm significant wave heights 4 hours later.

The observed rough seas could not be associated with a particularly intensive weather pattern. At the time of the rough seas, the synoptic maps showed the presence of a trough extending from the Arabian Peninsula toward the coast of Israel. During the periods of rough seas, the winds measured on shore at En Gedi were moderate (4–8 m/s) northerlies. Given a wind of 7 m/s, a fetch of about 35 km, and a duration of about 4 hours, the CERC-SPM (Coastal Engineering Research Center, 1984) nomogram predicted significant wave heights of about 75 cm and significant wave periods of about 4 seconds (a significant wave length of about 25 m). As a corollary, we looked at the sea state on the Dead Sea during the time that fronts and depressions crossed Israel (for example, December 5, 8, and 31, 1982). We could expect some fairly strong westerly winds to be associated with these weather patterns. However, as we pointed out previously, western winds do not have much of a fetch on the Dead Sea. Indeed, as expected, at the site of the wave gauge, these depressions did not seem to have a significant effect on the wave heights of the Dead Sea.

By averaging the data for a particular time, regardless of the date (e.g., average of all the data measured at 1400 hours), we attempted to find out whether there is a consistent diurnal pattern in the wave regime. Indeed, the results (Table 10-2) appear to indicate a diurnal cycle with a peak in the morning and significantly lower waves in the afternoon and evening.

Digital Measurements

In general, the second series of measurements (Table 10-3), though still incomplete, provided a better sample of the waves on the Dead Sea. Climatologically, these data complement the first series of measurements. Essentially, the digital data provided information on the wave climate of the Dead Sea during parts of the winter and spring of 1983.

During the winter and spring of 1983, the Dead Sea appeared to be very active (Fig. 10-6). No calm seas (i.e., waves lower than 10 cm) were observed, and almost 50% of the measurements consisted of waves with significant heights between 20 and 40 cm. About 19% of the records indicated waves with significant heights above 1 m, 3% above 2 m, and 0.75% above 2.5 m. The last group includes individual waves with maximum

Table 10-2 Diurnal variations of analog wave data (significant wave heights in cm)

Hour	N	Average ± St. Dev.
0000	65	23 ± 20
0400	68	28 ± 20
0800	65	30 ± 20
1200	67	25 ± 20
1600	63	13 ± 17
2000	64	14 ± 15

Table 10-3 Number of 10-minute digital records per day, 1983

Day	Feb	Mar	Apr	May	Jun	Jul	Aug	Sep	Oct	Nov
01		24	24	23						
02		24	24	24						
03	24	24	24	17						
04	24	24	24	24			07			
05	24	24	23	24			11			
06	24	24	11	24				05		
07	24	24		24						
08	24	24		24						12
09	24	24		24						24
10	24	24		24						03
11	24	24		10						
12	24	13								
13	24	24								
14	24	24								
15	24	24								
16	24	24								
17	24	24								
18	24	24				08				
19	24	24				24				
20	24	24				24				
21	24	24	07			06				
22	24	24	24							
23	24	24	24							
24	24	24	24							
25	24	24	24							
26	24	24	24							
27	24	24	24							
28	24	24	24							
29		24	20							
30		24	24							
31	24									

heights of as much as 4 m. The average crest periods (Fig. 10-7) show that most of the waves were short—2.5 to 3 s (10 to 14 m long)—although some longer waves were present. The average zero crossing periods (Fig. 10-8) indicate that 83% of the wave periods were between 2.5 and 5 s (10 to 39 m long), but there were also a relatively large number of longer waves, some even as long as 8 s (100 m). Thus, we observe a sea combined of short waves riding on top of some long waves, a situation that is also indicated by the spectral width distribution (Fig. 10-9). In the zero crossing distribution, a number of records (90 out of 1,890) indicated even longer waves; however, these were discarded as unreliable since, as shown previously, no such waves could develop on the Dead Sea.

In view of the data distribution (Table 10-3), we investigated the wave climate during different months, i.e., February, March, April, and May. We did not find significant differences between the months. An investigation of the diurnal variation of the wave heights (Table 10-4) showed once more that the highest waves occurred after midnight, between 0100 and 0300 hours, whereas the lowest waves occurred during the afternoon, between 1400 and 1600 hours.

The amount of digital data enabled us to test in greater detail another feature that appeared in the analog data, namely, the abrupt change from calm to rough seas. Table 10-5 shows selected cases of wave development in the Dead Sea. For every occurrence, the table shows the hour by hour significant wave height increase from the minimum to the maximum height. Thus, 00 hours is not the time of day but the starting point of the minimum wave field. Apparently, very rough seas characterized by significant waves higher than 2 m (maximum individual waves higher than 3 m) could develop in as short a time as 2 or 3 hours (e.g., Table 10-5, February 10, March 13, April 1).

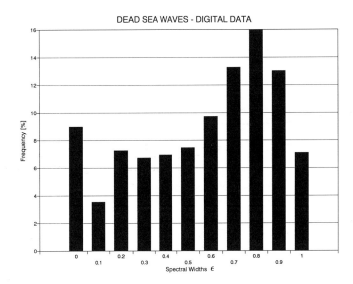

Figure 10-9 Digital data measurements; histogram of spectral width (e). To complete the histogram to 100%, the value for calm seas was included.

SUMMARY AND CONCLUSIONS

The wave measurements we analyzed comprise a unique series in the sense that, on the Dead Sea, no wave measurements have previously been recorded. However, since they are a very limited series, they could, at most, be viewed as a display of the phenomena that occurred during a limited period of time and an indicator of what can occur. The analog and the digital series complement each other in time. They present us with a relatively low wave field in the autumn and a far higher one during the winter and the spring. In view of the connection between the waves and the weather pattern, this is hardly surprising, and indeed it was already reported as such by Neev and Emery (1967). Our measurements showed that the size of the waves is much higher than that reported by Neev and Emery (1967). Whereas, they reported the largest waves as 1.1 m high and 10 m long (i.e., 2.5 s), we found waves that can reach a height of at least 3 m and have periods of up to 8 s (i.e., 100 m long). Therefore, their statement that the waves on the Dead Sea are lower than those on Lake Kinneret is incorrect.

The CERC-SPM (Coastal Engineering Research Center, 1984) nomogram predicted significant wave heights of 1.7 m, significant periods of 5.5 s, and wave lengths of 45 to 47 m. Thus, the nomogram-computed waves were much lower and much shorter that the ones we measured directly. The same nomogram was also used to predict the waves on Lake Kinneret (Boguslavski et al., 1988), which had a maximum height of 1.4 m, periods of 4.3 s, and lengths of 29 m. Once more, these were far below the wave heights of 3 m reported by Neev and Emery (1967). However, there appear to be no records of wave measurements of Lake Kinneret. Because the population on the shores and the surface of Lake Kinneret is far larger than that on the shores and the surface of the Dead Sea, the visual reports from Lake Kinneret are far more frequent. Introducing corrections for the "closed sea effect" (e.g., Leenknecht et al., 1992) resulted in a marginal increase in the Dead Sea wave heights and lengths. Thus, the nomograms appear to underestimate the wave field on the Dead Sea as well as on Lake Kinneret. Incidentally, this makes our criterion for the exclusion of certain wave measurements more conservative.

A number of factors are not taken into account by either the CERC-SPM (Coastal Engineering Research Center, 1984) or the Leenknecht et al. (1992) nomograms. One such factor is the orography of the two lakes and, in particular, the vertical components of the winds, which could be significant but find no expression in the wind measurement records. Other factors could be short-period wind variations (wind measurements are usually integrated over a time span), gusts, or currents at sea. One particularly well-known factor, and the one mentioned by Neev and Emery (1967), is the density difference between the two lakes (Lake Kinneret, essentially 1; the Dead Sea, essentially 1.3) and the open sea (essentially 1.02). Huss et al. (1986) show that, for a given wind, the transfer of energy to the surface currents results in a slower current in the Dead Sea than in the open ocean. Similarly, we may expect intuitively that, for the same conditions wind force, fetch, duration, and water depth), the waves developed in the Dead Sea should be lower than those developed in the open ocean or in Lake Kinneret, since the potential energy for a given wave height depends on the water density. Neev and Emery (1967) attempt to relate their observation that the waves on the Dead Sea are lower than those on Lake Kinneret to the higher density of the waters of the Dead Sea. However, as we have shown previously, the waves in the Dead Sea are in fact higher than those on Lake Kinneret. Moreover, Neev and Emery (1967) did not take into account all the factors related to the growth of the wave field (fetch and water depth, in particular), which, as a rule, would result in smaller waves in Lake Kinneret. Thus, although the density of the waters may affect wave development in the Dead Sea, this was not proven by Neev and Emery (1967).

The abrupt development of the rough seas, also reported by Neev and Emery (1967), as well as a similar feature of the wave field on Lake Kinneret, was confirmed quantitatively by our measurements. This effect can be related to the wind speed and wave frequency (e.g., Phillips, 1957; Miles, 1957; Inoue, 1967), as well as to specific local factors, such as the vertical component of the winds, the density of the waters, and the currents in the sea.

Finally, the periodicity of the wave field was also reported by Neev and Emery (1967). They report a mirror-like sea until

Dead Sea Diurnal Wind Speed Variations

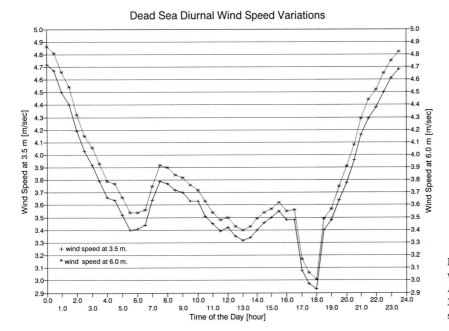

Figure 10-10 Average diurnal wind variation from winds measured at sea November 8, 1983 to August 19, 1984 at the position in Figure 10-1: +, at 3.5 m above the sea surface; *, at 6 m above the sea surface.

Table 10-4 Diurnal variation of digital wave data (significant wave heights in cm)

Hour	Average ± St. Dev.
00	77 ±62
01	81 ±66
02	81 ±65
03	81 ±64
04	77 ±60
05	77 ±62
06	71 ±56
07	66 ±56
08	59 ±51
09	62 ±54
10	62 ±57
11	58 ±50
12	51 ±45
13	44 ±45
14	38 ±38
15	34 ±33
16	38 ±38
17	40 ±42
18	40 ±40
19	47 ±42
20	50 ±45
21	53 ±40
22	57 ±44
23	66 ±57
All	59 ±54

from their report whether the sea stays calm throughout the night, i.e., between 2000 hours and 1000 hours the following day. The questions is, what is the reason for this periodicity? One obviously periodic driving force could be the sea breeze. Neev and Emery (1967) report that the winds over the Dead Sea are diurnally periodic: calm in the morning until 1000 hours, increasing to 1500 hours, and dying out by 2000 hours. According to Neumann and Stanhill (1978, p. 53), "One of the outstanding meteorological features of the Jordan Rift in summer is the arrival 'in force' in the afternoon hours of the Mediterranean sea breeze." Their investigation shows that the Mediterranean sea breeze arrives at Lake Kinneret in the afternoon hours and because of the orography of the region surrounding Lake Kinneret, it is amplified to twice and even two-and-a-half times the velocity near the Mediterranean shores. Thus, the breeze can reach speeds of 10 to 15 m/s and is the driving force behind the daily summer afternoon storms on Lake Kinneret. Weiss et al. (1987), as well as Weiss and Cohen (1988), mention that the high evening air temperatures on the Dead Sea, as well as the wind peaks occurring between 1700 and 1900 hours, are due to the Mediterranean breeze, but present no analysis to demonstrate their statements.

However, as we have seen, although our measurements show that the wave field is diurnally periodic, we observed the calmest seas in the afternoon and the highest seas past midnight at about 0200 hours, contrary to Neev and Emery (1967). Therefore, we decided to analyze the available wind measurements on the Dead Sea (i.e., November 8, 1983 to August 19, 1984). The data suffer from many gaps, and as a time series, they are not suitable for spectral analysis. A simpler and more superficial approach is to average the wind velocity for every half hour of the day, regardless of the date of the measurement (similar to the procedure we applied to the wave measurements). Our results (Fig. 10-10) show that the wind speed reaches a maximum around midnight and a minimum during the late afternoon hours. Thus the diurnal variations in the wind speed over the Dead Sea do not resemble those reported for Lake Kinneret, and, climatologically, fit in with the diurnal wave height as reported herein.

The present investigation has indicated a number of discrepancies between the wave field as reported by Neev and Emery (1967) and our own measurements. Far from trying to belittle

about 1000 hours, followed by a gradual intensification of the wave field, which reaches its peak at about 1500 hours and then dies out to become calm again at about 2000 hours. It is not clear

Table 10-5 Examples of the rate of wave development in the Dead Sea (significant wave heigths in cm)

Date	Time from Minimum to Maximum (hours)												
	00	01	02	03	04	05	06	07	08	09	10	11	12
09 Feb	14	15	15	24	50	92	112	201	197	236	240		
10 Feb	12	15	20	66	208	185	215	281	245	219	283	219	286
13 Mar	16	19	197	234	240	243							
20 Mar	14	19	23	53	143	183	262						
29 Mar	14	17	105	169	195	220							
30 Mar	15	20	20	110	155								
31 Mar	17	20	71	160	157	219							
01 Apr	17	17	120	216									
30 Apr	25	89	92	98	228								
01 May	27	58	44	84	117	151							
02 May	16	18	40	90	115	183							
05 May	19	19	26	26	54	58	168	231					
09 May	22	23	50	43	168	255							

the pioneering work of Neev and Emery (1967), to whom we are deeply indebted, we intend to stress the importance of objective prolonged measurements as opposed to sporadic observations. Therefore, we expect that an even longer and more precise series of wave measurements on the Dead Sea may yield different results than those we have presented.

Acknowledgments

The measurements for this investigation were carried out with the help of a dedicated group of technicians: G. Brookman, R. Sela, M. Udel, and J. Mouwes. We thank them for their hard work and perseverance. Our thanks are also due to M. Gonen, the skipper of the *Tiulit*, and to his crew for their help and assistance at sea; as well as to engineer D. S. Rosen for his help and advice. We would like to thank the referees, Dr. Wuest and Dr. Gloor, for their help in improving this chapter. The wind measurements ashore at En Gedi were kindly provided by the Meteorological Department of the Israeli Navy. The investigation was supported by a grant from the Ministry of Energy and Infrastructure.

REFERENCES

Boguslavski, A., Rosen, D. S., and Boguslavski, I., 1988, An opinion on the safety of cruising (on the Kinneret) in calm and rough seas with two interconnected passenger vessels Gamla and Magdala: Coastal Marine Research Institute of the Technion, 80 p., (in Hebrew).

Brainard, E. C., II, 1980, Wave orbital following buoy, *in* Proceedings, Marine Technology 80 Conference: Washington, D.C., October 1980, p. 473–478.

Coastal Engineering Research Center, 1984, Shore Protection Manual: Department of the Army, Waterways Experiment Station, Corps of Engineers, Coastal Engineering Research Center, Washington, D.C. chapter 3, 143 p.

Draper, L., 1963, Derivation of a "design wave" from instrumental records of sea waves: *Proceedings of the Institution of Civil Engineers*, v. 26, p. 291–316.

Draper, L., 1966, The analysis and presentation of wave data—A plea for uniformity, *in* Proceedings, 10th Conference on Coastal Engineering: Tokyo, Japan, September 1966, p. 1–11.

Draper, L., 1976, Revisions in wave data presentation, *in* Proceedings, 15th Conference on Coastal Engineering: Honolulu, Hawaii, July 1976, p. 38–48.

Huss, A., Ezer, T., and Hecht, A., 1986, How sluggish are the waters of the Dead Sea? *Israel Journal of Earth Sciences*, v. 35, p. 207–209.

Inoue, T., 1967, On the growth of the spectrum of a wind generated sea according to a modified Miles-Philips mechanism and its application to wave forecasting: New York University Department of Meteorology and Oceanography, Report TR–67–5, 51 p.

Leenknecht, D. A., Szuwalski, A., and Sherlock, A. R., 1992, Automated Coastal Engineering System, User's Guide, version. 1.06: Department of the Army, Waterways Experiment Station, Corps of Engineers, Coastal Engineering Research Center, Washington, D.C., 333 p.

Longuet-Higgins, M. S., 1952, On the statistical distribution of the heights of the sea waves: *Journal of Marine Research*, v. 2, p. 245–266.

Middleton, F. H., LeBlanc, L. R., and Czarnecki, F., 1976, Spectral tuning and calibration of a wave follower buoy, *in* Offshore Technology Conference, Paper 2597: Houston, Texas, April 1976, p. 753–762.

Miles, J. W., 1957, On the generation of surface waves by shear flows: *Journal of Fluid Mechanics*, v. 3, p. 185–204.

Neev, D., and Emery, K. O,, 1967, The Dead Sea: Depositional processes and environments of evaporites: Jerusalem, Geological Survey of Israel Bulletin 41, 147 p.

Neumann, J., and Stanhill, G., 1978, The general meteorological background, *in* Serruya, C., ed., Lake Kinneret: Monographiae Biologicae, Dr. W. Junk bv Publishers, v. 32, p. 49–58.

Phillips, O. M., 1957, On the generation of waves by turbulent wind: *Journal of Fluid Mechanics*, v. 2, p. 417–445.

Tucker, M. J., 1963, Analysis of records of sea waves, *in* Proceedings of the Institution of Civil Engineers, v. 26, p. 305–316.

Weiss, M., Cohen, A., and Mahrer, Y., 1987, Upper atmosphere measurements and meteorological measurements on the Dead Sea: Department of Meteorology, The Hebrew University of Jerusalem, Research Report for 1986–1987, 19 p. (in Hebrew).

Weiss, M., and Cohen, A., 1988, The atmosphere in the region of the Dead Sea—Measurements in the middle of the lake: Department of Meteorology, The Hebrew University of Jerusalem, Final report, 4 p. (in Hebrew).

11. EVAPORATION ESTIMATE FOR THE DEAD SEA: ESSENTIAL CONSIDERATIONS FOR SALINE LAKES

Ilana Steinhorn

This chapter presents an integrated view of the question: What is the evaporation rate from the Dead Sea and other saline lakes? It suggests that salinity-related aspects will be taken into consideration during all stages of evaporation estimate, from preparation and measurements to interpretation and application. To demonstrate the inconsistency associated with the use of traditional methods that were developed for fresh and sea water, the evaporation rate from several reasonable models of the Dead Sea is evaluated within the limitation of the available data.

QUESTIONS AND CONCEPTIONS

Most people are familiar with the fact, that by evaporation, saline water becomes more concentrated and may deposit salts. It is also evident, that evaporation from highly saline water like the Dead Sea water is slower than evaporation from fresh water under similar conditions. When evaporation from a whole freshwater body is concerned, professionals from various disciplines are struggling to overcome the difficulties associated with the complex response of a lake to changes in its environment, such as diurnal and seasonal climatic variations and various man-made effects. Additional difficulties exist when considering evaporation from a hypersaline instead of freshwater lake.

Because of the many applications of evaporation, the terms *evaporation* and *evaporation estimate* may mean different things to different people. Therefore, attempts to answer this and other questions may require an agreed upon definition of the basic concepts. In theory, the evaporation is known as the process in which water molecules transfer from the liquid to the gas phase. Yet the net flux of these molecules (the evaporation rate) is hard to measure even under controlled conditions, and by itself, it is of lesser interest. For many practical purposes, therefore, evaporation is defined in terms of the parameters of interest, among them the water loss or the decrease of water level caused by evaporation. Throughout this chapter, the term *evaporation* is used in its original meaning, while the practical evaporation terms are called *evaporation related parameters*. Unlike freshwater bodies, in the case of the Dead Sea, the difference between the theoretical and the practical terms is not only conceptual, but also quantitative.

Yet, regardless of what is meant by evaporation, the question underlying most attempts to evaluate the evaporation rate is the same: What does one wish to know about the lake by evaluating its evaporation rate? Since evaporation is a turbulent phenomenon, this question is considered with that of the time scale required. The answer for both would dictate, in some cases, the technique chosen to estimate evaporation.

For many freshwater lakes and reservoirs (Lake Kinneret, for example), it is important to know the amount of irrigation or drinking water lost by evaporation, and perhaps to investigate ways to reduce it. The time scale required is on the order of a week or month. Cooling of the water when latent heat is supplied for the evaporation process is also of practical importance. One example of this would be in power plants, where we want to know how long it would take to cool the heated

water by evaporation and other air-water interaction processes. In this case, the results of evaporation estimate may be needed within days. Other questions relate to the production of salt from concentrated saline water. In the evaporation ponds of the Dead Sea Works, higher rates of evaporation often mean that larger amount of minerals can be extracted within shorter periods of time. In every water body, dynamic processes in the water body and in the air above it depend on evaporation and are especially important in large lakes and seas. To estimate the vertical mixing and stability of the air and water layers, for instance, we need to measure turbulent fluxes. The time scale required for proper analysis of fluxes is within hours to seconds.

When the Dead Sea was first examined for the production of minerals in the 1930s, the question was, What is the order of magnitude of its monthly and yearly evaporation rates? In general, the subject of evaporation was important in the early days of meteorological research in the water-short Land of Israel, and measurements of the potential evaporation were taken in all areas, including the vicinity of the Dead Sea (Rozenan and Shore, 1952). Potential evaporation, however, depends on the meteorological conditions but not on the evaporating surface. The works of Neumann on the evaporation rate from the (then three) lakes of the Jordan Valley are (Neumann, 1953, 1958) were done shortly after the energy balance method was first applied to a lake (Anderson, 1952).

Today the questions are more complex. Everyone involved in this subject is familiar with the questions that both laymen and professionals ask. Among them are, How long would it take the Dead Sea to dry if its water supply were cut off completely? What is the role of evaporation in the water balance of the Dead Sea? How much sea water could be transferred to the Dead Sea during operation of the proposed Hydroelectric Seas Project to compensate for excess evaporation over inflow? The questions related to the current evaporation rate or those concerning observations could be answered, in principle, with proper measurements. However, the majority of the questions are prognostic. They cannot be answered without a better understanding of evaporation and its relations with other air and sea processes.

Although our present knowledge of the Dead Sea by far exceeds that of the 1950s, the same question regarding its annual and monthly evaporation rates is still being asked and has not yet been properly answered. The main reasons are that calculations of evporation rates have been based on a single method, the energy balance in its basic form, using limited data. The effects of statistical and other errors on the final results were not determined. The results were not verified by application of other empirical methods, such as the mass and salt balances. Moreover, modern measuring techniques, that could answer more specific questions on shorter time scales, were not used.

The problems concerning data collection and those involved in choosing parameters to represent the true conditions over the whole lake system are not unique to the Dead Sea or other saline lakes, and the interested readers may wish to consult other publications dealing with freshwater systems (see Brutsaert, 1982; Morton, 1986; Jones, 1992). However, selecting the parameters to measure in a hypersaline lake and interpreting the information do involve a critical view of the subject of evaporation estimate, which I do, in part, in this chapter.

Only a limited number of considerations are introduced here. A few quantitative examples are also given and are presented in Tables 11-1 to 11-3. The equations are all presented in the Appendix. Readers who are familiar with the classic formulations and those who are less interested in details may concentrate on the descriptive parts only, although some of the equations are presented in a modified form.

Although this article presents some calculations of evaporation rate, it provides no conclusive estimate for the elusive rate of evaporation from the Dead Sea. Yet, you may be able to get some feeling for the interrelations between evaporation rate and the air and water properties, particularly when changes in salt concentration and water level are concerned.

EVAPORATION FROM SALINE WATER

Is there any difference between evaporation from fresh and saline water? One of the ways to describe evaporation under natural conditions is to express the mass transferred from the surface as directly related to the wind speed, the air density, and the difference in vapor pressures between air and water. This is known as Dalton's Law or the mass transfer equation (see eq. 1 in Appendix). According to this law, there is basically no theoretical difference between evaporation from pure and saline water. There is, however, a quantitative difference mainly because of the reduction in saturated vapor pressure over saline water. This reduction has led to the common knowledge that the evaporation rate from saline water is lower than that from fresh water under similar meteorological and water temperature conditions.

But from the water balance aspect, the difference between evaporation from fresh and saline water lies in the fact that pure water can be fully evaporated, whereas saline water cannot. Part of the saline water, the salts, may remain in the solution and increase its concentration. When dry, the solution may still contain some water because of its hygroscopic nature. For saline water, then, the term *evaporation from* may be used, whereas the term *evaporation of* has a different meaning. The term *water balance* may also be misleading. Rather, the more accurate *mass balance* may be used, accompanied by the *salt balance* (or the *pure water balance*).

For hypersaline water, the evaporating (saline) water must have a bigger mass than the evaporated (pure) water molecules. For the Dead Sea, with salinity as high as 280 g/kg, the evaporating mass is considerably larger than the evaporated mass (eq. 2). The use of accurate evaporation definitions is important when we choose, for convenience, to express the evaporation rate in volume rather than in mass units (Steinhorn, 1991a). For relatively low-saline water, this distinction between the two masses is usually not made. For the ocean, there exist several classic equations that may not be used for the Dead Sea without modification (Steinhorn, 1991b).

By evaporation, a saline mass reduces its volume, yet more slowly than a pure water mass would. This is the same as saying that the density increases. The salt in the water, therefore, leads to contraction. Similarly, the water also reduces volume by cooling. The evaporation process therefore tends to reduce the volume of saline water in three ways: by reducing the mass, by cooling, and by increasing the salinity.

Because of these changes in volume, we should distinguish between several quantities expressed in volume units. All of these have the same value for pure water, but they are quantitatively different for Dead Sea water (for details see eqs. 3-9).

These quantities are, among others, the reduction of water level and the volume of evaporated water molecules if condensed to liquid at the lake temperature. The latter has conceptually the same value as would the volume of pure water needed to add to the lake to make up for the evaporation. The volume of saline water of the same salinity as the lake's needed to keep the lake mass constant is, however, much smaller. The reduction of the water level depends on the equation of state of the saline water, and it is less than, but very close to, the volume of evaporated water in liquid form.

Hypersaline water is often saturated with respect to one or more salts. By evaporation, the surface becomes more concentrated, and saturation or oversaturation may occur. The formation of salt on the water surface is perhaps likely to occur when the wind is calm and mixing with deeper water is temporarily reduced. Although salt can be formed in several ways, we do have evidence that it forms at the surface of the Dead Sea (Steinhorn, 1983). This subject, however, needs to be further investigated before it can be incorporated into evaporation studies.

METHODS OF EVAPORATION ESTIMATION FOR SALINE LAKES

There are two major ways to estimate evaporation based on in-situ measurements. One is to derive the fluxes at the air-sea interface and calculate the evaporation rate directly. The other is to apply conservation laws of energy and mass, and deduce the part of the evaporation in the balance. Since the most common parameter of interest is the water loss, a direct practical method is the water balance. It involves measurements or estimations of the lake level and all sources and sinks of water. This method may be applied to saline water with some adjustment: It requires that the conservation of mass (eq. 10) be used, rather than the conservation of volume. (As shown in the section "Evaporation from saline water," volume is not a conserved quantity and is affected by salinity and temperature changes).

A unique method for estimating evaporation from saline water is the salt balance method (eq. 11), and it refers to the evaporation of pure wter from the saline solution. For the Dead Sea, the measured parameter could be the density at a constant temperature rather than the total salt concentration. (The salt concentration is measured with greater difficulty at a lesser accuracy). Under controlled conditions, this method may be more accurate than the mass balance method (Steinhorn, 1992). If both salt and mass methods are combined, one may also be able to get information on some of the unmeasurable sources of mass and salt.

The energy budget method (eqs. 12-17) provides estimation of the heat lost by evaporation. The advantage of this method is that it relies on standard meteorological data with only occasional visits to the lake. However, it requires vertical profiles of the water column that may be difficult to obtain in deep lakes. Also, in most cases, measurements of the lake parameters and the air above the lake do not provide statistically true mean values of the water column or surface. These mean values are based on monthly, or less regular, visits to limited locations in the lake and on measurements by the lake shore. The difficulty in arriving at representative mean vertical values is more pronounced in the case of a deep and relatively large lake like the Dead Sea. In a very saline lake, the changes in heat content need to be calculated based on both temperatures and salinity measurements. (Modified equations are presented in the Appendix, eqs. 20-23).

Both practical methods, the energy balance and the mass (water) balance, are used in most cases for monthly or yearly estimation of evaporation rates. This time scale may be too large. Evaporation depends on the physical properties of air and water, and the turbulent conditions of these media yield fluctuating meteorological and marine conditions, as well as fluctuating evaporation values, on a very small scale. Averaging these and other parameters over a certain time period and calculating the evaporation rate based on these averages may eventually result in statistical errors. The errors are higher for longer time periods and may be different for different lakes. They also depend on the parameters measured and therefore on the method used. For instance, it has been shown for freshwater bodies that averaging the wind speed over a period longer than 2 or 3 days yields unsatisfactory results in the case of the mass transfer method (Jobson, 1972).

Modern techniques enable scientists to measure turbulent components of temperature, momentum, and humidity fluxes or to estimate fluxes from meteorological and water surface data. The evaporation rate can be evaluated without reference to hydrological, water column, or radiation measurements (eq. 1). If instruments are installed at a suitable location in the lake, the results may adequately represent the physical conditions at that point on the lake surface. There may be, however, some difficulties in applying these methods for the Dead Sea. In particular, there is no way at present to directly measure salt fluxes. In the ocean, salt fluxes are calculated from salinity, which can be measured using indirect parameters, such as conductivity. For hypersaline water, however, there is no way to measure the salinity in situ or to perform any continuous measurements of salinity-related parameters. Therefore, in the case of the Dead Sea, the effect of salinity cannot be included in these methods, although this effect is by far greater in the Dead Sea than it is in the ocean. The effect of evaporation on salt precipitation is an additional unmeasurable factor.

We should, therefore, be aware of the difficulties involved in measuring the required parameters. Some of these difficulties are practical, for example, where tools and techniques are available but difficult or expensive to use. Other difficulties lie in waiting for the development of techniques appropriate for use in saline and hypersaline water. The available information may also require modified analysis and interpretation.

EVAPORATION RATE FROM THE DEAD SEA: GOALS AND PITFALLS

I have stated that the basic question of the amount of yearly or monthly evaporation from the Dead Sea has not yet been resolved. On the other hand, our knowledge of the Dead Sea is greater now than it was when first attempts to answer this question were made, and some of this new knowledge has been incorporated into recent evaporation estimates.

The main criticism about the result of Neumann's (1958) calculations has been directed toward the quality of the data that were available at the time, for example, calculated solar radiation and estimated water temperature based on Lake Kinneret data. Much of the effort in improving the evaporation estimate from the Dead Sea was then devoted to obtaining more information on atmospheric and limnological parameters. In doing only that, however, the result would be simply an improvement of the evaporation estimate from Neumann's hypothetical model of the Dead Sea, not necessarily a more reliable estimate of the evaporation from the Dead Sea. When estimating evaporation rates for a lake, it is important to recognize the limitations of the model used, and to adjust the model to the changes in the lake, if appropriate.

Neumann (1958) calculated the evaporation rate to be 1,720 km/(m^{2}yr) from the northern basin and 2,133 km/(m^{2}yr) for the southern basin. Final values of 147 cm/yr for the northern basin and 180 cm/yr for the southern basin were obtained by dividing the evaporation rates (in mass units) by the densities of Dead Sea water (Neumann, 1958). However, as noted previously in the section entitled "Evaporation from saline water," this calculation would instead give the volume of the saline Dead Sea water needed to make up for the evaporated pure water mass. An appropriate measure for evaporation in volume units would be the volume of the evaporated pure water mass if condensed to liquid (obtained by dividing the evaporation rate by the density of the pure water) or the drop of the water level (obtained using eq. 6). These values are, respectively, 171.7 cm/yr and 172.5 cm/yr for the northern basin, and 212.7 cm/yr and 213.9 cm/yr for the southern basin. All are considerably higher than the values originally reported for the same evaporation rates.

In light of these higher-than-published evaporation values, Stanhill's (1994) results of 127 cm/yr for the northern basin, derived for the same period, may seem quite low. The differences between the two studies are mainly due to the different data set used for both air and lake properties. Stanhill used radiation and meteorological conditions as measured in the 1980s and suggested different lake temperatures than those used by Neumann. Their model lakes are similar, with one difference. Neumann calculated the annual energy budget and arrived at the monthly rates based on hydrological and other considerations. Stanhill applied the energy balance for each month and combined the sum of the monthly values to obtain the yearly evaporation.

Neumann's hypothetical Dead Sea is composed of two separate lakes. One represents the northern basin, and the other represents the southern basin; each assumes a different set of water and air properties. The lakes do not change from year to year and therefore are in equilibrium. The model lakes used by Stanhill (1994) include several diagnostic cases (for pre-1951, 1959–1960, and 1983–1987) and one prognostic case, which assumes the same surface density as freshwater, for the proposed Dead Sea–Mediterranean or Dead Sea–Red Sea Canal. Each one of those lakes represents only the northern basin. They assume no connection to the southern basin (either the natural or the artificial evaporation ponds). They all share the same overlying air but each has different water properties. All these model lakes, therefore, are similar to each other.

It is important to distinguish between the model lake used and the real lake, and between the model lake and the same model as combined with the data set. Because there are no sufficient measurements and observations, the models are relatively simple and do not fully represent the real lake. In addition, some of the data used are actually part of the conceptual models.

As an example, let us compare the two model lakes used by Stanhill (1994) for the pre-1950s and for the 1980s. Assuming that the data used are perfectly correct, to what extent does the evaporation rate from these model lakes truly represent the appropriate values from the Dead Sea during those time periods? Although we cannot answer this question, we can provide some comparison between the two lakes; both are examples of the same model lake, but each represents a different real lake. In the first period, the Dead Sea had a deeper basin connected to a smaller and much shallower one. In the second period, it had a deep basin, slightly smaller than the first one, with some seasonal outflow and also with inflow of heavier saline water. In this example, the same (one-dimensional) model is used for two different (multidimensional) situations of the changing lake.

In the following example, let us again compare the two previous model lakes. This time we will assume that the Dead Sea did not change size or any other property except for the water level and the water characteristics. We have reasonable infor-

Table 11-1 Annual meteorological conditions in the Dead Sea and related energy balance terms

No.	Air temperature (°C)	Relative humidity (%)	Short-wave radiation kgcal/(cm²yr)	Long-wave radiation kgcal/(cm²yr)	Vapor pressure (mb)	Cloud cover (tenths)
1a	23.6	55.5	175	268	15.9	2.9
1b	24.2	45.3	157	272	13.4	2.7
1c	25.4	44.2	157	276	13.4	2.7

Data collected at various locations near Kalia, Kidron, and En Gedi.
1a. Data based on Ashbel's measurements in the 1940s (Asbel, 1944), as used by Neuman (1958).
1b. and 1 c. Data from the 1970s and 1980s, based on the Israel Meteorological Service archives and Stanhill's 1987 measurements (Stanhill, 1987).

mation on the surface salinity of the Dead Sea during both periods of time. But for the first period, the meteorological data are less reliable than those for the second period, and water column data are practically nonexistent. To estimate the evaporation rate, we must complete the data set. How we do that will depend on the conceptual model we have in mind. We may wish to assume, like Stanhill (1994), that the meteorological conditions are the same for both model lakes. In this case, the model we adopt is of a lake that changes only in response to water level changes. We may instead accept past meteorological measurements as reliable, in which case our model lake may also change with the climate.

We also want to have our assumptions (and the results of our calculations) consistent with each other. For instance, we know that as a terminal lake, the average salinity of the Dead Sea increases with a decrease in water level. Therefore, the model lake for the 1940s should have lower salinity than the model lake for the 1960s, and the latter has lower salinity than the one representing the 1980s lake. In the past, consistency has not always been the rule, notably for seasonal calculations. Because many of the parameters used in evaporation estimation are related to each other, consistency is essential.

In many cases, the distinction between the model assumptions, the lake, and the data set may be less obvious than in the previous examples. Yet as long as there are limitations concerning data collecting and measuring techniques, it is important to be aware of possible inconsistencies and to recognize hidden assumptions.

ANNUAL EVAPORATION RATE FROM THE DEAD SEA

I now present simple energy balance calculations of evaporation rate from the Dead Sea. First, I would like to define the annual evaporation to include condensation. Condensation may occur when the vapor pressure at the water surface is lower than the vapor pressure in the air above. This could happen (eq. 1) when the humidity of the air is high and the water temperature is relatively low, as is sometimes the situation in winter. Because of the lower vapor pressure over saline water, condensation can occur more frequently and at much higher temperatures than in freshwater lakes. When long-term methods like the energy balance are used, the term *mean evaporation rate* may then be interpreted as the mean evaporation minus condensation rate.

The characteristics of the first hypothetical lake are the following: (1) The model includes only the northern part of the lake, from Kalia to Masada; (2) the air and sea conditions are

assumed to be the same all over the lake, although measured at several locations of different microclimates (by Kalia, Kidron, or En Gedi); (3) there is no heat, salt, or mass exchange between these and other parts of the lake; (4) inflow water has the same temperature as the lake water; (5) there is no net annual heat gain or loss; and (6) the model assumes annual averages. The properties of this model lake are the same as Neumann's (1958). The only difference is the empirical data on the saturated vapor pressure over Dead Sea water.

Why not just use Neumann's data? Neumann (1958) examined data obtained at the Dead Sea Works Laboratories. He found that for the density 1.17 g/cm³ (which, if it matches our definition of density, would correspond to salinity 210 g/kg), the vapor pressure lowering was close to that measured by Arons and Kientzler (1954) for sea water at chlorinity 13.5% (salinity 24.4 g/kg). The chemical composition of Dead Sea water is different from that of sea water, and it is difficult to find a match for other salinities. To my knowledge, the unpublished data mentioned by Neumann are not available today. Stanhill (1994) used two values for the activity coefficient of the Dead Sea water. One provides vapor pressure data close to those used by Neumann, and the other yields much lower values and is based on several unpublished recent measurements of Dead Sea water. The data I have chosen to use are based on several unpublished measurements performed by the Dead Sea Laboratories in the early 1970s (Table 11-1 compares the 1940s and 1970s datasets). The main advantage is that these data include several sets of vapor pressure measurements at a relatively large range of salinities and temperatures, with reasonable rate of change in both salinity and temperature. The activity coefficients thus show some dependence on temperature, as in Arons and Kientzler's (1954). But there is one important and somewhat puzzling difference: The vapor pressure values are lower than the earlier data reported by Neumann. The results of my calculations, therefore, yield a lower evaporation rate of 1,660 kg/(m²yr) (Table 11-2b). The volume of evaporated water is 166.5 cm/yr, and the resulting lowering of the water level is 165 cm/yr.

These evaporation estimates refer to climatic conditions that may have existed during the 1940s. A comparison between old and new data sets (Table 11-1) shows some changes in climate that may or may not represent the real situation. There is, however, no doubt about the changes in salt concentration caused by the reduction in inflow rate to the Dead Sea. It may be reasonable, then, to first examine the response of the lake to the changes in salt concentrations under unchanged climatic conditions.

Table 11-2 Dead Sea water characteristics, energy budget terms, and evaporation related quantities

No.	Climate (Table 11-1)	Water temperature (°C)	Water density (g/cm³)	Water salinity (g.kg)	Bowen ratio	Long-wave radiation [kgcal/(cm²yr)]	Sensible heat [kgcal/(cm²yr)]	Latent heat [kgcal/(cm²yr]	Evaporation rate kg/(m²yr)	Level lowering (cm/yr)	Evaporation volume (cm/yr)
2a	1a	24.7	1.17	210	0.065	333	6.7	103.3	1,720	171.7	172.5
2b	1a	24.7	1.17	214	0.125	333	12.2	97.8	1,660	165.0	166.5
2c	1a	24.7	1.23	276	0.711	333	45.7	64.3	1,089	107.0	109.2
2d	1a	26.9	1.23	276	0.538	339	36.6	67.4	1,141	112.1	114.4
2e	1b	26.9	1.23	276	0.270	339	19.0	71.0	1,202	118.1	120.6
2f	1c	27.25	1.23	276	0.172	340	13.1	76.4	1,294	127.1	129.8
2g	1c	27.25	1.23	276	0.172	340	14.2	82.8	1,403	140.0	140.7

2a. Neumann's 1958 estimate, with correction for evaporation volumes.
2b. Estimate based on data used by Neumann (1958), with vapor pressure as measured by the Dead Sea Works Laboratories in the 1970s.
2c. The same surface temperature as in 2b, with higher salinity. The sensible heat is different than before, in contradiction to the basic assumption of no change in climate.
2d. Increase of surface salinity requires increase of temperature when climatic conditions remain the same.
2e. The waer conditions are as in 2d, yet the evaporation rate is different because of different climatic conditions.
2f. With initial water conditions as in 2d and 2f, the changes in the climatic conditions from those described in Table 11-1b to those in Table 11-1c result in the increase of surface temperature and evaporation rate.
2g. Additional increase of evaporation rate is caused by advection of 3.6 kgcal/(cm²yr) from the southern parts of the lake.

EFFECT OF SALINITY CHANGES ON EVAPORATION

During the 1940s, the Dead Sea has a salinity of about 210 g/kg and a density of 1.17 g/cm³. Some 40 years later, it has reached 275 g/kg and 1.23 g/cm³. It is reasonable to assume that the evaporation rate from its surface has changed too.

The following calculation is based on similar assumptions as in the previous section. In particular, the lake has achieved equilibrium during at least 1 year of the second period (1980s), as it is assumed to have done during the first period (1940s). The calculated evaporation rate for the second period is only 109.2 cm/yr (Table 11-2c), compared to 166.5 cm/yr for the first period (Table 11-2b). This value may seem reasonable because of the expected decrease of vapor pressure with the increase of salinity. However, the sensible heat in the second period is calculated to be 45.7 kgcal/(cm²yr), which is vastly different from that calculated for the first period, 12.2 kgcal/(cm²yr). Why is there a difference? From equation 15, we conclude that this could be due either to changes in air temperature or to changes in winds and atmospheric stability. But our basic assumption was that the climatic conditions remain the same, and therefore no change of air temperature or wind speed are allowed. It follows that changes of sensible heat cannot be accepted. The water temperature must have changed to allow for changes of sensible heat.

Why does the temperature change? When the evaporation rate decreases, the total latent heat required to supply the evaporating water decreases too (although the latent heat per unit mass slightly increases with salinity). As a result, the amount of heat transferred to the lake is larger than the amount released from the lake to the air, and the surface temperature increases. The increase of surface temperature enables more heat to be transferred to the air through higher sensible heat and infrared radiation, and also through a slight increase of evaporation with temperatures.

What is the new surface temperature? To calculate this, I use the set of implicit equations 12-13 and 15-16. The result is 26.9°C. This is a dramatic increase of 2.2°C attributed solely to changes in surface salinity. Indeed, an increase in temperature with salinity has long been observed in the series of evaporation ponds of the Dead Sea Works and was systematically examined in a series of experimental pans. Salhotra et al. (1985, 1987), who analyzed the pond data, provided clear explanation of the feedback effect of salinity and temperature on evaporation rate. The temperature calculated with equations 12–13 and 15–16 for Dead Sea water is close to that actually measured in the 1980s. It is evident that the Dead Sea temperature continues to rise with the average increase of salinity (Anati and Stiller, 1991). The increase in Dead Sea temperature is supported by changes in the stratification pattern that accompany salinity increases in this lake (Steinhorn, 1985).

With the higher calculated surface temperature, the evaporation rate is 1,141 kg/(m²yr), or 114.4 cm/yr in volume units (Table 11-2d). This calculation for the model lake of the 1980s is consistent with the salinity changes of the model lake of the 1940s (Table 11-2b) because of the choice of surface temperature and surface vapor pressure, which change consistently with the salinity.

RESPONSE TO CLIMATE AND METEOROLOGICAL CHANGES

The meteorological data from the 1970s and 1980s (Table 11-1b, 11-1c) indicate higher air temperatures compared to the 1940s (Table 11-1a). Solar radiation was first measured in the 1980s (Stanhill, 1987), indicating lower values than those used by Neumann (1958). With the meteorological conditions of the 1940s and the density value changing to that of the 1980s (see previous section), the lake must have raised its temperature to 26.9°C and changed its evaporation rate to 1,141 kg/(m²yr) or 114.4 cm/yr in evaporation volume (Table 11-2d). What would be the evaporation rate in the case of a change in climate?

First, let us demonstrate the difference in evaporation, assuming changes in climatic conditions, while the lake salinity and temperature remain the same. In this example, I assume the lake properties are as in Table 11-2d, but the climatic conditions are those of the 1980s (as in Table 11-1b, instead of Table 11-1a). The evaporation rate is calculated to be 1,202 kg/(m²yr) or 120.6 cm/yr (Table 11-2e). Although the incident radiation is lower than before, the higher air temperature and the lower humidity bring a higher evaporation rate for a lake having similar surface temperature and salinity.

The following calculations demonstrate the change in lake temperature in response to climatic variations. The initial lake conditions are assumed to be as in Table 11-2d, whereas the meteorological conditions change from Table 11-1b to Table 11-1c. As a result, the lake changes its temperature from 26.9°C to 27.25°C, and its evaporation rate in volume units increases to 129.8 cm/yr (Table 11-2f).

The Dead Sea as a system also includes the air above it. Changes or variations in regional climate are likely to affect the Dead Sea climate. Also, because of the relatively large area of the lake, changes in surface temperature may induce changes in the air above the water. The Dead Sea is known to affect the climate and meteorological pattern of the air in its neighborhood (Ashbel, 1939; Segal, 1983). Estimating evaporation from the Dead Sea may therefore require investigation of both lake and air conditions, using a coupled air-sea model.

EVAPORATION RATE IN NONEQUILIBRIUM CONDITIONS

The evaporation rates presented in Table 11-2 are calculated assuming annual equilibrium; that is, each year the Dead Sea attains the same physical properties as it had in the previous year. In freshwater lakes, the transition between two periods of equilibrium conditions involves heating or cooling of the water column. This may take at least 1 year of nonequilibrium conditions, where there is net heat transfer to or from the lake.

In warm freshwater lakes, the month of minimum temperature is also the month of maximum mixing. This is not necessarily so in the Dead Sea, where both temperature and salinity patterns contribute to the vertical mixing. Vertical measurements in the Dead Sea indicate that minimum temperature is generally achieved in mid-February. Neumann (1958) already mentioned this, based on his experience with the freshwater Lakes Kinneret and Hula. In the summer, when evaporation is greater than inflow, surface salinity increases. Cooling of the water in later summer causes mixing of the more saline water within the deeper layers. It is reasonable to assume that in years when the water level declines or remains in relative annual equilibrium, the maximum depth of mixing coincides with the minimum temperature in midwinter. In this case, the seasonal stratification that results from the inflow of freshwater into the Dead Sea occurs in the spring. In other years, the winter inflow may be significant enough to cause stratification, and the maximum rate of mixing is achieved earlier, in late fall or early winter.

When the energy balance method is used, knowledge of the heat content of the lake is required. Because this information was unavailable, Neumann (1958) assumed annual equilibrium. Monthly calculations should always include changes of heat content of the lake, because the lake is under obvious nonequilibrium conditions between months. Stanhill (1994) based his calculations on mean monthly temperature profiles of the Dead Sea, but he did not mention whether his model lake assumes equilibrium at the end of the year. For the 1980s period, however, an average heating of the Dead Sea is reported elsewhere (Stanhill, 1990).

The traditional formula for evaluating the changes in heat content relies on temperature profiles of the water column (eqs. 18, 19). Between peak heating and peak cooling, these changes usually signify the annual heat budget of the lake. Year-to-year changes reflect long-term heating or cooling of the lake. In hypersaline lakes, changes in salinity and mass must be considered as well. Obviously, a lake loses enthalpy by losing mass. In freshwater lakes, this often means that the net heat transfer from the surface is redistributed within a smaller mass (if shallow) or that deeper layers of the lake are now involved in the seasonal processes. In a saline lake, however, the heat loss is due to the need to supply energy for redissolution of the salt left by evaporation. The changes in the thermal properties may

therefore include changes in the specific heat of the water. Equations 20-22 take these effects into consideration.

Why is it important to consider the changes in specific heat capacity? One of the reasons is that the high salinity of the Dead Sea implies relatively large salinity changes. From a physical point of view, the amount of heat required to raise the temperature of a water column is reduced with an increase in salinity. If there is a net income of heat, the same amount of heat would cause a greater rise in temperature. Therefore, for a steady income of heat and an increase in salinity, there is an acceleration of temperature increase with time. During the 40-year period (1940s to 1980s), the specific heat of Dead Sea water actually changed from approximately 3,085 J/(kgK) at salinity 210 g/kg to 2,985 J/(kgK) at salinity 275 g/kg. (This is based on interpolation of unpublished data, measured at the Dead Sea Works Laboratories in the early 1970s.) It follows that the amount of heat required in the past to raise the upper part of the water column by 1°C is now sufficient to raise it by 1.03°C.

Furthermore, using equation 23, we can tell that since the 1940s, the mean temperature of the upper layers of the Dead Sea has risen by 0.72°C solely as a result of changes in specific heat. This increase does not require heat income. In the section entitled "Effect of salinity changes on evaporation," I stated that the surface temperature could increase by 2.2°C to compensate for the reduced latent heat by reducing evaporation resulting from the salinity increase. This value refers to surface conditions. Therefore, heat must have been transferred to the lake surface to raise its temperature by an additional 1.5°C. The actual increase in the temperature of the water column is determined by the mixing processes in the Dead Sea, and it depends also on the changes of both mass and area of the lake.

When heat is transferred from the surface to deeper parts of the water column, less heat is available for evaporation. Therefore, for the same water and air conditions as reported above, the calculated evaporation rates from the Dead Sea would be lower than those presented in Table 11-2. Yet this does not necessarily mean that the actual evaporation rate is lower than that presented previously, because the data used for the calculations are based on averages over several years. But under nonequilibrium conditions, surface water may have had different characteristics than the average. For instance, higher than average surface salinity may be reflected in the temperature, resulting in underestimation of evaporation rates.

It is reasonable to assume that the Dead Sea has been under nonequilibrium conditions most of the years because of changes in inflow rate and the seasonal pattern, and perhaps also because of variations in climatic conditions. A detailed account of this subject requires analysis of the vertical structure of the Dead Sea.

MONTHLY EVAPORATION RATES

Seasonal changes in the physical properties of a lake, and therefore of its evaporation rate, are basically changes that occur in a nonequilibrium state. These changes are, however, larger than the year-to-year variations. During each month, a certain amount of heat is transferred through the lake layers. Overestimation of this heat (for instance, by neglecting to consider the seasonal increase of salinity) may result in underestimation of evaporation rates.

Monthly evaporation rates depend on the vertical structure of the water column. In the Dead Sea, this has changed dramatically over the past 50 years. In general, meromictic stratification, where mixing is limited to the upper part of the water column, may result in lower winter and spring surface temperatures than in other years. This is because the water at a lower salinity could be cooler and still lighter than the warmer, more saline water below. When vertical mixing is more pronounced, the mean surface temperature may increase because of the lower salinity gradient and the relatively high temperature of the Dead Sea water column. Because the vertical structure of the Dead Sea is largely dependent on the amount of seasonal income of fresh water, the year-to-year variations of the vertical structure of the water column are large for most months. Therefore, averaging any monthly property over several years should be done cautiously.

Several other considerations may be taken into account when estimating monthly evaporation rates. For instance, the energy balance method introduces some bias in the calculation of monthly evaporation rates because of its basic assumption: The Bowen Ratio (eq. 16) assumes the mass transfer and the heat transfer coefficients to be equal in value. The ratio is, in effect, largely dependent on the air stability. During inversion conditions, the heat transfer coefficient in some lakes is 20-40% smaller than the mass transfer coefficient. This may result in overestimation of evaporation rates, especially during the summer.

Another bias is derived from neglecting to consider thermal expansion or salinity contraction of the water column. This is particularly important for estimating evaporation using the mass balance method. Monthly estimates by Neumann (1958), based on various considerations, arrive at lower evaporation rates in the first half of the "Dead Sea year" (beginning in mid-February) compared to those in the second half. At least part of this difference could be the result of underestimation of summer evaporation and overestimation of winter evaporation, caused by neglecting the thermal expansion (or contraction) of the Dead Sea water. (These were calculated to be 5-10 cm during 1977-1979; Steinhorn, 1985).

This discussion demonstrates the importance of using a modified approach to monthly evaporation estimates. With average monthly evaporation rates of 12 to 15 cm a month and summer rates several times higher than those in the winter, error or bias of a few centimeters is therefore significant.

EFFECT OF THE SOUTHERN AREA ON EVAPORATION FROM THE CENTRAL NORTHERN PART OF THE DEAD SEA

The calculations done so far assume the Dead Sea to be horizontally uniform, with physical properties of sea and air measured in its northern part. A point at the center of the lake will receive cooler, fresher water from the north (Jordan River and springs) and from the east (Arnon River), and warmer and more saline water from the south. The amount of inflow from all directions was perhaps vastly different in the past, when the southern basin was in its natural (wet or dry) condition and the flow in the rivers was naturally controlled and dependent on winter precipitation and spring snow melt.

The southern parts of the lake, which include the southern part of the northern basin, are warmer and more saline than the northern areas because of the warmer climate, lower cloud cover, and shallower water. In the southern part of the northern basin, the whole water column (30- to 50-m depth) is consistently warmer than the water column in northern areas by about 2°C (Steinhorn, 1985). In its natural condition, the southern basin of the Dead Sea was only a few meters deep, and it exchanged water with the northern basin, over 300 m deep, through the Lisan straits. In the late 1960s, the western half of the southern basin was disconnected from the lake to serve as evaporation ponds for the production of salts. By the late 1970s, the water level had decreased below the straits level.

Since then there has been, on the average, a further decrease in the water level, resulting in further recession of the lake in the shallower southern parts.

How would the advected heat from the south affect the annual evaporation rate? As an example, I assume the mass coming from the south to be equivalent to that of the upper 25-m layer. This assumption means that the mixed layer is exchanging between northern and southern parts during the year. With a 2°C difference in temperature, the advected heat is approximately 3.6 kgcal/(cm²yr). Under these conditions, the volume of the evaporated water is calculated to be 140.7 cm/yr (Table 11-2g). Advection from the south therefore increases the evaporation rate from the main parts of the Dead Sea.

The effects of the Dead Sea Works activities on evaporation from the Dead Sea can be looked at in several ways. One is to consider the whole system of the Dead Sea–evaporation ponds. This system has no outflow, and the evaporation ponds are regarded as a tool to extend the evaporating area of the Dead Sea. This additional area increases the total evaporation, whereas the evaporation rate from the Dead Sea as part of the Dead Sea–evaporation ponds system is reduced because of the higher salinity of the ponds. The other way is to consider the Dead Sea alone. This lake is not entirely terminal, as water is pumped from it into the evaporation ponds. The ponds, in turn, provide sources of heat and salt (heavy "end brine"). The pumped (outflow) water does not change the evaporation rate of the Dead Sea directly, because it has no effect on the energy balance at the surface. Yet, the mass of the lake is reduced. The inflow of more concentrated brines into the lake increases its salinity and contributes to the radiative transfer of evaporation rate. Therefore, because of the ponds, there is a decrease in both the evaporation rate and the total evaporation from the Dead Sea. From both points of view, the evaporating ponds accelerate the rate at which the Dead Sea reduces evaporation rate.

PROGNOSTIC ESTIMATIONS OF EVAPORATION RATE

Diagnostic evaporation estimates are based on a wealth of information on the physical properties of the lake and the air above it. Prognostic estimates require knowledge of the physical characteristics of the lake water and their feedback effects on evaporation. In particular, the prognostic mixing pattern of the lake should be modeled because it largely determines the surface properties of salinity and temperature.

What would be the evaporation rate from lakes of different salinity if they were located at the site of the Dead Sea? There are at least three ways of approaching this problem: The "classic" way is to assume similar water characteristics (with the exception of salinity) and the same climatic conditions as for the Dead Sea. Because of the difference in vapor pressure, a lower-salinity lake would be expected to have a higher evaporation rate. In this case, because of the various heat processes involved, the differences in vapor pressure cannot be the only factor responsible for the difference in evaporation rates (as discussed in "Effect of salinity changes on evaporation"). If the lake temperature remains the same, there must be some difference in the wind speed to adjust the sensible heat transfer to compensate for the changes in latent heat. The classic way, therefore, implies some changes in climate. With this in mind, we find that a freshwater lake with the same temperature as the Dead Sea in the 1940s would have an evaporation rate of 182 cm/yr for climate 1a (Table 11-3a), whereas the Dead Sea would have 166 cm/yr (Table 11-3e, as in Table 11-2b). For the conditions of the 1980s, a freshwater lake would evaporate 153 cm/

Table 11-3 Evaporation rates and water temperatures of hypothetical lakes of various salinitites.

No.	Air temperature (°C)	Relative humidity (%)	Short-wave radiation [kgcal/(cm²yr)]	Water salinity g/kg	Water temperature (°C)	Water density (g/cm³)	Evaporation volume (cm/yr)
3a	23.6	55	175	0	24.7	1.00	182
3b	23.6	55	175	0	21.1	1.00	271
3c	20.0	55	175	0	21.1	1.00	204
3d	23.6	55	175	40	21.3	1.02	265
3e	23.6	55	175	214	24.7	1.17	166
3f	23.6	55	175	244	25.0	1.20	147
3g	23.6	55	175	276	26.9	1.23	116
3h	25.4	45	157	0	27.2	1.00	153
3i	25.4	45	157	0	22.0	1.00	240
3j	20.9	45	157	0	22.0	1.00	163
3k	25.4	45	157	276	27.2	1.23	130

3a. Evaporation from a pure water lake having the same temperature as the Dead Sea in 3e. The radiation, air temperature, and humidity are as in 3k, but the wind speed should be different.

3b, 3d, 3f, and 3g. The temperatures of several lakes of various salinities are adjusted with salinity to fit all climate conditions as assumed for the 214 g/kg Dead Sea in 3e.

3c. Evaporation from the pure water lakes having the physical characteristics as in 3b and 3i. The air temperature is 1.1°C lower than the lake temperature.

3e. Evaporation from the Dead Sea, where air and lake conditions are as in the 1940s (Table 11-2b).

3h. Evaporation from a pure water lake having the same temperature and climate conditions (except the wind) as the Dead Sea in 3k.

3i. The temperature of the pure water lake is adjusted with salinity to fit all climate conditions as for the 276 g/kg Dead Sea in 3k.

3j. Evaporation from the pure water lakes having the physical characteristics as in 3i. The air temperatures is 1.1°C lower than the lake temperature.

3k. Evaporation from the Dead Sea, where air and lake conditions are as in the 1980s (Table 11-2f).

yr at climate 1c (Table 11-3h). The Dead Sea, with higher salinity and temperature than before, would evaporate 130 cm/yr (Table 11-3k).

The second way of comparing evaporation rates from lakes of different salinities is to calculate the salinity-induced changes in both lake evaporation and temperature while assuming no change in the air above the lake. With the climate of the 1940s, a freshwater lake would have a temperature of 21.1°C (compared to 24.7°C for the Dead Sea). Although its temperature is lower, its evaporation rate is significantly higher (271 cm/yr in Table 11-3b, compared to 182 cm/yr in Table 11-3a). A sea water lake would have a slightly higher temperature of 21.3°C and evaporation rate of 265 cm/yr (Table 11-3d). Higher temperatures and lower evaporation rates are calculated for Dead Sea water of higher salinities (Table 11-3f, 11-3g). The second way of comparison produces higher evaporation rates for fresh or low-saline water because of the gain in sensible heat.

The air above the lake is in constant contact with the water at the surface. In large water bodies, the air may follow the temperature changes of the lake surface. The third way of comparison would, therefore, require coupled air-water calculations. If, for instance, the air adjusts its temperature to be always 1.1°C cooler than the lake, the calculated evaporation rates of a freshwater lake would be 204 cm/yr and 163 cm/yr for the two climates considered (Table 11-3c, 11-3j). Such adjustment of the air temperature may keep the lake evaporation within a moderate range of change.

The changes in the physical properties of the lake and the overlying air are linked through various heat transfer mechanisms, including evaporation. Attempts to evaluate future and past evaporation rates from the Dead Sea should therefore be based, if possible, on a comprehensive model of the lake and of the air above it.

CONCLUSIONS

Methods of estimating evaporation have been developed and applied for both fresh water (lakes and reservoirs) and saline water (oceans). In fresh and low-saline water bodies and in the ocean, the water temperature plays a major role. Most physical properties of the water, including vapor pressure and density, are largely dependent on temperature. Therefore, the evaporation rate is largely determined by the temperature variations, both directly (at the surface) and indirectly (through the density-driven mixing and heat transfer within the water body). The effect of dissolved salts on the evaporation rate is generally taken into consideration by assuming the appropriate values of vapor pressure and latent heat, which are both dependent on salinity. In this chapter, I suggest the importance of another physical parameter-changes in salinity.

Why should salinity changes, which in most situations have a minor effect on the vapor pressure, be considered? One reason that salinity changes are important is their indirect effect on the temperature changes. The other main reason is that salinity changes largely contribute to the changes in the mixing pattern. Both effects are more pronounced in highly saline water or when there is inflow of fresh or lower-salinity water.

Evaporation estimates for very saline water should therefore rely on salinity measurements. Depending on the specific method used, evaporation estimates should be based on knowledge of the relevant lake and water properties and their feedback effects on evaporation. Some limitations exist because of the limited knowledge of several of the mechanisms involved (like salt precipitation), the lack of sufficient information on physicochemical properties (like activity coefficient), and the status of measuring techniques (in particular, continuous, in-situ salinity).

A reliable estimate of the evaporation rate from the Dead Sea should be based on a combination of several methods for comparison and verification of the results. It is also reasonable to expect the evaporation rate to change with the physical conditions of the lake and its water. At this time, approximate estimate of the evaporation rate from the Dead Sea can be obtained based on the energy balance method. As shown in this article, with the traditional use of the energy balance, it is possible to arrive at quite a large range of numbers. Yet, moderate values are achieved with a choice of a model lake and overlying air that consistently change in response to salinity changes.

REFERENCES

Anati, D. A., and Stiller, M., 1991, The post-1979 thermohaline structure of the Dead Sea and the role of double-diffuse mixing: *Limnology and Oceanography*, v. 36, p. 342–354.

Anderson, E. R., 1952, Energy budget studies, *in* Water loss investigation, Lake Hefner studies: U.S. Geological Survey Technical Report, v. 1, p. 71–119.

Arons, A. B., Kientzler, C. F., 1954, Vapor pressure of sea salt solutions: *Transactions of the American Geophysical Union*, v. 35, p. 722–726.

Ashbel, D., 1939, The influence of the Dead sea on the climate of its neighborhood: *Quaternary Journal of the Royal Meteorology Society*, v. 65, p. 185–194.

Ashbel, D., 1944, Fifteen years' observations on the climatology and hydrography of the Dead Sea: Jerusalem, The Hebrew University, 114 p.

Brutsaert, W. H., 1982 Evaporation into the atmosphere: Theory, history, and applications: Boston, D. Reidel, 299 p.

Jobson, H. E., 1972, Effect of using averaged data on the computed evaporation: *Water Resources Research*, v. 8(2), p. 513–518.

Jones, F. E., 1992, Evaporation of water with emphasis on applications and measurements: Michigan, Lewis Publications, 188 p.

Morton, F. I., 1986, Practical estimates of lake evaporation: *Journal of Climate and Applied Meteorology*, v. 25, p. 371–387.

Neumann, J., 1953, Energy balance of and evaporation from sweet-water lakes of the Jordan rift: *Israel Research Council Bulletin*, v. 2, p. 337–357.

Neumann, J., 1958, Tentative energy and water balances for the Dead Sea: *Israel Research Council Bulletin*, v. 7G, p. 137–163.

Rozenan, N., and Shore, Z., 1952, Measuring evaporation with Piche's technique: Meteorological Notes, No. 1, Israel Meteorological Service, Office of Transportation, 16 p.

Salhotra, A. M., Adams, E. E., and Harleman, D. R. F., 1985, Effect of salinity and ionic composition on evaporation: Analysis of Dead Sea evaporation pans: Water Resources Research, v. 21, p. 1,336-1,344.

Salhotra, A. M., Adams, E. E., and Harleman, D. R. F., 1987, The alpha, beta, gamma of evaporation from saline water bodies: *Water Resources Research*, v. 23(9), p. 1,769–1,774.

Segal, M., 1983, A study of meteorological patterns associated with a lake confined by mountains—The Dead Sea case: *Quaternary Journal of the Royal Meteorology Society*, v. 109, p. 549–564.

Stanhill, G., 1987, The radiation climate of the Dead Sea: *Journal of Climatology*, v. 7, p. 247–265.

Stanhill, G., 1990, Changes in the surface temperature of the Dead Sea and its heat storage: *International Journal of Climatology*, v. 10, p. 519–536.

Stanhill, G., 1994, Changes in the rate of evaporation from the Dead Sea: *International Journal of Climatology*, v. 14, p. 465-471.

Steinhorn, I. 1980, The density of Dead Sea water as a function of temperature and salt concentration: *Israel Journal of Earth Science*, v. 29, p. 191–196.

Steinhorn, I., 1983, In-situ sale precipitation at the Dead Sea: *Limnology and Oceanography*, v. 28(3), p. 580–583.

Steinhorn, I., 1985, The disappearance of the long term mermictic stratification of the Dead Sea: *Limnology and Oceanography*, v. 30(3), p. 451–472.

Steinhorn, I., 1991a, On the concept of evaporation from fresh and saline water bodies: *Water Resources Research*, v. 27(4), p. 645–648.

Steinhorn, I., 1991b, Salt flux and evaporation: *Journal of Physical Oceanography*, v. 21(11), p. 1,681–1,683.

Steinhorn, I., 1992, Reply to Comment by Salhotra, Adams and Harleman on "On the concept of evaporation from fresh and saline water bodies": *Water Resources Research*, v. 28(5), p. 1,493–1,495.

APPENDIX

By evaporation, water molecules transfer from the liquid to the gas phase. The water vapor is removed from the lake surface by the vertical movements of the air and transported to higher levels. The evaporation rate E_w depends on the ability of the air to absorb and transport the added vapor. Dalton's Law gives a semiempirical, quantitative way to express this rate:

$$E_w = C_E \rho_a U \left(\frac{M_V}{M_d} \right) \frac{(e - e_a)}{P} \qquad (1)$$

The evaporation rate E_w is given in mass per unit are per unit time; C_E is the mass transfer coefficient; ρ_a is the air density; U is the wind speed at a level above the surface; $M_v/M_d \approx 0.622$ is the ratio of the molecular weights of water and dry air; P is the air pressure; e_a is the vapor pressure of the air and it depends on the relative humidity. The saturated vapor pressure of water at the surface temperature T and salinity S, $e(S, T)$, is lower for higher salinities. C_E may also depend on salinity because of its dependency on the air stability. Equation 1 and other similar forms are used for evaporation estimate using the mass transfer, profile, and bulk aerodynamics methods, where the existence of a vertical gradient of the mean value of the water vapor in the turbulent medium is assumed. The fluctuating values of water vapor are correlated with the vertical component of the eddy motion (product of the density and velocity fluctuations). Short-time scale eddy correlation methods are based on this property.

The evaporated mass, E_w, is initiated from mass E_s of saline water that has originally also contained a mass of dissolved salts, M_s:

$$E_S = E_W + M_S \qquad (2)$$

When E_w is expressed per unit time per unit area, it is equivalent to the evaporation rate. For the Dead Sea (S = 0.28 g/g), the evaporating mass E_s is considerably larger than the evaporated mass E_w because $E_w = E_s(1 - S)$.

The volume of the evaporated pure water if condensed to water at the lake temperature is:

$$V_w = \frac{E_w}{\rho_w} \quad (3)$$

where ρ_w is the pure water density at the lake temperature T. For convenience, V_w is taken, in this chapter, as the evaporation rate in volume units, or the evaporation volume. This quantity is different from several others that give similar values in the case of pure water. One of them is the volume of evaporating saline water, V_s:

$$V_S = \frac{E_S}{\rho} \quad (4)$$

The volume reduction of the lake, ΔV, may be subdivided as follows:

$$\Delta V = V_e + V_T \quad (5)$$

where V_e is the reduction in volume that results from water loss and contraction by salinity increase, and V_T is the volume reduction that results from the temperature decrease following evaporation.

The relations between the quantities may be derived from conservation laws and the equation of state. For example,

$$V_e = E_W \frac{(1 + \beta S)}{\rho} \quad (6)$$

and

$$V_T = \frac{\alpha L E_W}{\rho c_\rho} \quad (7)$$

V_e may be expressed as $V_w - V_c$, where V_c may be looked at as the difference between the combined volumes of remaining salts and lake water before and after their remixing. Because V_c is relatively small, the reduction of level resulting from water loss, V_e, is a little less than, but closely related to, V_w. However, V_e is quite different in value from

$$V_{WS} = \frac{E_W}{\rho} \quad (8)$$

which is the volume of saline water of the same salinity as the lake's required to keep the mass of the lake constant during evaporation.

In equations 6 and 7, the terms α and β are derived from the equation of state in its linear form:

$$\Delta \rho = -\alpha \rho \Delta T + \beta \rho \Delta S . \quad (9)$$

For the Dead Sea, we define the salinity in terms of total dissolved salts: $\alpha = (1/\rho)(\partial \rho / \partial T)$ and $\beta = (1/\rho)(\partial \rho / \partial T)$. According to Steinhorn (1980), $\delta \rho / \delta S$ is 0.91–0.92 · 10^{-3} g/cm³, $\partial \rho / \partial T$ is 0.40–0.43 · 10^{-3} g/(cm°K), and a reference value for ρ is chosen as 276 g/kg for ρ = 1.2318 g/cm³ at 25°C. The equation of state of Dead Sea water indicates the thermal expansion parameter to be higher than that for fresh or sea water.

The mass balance equation in its correct form is

$$\rho_2 V_2 = \rho_1 V_1 + \sum \rho_{in} V_{in} - \sum \rho_{out} V_{out} + P_r - E_W - M_{slt} \quad (10)$$

where terms for precipitation, P_r, and salt deposition, M_{slt}, are also included.

The salt balance equation may be expressed in a more general form as

$$\Delta \int S\rho V + S_{in}\rho_{in}V_{in} - S_{out}\rho_{out}V_{out} - M_{slt} \quad (11)$$

One can use equation 11 in combination with equation 10 to derive the amount of saline water flowing into or out of the lake; or a modified form, $\Delta(\int S\rho V) = 0$ (assuming no massive salt precipitation or salt sources), may be used to estimate the amount of freshwater that left the system. The latter approach can give much more accurate results than an estimate based on water level measurements (Steinhorn, 1992).

The energy balance method estimates the heat loss by evaporation, $Q_{EVP} = L E_w$ where L is the latent heat. It is derived from the energy balance equation

$$Q_{NET} = (Q_S + Q_A) - (Q_{WR} + Q_{CON} + Q_{EVP} + Q_{ADV}) \quad (12)$$

The lake absorbs heat during the warming period until midsummer and loses it during the cooling period. Q_{NET} is the resultant of the heat exchange processes; Q_S is the heat absorbed in the lake as a result of short-wave solar radiation (incident total short-wave radiation minus the reflected); Q_A is the atmospheric long-wave radiation minus the part of this radiation reflected from the lake surface; Q_{WR} is the heat lost by the water long-wave radiation; Q_{CON} is the heat lost by conduction to the atmosphere; and Q_{ADV} is the heat lost or gained by advection. There are several ways to estimate, calculate, or measure the various terms in this equation. For instance, a basic empirical equation for infrared radiation from the air or from the water surface is:

$$Q_{WA} = \varepsilon \sigma (T + T')^4 \quad (13)$$

where T is the air or water temperatures in degrees Celsius, T' is the temperature 0°C in degrees Kelvin (273.15°K), ε is the emissivity in relation to black body, and σ is Stephan-Bolzmann constant. The advected heat may be expressed as

$$Q_{ADV} = \sum (c_p T - c_{pin} T_{in}) M_{in} \quad (14)$$

where c_{pin}, T_{in}, and M_{in} are the specific heat, temperature, and mass of the incoming water, respectively.

The sensible heat is

$$Q_{CON} = C_H c_{pa} \rho_a (T - T_a) U \quad (15)$$

where c_{pa} is the specific heat of air at a constant pressure and C_H is the heat transfer coefficient. To avoid difficulty of estimating the mass transfer coefficient, C_E (eq. 1), and the heat transfer coefficient, C_H (eq. 15), it is assumed that C_E is equal to C_H. The evaporation rate may then be calculated using the Bowen Ratio, the ratio between the heat transferred by conduction and the heat lost be evaporation:

$$R_B = \frac{Q_{CON}}{Q_{EVP}} = 0.622 \frac{(e - e_a)}{c_{pa} P (T - T_a)} \quad (16)$$

where the air pressure, P, is high in the Dead Sea area, about 1,060 mb.

When there is no net gain or loss, $Q_{NET} = 0$, and the energy balance equation, therefore, takes the form

$$Q_{EVP} = \frac{Q_S + Q_A - Q_{WR} - Q_{ADV}}{1 + R_B} \qquad (17)$$

When there is net gain or loss of heat, it is assumed that heat is transferred to or from the deeper layers of the lake to increase the heat content of the lake, Q_L. The change in heat content per unit time is assumed to represent Q_{NET} when no apparent other heat sources exist:

$$Q_L = Q_{NET} \qquad (18)$$

With the traditional formula, Q_L is proportional to ΔT:

$$Q_L = c_p \rho \int \Delta T A \, dh \qquad (19)$$

where the integration is over the lake layers of height h having different areas, temperatures, etc. It is assumed that c_p and p are constant, and there is no change in water balance. For the period between peak summer and peak winter, usually Q_{NET} signifies the annual heat budget (heat storage) of the lake.

It is my opinion that, because of changes in the specific heat capacity, salinity, and water budget, the traditional calculation may not be sufficient for correctly presenting conservation of enthalpy. I suggest the following alternative

$$Q_L = \Delta \int c_p T \, dM \quad \text{where } Q_L \neq Q_{NET} \qquad (20)$$

The integration is on all lake layers, where, for each one,

$$\Delta (M c_p T) = M T \Delta c_p + M c_p \Delta T + c_p T \Delta M \qquad (21)$$

The specific heat depends on both temperature and salinity. In linear form,

$$\Delta Cp = \frac{\partial c_p}{\partial T} \Delta T + \frac{\partial c_p}{\partial S} \Delta S \ . \qquad (22)$$

Q_{NET} is a major, but not sole, contributor to ΔT, and it may be derived for Q_L, where the effect of ΔM on the heat content is extracted, and the salinity effect on ΔT and ΔCp are taken into consideration. The relative effect of salinity changes may be estimated as:

$$r_S = \frac{T \Delta cp}{cp \Delta T} \ . \qquad (23)$$

If $r_s > 1$, the temperature changes of the water column are mainly the result of salinity changes.

12. EVOLUTION OF THE DEAD SEA BRINES

ISRAEL ZAK

Since the beginning of its subsidence in late Neogene times, the Dead Sea rift has accumulated up to 10,000 m of evaporites and clastics (the Dead Sea Group). The thick rock salt sequence of the Sedom Formation represents the marine evaporites deposited in the subsiding distal part of the Gulf of Sedom. Overlying it are lacustrine deposits, including evaporites, up to and including those of the present Dead Sea itself.

The Dead Sea Group and the surrounding and underlying prerift formations contain a great volume of interstitial and effluent brines. Commonest among the brines is the Dead Sea brine, of Ca-chloridic composition, with over 332 g/l dissolved salts and with a uniquely high Br content (over 5 g/l). The lake, composed of this brine, fills a deep trough in the northern Dead Sea basin and is considered an outcrop of the surrounding underground fossil brine. This, and the brines known from springs and boreholes around the lake, can all be derived from the mixing and chemical evolution of three basic brines. The first is a diagenetic brine, originally evaporated seawater of the Gulf of Sedom and later modified by Mg÷Ca exchange into Ca-Mg-Na-K-Cl-Br water (with $Ca^{2+} > HCO_3^- + SO_4^{2-}$). In this brine, the original (marine) Mg has been exchanged for the Ca of carbonate sediments of the Dead Sea Group and of the surrounding prerift Cretaceous and older carbonate rocks, and most of the original sulfate has been lost by precipitation with this sequestered Ca. The second type of brine consists of meteoric waters, dilute to mesohaline, whose composition has seawater affinity (airborne seasalts). Metamorphic brine, the third type, is the result of incongruent alteration and dissolution of hydrous evaporite minerals (e.g., carnallite and gypsum) of the Dead Sea Group. The highly soluble salts are selectively and alternately retained in solution, in successive lakes, or in temporary solid deposits during periods of desiccation.

Dead Sea brine, in the strictest sense, is modified and retained by interacting with these solutions and by multiple progressive-regressive evaporation cycles, with occasional desiccation. The processes are also controlled by mixing along hydraulic pathways created by the rift's marginal faults, which provide porosity and enable brines to sink to depths of geothermal heating. Brine properties and behavior are explained, illustrated, and predicted by application of Jaenecke's Gibbs' diagrams. The evolution of the Dead Sea brines is a characteristic model for the hydrochemistry of continental rift basins temporarily connected with the ocean.

THE DEAD SEA TODAY

The Dead Sea, a hypersaline terminal lake occupying the lowest part of the continental Dead Sea rift, succeeds a series of Pleistocene to Recent lakes, which in turn succeeded an evaporative marine gulf of Pliocene age—the Gulf of Sedom. Altogether, these water bodies deposited up to 10,000 m of rock salt (and some carnallite and sylvite), with associated clastics, together known as the Dead Sea Group. In places, salt bodies from this sequence rise in diapirs, one of which has broken to the surface, forming the monumental salt ridge of Mount Sedom (Figs. 12-1 and 12-2).

The Dead Sea itself consists of about 150 km³ of hypersaline

brine. With over 330 g/l dissolved ions, it is 10 times more concentrated than seawater. However, unlike the sulfatic seawater, the Dead Sea brine has essentially a Ca-chloridic composition ($Ca^{2+} > HCO_3^- + SO_4^{2-}$), saturated with respect to carbonate and sulfate, both present in very small concentrations, and is close to halite saturation. Carnallite is the next mineral to precipitate, after evaporation to about half the volume (Epstein, 1976; Menczel et al., 1980). The lake, which reaches a depth of about 325 m, may be regarded as an outcrop of brine, partly fossil and partly of recent derivation, that pervades both the Dead Sea Group and the downfaulted Mesozoic formations flanking and underlying it.

Basically, the Dead Sea brine (hence DS brine) is of an Mg-Ca-K-Na and Cl-Br type, derived from the highly evaporated seawater of the Gulf of Sedom. This residue underwent modifying and mixing processes, combined with multiple progressive and regressive evaporation cycles, with periodic precipitation of halite and occasionally of carnallite.

This chapter discusses the origin of the brines and the hydrochemical pathways by which they evolved into the present DS brine and into other derivatives found in the Dead Sea basin.

GEOLOGICAL BACKGROUND

The Dead Sea basin, despite its barrenness, has been inhabited continuously since prehistoric times. Jericho to the north of the Dead Sea and Tso'ar at the south end (Fig. 12-1) were thriving settlements. Disastrous geological events are probably background to the biblical story of the destruction of Sodom and Gomorrah (Sedom and 'Amora in Hebrew), viz. the nomenclature chosen for units of the Dead Sea Group (Fig. 12-2). Salt obtained from the DS brine (salt of Sodomith), pure and sulfate-free, was used for sacerdotal purposes in the Second Temple (sixth century B.C.–first century A.D.; Rozenson and Zak, 1987). Rabbi Eli'ezer ben Hyrcanos (first century A.D.) is reputed to have described the daily rising of the salt of Mount of Sedom; this is probably the first description, admittedly homiletic, of a rising salt diapir (Zak, 1967).

Geographically, the Dead Sea is a significant feature that accompanies major crustal rifting. The northern basin of the Dead Sea (today's remaining lake) fills a tectonic depression, a pull-apart basin, associated with the Dead Sea-Jordan transform, which separates the Sinai-Israel subplate from the Arabian plate (Garfunkel et al., 1981; Zak, 1967; Zak and Freund, 1981; Garfunkel, chapter 4, this volume). The northern Dead Sea trough is about 45 km long and 16 km wide, and its level floor is about 730 m below mean sea level (MSL). The western shoulder of the rift, which constitutes the Judean Desert, reaches a few hundred meters above MSL, and the eastern shoulder, the highland of Moab, rises to 1,000–1,200 m above MSL. The shoulders and Dead Sea valley are separated stepwise by long, normal-faulted blocks, the lowest of which are buried under thousands of meters of rift fill (Figs. 12-1 and 12-2). The faulted blocks on the west side expose mostly Cretaceous carbonates and clastics (of the Kurnub and Judea groups, Figs. 12-2 and 12-3). Along the eastern side of the Dead Sea, older and deeper layers are exposed, and toward the south the

133

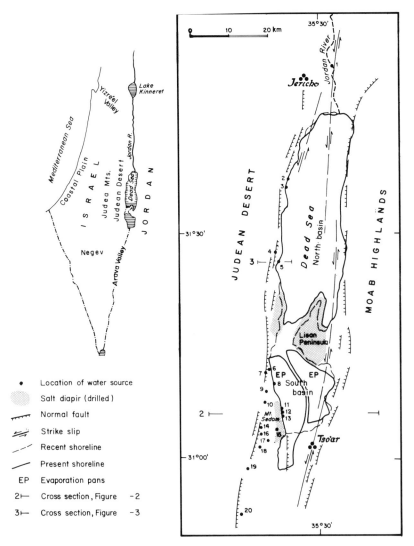

Figure 12-1 Tectonic configuration of the Dead Sea basin, and location of selected water sources (faults according to Bentor et al., 1965; Zak, 1967).

List of water sources; analysis number (Table 12-1, 12-2, and Fig. 12-4) in parantheses.

1	Jordan River	(31, 32)
2, 3	'Ein Kaneh-"Ein Samar	(19-21, 28-30)
4	Nahal "Arugot	(27)
5	Hammei Yesha'	(14)
6	'Ein No'it	(25)
7	Nahal Boqeq	(23, 24)
"	'Ein Boqeq	(26)
8	Hammei Zohar	(15)
9	Zohar-1	(16)
10	Lot-1	(5, 6)
11	'Ein Ashlag	(10)
12	Tamilleh	(13)
13	'Ein Siddim	(11)
14	Sedom-1	(12)
15	Amiaz-1	(17)
16	Amiaz-2	(18)
17	Sedom-2	(8)
18	Heimar-1	(9)
19	Tamar-3	(22)
20	Arava-1	(7)
	Dead Sea	(1-4)
	Ocean	(33)

igneous Precambrian basement rises to form the eastern wall of the rift (Fig. 12-2).

The basal strata, which are buried and downfaulted, of the Dead Sea Group are not well known. They are overlain by Pliocene brackish-marine sediments, deposited in the marine gulf that entered the rift valley from the north, followed by a very thick (over 4,000 m) marine evaporite series, the Sedom Formation (Fig. 12-2). This formation consists of bedded rock salt with interlayers of anhydrite, gypsum, and dolomitic shales, as well as local sequences with carnallite, sylvite, polyhalite, and kieserite. Polygonal cracks (tepee structures) in the Sedom rock salt, indicate periods of exposure and desiccation. The higher parts of the Dead Sea Group, the 'Amora and Lisan Formations, are lacustrine, consisting of alternating or varve-like series of thinly bedded aragonite and detrital carbonates, some gypsum and halite, and conglomerates.

Both the marine evaporitic Sedom Formation and the lacustrine 'Amora and Lisan Formations contain occasional highly soluble connate K-Mg-Ca-chloride salts within their pore spaces, in places amounting to 10% of the rock weight. Their presence testifies to episodes of advanced evaporation following primary diagenesis (Zak 1967, 1974, Zak et al., 1992).

The Sedom Formation has risen in salt diapirs, piercing through the overlying strata. These include (1) the large Lisan Peninsula salt diapir (found under an area of about 8 x 15 km²), a pillow structure that has pierced through 'Amora and Lisan sediments up to 100 m from the surface; (2) the exposed diapir of Mount Sedom, a salt wall structure (11 km long and 2 km wide exposed at the surface, Fig. 12-2); and (3) the Tse'elim structure (west of the Lisan Peninsula), still covered by several hundred meters of younger sediments (Zak, 1967; Zak and Freund, 1981). Neev and Hall (1979), using geophysical methods, found numerous smaller diapirs rising through layered Dead Sea sediments under the northern basin.

The history of the Dead Sea Group, its sediments, and its fluids, is marked by alternating phases of progressive and regressive evaporation (i.e., salt of more evaporated brine alternating with salt of less evaporated brine). Progressive evaporation reached the stage of halite precipitation throughout most of the Sedom Formation, at times progressing to carnallite precipitation and at times reaching virtual desiccation, as indicated by mineral associations, desiccation structures, and the hydrochemistry of the interstitial connate brines (Zak, 1974). The long history of deposition of chloride minerals has left its mark in the uniquely high bromine content and high Br/Cl ratio of the lake's waters. Today's Dead Sea may be regarded as a body

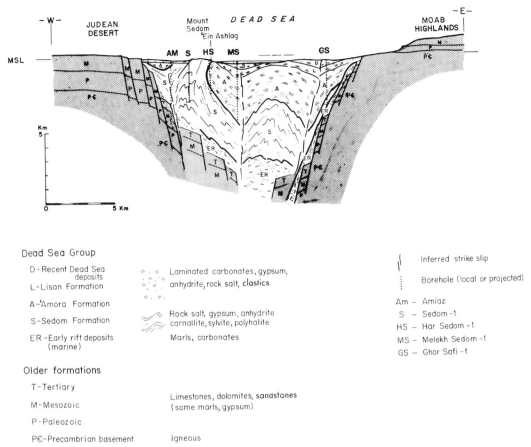

Figure 12-2 Generalized cross section through the Dead Sea rift, at the latitude of Mount Sedom (modified from Zak, 1967; Zak and Freund, 1981).

Dead Sea Group

D - Recent Dead Sea deposits
L - Lisan Formation
A - 'Amora Formation
S - Sedom Formation
ER - Early rift deposits (marine)

Laminated carbonates, gypsum, anhydrite, rock salt, clastics

Rock salt, gypsum, anhydrite carnallite, sylvite, polyhalite

Marls, carbonates

Older formations

T - Tertiary
M - Mesozoic
P - Paleozoic
P€ - Precambrian basement

Limestones, dolomites, sandstones (some marls, gypsum)

Igneous

Inferred strike slip
Borehole (local or projected)

Am — Amiaz
S — Sedom -1
HS — Har Sedom -1
MS — Melekh Sedom -1
GS — Ghor Safi -1

of diluted carnallitic brine, which has reached its present state through a stage of progressive carnallite deposition followed by dilution and regressive evaporation. At present, the Dead Sea is again undergoing progressive evaporation, considerably influenced by human activities.

BRINES OF THE DEAD SEA BASIN: PREVIOUS STUDIES

The chemical and physical characteristics of the Dead Sea and related saline waters have aroused curiosity and interest for centuries. The first analysis of Dead Sea water is imputed to Lavoisier in 1772 (for a review see Nissenbaum, 1970), but systematic scientific study began only in latter decades. M. A. Novomeysky and M. R. Bloch (of the Palestine Potash Company, precursor to the present Dead Sea Works) were pioneer investigators of the evaporational path of the Dead Sea waters. Bloch and Schnerb (1953) were among the first to use bromine as an indicator for the history of evaporation and precipitation from seawater and subsequent brines, as well as from the Dead Sea. An outstanding systematic study of the Dead Sea was made by Neev and Emery (1967), who presented a comprehensive limnological review of all the lake's aspects: physiography, hypsometry, temperatures, currents and stratification, sediments, and hydrochemistry. Lerman (1967), Starinsky (1974), and Gavrieli et al. (1989) studied some physicochemical aspects of the Dead Sea brines, and Menczel et al. (1980) studied the $NaCl-KCl-MgCl_2-CaCl_2-H_2O$ system and the stability fields (between 10° and 40° C) of the pertaining minerals. Mazor and Rosenthal (1967), Mazor et al. (1969), Nissenbaum (1969), and Starinsky (1974) studied the hydrochemistry of numerous springs, wells, and deep boreholes in the Dead Sea basin and in

the Mesozoic rock formations along the western boundary of the rift, and defined water groups or water types.

Neev and Emery (1967), established the meromictic character of the Dead Sea, distinguishing an upper water mass down to about 40 m, a transitory zone down to about 70 m, and a denser, anoxic lower water mass down to the bottom at 330 m. The lake level dropped 0.36 m/y during the years 1969–1977 and about 0.8 m/y during 1981–1988 (Anati and Shasha, 1989), coinciding with a steady decrease in average rainfall, decrease of recharge, and an increase diversion of water for human consumption. As levels continued to drop, the lake's southern basin became separated (as it had done at times in the past), thus eliminating the main source of intensely evaporated brines that used to produce the long-term, three-layered meromictic structure of the northern basin (Neev and Emery, 1967; Steinhorn et al., 1979; Beyth, 1980).

Quantitative balance is a commonly used approach for explaining the evolution of the DS brines (Bentor, 1961; Loewengart, 1962; Neev and Emery, 1967). This approach is based on a quantitative evaluation of the ion input by the various water sources, compared to the actual composition of the DS brines.

Loewengart (1962), on the basis of quantitative data, considered airborne sea salts as an adequate source, which over a period of 1.5 Ma could have supplied the mass of salts dissolved in the Dead Sea and could also account for the salt deposits, as they were then known, in the sediments of the Dead Sea Group. Loewengart proposed the still-accepted basic mechanism for changing the Mg-chloridic-sulfatic marine-type brine into a Ca-Mg-chloridic type of brine, through Mg+Ca exchange between the brine and the solid Ca-carbonate of the

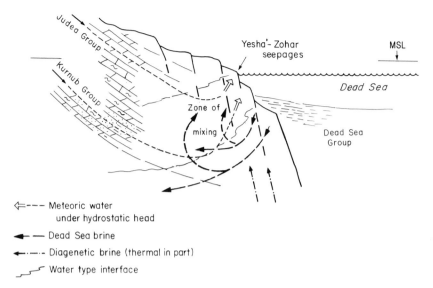

Figure 12-3 Schematic cross section, showing zone of mixing between Dead Sea brine, diagenetic (thermal) brine, and meteoric waters, forming (among others) Yesha'-Zohar type brine.

host rock, and subsequent precipitation of the dissolved sulfate ion in the form of $CaSO_4$ minerals (Loewengart, 1962; also cited by Braitsch, 1971; Loewengart and Zak, 1972, 1983; Starinsky, 1974). Loewengart explains the depletion in marine potassium by absorption and exchange in clays throughout the catchment area ($40,000$ km^2) of the Dead Sea. The clays filter the rainwater and its dissolved airborne salts as they flow lakeward. Zak (1974, 1980) accepted the mechanism of Mg÷Ca exchange and $CaSO_4$ precipitation, but showed that rather than arriving only as airborne salts (insufficient to account for the great amount of salt in the Dead Sea Group), seawater had entered the ancestral Dead Sea basin through a late Neogene gulf (the Gulf of Sedom), and in the course of long periods of evaporation, accumulated a thick marine evaporite deposit. Marine to brackish sediments of this gulf (predating the evaporitic sequence) can be traced, mainly in boreholes, from the Mediterranean Sea through the Valley of Yizre'el, southward along the Jordan Valley, and into the Dead Sea basin itself (Fig. 12-1; Zak, 1967). Also, the amounts of halite and K-chloride minerals making up the Sedom Formation (of the Dead Sea Group) account for much of the Na and K depletion in the DS brine compared to seawater (Zak et al., 1992). Some potassium may be lost to the synthesis of K-aluminosilicates at depth, a possible mechanism proposed by Carpenter (1978) for deep-seated brines.

Dense evaporational brines produced in the Gulf of Sedom could have escaped seaward during less progressive evaporation stages, but this path was repeatedly barred and eventually terminated by the dropping of ocean levels or by tectonism (Zak et al., 1992). Brines produced in the Gulf of Sedom and in the later 'Amora and Lisan rift lakes, are, however, present in the subsurface under the Dead Sea and within the surrounding older formations, as is indicated among other things by the composition of pore fluids and Br/Cl ratios (Zak, 1967, 1974). Starinsky (1974) identified brines of highly evaporated seawater, which he considers to have originated in the Dead Sea basin, as far as 100 km from the Dead Sea basin, along a wide belt through the northern Negev and the southern coastal plain of Israel.

WATER ANALYSIS DATA AND THEIR PRESENTATION

Tables 12-1 and 12-2 present selected analyses of water samples from the Dead Sea basin, and Figure 12-1 shows their geo-graphical locations. Samples were selected to represent the various water types and their characteristic variability. Wells and springs offer a wide choice of samples, though uncontaminated samples from deep boreholes are scant. Bromine data are often not available.

When low-salinity waters mix, the resulting solution shows the weighted average ionic composition, and only the ionic activities may change slightly. The quantitative evaluation of such unsaturated solutions and processes of mixing are best presented and studied by the classic methods of research and presentation. This is not the case with brines of high salinity and under evaporation, for which phase diagrams are useful and, where precipitation is involved, a necessity. When a wide spectrum of salinities is under study, as here, both approaches should be used. The present study, however, makes use of the research by Mazor et al. (1969), Mazor (chapter 25, this volume), and Starinsky (1974), for the low-salinity water types of the Dead Sea basin, and refers to the Gibbs' ternary diagrams (Fig. 12-4) for high-salinity brines.

As mentioned, all brine and water types went through histories of mixing, brine-rock interaction both before and after mixing, open-lake mixing between DS brine and incoming waters, and massive atmospheric evaporation. The processes are most clearly illustrated, following Braitsch 1971, by Jaenecke's use of Gibb's ternary diagrams (Fig. 12-4). This method enables portrayal and even prediction of processes, some of which have already been observed and described by Mazor et al. (1969), Mazor (chapter 25, this volume), Starinsky (1974), and Zak (1967, 1974, 1980).

On each such ternary diagram, three principal ions in the solution are considered as of 100% mole concentration, after subtraction of the amount of potential precipitate, e.g., $CaCO_3$ and $CaSO_4$ (concentration of other ions in the solution is calculated in relation to the sum of the three chosen ions). In our case, systems at saturation point for halite, with practically no dissolved $CaCO_3$ and $CaSO_4$ left, are considered. Thus, for the Mg-Ca-K$_2$ and for the Mg-K$_2$-SO$_4$ ternary diagrams (in Fig. 12-4), Ca^{2+} balanced by HCO_3^- and by SO_4^{2-} was subtracted from the total Ca^{2+} concentration before calculating the percentage of each in the sum of the three ions. For Ca-chloridic brines, defined by $Ca^{2+} > HCO_3^- + SO_4^{2-}$, the subtraction leaves some Ca^{2+} in solution. For chloridic-sulfatic brines, where $Ca^{2+} \leq HCO_3^- + SO_4^{2-}$, excess sulfate remains after Ca^{2+} is removed by precipitation. In Tables 12-1 and 12-2, analyses 1–21 are of the

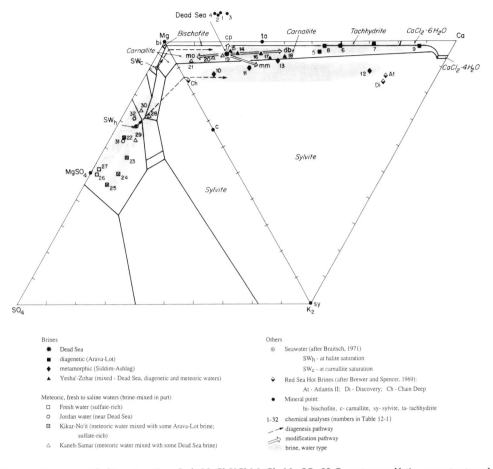

Figure 12-4 Gibbs' ternary diagrams at halite saturation. Left, NaCl-KCl-MgCl$_2$-Na$_2$SO$_4$-H$_2$O system; sulfatic, seawater-type brines; Right, NaCl-KCl-MgCl$_2$-CaCl$_2$-H$_2$O system; Ca-chloridic, Dead Sea type brines. (Mineral stability fields after Assarsson and Balder, 1955; Braitsch, 1971; Wood, 1975). Diagenesis pathways of evaporated seawater: Left triangle, the brine loses Mg and SO$_4$ through Mg÷Ca exchange and subsequent precipitation of Ca-sulfate. The brine shifts upward to the right, from the SW$_h$ point, if at saturation to halite, and from the SW$_c$ point, if at saturation to carnallite (away from the MgSO$_4$ mineral point). With progressive Mg and Ca exchange, the brine shifts into and along the right triangle; Right triangle, the brine loses Mg and gains more Ca, leading to a Ca-chloridic, Dead Sea type brine and to a diagenetic ('Arava-Lot) type brine. Modification pathways of Dead Sea brines, along arrows: mo—mixing with meteoric waters (tributaries and springs); db—mixing with diagenetic brines ('Arava-Lot type); mm—mixing with metamorphic brines (Siddim-Ashlag type); cp—the brine shifts toward the Mg-Ca side by precipitation of carnallite, away from carnallite point c on the Mg-K$_2$ side.

Ca-chloridic type and analyses 22–32 are of the chloridic-sulfatic type.

DESCRIPTION OF THE WATER TYPES

Typical Dead Sea water, with about 332 g/l dissolved ions (Table 12-1, analysis 3; 27 wt% in a solution of specific gravity 1.242) is a natural brine of high salinity, consisting of water with Mg>Na>Ca>K and Cl>Br and only traces of SO$_4$ (about 0.1 wt%). It is unique in being an open body of water with a Ca-chloridic composition and with over 5g/l of bromine (analyses 1–4 in Tables 12-1 and 12-2 and Fig. 12-4).

Dead Sea brine is believed to be the result of mixing between brines of three basic source types (Table 12-1, 12-2, and Fig. 12-

4): (1) the 'Arava-Lot Ca-chloridic diagenetic brine (Starinsky's R$_S$1), which is evaporated seawater modified by Mg÷Ca exchange with the surrounding carbonate rock (analyses 5–9); (2) the chloridic-sulfatic (and bicarbonatic) meteoric water, which is rainwater and surface runoff that bears the chemical signature of the ambient marine aerosol, and groundwater that is periodically replenished by such rainwater, issuing from natural springs around the Dead Sea (analyses 26–32);and (3) the Siddim-Ashlag metamorphic brine of Mg-Ca-K-chloridic composition resulting from the incongruent alteration and dissolution of hydrous evaporite minerals (analyses 10–13). These three brine types, although occasionally found in a close to pure state, are end members, whose mixing has produced the widespread seepages of water types of, for example, Yesha'-

Table 12-1 Chemical composition of saline waters and brines, Dead Sea region (mg/l)

Analysis	Location	Sampling notes	Na	K	Mg	Ca	Cl	Br	SO$_4$	HCO$_3$	TDI*	1000Br/Cl	Cl/Br
1	Dead Sea -halite	(1) close to surface	41300	7600	42120	17600	224200	4500	1		337321	20.1	50
2	Dead Sea -carnallite	(1) close to surface	7870	14160	75310	30700	298980	9000			436020	30.1	33
3	Dead Sea -upper mass	(2)	38510	6500	36150	16380	196940	4600	580	230	299890	23.4	43
4	Dead Sea -lower mass	(2)	39700	7590	42430	17180	219250	5270	420	220	332060	24.0	42
5	Lot-1	(3) artesian, 717 m	6000	800	2991	6192	28356	510	1225	366	46440	18.0	56
6	Lot-1	(3) 1269-79 m	15116	1679	13959	34677	130985	3040	1004	61	200521	23.2	43
7	Arava-1	(3) 1935-45 m	50483	1033	12039	50636	202245	3670	122	61	320289	18.1	55
8	Sedom-2	(4) artesian	33200	990	8510	17330	105600	1790	905	121	168446	17.0	59
9	Heimar-1	(3) 2079-94 m	24000	1600	3630	43006	137841	2750	1284	18	214129	20.0	50
10	'Ein-Ashlag, Mt Sedom	(4) spring	21620	28780	56220	12860	240170	3830	134	376	363990	15.9	63
11	'Ein-Siddim, Mt Sedom	(4) spring	33450	18355	38360	22810	228600	2330	263		344168	10.2	98
12	Sedom-1	(4) artesian	17930	27360	18430	80530	255000	3870	108		403228	15.2	66
13	Tamileh, Mt Sedom	(4) close to surface	26600	14900	41000	40900	259030	3310		380	386120	12.8	78
14	Hammei Yesha'	(5) mre-93, well	27361	3520	18086	9813	114510	835	1078	107	175310	7.3	137
15	Hammei Zohar	(5) mre-69	15850	2120	14200	6660	79400	1196	713	179	120318	15.1	66
16	Zohar -1	(5) mre-58, 158 m	4831	342	1457	1453	13943	221	658	183	23088	15.9	63
17	Amiaz -1	(5) mre-52, 70 m	2698	165	519	830	6578	103	788	146	11827	15.7	64
18	Amiaz-2	(5) mre-48	3537	230	784	1309	9440	140	905	171	16516	14.8	67
19	Kaneh-Samar	(6) m-90, well	1250	250	1150	590	6428	134	166	279	10247	20.8	48
20	Kaneh-Samar	(6) m-83, spring (a)	460	87	375	260	2218	42	122	322	3886	18.9	53
21	Kaneh-Samar	(6) m-80, spring (b)	350	76	290	175	1390	26	116	294	2717	18.7	53
22	Tamar-3	(5) mre-16	640	32	152	250	1115	7	663	286	3145	6.5	153
23	Nahal Boqeq -1	(5) mre -74, 42 m	1902	144	288	577	3395	19	1835	280	8440	5.6	179
24	Nahal Boqeq -1	(5) mre -76, 66 m	1810	111	193	623	3000	10	1860	286	7893	3.3	300
25	'Ein-No'it	(5) mre-83, spring	1340	62	122	600	2030	8	1710	245	6117	3.9	254
26	'Ein Boqeq	(5) mre-72, spring	314	9	70	100	550	1	351	180	1575	1.8	550
27	Nahal 'Arugot	(5) mre-102, surface	63	3	30	44	113	<1	53	195	501	3.5	283
28	Kaneh-Samar	(6) m -79, well, shallow	107	31	82	80	385	3	53	290	1031	7.8	128
29	Kaneh-Samar	(6) m -77, well 2, shallow	43	19	44	63	151		13	277	610		
30	Kaneh-Samar	(6) m -75, well 3, shallow	42	7	38	62	118		10	233	510		
31	Jordan R.-winter	(7) near Dead Sea	219	13	55	86	442	4	143	219	1181	8.6	116
32	Jordan R.-summer	(7) near Dead Sea	417	22	113	147	902	8	279	284	2172	8.3	120
33	Seawater	(8)	10556	380	1272	400	18980	65	2649	140	34441	3.4	294

*Total dissolved ions.

Sources of data: (1) Epstein, 1976; (2) Neev and Emery, 1967; (3) Starinsky, 1974; (4) Zak, 1960; 1967; and unpubl.; (5) Mazor, et al., 1969; (6) Mazor, this volume; (7) Average from (2), (3), (4), and Bentor 1961; Nissenbaum, 1969; (8) Sverdrup et al., 1942.

Table 12-2 Ratios of ions in saline waters and brines, Dead Sea region (mole/l)

Analysis	Location	Sampling notes	RE**	Na/Cl	K/Cl	Mg/Cl	Ca/Cl	Ca/Mg	Ca/(HCO₃+SO₄) > 1			Ca/(HCO₃+SO₄) ≤ 1		
									%K₂	%Mg	%Ca	%K₂	%Mg	%SO₄
1	Dead Sea -halite	(1) close to surface	-0.4	0.28	0.03	0.27	0.07	0.25	4.3	76.4	19.4			
2	Dead Sea -carnallite	(1) close to surface	-0.7	0.04	0.04	0.37	0.09	0.25	4.5	76.6	18.9			
3	Dead Sea -upper mass	(2)	<0.1	0.30	0.03	0.27	0.07	0.27	4.2	75.5	20.3			
4	Dead Sea -lower mass	(2)	0.1	0.28	0.03	0.28	0.07	0.25	4.3	77.1	18.6			
5	Lot-1	(3) artesian, 717 m	-0.1	0.33	0.03	0.15	0.19	1.26	3.8	45.2	51.0			
6	Lot-1	(3) 1269-79 m	-2.4	0.18	0.01	0.16	0.23	1.51	1.5	39.6	58.9			
7	Arava-1	(3) 1935-45 m	-0.1	0.38	<0.01	0.09	0.22	2.55	0.7	28.0	71.3			
8	Sedom-2	(4) artesian	0.2	0.48	0.01	0.12	0.15	1.23	1.6	44.6	53.8			
9	Heimar-1	(3) 2079-94 m	-5.6	0.27	0.01	0.04	0.28	7.18	1.7	12.1	86.2			
10	'Ein-Ashlag, Mt Sedom	(4) spring	0.8	0.14	0.11	0.34	0.05	0.14	12.3	77.2	10.6			
11	'Ein-Siddim, Mt Sedom	(4) spring	-2.1	0.23	0.07	0.24	0.09	0.36	9.9	66.3	23.8			
12	Sedom-1	(4) artesian	-1.6	0.11	0.10	0.11	0.28	2.65	11.2	24.3	64.4			
13	Tamileh, Mt Sedom	(4) close to surface	-2.8	0.16	0.05	0.23	0.14	0.60	6.6	58.3	35.1			
14	Hammei Yesha'	(5) mre-93, well	-0.1	0.37	0.03	0.23	0.08	0.33	4.4	72.8	22.8			
15	Hammei Zohar	(5) mre-69	-0.6	0.31	0.02	0.26	0.07	0.28	3.5	76.0	20.5			
16	Zohar -1	(5) mre-58, 158 m	-0.2	0.53	0.02	0.15	0.09	0.60	4.7	65.0	30.3			
17	Amiaz -1	(5) mre-52, 70 m	<0.1	0.63	0.02	0.12	0.11	0.97	6.1	61.4	32.5			
18	Amiaz-2	(5) mre-48	<0.1	0.58	0.02	0.12	0.12	1.01	5.2	56.6	38.3			
19	Kaneh-Samar	(6) m-90, well	-1.6	0.30	0.04	0.26	0.08	0.31	5.2	77.3	17.5			
20	Kaneh-Samar	(6) m-83, spring (a)	-3.5	0.32	0.04	0.25	0.10	0.42	5.8	80.7	13.5			
21	Kaneh-Samar	(6) m-80, spring (b)	3.1	0.39	0.05	0.30	0.11	0.37	7.1	87.4	5.5			
22	Tamar-3	(5) mre-16	3.5	0.89	0.03	0.20	0.20	1.00				4.2	64.7	31.1
23	Nahal Boqeq -1	(5) mre-74, 42 m	<0.1	0.86	0.04	0.12	0.15	1.21				8.9	57.3	33.8
24	Nahal Boqeq -1	(5) mre-76, 66 m	0.1	0.93	0.03	0.09	0.18	1.96				9.1	51.2	39.7
25	'Ein-No'it	(5) mre-83, spring	1.5	1.02	0.03	0.09	0.26	2.98				7.4	47.1	45.4
26	'Ein Boqeq	(5) mre-72, spring	-2.3	0.88	0.01	0.19	0.16	0.87				2.0	51.2	46.8
27	Nahal 'Arugot	(5) mre-102, surface	-0.1	0.86	0.02	0.39	0.34	0.89				1.7	53.1	45.3
28	Kaneh-Samar	(6) m-79, well, shallow	-1.7	0.43	0.07	0.31	0.18	0.59				8.4	71.7	19.8
29	Kaneh-Samar	(6) m-77, well 2, shallow	0.3	0.44	0.11	0.43	0.37	0.87				8.4	62.7	28.9
30	Kaneh-Samar	(6) m-75, well 3, shallow	5.6	0.55	0.05	0.47	0.46	0.99				4.2	73.8	22.0
31	Jordan R.-winter	(7) near Dead Sea	-1.0	0.76	0.03	0.18	0.17	0.95				4.5	63.8	31.7
32	Jordan R.-summer	(7) near Dead Sea	-0.9	0.71	0.02	0.18	0.14	0.79				4.4	71.5	24.1
33	Seawater	(8)	<0.1	0.86	0.02	0.10	0.02	0.19				6.4	68.9	24.7

RE**: Reaction Error = 100 * (sum cations - sum anions)/(sum cations + sum anions) (equivalent values).
Sources of data: (1) Epstein, 1976; (2) Neev and Emery, 1967; (3) Starinsky, 1974; (4) Zak, 1960, 1967, and unpubl.; (5) Mazor et al., 1969; (6) Mazor, this volume; (7) Average, from (2), (3), (4), Bentor, 1961; Nissenbaum, 1969; (8) Sverdrup et al., 1942.

Zohar, Kikar-No'it and Kaneh-Samar (defined by Mazor et al., 1969; Mazor, chapter 25, this volume), found along the western shore of the Dead Sea. A discussion of each type of brine, its history, and its role in the evolution of today's DS brine follows.

The 'Arava-Lot diagenetic brine

This is a highly saline, deep-seated groundwater with up to 320 g/l dissolved solids. The brine is Ca-chloridic, with $Ca^{2+} \gg HCO_3^- + SO_4^{2-}$ (equivalent values), a very low Na/Cl ratio compared to seawater (but similar to, or higher than, the DS brine), and a K/Cl ratio lower than either seawater or DS brine (Tables 12-1 and 12-2 and Fig. 12-4, analyses 5–9). It occurs as a pore fluid within the Dead Sea Group sediments and in the laterally contiguous permeable formations of prerift age around the Dead Sea. Diagenetic brine may enter the lake from springs and seepages, some of them on the lake bottom. The diagenetic brine is generally mixed with meteoric waters from present-day recharge. When also mixed with DS brine, it becomes the typical shoreline seepage brine of the Yesha'-Zohar type. The water is geothermally hot, up to 42° C at point of issue (where it is already mixed with meteoric water and with DS brine). Mixing is facilitated by the high permeability of breccious fault zones along the shores.

The diagenetic brine is essentially a relict of the evaporated water of the Gulf of Sedom, which accumulated on the bottom of the then-deepest part of the rift during the Pliocene to (?)Early Pleistocene. Less-concentrated brines were probably returned seaward, flowing to the Mediterranean Sea atop the denser bottom brine (Zak, 1967, 1974; Starinsky, 1974). The denser, more evaporated brines remained in the basin, filling pore space in the contemporaneous carbonate sediments (primary aragonite), and finding their way into the aquifers of the surrounding limestone formations (Starinsky, 1974; Zak, 1974).

During its residence within the carbonate sediments and host rock, the Mg-chloridic-sulfatic marine brine was modified to Ca-Mg-chloridic brine through replacement of calcium in the carbonate minerals (aragonite and calcite) by magnesium from the brine. The process was possibly sped up in the hot environments that the brine reached along fault-brecciated zones, producing the characteristic dolomitization and cavernous texture that marks these faults at many localities in the Judean Desert. Shortly after its dissolution by exchange, the calcium precipitated with the available sulfate, forming pore fillings and veins of gypsum, and the excess calcium remained in the now Ca-chloridic solution. The net effect of this diagenetic exchange on the marine-derived brine is loss of magnesium and enrichment in calcium, and the loss of nearly all its sulfate by precipitation with rock-derived calcium (Loewengart, 1962, and in Braitsch, 1971; Loewengart and Zak, 1972, 1983; Starinsky, 1974; Zak 1980; Fig. 12-4, diagenesis pathways, from points SW_h and SW_c).

Sulfate is also removed from the diagenetic brine, and probably also from the original marine brine, by the activity of sulfate-reducing bacteria (Neev and Emery, 1967; Carpenter, 1978). The presence of reduced sulfur is felt in the H_2S emanations along the western fault escarpment of the Dead Sea shore, especially in the vicinity of Yesha'-Zohar seepages and in the pyrite mineralization that forms in odorous mud in and around springs and spring-fed marshes (Mazor and Rosenthal, 1967).

The high temperatures of the mixed spring waters indicate that the diagenetic brine rises from great depths and through fast-flowing conduits. Deep depths of origin are implied given the low geothermal gradient (20°C/km) in the Dead Sea basin (Maurath, 1989). The high temperatures may be an important factor in promoting the Mg÷Ca exchange, even today, and the

breccia-filled tectonic fissures may provide effective conduits for the convective rising of large volumes of brine.

Meteoric waters

These are rain-derived groundwater and surface runoff waters, marked by close-to-marine ionic composition. These waters gain their primary salinity from airborne seasalts (Loewengart, 1961). They become more saline and modified (e.g., an increase of the SO_4/Cl ratio) by evapotranspiration processes in the vadose zone, as is typical for arid regions (Loewengart, 1961; Arad, 1964). They reach the Dead Sea basin by seasonal rain and associated creek flow through springs from active (phreatic and confined) aquifers, as well as from the Jordan River, the lake's main tributary, and other rivers in the catchment area. Being of very low salinity, the meteoric waters are easily marked by contamination, even by small amounts of other water types or brines (e.g., Kikar-No'it and Kaneh-Samar types, described subsequently). Springs near the Dead Sea as well as the Jordan River have salinities up to about 2,000 mg/l (Table 12-1 and Fig. 12-4, analyses 26–32). Compared to seawater, some spring waters have higher sulfate ratios (after eliminating sulfate balanced by calcium; analyses 26–27), whereas others meteoric waters have higher potassium ratios (analyses 28–29); their Br/Cl ratio is generally higher, and their Na/Cl ratio is the same as or lower than that of seawater.

When entering the aquifer in the recharge zones of the Judea Mountains and the northern Negev, meteoric waters are bicarbonatic-sulfatic and chloridic. Farther away from the coastal plain and Mediterranean Sea toward the Dead Sea, the meteoric waters become increasingly chloridic-sulfatic and bicarbonatic, to sulfatic-chloridic, and Na and K-rich in the northeastern Negev aquifer (Arad, 1964). The Jordan River (analyses 31–32), fed mainly by snow meltwater from Mount Hermon in the north and by numerous springs issuing from the bottom of Lake Kinneret and along the Jordan Valley, is slightly influenced by contributions of connate fossil brines and dissolved evaporitic salts, which, however, do not change the basically meteoric nature, that is, seawater affinity, of its waters.

The Siddim-Ashlag metamorphic brines

These are highly concentrated Mg-Ca-K-Na-chlorides, with over 340 g/l salts and with a Br/Cl ratio lower than the DS brine (analyses 10–13 in Tables 12-1, 12-2, and in Fig. 12-4). They issue from valley floor springs near the Mount Sedom diapir, such as 'Ein Siddim and 'Ein Ashlag, where they have a very high radon content, and from shallow, near-shore groundwater, such as the Tamileh pool. Metamorphic brine is also encountered under artesian pressure in deep boreholes (e.g., Sedom 1). The brines, which are generally saturated with respect to halite, sylvite, and carnallite, have apparently evolved through incongruent alteration and dissolution of hydrous evaporite minerals by Ca-chloridic brines. The main reaction is probably between Ca-chloridic brine and carnallite (within the Sedom evaporites), which is incongruently altered to sylvite, possibly with the alteration of hydrous sulfates like gypsum to anhydrite, and a newly formed brine (see also Braitsch, 1971, p. 212). Similar hypersaline brines may have formed repeatedly throughout the continental history of the Dead Sea basin whenever evaporation approached desiccation. The same is also argued for the early phases of today's Dead Sea (Zak, 1974, 1980).

Bentor (1961), attempting to evaluate the age of the Dead Sea from yearly salt input, partial salt precipitation, and its present salt content and composition, proposed that the lake has been accumulating salt for 12,000 years (the lake's proposed age), and for certain ions, up to 70,000 years. Bentor suggested that about one third of the salts were supplied by Jordan River waters and that the two thirds missing from the balance were derived from Mg-Ca-K-Na-chloridic brine seepages of the Siddim-Ashlag type. However, borehole data that have been obtained since indicate that the total available volume of such metamorphic brines, as well as their rate of flow, are insufficient to account for the required mass and composition of the Dead Sea's salt. More salts would be missing if it were accepted (Zak, 1974, 1980) that the Dead Sea underwent a phase that was close to desiccation with precipitation of carnallite at its bottom. Thus, a bigger source is required for the necessary supply of Mg-Ca-K-rich brine.

Desiccation of DS brines at temperatures above 22°C coincides with a triple point of carnallite, bischofite, and tachhydrite (or antarcticite instead of bischofite at higher Ca/Mg ratios; see Assarsson and Balder, 1955; Braitsch, 1971). Such highly soluble hydrous chloride salts were probably retained in heliothermal puddles that were left from Lake Lisan, an inheritance of terminal pools from the Gulf of Sedom. Such a recurrence of bittern salts, increasingly depleted of their penultimate components, are probably a major depository of the metamorphic brine component of the DS brine. These bittern salts throughout history resided in solution (during less progressive evaporation periods) or occasionally in hygroscopic solids, before redissolving (upon redilution) as metamorphic brines. These salts are inherited from a long history of salt accumulation that began long before the present Dead Sea, and could not provide the salt clock proposed by Bentor (1961).

Mixtures of brine and saline water

The diagenetic brine and meteoric water mix with each other and with the DS brine and in groundwater aquifers to produce the various local types described by Mazor et al. (1969), Mazor (present volume), and Starinsky (1974). They include the Yesha'-Zohar and Kaneh-Samar types, which have a DS brine affinity, and the Kikar-No'it type, of more meteoric (seawater) affinity. The diagenetic and meteoric end members rarely, if at all, reach the Dead Sea in pure condition. It is through inflow of these intermediate water types that the Dead Sea composition is maintained.

The Yesha'-Zohar brine group

This group is defined by Mazor et al. (1969), as type R_S2 by Starinsky (1974), and as the "Dead Sea affinity Zohar-Yesha" group, by Mazor, present volume). Waters in this group (analyses 14–18, in Tables 12-1, 12-2, and in Fig. 12-4) are considerably saline (≤ 175 g/l), up to about half that of the Dead Sea, are often thermal (up to 32°C), and contain radon, radium, and H_2S. The Br/Cl ratio is lower but quite close to that of the DS brine, Ca/Mg and Na/Cl ratios are higher, and K/Cl ratio is lower than those of the DS brine. Mazor et al. (1969) consider the Zohar-Yesha' group to be recycled Dead Sea water mixed with Kikar-No'it waters (see subsequent discussion) and diluted by rising confined-aquifer water of meteoric affinity. Starinsky (1974), on the contrary, points out that these waters, through evaporation, would become quite similar to DS brine, differing in higher Ca concentration and lower Mg, K, and Br. Starinsky suggests that these brines are not recycled DS brine but a remnant of the parent brine type from which the DS brine is derived. The evidence that points to consecutive series of mixing favors Mazor's model, according to which the Yesha'-Zohar brine is Dead Sea brine that percolated into the sur-

rounding rock and there mixed with diagenetic ('Arava-Lot) brine ascending along fault-brecciated zones, where it was entrained by meteoric waters from deep aquifers flowing up under hydrostatic head (Fig. 12-3). Such mixing is also indicated by isotopic concentrations of deuterium and ^{18}O (Gat et al, 1969; Mazor, chapter 25, this volume).

The Kikar-No'it saline water

Defined by Mazor et al. (1969), these waters (analyses 22–25 in Tables 12-1, 12-2, and Fig. 12-4) are of low to medium salinity (about 5-50 % that of seawater), often thermal (up to 42°C), radioactive (mainly radon), and contain H_2S. Mazor et al. (1969) consider this group to be unevaporated seawater that invaded the Dead Sea rift, forming a precursor of the Lisan Lake in the late Quaternary, and that percolated to depths of high temperature. Nevertheless, its composition, higher sulfate and K content, higher Ca/Mg and Br/Cl ratios, and varying higher and lower Na/Cl ratios compared to seawater, indicate mixing with sulfatic, Na- and K-rich, meteoric waters (Arad, 1964). The high temperature suggests mixing with some uprising diagenetic ('Arava-Lot) brines and with small and varying amounts of DS brine.

The Kaneh-Samar group

This group as defined by Mazor (present volume) consists of freshwater with a dissolved salt content of 500 to 10,000 mg/l (analyses 19–21 and 28–30 in Tables 12-1 and 12-2 and Fig. 12-4). Like the Yesha'-Zohar and Kikar-No'it waters, it has elevated H_2S and radon contents and is somewhat thermal. The Kaneh-Samar group consists of two subtypes, sulfatic and Cachloridic, which reveal nicely the mixing of meteoric waters with brines of the Dead Sea basin. The sulfatic subtype (28–30) has low salinity (ca. 500-1,000 mg/l), higher Na/Cl, K/Cl and Mg/Cl ratios, and a lower Br/Cl ratio, compared to the Cachloridic subtype. The bromine content is very low, within the analytical error, though the low Br/Cl ratio is similar to that of most meteoric waters contaminated by DS brines. Ca/Cl and Ca/Mg ratios are somewhat high compared to seawater because of the presence of relatively high concentrations of dissolved $CaCO_3$ and $CaSO_4$. (The ratios are calculated for original composition, without deduction of Ca carbonate and sulfate, as done in the ternary diagrams.) The Ca-chloridic subtype waters (19–21) are more saline (up to about 10,000 mg/l), their ionic ratios are quite similar to those of the DS brine, though with somewhat lower Mg/Cl and Br/Cl ratios and a somewhat higher K/Cl ratio. The two subtypes have a major meteoric water component mixed with DS brines (Mazor, present volume) and with some 'Arava-Lot brine, which contributed to its salinity and its thermal heat.

BRINE EVOLUTION: POTENTIAL AND ACTUAL CHEMICAL PATHWAYS

As stated, these three end-member water types and their derivative mixtures, having gone through a history of surface processes (progressive/regressive evaporation), constitute the dynamic equilibrium system that is the lake's water body at any specified time, whether stratified or seasonally overturning as it is today.

The various waters of the Dead Sea basin have been plotted on two contiguous Gibbs' ternary diagrams (Fig. 12-4). The left ternary diagram has been constructed for the chloridic-sulfatic system, $NaCl-KCl-MgCl_2-Na_2SO_4-H_2O$, to which belong the chloridic-sulfatic meteoric waters and seawater. The right ternary diagram is for the chloridic system, $NaCl-KCl-MgCl_2-$

CaCl$_2$-H$_2$O, which does not contain sulfate and to which belong the Ca-chloridic brines. Both diagrams are designed for solutions at saturation with respect to halite, as is the case with the Dead Sea. Observed and theoretical paths of evolution of DS brines and their precipitates may thus be projected in relation to the stability fields of the minerals known, or assumed, to be present within the Dead Sea Group.

The DS brine (generalized by an asterisk within the right diagram of Fig. 12-4 and detailed outside the triangle by a group of asterisks for analyses 1–4, at fourfold the diagram's scale) is well within the carnallite stability field, with high Mg, less Ca, and low K concentrations. The diagenetic ('Arava-Lot) brines (5–9), which have higher Ca/Mg ratios and a lower K content than the DS brine, are spread toward the Ca corner, along the carnallite-tachhydrite fields. Closer to the Dead Sea point, within the carnallite field, are the Yesha'-Zohar brines (14–18), spread along arrow db, signifying the mixing of DS brine with diagenetic ('Arava-Lot) brine and meteoric waters. The metamorphic (Siddim-Ashlag) brines (10–13) spread from close to the Ca side to the Mg side, within the sylvite stability field (showing the composition of incongruently dissolved carnallite, with excess Mg or Ca).

The meteoric waters (unsaturated with respect to halite) are located on the sulfatic diagram (Fig. 12-4, left). The Jordan waters (31–32) are near the SW$_h$ point. They are of higher salinity and shifted toward the Mg corner in summer, and of lower salinity and shifted toward the SO$_4$ corner in winter. Those with higher SO$_4$ content (25–26) are spread from the SW$_h$ point down toward the SO$_4$ corner. The Kaneh-Samar type is found here, split into sulfatic and Ca-chloridic subtypes. The sulfatic subtype (28–30) is on the left side of the diagram, where it spreads up above SW$_h$ toward the Mg corner. Being of low salinity, this subtype may have evolved by mixing meteoric waters with a small proportion of Mg-Ca-chloridic brine. The Ca-chloridic subtype (19–21) spreads between the Dead Sea point and the boundary line Mg-K$_2$, which is common to the two ternary diagrams. Waters of this subtype may have evolved by mixing meteoric water with more Mg-Ca-chloridic brine. On the left side are also the peculiar Kikar-No'it saline waters (22–25), with higher sulfate and K and somewhat higher Na compared to seawater. This is apparently due to the presence of sulfate and K-rich meteoric waters (Arad, 1964), and, judging by the somewhat high Br/Cl ratio, there was some mixing with Dead Sea basin brine (DS and/or diagenetic brine).

The dissimilarity between DS brine and normal seawater raises apparent difficulties in deriving one from the other. These difficulties disappear when the Mg÷Ca exchange (seawater-rock interaction) is considered and when seawater and Dead Sea brine, both at halite and at carnallite saturation are compared (SW$_h$, SW$_c$ and Dead Sea points), that is, after seawater has passed through a history of evaporation processes. Such events are independently compatible with the geological history of the Dead Sea Rift and its water bodies.

The trends and products of the Mg÷Ca reaction are shown by two diagenesis pathways (the dashed lines on Fig. 12-4), starting from SW$_h$ and SW$_c$, respectively. The two diagenesis lines stand for two pathways: seawater saturated with halite (SW$_h$) exchanging its Mg with rock Ca, and, alternatively, for the same interaction by seawater saturated with carnallite (SW$_c$). Through exchange, Mg is lost from the solution, and as long as the sulfate ion is present, an equivalent amount of SO$_4$ is lost to form CaSO$_4$ precipitate. Along either path, seawater shifts away from the point for MgSO$_4$, which is lost from seawater as the reaction progresses toward the boundary between the two ternary diagrams (line Mg-K$_2$). This boundary is reached when

virtually all the available SO$_4$ has precipitated. From this boundary onward, the brines move into the Ca-chloridic (right) diagram, with exchanged Ca remaining in solution and increasing at the expense of Mg. The naturally occurring DS brine is located between the two path lines, within the carnallite field (close to the carnallite-bischofite boundary), indicating the more probable path of evolution of DS brine from seawater.

The composition of the DS brine is a dynamic balance controlled by the rates of input and output of the various components. Input of meteoric water types (chloridic-sulfatic-bicarbonatic) will cause the DS brine to shift along arrow mo, toward the sulfatic diagram. Input of diagenetic ('Arava-Lot) brine (upper carnallite and tachhydrite fields, more Ca-chloridic) will shift it along the db arrow. Addition of metamorphic (Siddim-Ashlag) brines (in the upper sylvite field) will cause a shift along arrow mm in a wide arc. Precipitation of carnallite will cause it to shift along arrow cp, away from the carnallite point (c) on the boundary line Mg-K$_2$. Long-term stratification in the Dead Sea (as was the case for centuries, until recently; Steinhorn et al., 1979) will cause separation of the Dead Sea water into masses of different concentrations and even composition (analyses 3–4 in Tables 12-1 and 12-2, and asterisks above the right triangle in Fig. 12-4). With this separation, the capacity for dissolving halite increases (as solubility of NaCl in a chloridic solution, with high concentrations of Mg and Ca, is nonlinear; Lerman, 1967). Mixing of the water masses again lowers the solubility product. The water's capacity for holding NaCl in solution decreases, and eventually halite may crystallize.

The input of diagenetic ('Arava-Lot) brine to the DS brine adds more Ca than Mg and adds K and Br (although in low concentration, remaining in solution). Introduced Na, on the other hand, would precipitate in halite. The input of meteoric water of seawater affinity (with dissolved Ca-carbonate and Ca-sulfate) adds Ca, bicarbonate, and sulfate ions. Consequently, CaCO$_3$ and CaSO$_4$, with which DS brines are already saturated, precipitate quickly, as eventually does NaCl. Adding meteoric water also brings sulfate balanced by Mg, which will take Ca from the DS brine to precipitate more CaSO$_4$ (the Mg remaining in solution, balanced by Cl ions). Thus, the net effect of meteoric water input on the DS brine will be the loss of Ca and the gain of Mg, K, and Br. The input of metamorphic (Siddim-Ashlag) brine results in the precipitation of halite (for which both are already saturated), the addition of K and Br, and an increase of the Ca/Mg ratio.

The K/Cl ratio in the Dead Sea is high compared to seawater, because of the loss of Cl in halite precipitation. Nevertheless, K concentration in the Dead Sea is comparable to that of evaporated seawater, which has lost K through carnallite precipitation. Thus, the DS brine has gone through a carnallite precipitation stage, either when it had the composition of seawater (in the evaporating Gulf of Sedom) or when it became a Ca-chloridic type brine ('Amora, Lisan, or Dead Sea lakes). Carnallite is present as beds in the Sedom Formation, as connate salt in sediments of the successive lakes, and probably among the basal sediments of the Dead Sea. Thus, K has been lost in this way several times in the history of the Dead Sea basin. Carnallite precipitation is not the only way to lose K. 'Arava-Lot brine has even less K than DS brine, but according to mechanisms described by Carpenter (1978), this can be the result of building authigenic K-aluminosilicates in the subsurface.

The high concentration of Br and the high Br/Cl ratio in the Dead Sea are also strong evidence for multiple evaporation events that went beyond the carnallite precipitation phase. Moreover, DS brine is enriched in Br even though all contributing waters have Br/Cl ratios lower than those of the Dead Sea. As a rule, evaporating solutions are enriched in Br, because it does not form its own minerals and it substitutes for Cl in chloride minerals in traces only. The uniquely high Br concentration in the Dead Sea, more than in any other known natural exposed body of water, testifies to a long history of precipitation of chloride minerals.

GEOLOGICAL ANALOGUES

The geochemical evolution of the Dead Sea may arguably be a meaningful model for continental rift basins that are formed in the course of the early, preoceanic rifting of a continental plate. The geochemistry is comparable to the Ca-chloride evaporites of the proto-Atlantic Sergipe basin (northeast Brazil) and the Congo-Gabon basin (west Africa) (Bonatti et al., 1970; Wardlaw, 1972) and with the recent Ca-chloridic Hot Brine pools of the Red Sea (Fig. 12-4; Brewer and Spencer, 1969). Indeed, continental rifts are predisposed to endorheic drainage, and as such they may be regarded as typical, morphotectonic evaporite environments. The anomalous magnitudes of subsidence of the continental rift basins are a driving force of evaporite accumulation and water-rock interaction, and their fault systems enable strong local thermal influences.

Acknowledgments

I thank Drs. J.R. Gat, E. Mazor and A. Starinsky for stimulating discussions and also I. Perath, whose critical review and editorial assistance are greatly appreciated. I thank Dr. T. M. Niemi and the Oxford University Press staff for making editing improvements in the manuscript. I owe deep gratitude to the late Dr. S. Loewengart and the late Dr. M. R. Bloch, true scholars from whom I have learned much about the Dead Sea.

REFERENCES

Anati, D. A., and Shasha, S., 1989, Dead Sea surface-level changes: *Israel Journal of Earth Sciences*, v. 38, p. 29–32.

Arad, A., 1964, The geology and hydrogeology of the Lower Cretaceous of the northern Negev and Judea Mountains [Ph.D. thesis]: Jerusalem, Israel, The Hebrew University, 160 p. (in Hebrew, English summary).

Assarsson, G. O., and Balder, A., 1955, The poly-component aqueous systems containing the chlorides of Ca^{2+}, Mg^{2+}, Sr^{2+}, K^+ and Na^+ between 18° and 93°C: *Journal of Physical Chemistry*, v. 59, p. 631–633.

Bentor, Y. K., 1961, Some geochemical aspects of the Dead Sea and the question of its age: *Geochimica et Cosmochimica Acta*, v. 25, p. 239–260.

Bentor, Y. K., Golani, U., Picard, L., Vroman, A., and Zak, I., 1965, The geological map of Israel, 1:250,000 (North and South sheets): Tel Aviv, Israel, Survey of Israel.

Beyth, M., 1980, Recent evolution and present stage of Dead Sea brines, *in* Nissenbaum, A., ed., *Hypersaline brines and evaporitic environments*: Amsterdam, Holland, Elsevier Scientific Publication, p. 155–165.

Bloch, M. R., and Schnerb, I., 1953, On the Cl^-/Br^- ratio and the distribution of Br^- ions in liquids and solids during evaporation of bromide-containing chloride solutions: *Research Council of Israel Bulletin*, v. 3, p. 151–158.

Bonatti, E., Ball, M., and Schubert, C., 1970, Evaporites and continental drift: *Naturwissenschaft*, v. 57, p. 107–108.

Braitsch, O., 1971, *Salt deposits, their origin and composition*:

(Translation of Burek, P. J., and Nairn, A. E. M., 1962, *Entstehung und Stoffbestand der Salzlagerstaetten*: Berlin, Germany, Springer-Verlag): Berlin, Germany, Springer-Verlag, 232 p.

Brewer, P. G., and Spencer, D. W., 1969, A note on the chemical composition of the Red Sea brines, *in* Degens, E. T., and Ross, D. A., eds., *Hot brines and recent heavy metal deposits in the Red Sea*: New York, N.Y., Springer-Verlag, p. 174–179.

Carpenter, A. B., 1978, Origin and chemical evolution of brines in sedimentary basins: Oklahoma Geological Survey Circular 79, p. 60–77.

Epstein, J. A., 1976, Utilization of the Dead Sea minerals (a review): *Hydrometallurgy*, v. 2, p. 1–10.

Garfunkel, Z., Zak, I., and Freund, R., 1981, Active faulting in the Dead Sea Rift: *Tectonophysics*, v. 80, p. 1–26.

Gat, J. R., Mazor, E., and Tzur, Y., 1969, The stable isotope composition of mineral waters in the Jordan rift valley, Israel: *Journal of Hydrology*, v. 7, p. 334–352.

Gavrieli, I., Starinsky, A., and Bein, A., 1989, The solubility of halite as a function of temperature in the highly saline Dead Sea brine system: *Limnology and Oceanography*, v. 34(7), p. 1,224–1,234.

Lerman, A., 1967, Model of chemical evolution of a chloride lake—The Dead Sea: *Geochimica et Cosmochimica Acta*, v. 31, p. 2,309–2,330.

Loewengart, S., 1961, Airborne salts—The major source of the salinity of waters in Israel: *Research Council of Israel Bulletin*, v. 10G, p. 183–206.

Loewengart, S., 1962, The geochemical evolution of the Dead Sea Basin: *Research Council of Israel Bulletin*, v. 11G, p. 85–96.

Loewengart, S., and Zak, I., 1972, Ca-Mg exchange between subsurface waters and carbonate sedimentary rocks: Jerusalem. Department of Geology Report, The Hebrew University, 47 p.

Loewengart, S., and Zak, I., 1983, Ca-Mg equilibria in water-carbonates and silicates interaction [abs.]: Water-Rock Interaction, Fourth International Symposium, Misasa, Japan, p. 297–300.

Maurath, G., 1989, Thermal constraints on tectonic processes along the Jordan–Dead Sea transform [Ph.D. thesis]: Kent, Ohio, Kent State University, Graduate College, 278 p.

Mazor, E., and Rosenthal, E., 1967, Notes on the sulfur cycle in the mineral waters and rocks of the Lake Tiberias-Dead Sea rift valley, Israel: *Israel Journal of Earth Sciences*, v. 16, p. 198–205.

Mazor, E., Rosenthal, E., and Ekstein, J., 1969, Geochemical tracing of mineral water sources in the southwestern Dead Sea basin, Israel: *Journal of Hydrology*, v. 7, p. 246–275.

Menczel, E, Apelblat, A., Roy, A., and Korin, E., 1980, Thermodynamic description of the system $NaCl$-KCl-$MgCl_2$-$CaCl_2$-H_2O in the 10–40°C temperature range: *Revue de Chimie Minerale*, v. 17, p. 508–516.

Neev, D., and Emery, K. O., 1967, The Dead Sea: Depositional processes and environments of evaporites: Geological Survey of Israel Bulletin 41, p. 1–147.

Neev, D., and Hall, J.K., 1979, Geophysical investigations in the Dead Sea: *Sedimentary Geology*, v. 23, p. 209–238.

Nissenbaum, A., 1969, Studies in the geochemistry of the Jordan River–Dead Sea system [Ph.D. thesis]: Los Angeles, California, University of California, 288 p.

Nissenbaum, A., 1970, Chemical analyses of Jordan River and Dead Sea water, 1778–1830: *Israel Journal of Earth Sciences*, v. 8, p. 281–287.

Rozenson, I., and Zak, I., 1987, Salt of Sedomith—Its nature, extraction and Halacha: *Techumin*, v. 8, p. 417–428, (in Hebrew).

Starinsky, A., 1974, Relationship between Ca-chloride brines and sedimentary rocks in Israel [Ph.D. thesis]: Jerusalem, Israel, The Hebrew University, 176 p., appendices, (in Hebrew, English summary).

Steinhorn, I., Assaf, G., Gat, J. R., Nishri, A., Nissenbaum, A., Stiller, M., Beyth, M., Neev, D., Garber, R., Friedman, G. M., and Weiss, W., 1979, The Dead Sea: Deepening of the mixolimnion signifies the overture to overturn of the water column: Science, v. 206, p. 55–57.

Sverdrup, H. U., Johnson, M. W., and Fleming, R. H., 1942, The Oceans, their physics, chemistry and biology: Englewood Cliffs, New Jersey, Prentice-Hall, 1087 p.

Wardlaw, N. C., 1972, Unusual marine evaporites with salts of calcium and magnesium chloride in Cretaceous basins of Sergipe, Brazil: Economic Geology, v. 67, p. 156–168.

Wood, J. R., 1975, Thermodynamics of brine-salt equilibria—1. The System $NaCl-KCl-MgCl_2-CaCl_2-H_2O$ and $NaCl-Na_2SO_4$-H_2O at 25°C: Geochimica et Cosmochimica Acta, v. 39, p. 1147–1163.

Zak, I., 1960, Mount Sedom area, a general geological survey and geochemical prospecting for potassium: Jerusalem, Geological Survey of Israel Report, 22 p., appendices, maps, (in Hebrew).

Zak, I., 1967, The geology of Mount Sedom [Ph.D. thesis]: Jerusalem, The Hebrew University, 208 p., appendices, (in Hebrew, English summary).

Zak, I., 1974, Sedimentology and bromine geochemistry of marine and continental evaporites in the Dead Sea basin, in Coogan, A. H., ed., Fourth Symposium on Salt: Cleveland, Ohio, Northern Ohio Geological Society, v. 1, p. 349–361.

Zak, I., 1980, The geochemical evolution of the Dead Sea, in Coogan, A. H., and Hauber, L., eds., Fifth Symposium on Salt: Cleveland, Ohio, Northern Ohio Geological Society, v. 1, p. 181–184.

Zak, I., and Freund, R., 1981, Asymmetry and basin migration in the Dead Sea Rift: Tectonophysics, v. 80, p. 27–38.

Zak, I., Starinsky, A., Metcalfe, J., and Goor, A., 1992, Thoughts on the origin of the brines and salts in the Dead Sea basin, in Ginzburg, D., ed., The Israeli Association for the Advancement of Mineral Engineering, 11th Conference, Nahariya, December 1992, p. E-300–307.

13. ION INTERACTION APPROACH TO GEOCHEMICAL ASPECTS OF THE DEAD SEA

Boris S. Krumgalz

The Dead Sea is a very interesting natural formation with waters whose ionic strength is approximately 12 times that of ordinary seawater. Dead Sea waters are enriched in calcium, magnesium, potassium, and bromide and are depleted in sodium, sulfate, and carbonate in contrast to other natural brines of marine origin. A better understanding of the formation mechanism of oversaturated solutions in natural hypersaline brines and of the mechanism of mineral crystallization from such brines would be a significant contribution both to fundamental science and to the intensive development of the chemical and tourist industries in this area. The possible cooperation between Israel and the Kingdom of Jordan, after signing a peace treaty, in studies of human impact on the ecology of the Dead Sea would result in joint benefits for both countries. The future hydrographic and geochemical development of the Dead Sea will depend very strongly on its saturation with respect to various minerals, and hence in some cases on the precipitation.

ION INTERACTION MODEL

To predict the chemical behavior of dissolved ionic solutes in complex chemical solutions, it is necessary to know their chemical potential. Chemical potential is calculated using the formula,

$$\mu_i = -R \cdot T \cdot \ln(a_i) \tag{1}$$

where the thermodynamic activity (a_i) and the total concentration (m_i) of an ionic component are related by the equation:

$$a_i = \gamma_i \cdot m_i \tag{2}$$

with γ_i as the total or stoichiometric ionic activity coefficient related to the ionic interactions in the solution. The chemical potentials of seawater components, including both the major ions and water in simplified seawater, were calculated by Robinson and Wood (1972), Leyendekkers (1973), and Whitfield (1973, 1974, 1975a, 1975b, 1979) using a specific ionic interaction model. The calculated values were in reasonable agreement with thermodynamic data available for seawater. A very detailed review devoted to the application of ionic interaction models to natural waters, such as seawater, rivers, lakes, and brines, has been recently published by Millero (1982). The general conclusion of the reviewed studies has been that the specific interaction models yield reliable activity coefficients for the major ionic components.

The ion interaction model developed by Pitzer's scientific school (Pitzer, 1973, 1975, 1979, 1981a, 1981b, 1983, 1986, 1987, 1991; Pitzer and Mayorga, 1973, 1974; Pitzer and Kim, 1974; Downes and Pitzer, 1976; Pitzer and Silvester, 1976, 1978; Pitzer et al., 1977, 1978, 1984, 1985; Silvester and Pitzer, 1978; Pitzer and Peiper, 1980; Rogers and Pitzer, 1981, 1982; Peiper and Pitzer, 1982; de Lima and Pitzer, 1983a, 1983b; Phutela and Pitzer, 1983, 1986; Roy et al., 1983, 1984; Saluja et al., 1986; Pabalan and Pitzer, 1987, 1988a, 1988b, 1991; Phutela et al., 1987) for calculations of thermodynamic properties is based on a set of theoretically and empirically derived equations that account for the interactions between the particular ions present in solution and for indirect forces arising from the solvent. The Pitzer approach uses so-called ion interaction parameters, calculated from the appropriate experimental data for single and mixed (ternary) electrolyte solutions.

The equations developed by Pitzer's scientific school, though semiempirical in character, have been quite successful in predicting the thermodynamic properties of mixed electrolyte solutions over a wide range of solution concentrations, and they were found to be especially effective for the prediction of various thermodynamic properties of natural hypersaline brines of large complexity (Harvie and Weare, 1980; Harvie et al., 1984). The number of possible species combined from various ions in natural brines is much larger than the number of single ions. Therefore, for such complex mixtures, it is convenient to calculate the conventional single-ion activity coefficients, because these values can then be combined to obtain the activity coefficient for any species represented in the brines.

Pitzer's model avoids some of the restrictions or assumptions required by previous models, such as the intrinsic limitation to low or moderate concentrations or the extrapolation of some of the solution properties beyond the solubility limit of electrolytes in pure water. For example, to use Young's approach to calculate various thermodynamic properties of Dead Sea water at an ionic strength of about 9, we have to know the thermodynamic properties of all single-solute solutions for each salt existing in this water at the same concentration. However, neither NaCl or KCl solutions of 9 molal exist: the solubilities of NaCl and KCl in water at 25°C are equal to 6.16 and 4.81 molal (Linke, 1958), respectively. So these quantities could be calculated only by extrapolation of the equations (often highly nonlinear, such as the Debye-Hückel equation) used to represent the desired quantity in the measurable concentration range. Therefore, the results obtained were dependent on the shape of the mathematical expression and on the method chosen for the extrapolation. This problem was discussed in detail by Whitfield (1979) in his description of the use of Harned's rule to calculate activity coefficients in concentrated multicomponent electrolyte solutions, and by Krumgalz et al. (1992) in their calculations of heat capacities for multiple electrolyte solutions resembling Dead Sea water with ionic strength from 8.66 to 9.23 molal, using Young's rule.

The main goal of this paper is to apply the Pitzer system of equations to natural hypersaline brines, such as Dead Sea water. The possibility of using these equations in concentrated electrolyte solutions was demonstrated by Pitzer and Mayorga (1973) for single electrolytes and by Pitzer and Kim (1974) for mixed aqueous electrolytes. Harvie and Weare (1980), Harvie et al. (1982, 1984), Scrivner and Staples (1982), and Meissner and Manning (1983) have used the Pitzer system of equations to predict mineral solubilities, conventional single-ion activity coefficients, and osmotic coefficients in numerous concentrated multicomponent systems and in natural brines. Recently, Krumgalz and Millero (1982, 1983a, 1983b, 1989) used the Pitzer system of equations to calculate ionic activity coefficients of some major ions, as well as, gypsum and halite solubility in

Dead Sea brines, and Krumgalz et al. (1990) studied anthropogenic fluorite precipitation in estuarine waters.

PITZER'S SYSTEM OF EQUATIONS

The prediction of thermodynamic properties of concentrated multiple electrolyte solutions remains one of the outstanding problems of physical solution chemistry. All physicochemical processes, including the solubility or deposition of different salts, are known to be determined by the ionic composition of the solutions. The solubilities of solutes in complex systems can be calculated from thermodynamic considerations provided that equilibrium constants are known and activity coefficients can be obtained. The difficulty in predicting salt solubilities is largely due to the fact that the solubilities of many salts become exceedingly dependent on the solution composition as the concentrations of other solutes increase. The solubility equilibrium of a certain mineral of the formula $X_a Y_b \cdot n\mathrm{H_2O}$ may be written as

$$X_a Y_b \cdot n\mathrm{H_2O} \leftrightarrow a \cdot X^{b+} + b \cdot Y^{a-} + n\mathrm{H_2O}. \quad (3)$$

A thermodynamic solubility product ($K_{sp, X_a Y_b \cdot n\mathrm{H_2O}}$) of the mineral, if the activity of the solid phase is considered to be equal to 1, is defined as

$$
\begin{aligned}
K_{sp, X_a Y_b \cdot n\mathrm{H_2O}} &= a_X^a \cdot a_Y^b \cdot a_{H_2O}^n \\
&= (m_{X, sat} \cdot \gamma_X)^a \cdot (m_{Y, sat} \cdot \gamma_Y)^b \cdot a_{H_2O}^n \quad (4)
\end{aligned}
$$

where $m_{X,sat}$ and $m_{Y,sat}$ are molal concentrations of cation and anion, respectively, in the liquid phase saturated with respect to the solid phase and γ_X and γ_Y are conventional single-ion activity coefficients of cation and anion at proper concentration, respectively. The activity of a solvent ($a_{solvent}$) in any multicomponent system is conventionally defined as:

$$a_{solvent} = \frac{p}{p_o} = \exp\left(-\frac{M \cdot \phi}{1000} \cdot \sum_j m_j\right) \quad (5)$$

where p and p_o are the vapor pressure above a particular system and above a pure solvent at the same temperature, respectively; M is the solvent molecular weight; ϕ is the osmotic coefficient; and $\sum_j m_j$ is the summation of the molalities of all solute species, including ionic species and neutral substances.

The estimation of the degree of saturation (Ω) of a particular brine with respect to a certain mineral is very important in the study of the present state and geochemical evolution of the brine. In general, the degree of saturation of a brine with respect to a mineral of the formula $X_a Y_b \cdot n\mathrm{H_2O}$ is defined by

$$\Omega_{X_a Y_b \cdot n\mathrm{H_2O}} = \frac{m_{X,i}^a \cdot m_{Y,i}^b}{m_{X,sat}^a \cdot m_{Y,sat}^b} = \frac{m_{X,i}^a \cdot m_{Y,i}^b}{\dfrac{K_{sp, X_a Y_b \cdot n\mathrm{H_2O}}}{\gamma_X^a \cdot \gamma_Y^b \cdot a_{H_2O}^n}} \quad (6)$$

where the numerator of the equation is the product of the real ionic concentrations in a particular system.

The ionic activity coefficients in multiple electrolyte solutions were expressed, according to ionic solution theory (Mayer, 1950), in the form

$$\ln\gamma_i = D.H. + \sum_j B_{i,j} \cdot m_j + \sum_{j,k} C_{i,j,k} \cdot m_j \cdot m_k \quad (7)$$

where $D.H.$ is a Debye-Hückel term; and $B_{i,j}$ and $C_{i,j,k}$ are the virial coefficients, respectively, which are functions of ionic

strength. The virial coefficients ($B_{i,j}$) are related to the pairwise interactions of ions, and the viral coefficients ($C_{i,j,k}$) are related to ion triplet interactions that dominate at high concentrations.

The Pitzer approach for conventional single-ion activity coefficient calculations was based on equation 7. The original form of Pitzer's equations (Pitzer, 1973) has been transformed several times during the last 20 years. The most modern and complete set of Pitzer's equations for the calculation of conventional single ion activity and osmotic coefficients, which include the terms with pairwise interactions of like and unlike charged ions, ion triplet interactions, and interaction terms with neutral species is presented in equations 8-10:

$$
\begin{aligned}
\ln\gamma_X = &-z_X^2 \cdot A^\phi \cdot \left[\frac{\sqrt{I}}{1 + b \cdot \sqrt{I}} + \frac{2}{b} \cdot \ln(1 + b \cdot \sqrt{I})\right] + \\
&2 \sum_a m_a \cdot (B_{X,a} + E \cdot C_{X,a}) + \\
&z_X^2 \cdot \sum_c \sum_a m_c \cdot m_a \cdot B'_{c,a} + z_X \cdot \sum_c \sum_a m_c \cdot m_a \cdot C_{c,a} + \\
&\sum_c m_c \cdot \left(2\Phi_{X,c} + \sum_a m_a \cdot \Psi_{X,c,a}\right) + \\
&\sum_a \sum_{<a'} m_a \cdot m_{a'} \cdot \Psi_{a,a',X} + 2\sum_c m_c \cdot {}^E\Phi_{X,c} + \\
&z_X^2 \cdot \left(\sum_c \sum_{<c'} m_c \cdot m_{c'} \cdot {}^E\Phi'_{c,c'}\right) + \\
&z_X^2 \cdot \left(\sum_a \sum_{<a'} m_a \cdot m_{a'} \cdot {}^E\Phi'_{a,a'}\right) + 2\sum_n m_n \cdot \lambda_{n,X} + \cdots
\end{aligned} \quad (8)
$$

$$
\begin{aligned}
\ln\gamma_Y = &-z_Y^2 \cdot A^\phi \cdot \left[\frac{\sqrt{I}}{1 + b \cdot \sqrt{I}} + \frac{2}{b} \cdot \ln(1 + b \cdot \sqrt{I})\right] + \\
&2 \sum_c m_c \cdot (B_{c,Y} + E \cdot C_{c,Y}) + \\
&z_Y^2 \cdot \sum_c \sum_a m_c \cdot m_a \cdot B'_{c,a} + |z_Y| \cdot \sum_c \sum_a m_c \cdot m_a \cdot C_{c,a} + \\
&\sum_a m_a \cdot \left(2\Phi_{Y,a} + \sum_c m_c \cdot \Psi_{Y,a,c}\right) + \\
&\sum_c \sum_{<c'} m_c \cdot m_{c'} \cdot \Psi_{c,c',Y} + 2\sum_a m_a \cdot {}^E\Phi_{Y,a} + \\
&z_Y^2 \cdot \left(\sum_c \sum_{<c'} m_c \cdot m_{c'} \cdot {}^E\Phi'_{c,c'}\right) + \\
&z_Y^2 \cdot \left(\sum_a \sum_{<a'} m_a \cdot m_{a'} \cdot {}^E\Phi'_{a,a'}\right) + \\
&2 \cdot \sum_n m_n \cdot \lambda_{n,Y} + \cdots
\end{aligned} \quad (9)
$$

$$
\begin{aligned}
\phi = 1 + 2 \cdot \left(\sum_i m_i\right)^{-1} \cdot \Bigg[&-\frac{A^\phi \cdot \sqrt{I^3}}{1 + b \cdot \sqrt{I}} + \\
&\sum_c \sum_a m_c \cdot m_a \cdot \left(B^\phi_{c,a} + \frac{E}{|z_c \cdot z_a|^{1/2}} \cdot C^\phi_{c,a}\right) + \\
&\sum_c \sum_{<c'} m_c \cdot m_{c'} \cdot \left(\Phi^\phi_{c,c'} + \sum_a m_a \cdot \Psi_{c,c',a}\right) +
\end{aligned} \quad (10)
$$

$$\sum_a \sum_{<a'} m_a \cdot m_{a'} \cdot \left(\Phi_{a,a'}^{\phi} + \sum_c m_c \cdot \Psi_{a,a',c} \right) +$$

$$\sum_n \sum_a m_n \cdot m_a \cdot \lambda_{n,a} + \sum_n \sum_c m_n \cdot m_c \cdot \lambda_{n,c} +$$

$$\sum_n \sum_{<n'} m_n \cdot m_{n'} \cdot \lambda_{n,n'} + \frac{1}{2} \sum_n m_n^2 \cdot \lambda_{n,n} \Bigg]$$

Throughout this article, the subscripts X and c refer to cations; similarly, the subscripts Y and a refer to anions. In equations 8-10, z_X, z_Y, m_c, and m_a are the change and molal concentration of proper cation and anion, respectively. The indices $\sum_a$ and $\sum_c$ indicate summation of the properties of all anions and cations, respectively. The double summation indices denote the sum over all possible pairs of ions and ions with a neutral component. More detailed explanations related to the meanings and calculation procedures of all parameters in these equations can be found in Krumgalz and Millero (1982, 1983a, 1983b), Harvie et al. (1984), and Pitzer (1991). The terms $\lambda_{n,c}$, $\lambda_{n,a}$, $\lambda_{n,n}$, and $\lambda_{n,n'}$ which account for the interactions between neutral species and ionic species and between neutral species, respectively (Pitzer and Silvester, 1976), are especially important in natural carbonate systems, where the dissolved $CO_2(aq)$ must be taken into account. The neutral molecule-ion interaction parameters $\lambda_{CO_2,a}$ and $\lambda_{CO_2,c}$ for some anions and cations have been recently reported by Harvie et al. (1984).

ION INTERACTION PARAMETERS

The precise calculation of conventional single-ion activity coefficients and osmotic coefficients using the full form of Pitzer's equations (eqs. 8-10) is possible only when all interaction parameters, including the parameters $\Phi_{i,j}$ and $\Phi_{i,j}^{\phi}$ relating to the like-charged ion interactions and the parameters $\Psi_{i,j}$ relating to the interaction between three ions (cation$_i$-cation$_j$-anion$_k$, anion$_i$-anion$_j$-cation$_k$) are available. Recently, Millero (1982) has demonstrated the importance of the $\Phi_{Na,c}$, $\Phi_{Cl,a}$, $\Psi_{Na,c,Cl}$, $\Psi_{Cl,c,Na}$ terms for H^+, Li^+, K^+, Cs^+, OH^-, HCO_3^-, NO_3^-, $H_2PO_4^-$, SO_4^{2-}, and CO_3^{2-} ions in NaCl media. Since the higher-order terms $\Phi_{i,j}$ and $\Psi_{i,j,k}$ are not available for all ionic combinations, the precise calculation of osmotic and ionic activity coefficients using the full form of Pitzer's equations cannot be carried out at present except for mixed solutions containing H^+, Na^+, K^+, Mg^{2+}, Ca^{2+}, Cl^-, Br^-, HCO_3^-, CO_3^{2-}, and SO_4^{2-} ions.

When calculating conventional single-ion activity coefficients and osmotic coefficients by the Pitzer system of equations, we have to choose $\beta^{(0)}_{i,j}$, $\beta^{(1)}_{i,j}$, $\beta^{(2)}_{i,j}$, $C^{\phi}_{i,j}$, $\theta_{i,j}$, and $\Psi_{i,j,k}$ ion interaction parameters from different sources. During recent years, many of the ion interaction parameters were recalculated on the basis of new experimental data, and some new parameters have been published. This has made it very difficult for users to find and choose the appropriate ion interaction parameters from among the numerous parameters scattered in various publications, including the most trustworthy ones. The variations in these parameters arose primarily from the use of different sets of experimental data for their calculations. The most recent update of these parameters can be found in the thorough review of Pitzer (1991).

The problem of proper selection of the ion interaction parameters becomes very complicated, especially for hypersaline natural waters. Dealing with Dead Sea waters, it is worth reviewing all existing ion interaction parameters and selecting those determined for a high-solute concentration range. All ion interaction parameters at 25°C found in the literature for pure electrolytes and for mixed solutions combined from major

Dead Sea ions are summarized in Tables 13-1, 13-2, and 13-3 with the maximum dissolved solute concentrations for which these parameters are valid. In all cases where discrepancies exist between the parameters published by various authors, the recommended values are pointed out. The selection of the recommended ion interaction parameters in the cases when they differ from those given in Pitzer's review (1991) is discussed next.

Ion interaction parameters for single electrolyte solutions

The ion interaction parameters $\beta^{(0)}_{i,j}$, $\beta^{(1)}_{i,j}$, $\beta^{(2)}_{i,j}$, and $C^{\phi}_{i,j}$ for single electrolyte solutions of all possible combinations of major ions existing in Dead Sea water are presented in Tables 13-1 and 13-2. The method of evaluating these parameters from the osmotic and activity coefficients has been discussed by Pitzer and Mayorga (1973, 1974). In a few cases of bicarbonates and carbonates, $KHCO_3$, K_2CO_3, and $Mg(HCO_3)_2$, the ion interaction parameters recommended herein differ from those recommended in Pitzer (1991). In these cases, I favor the parameters suggested by Harvie et al. (1984) because their database is more relevant to hypersaline natural brines. The recommended parameters for $MgCO_3$ were taken from Millero and Thurmond (1983), but I could not recommend any of the rough estimates published for $CaCO_3$ (Millero, 1982, Harvie et al., 1984) for practical calculations. The ion interaction parameters included for $MgCO_3$ led to very considerable differences between γ_{CO_3} and degrees of saturation of the carbonate minerals calculated with the use of the ion interaction parameters presented herein and the ones recommended in Pitzer (1991).

Ion interaction parameters for mixed electrolyte solutions

The major effects of mixing electrolytes are known to arise from the differences in ion interaction parameters $\beta^{(0)}_{i,j}$, $\beta^{(1)}_{i,j}$, $\beta^{(2)}_{i,j}$ and $C^{\phi}_{i,j}$ for a single electrolyte. Nevertheless, the thermodynamic calculations for complex electrolyte solutions should also take into consideration the interactions between like-charged ions and the interactions between three ions, since such interactions have measurable effects in some cases. The latter types of interactions are described by the ion interaction parameters $\theta_{i,j}$ and $\Psi_{i,j,k}$ which are calculated from the differences between the experimental osmotic and activity coefficients for mixed solutions and those calculated with all single electrolyte terms. For unsymmetrical mixtures (containing like-charged ions with different charges), additional electrostatic terms should be considered. All $\theta_{i,j}$ and $\Psi_{i,j,k}$ parameter for various combinations of major Dead Sea ions, presented in Table 13-3, include the electrostatic terms. The parameters $\theta_{i,j}$ and $\Psi_{i,j,k}$ recommended in Table 13-3, are similar to those recommended in Pitzer's review (1991). However, I added the recently obtained ion interaction mixing parameters for the ionic combinations K, Ca, SO_4 (Harvie and Weare, 1980) and K, Mg, Br (Balarew et al., 1993), instead of the missing values of Pitzer's review (1991).

MINERAL SOLUBILITIES IN THE DEAD SEA

Since the early work of Van't Hoff, geochemists have been using chemical models to clarify field observations. This approach has provided the basis for interpreting field data in terms of the chemical behavior (Garrels and Thompson, 1962; Helgeson, 1969) that influenced the formation of evaporative sediments. This approach has been very fruitful, and it is now common practice to compare the results of natural water analysis with saturation levels computed from models to obtain information about mineral precipitation and dissolution. However, because of the lack of an accurate theoretical description

Table 13-1 Parameters $\beta^{(0)}_{i,j}$, $\beta^{(1)}_{i,j}$, $C^{\phi}_{i,j}$ at 25°C for the major Dead Sea salts (1:1, 1:2, and 2:1 electrolytes).

Cation	Anion	$\beta^{(0)}_{i,j}$ (kg/mol)	$\beta^{(1)}_{i,j}$ (kg/mol)	$C^{\phi}_{i,j}$ (kg^2/mol^2)	I_{max}[a]	References
Na	Cl	0.07670	0.26495	0.00122	6	Pitzer, 1973
		0.0765[b]	0.2664[b]	0.00127[b]	6	Pitzer and Mayorga, 1973
		0.07669	0.26461	0.0012193	6	Macaskill et al., 1978
		0.0754	0.2770	0.00140	6	Pitzer, 1986
		0.0762	0.2780	0.001281	3	Simonson et al., 1987
		0.07537	0.2770	0.001407	—	Venkateswarlu and Ananthaswamy, 1988
		0.07722	0.25183	0.00106	6.14	Kim and Frederick, 1988a
		0.0754	0.2770	0.0014	6	Greenberg and Moller, 1989
		0.07627	0.2804	0.001271	6.1	Spencer et al., 1990
		0.075443	0.027703	0.001373	—	Pabalan and Pitzer, 1991
Na	Br	0.0973[b]	0.2791[b]	0.00116[b]	4	Pitzer and Mayorga, 1973
		0.11077	0.13760	-0.00153	9	Kim and Frederick, 1988a
		0.09934	0.26202	0.00097	5	Kim and Frederick, 1988b
		0.09582[c]	0.27199[c]	0.0013678[c]	4.0	Esteso et al., 1991
Na	HCO$_3$	0.0277	0.0411	—	1	Pitzer and Peiper, 1980
		0.028[b]	0.044[b]	—	1.22	Peiper and Pitzer, 1982
		-0.04096	0.5062	0.005250	1.3	Sarbar et al., 1982b
Na	SO$_4$	0.019575[b]	1.1130[b]	0.0049745[b]	12	Pitzer and Mayorga, 1973
		0.0148	1.218	0.00634	12	Pitzer, 1973
		0.019575	1.1130	0.00570	12	Pitzer, 1979
		0.01869[b]	1.09942[b]	0.005549[b]	12	Rard and Miller, 1981a
		0.0181	1.0559	0.00381	5.4	Pitzer, 1986
		0.04604	0.93350	-0.00483	6.25	Kim and Frederick, 1988a
		0.018693	1.09941	0.006298	12	Greenberg and Moller, 1989
		0.012733	1.09297	0.006267	4.2	Spencer et al., 1990
Na	CO$_3$	0.18975	0.8460	-0.04803	4.5	Pitzer and Mayorga, 1973
		0.040822	1.46790	0.0042374	8.4	Robinson and Macaskill, 1979
		0.0362[b]	1.510[b]	0.0052[b]	8.4	Peiper and Pitzer, 1982
		0.0399	1.389	0.0044	8.4	Harvie et al., 1984
		0.05306	1.29262	0.00094	8.25	Kim and Frederick, 1988a
		0.07185	1.15645	-0.00835	4.5	Kim and Frederick, 1988b
K	Cl	0.04835[b]	0.2122[b]	-0.00084[b]	4.8	Pitzer and Mayorga, 1973
		0.04827	0.20887	-0.00082	6.0	Pitzer, 1973
		0.0481	0.2187	-0.000788	—	Pitzer, 1986
		0.0487	0.2203	-0.000853	3	Simonson et al., 1987
		0.04661	0.22341	-0.00044	4.80	Kim and Frederick, 1988a
		0.04680	0.22096	-0.00050	4.00	Kim and Frederick, 1988b
		0.04808	0.21803	-0.000788	4.8	Greenberg and Moller, 1989
		0.04834	0.2104	-0.000839	4.8	Spencer et al., 1990
		0.04808	0.21875	-0.000788	—	Pabalan and Pitzer, 1991
K	Br	0.0569[b]	0.2212[b]	-0.00180[b]	5.5	Pitzer and Mayorga, 1973
		0.05592	0.22094	-0.00162	5.5	Kim and Frederick, 1988a
		0.0559	0.2296	-0.0017	sat.	Balarew et al., 1993
K	HCO$_3$	-0.0005	-0.013	—	—	Pitzer and Peiper, 1980
		-0.022	0.09	0	1.0	Roy et al., 1983
		0.0296[b]	-0.013[b]	-0.008[b]	3.2	Harvie et al., 1984
		-0.0107	0.0478	0	2	Roy et al., 1984
K	SO$_4$	0.04995[b]	0.77925[b]	0[b]	2.1	Pitzer and Mayorga, 1973
		0.0927	0.603	-0.03661	—	Holmes and Mesmer, 1986
		0.0000	1.1023	0.01254	—	Pitzer, 1986
		0.07548	0.44371	—	2.08	Kim and Frederick, 1988a
		0.0555	0.7964	-0.0188	2.1	Greenberg and Moller, 1989

Table 13-1 Continued

Cation	Anion	$\beta^{(0)}_{i,j}$ (kg/mol)	$\beta^{(1)}_{i,j}$ (kg/mol)	$C^{\phi}_{i,j}$ (kg²/mol²)	I_{max}[a]	References
K	CO₃	0.1240	1.649	-0.0068	sat.	Sarbar et al.,[d] 1982a
		0.1288	1.433	0.00050	sat.	Roy et al., 1984
		0.1488[b]	1.43[b]	-0.0015[b]	sat.	Harvie et al., 1984
Mg	Cl	0.35235[b]	1.6815[b]	0.005192[b]	13.5	Pitzer and Mayorga, 1973
		0.35093	1.65075	0.006507	17	Rard and Miller, 1981b
		0.3511	1.6512	0.0064142	—	Lima and Pitzer, 1983b
		0.3509	1.651	0.00434	—	Pitzer, 1986
		0.3524	1.740	0.005192	—	Reddy et al., 1988
		0.35573	1.61738	0.00474	17.25	Kim and Frederick, 1988a
		0.35372	1.70054	0.00524	15	Kim and Frederick, 1988b
		0.3515	1.8206	0.006507	17.4	Spencer et al., 1990
Mg	Br	0.43268[b]	1.75275[b]	0.00312[b]	15	Pitzer and Mayorga, 1973
		0.4346	1.73184	0.00275	16.83	Kim and Frederick, 1988a
		0.4328	1.7457	0.0029	sat.	Balarew et al., 1993
Mg	HCO₃	0.18	0.49	—	—	Weare, pers. com.
		0.46	—	—	—	Millero, 1982
		0.0193	0.584	—	2.5	Millero and Thurmond, 1983
		0.329	0.6072	—	sat.	Harvie et al., 1984
		0.033[b]	0.8498[b]	—	—	Pitzer et al., 1985
Ca	Cl	0.3159[b]	1.6140[b]	-0.000339[b]	7.5	Pitzer and Mayorga, 1973
		0.5292	-3.9317	-0.02245	22.5	Ananthaswamy and Atkinson, 1982
		0.305325	1.7085	0.002153	12	Phutela and Pitzer, 1983
		0.3053	1.708	0.001435	12.9	Pitzer, 1986
		0.32579	1.38412	-0.00174	18	Kim and Frederick, 1988a
		0.30654	1.64278	0.00222	10.5	Kim and Frederick, 1988b
		0.312605	1.70813	0.002342	—	Greenberg and Moller, 1989
		0.3041	1.7081	-0.001985	22.2	Spencer et al., 1990
		0.3397	1.505	-0.02679	—	Atkinson et al., 1991
Ca	Br	0.3816[b]	1.6132[b]	-0.00257[b]	6	Pitzer and Mayorga, 1973
		0.3648	1.7269	-0.004572	9	Khoo et al., 1979
		0.3562	1.872	0.005291	8.4	Tialowska-Mocharla and Atkinson, 1985
		0.33899	2.04551	0.01067	18	Kim and Frederick, 1988a
		0.36272	1.81585	0.00349	7.5	Kim and Frederick, 1988b
Ca	HCO₃	0.12	—	—	—	Millero, 1982
		0.39975[b]	2.9775[b]	—	sat.	Harvie et al., 1984
		0.28	0.3	—	—	Pitzer et al., 1985

[a]In some publications, there was no indication of the maximum concentration for the validity of the ion interation parameters presented.
[b]Recommended values among various sets of ion interaction parameters.
[c]The authors used the value $\alpha_1 = 1.9455$ instead of the generally accepted value, $\alpha_1 = 2.0$ for the calculations of the ion interaction parameters.
[d]For calculating $\beta^{(0)}_{i,j}$, $\beta^{(1)}_{i,j}$, and $C^{\phi}_{i,j}$ values, the authors used an equation that includes the additional adjustable parameters D^{ϕ}, E^{ϕ}, F^{ϕ} (Sarbar et al., 1982a). In addition, the authors did not take into account the hydrolysis and disproportionate equilibria:

$$CO_3^{2-}(aq) + H_2O \longleftrightarrow HCO_3^-(aq) + OH^-(aq)$$
$$2HCO_3^-(aq) \longleftrightarrow CO_3^{2-}(aq) + CO_2(g) + H_2O$$

in the treatment of their experimental data, as emphasized by Roy et al. (1984).

of the thermodynamics of highly concentrated solutions, until recently these models were limited to dilute aqueous solutions, among them such natural waters as seawater. The existing limitations were discussed previously (Harvie and Weare, 1980; Harvie et al., 1980, 1984).

The chemical composition of most natural brines can be approximated by the major ions Na^+, K^+, Mg^{2+}, Ca^{2+}, Cl^-, Br^-, HCO_3^-, CO_3^{2-}, and SO_4^{2-}. These include seawater-derived brines, waters associated with continental evaporates, and formation brines of mixed origin, as well as various geochemical brines, deep-sea hydrothermal vents, etc. The ionic concentration of Dead Sea water varies from the mouth of the Jordan River (northern Dead Sea), where the salt concentration is somewhat lower because of the dilution of Dead Sea water with fresh river

Table 13-2 Parameters $\beta^{(0)}_{i,j}$, $\beta^{(1)}_{i,j}$, $\beta^{(2)}_{i,j}$, and $C^\phi_{i,j}$ at 25°C for the major Dead Sea salts (2:2 electrolytes).

Cation	Anion	$\beta^{(0)}_{i,j}$ (kg/mol)	$\beta^{(1)}_{i,j}$ (kg/mol)	$\beta^{(2)}_{i,j}$ (kg/mol)	$C^\phi_{i,j}$ (kg²/mol²)	I_{max}[a]	References
Mg	SO₄	0.2210[b]	3.343[b]	-37.23[b]	0.0250	12	Pitzer and Mayorga, 1974
		0.21499	3.3646	-32.743	0.02797	14.5	Rard and Miller, 1981a
		0.2150	3.3636	-32.74	0.001748		Phutela and Pitzer, 1986
		0.22438	3.3067	-40.493	0.02512	12	Kim and Frederick, 1988a
		0.25092	4.0696	—	0.028067	12	Spencer et al., 1990
Mg	CO₃	-6.8	—	—	—		Millero, 1982
		2.833[b]	-15.069[b]	-204.90[b]	0.970[b]	2.5	Millero and Thurmond, 1983
Ca	SO₄	0.20	2.65	-55.7	—	0.04	Pitzer and Mayorga, 1974
		0.20[b]	3.1973[b]	-54.24b	0[b]	0.04	Harvie et al., 1982
		0.20	3.7762	-58.388	—	0.08	Kim and Frederick, 1988a
		0.015	3.00	-10.011	0	0.056	Greenberg and Moller, 1989
		0.15	3.0	-10.011	—	0.08	Spencer et al., 1990
Ca	CO₃	-11.9	—	—	—		Millero, 1982
		—	—	-200	—	sat.	Harvie et al., 1984

[a]In some publications, there was no indication of the maximum concentration for the validity of the ion interaction parameters presented.
[b]Recommended values among various sets of ion interaction parameters.

water, to the southern extremity of the Dead Sea, where the salt concentration is the greatest as a result of intensive evaporation and end-brine discharges from Israeli and Jordanian chemical plants. The concentration of Dead Sea water varies also with depth because of stratification.

Until 15 years ago, the Dead Sea was considered to be a classical example of a permanently physicochemically stratified water body with two superimposed water masses: the upper water mass and the lower one. The ionic concentrations and various physicochemical properties of the upper water mass (from surface to 40 m) were shown (Steinhorn and Assaf, 1980; Krumgalz and Gat, 1984) to be very sensitive to all naturally occurring processes, such as water evaporation or the inflow of freshwater owing to winter floods. The water column below 40 m was very conservative, having constant density and salinity and being a "fossil," stagnant water body (Neev and Emery, 1967). Therefore, for further calculations, we could consider the water column from a depth of 50 to 320 m as a homogenous lower water mass with the average ionic concentrations presented by Krumgalz and Millero (1989).

Using the approach discussed above, we analyzed the mineral saturation in the low water mass of the Dead Sea sampled on various occasions. The calculated ionic activity coefficients for major ions, water activity, and degrees of mineral saturation for various minerals formed from these ions are reported in Tables 13-4 and 13-5 for water sampled in the central part of the Dead Sea (Fig. 13-1). The thermodynamic solubility products at 25°C are from Table 13-6. This table contains the recommended thermodynamic solubility products at 25°C of all possible minerals formed from various combinations of major ions of Dead Sea water. As can be seen from the results, these particular Dead Sea waters at 25°C were oversaturated with respect to dolomite, huntite, magnesite, calcite, aragonite, anhydrite, and gypsum, and were very close to saturation with respect to halite. It is possible that the HCO₃⁻ and CO₃²⁻ concentrations determined by the alkalinity titration method without the borate correction are shifted from the real concentrations (Stiller et al., 1985, this volume). To the best of my knowledge, there are no good experimental data HCO₃⁻ and CO₃²⁻ concentrations in Dead Sea waters. Therefore, the degrees of saturation for carbonate minerals summarized in Table 13-5 are only estimates.

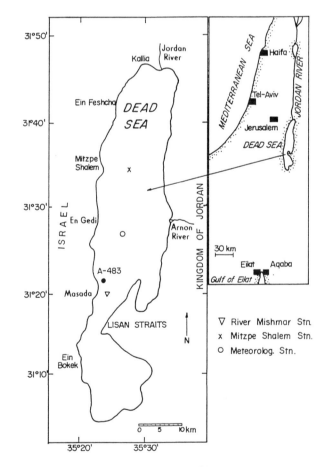

Figure 13-1 Sampling locations in the Dead Sea.

Even though Dead Sea waters were found to be highly oversaturated with respect to many calcium and magnesium carbonates and sulfates, no massive deposition of these minerals

Table 13-3 Pitzer interaction parameters $\theta_{i,j}$ and $\psi_{i,j,k}$ with higher order electrostatic interaction terms at 25°C for the major Dead Sea salts.

Ions i-j-k	$\theta_{i,j}$ (kg/mol)	$\psi_{i,j,k}$ (kg²/mol²)	I_{max}[a]	References
Na-K-Cl	-0.012[b]	-0.0018[b]	4.8	Pitzer and Kim, 1974
	-0.0102	-0.002	1	Whitfield, 1975b
	-0.0154	-0.00078		Holmes et al., 1979
	0.0070	-0.0098	4.30	Kim and Frederick, 1988b
	-0.003203	-0.003691	12	Greenberg and Moller, 1989
	-0.00948	-0.002539	7	Spencer et al., 1990
Na-K-Br	-0.012[b]	-0.0022	4	Pitzer and Kim, 1974
Na-K-HCO₃	-0.012[b]	-0.003[b]	sat.	Harvie et al., 1984
Na-K-SO₄	-0.012[b]	-0.010[b]	3.6	Pitzer and Kim, 1974
	-0.00948	0.007253	6	Spencer et al., 1990
Na-K-CO₃	-0.012b	0.003b	sat.	Harvie et al., 1984
Na-Mg-Cl	0.070	-0.010		Pitzer, 1975
	0.070[b]	-0.012[b]	10	Harvie and Weare, 1980
	0.097	-0.0517	7.14	Kim and Frederick, 1988b
	0.0594	-0.0128	3	Rao and Ananthaswamy, 1989
	0.07	-0.008172	18	Spencer et al., 1990
	0	0		Kurilenko et al., 1990
Na-Mg-SO₄	0.070	-0.023		Pitzer, 1975
	0.070[b]	-0.015[b]	10	Harvie and Weare, 1980
	0.070	-0.010		Pitzer, 1983
	0.0970	-0.0352	8.83	Kim and Frederick, 1988b
	0.07	-0.02332	6	Spencer et al., 1990
	0	-0.0056		Kurilenko et al., 1990
Na-Ca-Cl	0.070[b]	-0.007[b]		Pitzer, 1975
	0.070	-0.014	10	Harvie and Weare, 1980
	0.070	-0.007		Harvie et al., 1982
	0.05	-0.003	15	Moller, 1988
	0.05	-0.003		Greenberg and Moller, 1989
	0.0725	-0.0108	5	Tishchenko, 1989
	0.05	-0.003297	15	Spencer et al., 1990
Na-Ca-SO₄	0.070	-0.067		Pitzer, 1975
	0.070	-0.023	6	Harvie and Weare, 1980
	0.070[b]	-0.055[b]	6.4	Harvie et al., 1982
	0.05	-0.012		Moller, 1988
	0.05	-0.012	6	Spencer et al., 1990
K-Mg-Cl	0.0[b]	-0.022[b]	8	Harvie and Weare, 1980
	0.1167	-0.04948	15	Spencer et al., 1990
K-Mg-Br	0[b]	-0.0265[b]	sat.	Balarew et al., 1993
K-Mg-SO₄	0[b]	-0.048[b]	8	Harvie and Weare, 1980
	0.1167	-0.1037		Spencer et al., 1990
K-Ca-Cl	0.032[b]	-0.025[b]		Pitzer, 1975
	0.1156	-0.043189	11	Greenberg and Moller, 1989
	0.05642	-0.02856	20	Spencer et al., 1990
K-Ca-SO₄	0.032[b]	0[b]	0.7	Harvie and Weare, 1980
Mg-Ca-Cl	0.010	0	7.7	Pitzer and Kim, 1974
	0.007[b]	-0.012[b]	26	Harvie and Weare, 1980
	0.1244	-0.02381	25	Spencer et al., 1990
Mg-Ca-SO₄	0.010	0.020		Pitzer, 1975
	0.007	0.05	4	Harvie and Weare, 1980
	0.007[b]	0.024[b]	9	Harvie et al., 1982
Cl-Br-Na	0.000[b]	0.000[b]	4.4	Pitzer and Kim, 1974
Cl-Br-K	0.000[b]	0.000[b]	4.4	Pitzer and Kim, 1974
Cl-HCO₃-Na	0.0359	-0.0143	3	Peiper and Pitzer, 1982
	0.03b	-0.015b	sat.	Harvie et al., 1984
	0.0735	0.0989	1.10	Kim and Frederick, 1988b
Cl-HCO₃-K	0.0359	-0.0037		Roy et al., 1984
Cl-HCO₃-Mg	0.03[b]	-0.096[b]	sat.	Harvie et al., 1984
Cl-SO₄-Na	0.020	0.0014		Pitzer, 1975
	0.020	0.0014	8	Harvie and Weare, 1980
	0.030[b]	0.000[b]		Pitzer, 1983
	0.0380	0.0081	6.00	Kim and Frederick, 1988b
	0.030	-1.04E-08	7	Palaban and Pitzer, 1988b
	0.07	-0.009		Moller, 1988
	0.07	-0.00481	5.2	Spencer et al., 1990
	-0.0450	0.0070		Kurilenko et al., 1990
	0.0452	-0.0054	4.2	Tishchenko et al., 1992

Table 13-3 Continued

Ions i-j-k	$\theta_{i,j}$ (kg/mol)	$\psi_{i,j,k}$ (kg^2/mol^2)	I_{max}[a]	References
Cl-SO$_4$-K	0.020	0.000		Pitzer, 1975
	0.020	0.000	4.8	Harvie and Weare, 1980
	0.030[b]	-0.005[b]		Pitzer, 1983
	0.07	-0.001615	3.3	Spencer et al., 1990
Cl-SO$_4$-Mg	0.020	-0.014		Pitzer, 1975
	0.020	-0.004	17	Harvie and Weare, 1980
	0.030	-0.020		Pitzer, 1983
	0.030[b]	-0.008[b]	15	Pabalan and Pitzer, 1987
	0.0380	-0.0062	7.71	Kim and Frederick, 1988b
	0.07	-0.03186		Spencer et al., 1990
	-0.045	0.0070		Kurilenko et al., 1990
Cl-SO$_4$-Ca	0.020	0	6	Harvie and Weare, 1980
	0.030	-0.027		Pitzer, 1983
	0.020	0.018[c]	6.4	Harvie et al., 1982
	0.020	-0.018	sat.	Harvie et al., 1984
	0.07	-0.001615		Greenberg and Moller, 1989
	0.07	-0.018		Spencer et al., 1990
	-0.030[b]	-0.002[b]		Pitzer, 1991
Cl-CO$_3$-Na	-0.053	—	3	Peiper and Pitzer, 1982
	-0.053	0.016		Thurmond and Millero, 1982
	-0.02[b]	0.0085[b]	sat.	Harvie et al., 1984
	-0.0630	0.0025	5.7	Kim and Frederick, 1988b
Cl-CO$_3$-K	-0.02[b]	0.004[b]	sat.	Harvie et al., 1984
	-0.053	0.024	7	Roy et al., 1984
HCO$_3$-SO$_4$-Na	0.01b	-0.005[b]	sat.	Harvie et al., 1984
HCO$_3$-SO$_4$-Mg	0.01[b]	-0.161[b]	sat.	Harvie et al., 1984
HCO$_3$-CO$_3$-Na	-0.04[b]	0.002[b]	sat.	Harvie et al., 1984
HCO$_3$-CO$_3$-K	-0.04[b]	0.012[b]	sat.	Harvie et al., 1984
	-0.089	0.036	8	Roy et al., 1984
SO$_4$-CO$_3$-Na	0.02[b]	-0.005[b]	sat.	Harvie et al., 1984
SO$_4$-CO$_3$-K	0.02[b]	-0.009[b]	sat.	Harvie et al., 1984

[a]In some publications, there was no indication of the maximum concentration for the validity of the ion interaction parameters presented.
[b]Recommended values among various sets of ion interaction parameters.
[c]The value of $\psi_{Cl, SO_4, Ca} = 0.018$ is a misprint, since Harvie's thesis (which is a later source) and Harvie et al. (1984) give this value as -0.018.

Table 13-4 Some mineral solubility in Dead Sea water[a].

Ion	Na$^+$	K$^+$	Mg^{2+}	Ca^{2+}	Cl$^-$	Br$^-$	HCO$_3^{-}$ [b]	SO$_4^{2-}$	CO$_3^{2-}$ [b]
Molality	1.8875	0.2194	2.0181	0.4744	6.9863	0.07630	0.00550	0.00572	0.00027
γ_i	0.872$_5$	0.281$_9$	1.813$_4$	0.826$_1$	3.038$_3$	5.118$_8$	0.2057	0.043$_3$	0.00018$_3$

$I = 9.58488$, $\phi = 1.84760$, $a_{H_2O} = 0.6780$, $p = 16.11_5$ mm Hg
[a]Calculations were made at 25°C for low Dead Sea water mass sampled January–April,1983 at River Mishmar Stn (see Table 13-7 and Fig. 13-1).
[b]Bicarbonate and carbonate concentrations were taken according to S. Ben-Yaakov (pers. com., cited from Krumgalz and Gat, 1984).

was observed under present conditions (Levy, 1988b). The spontaneous precipitation of some of these minerals in Dead Sea water has been observed only in very rare cases. Bloch et al. (1944), Shalem (1949), and Rasumny (1962) described a whiting (whitening) phenomenon observed in the Dead Sea from August to December 1943. Bloch et al. (1944) estimated that about 80% of the white solid was calcium carbonate, 15% was insoluble in dilute HCl, and 5% was insoluble even in concentrated HCl. According to their rough estimation, one million tons of calcium carbonate seem to have spread over the whole Dead Sea during this phenomenon. Bloch et al. (1944), Shalem (1949), and Rasumny (1963) did not find any mention of such

an outstanding event in the previous history of the Dead Sea except a quotation from the famous Greek physician Galen, who visited the Dead Sea at about 158 C.E. and stated that the waters of the Dead Sea "appear at first glance whiter and heavier than all Seas." Neev (1963, 1964) again observed a whiting in the Dead Sea in August 1959. He suggested that aragonite was precipitated during whitings when the Dead Sea water temperatures reach their annual maximum. Friedman (1965, 1993) concluded that the mechanism of whiting in the Dead Sea was physicochemical precipitation of aragonite in contrast to the biologically induced mechanism for Bahaman whitings (Robbins and Blackwelder, 1992). The mechanism of whiting in the Dead Sea has not yet been explained completely, even though it is obvious that kinetic factors are responsible.

Kinetic factors related to crystal nuclei formation, chemical reactions on the surface of the crystals (Mucci and Morse, 1985; Mucci et al., 1985), and some physicochemical properties of the oversaturated liquid phase are responsible for the supersaturation of natural hypersaline waters with respect to various minerals. In my opinion, some of the possible reasons for the high degree of supersaturation of Dead Sea waters with respect to calcium-containing minerals are either inhibitor effects of magnesium ions or of some microcomponents and organic compounds in water. An unexpected occasional precipitation of minerals in natural waters, sometimes caused by human impact, can lead to negative environmental changes influencing the local economy. For example, an unexpected precipitation of minerals in the Dead Sea, such as a whiting phenomenon, can lead to undesirable environmental changes influencing technology processes of mineral production at the Dead Sea Works Ltd. and harming the tourist industry in this area. Therefore, the ability to predict the conditions when such a phenomenon can occur is very important for Israel's national economy. The ion interaction approach can be a powerful means to explain the formation mechanism, nature, and development of natural water bodies under various conditions. As examples of the application of this approach to the Dead Sea, the saturation and precipitation of gypsum and halite have been considered (Krumgalz and Millero, 1982, 1983a, 1989).

Gypsum solubility

Neev and Emery (1967) have shown that one of the sediment components in the Dead Sea is gypsum, which has precipitated from Dead Sea waters during at least the last 100 years. Gypsum is also known to be very important in the present sedimentational stage. The solubility equilibrium of gypsum may be written as:

$$CaSO_4 \cdot 2H_2O \Leftrightarrow Ca^{2+} + SO_4^{2-} + 2H_2O \qquad (11)$$

A thermodynamic solubility product of gypsum ($K_{sp,gypsum}$), if the activity of the solid phase is considered to be equal to 1, is defined as:

$$K_{sp,\,gypsum} = a_{Ca^{2+}} \cdot a_{SO_4^{2-}} \cdot a_{H_2O}^2 = \qquad (4a)$$

$$(m_{Ca^{2+},\,sat} \cdot \gamma_{Ca^{2+}}) \cdot (m_{SO_4^{2-},\,sat} \cdot \gamma_{SO_4^{2-}}) \cdot a_{H_2O}^2 =$$

$$K_{sp,\,gypsum}^* \cdot \gamma_{Ca^{2+}} \cdot \gamma_{SO_4^{2-}} \cdot a_{H_2O}^2$$

where a_{H_2O} is the activity of water; $m_{Ca^{2+},sat}$ and $m_{SO_4^{2-},sat}$ are molal concentrations of Ca^{2+} and SO_4^{2-} ions, respectively, in the liquid phase, saturated with respect to gypsum; $\gamma_{Ca^{2+}}$ and $\gamma_{SO_4^{2-}}$ are conventional single-ion activity coefficients of Ca^{2+} and SO_4^{2-}, respectively, and $K_{sp,gypsum}^*$ is the stoichiometric solubility product.

Gypsum solubility has been experimentally studied in various natural brines: Hara et al. (1934), Posnjak (1940), Shaffer (1967) and Marshall and Slusher (1968) in seawater concentrates and dilutents; Schnerb and Yaron (1952), in Dead Sea brines; Madgin and Swales (1956) and Marshall and Slusher (1966) in NaCl-H$_2$O solutions with MgCl$_2$ additions; and Katz et al. (1981) in mixtures of Dead Sea and Mediterranean Sea waters and their concentrates. A comparison of the theoretically calculated, apparent solubility product of gypsum, and

Table 13-5 Mineral saturation in Dead Sea water

Mineral		Degree of saturation[a]
$CaCO_3$	(aragonite)	3.2
$CaCO_3$	(calcite)	4.92
$CaCO_3 \cdot H_2O$	(monohydrocalcite)	0.456
$CaMg(CO_3)_2$	(dolomite)	421
$CaMg_3(CO_3)_4$	(huntite)	330
$CaSO_4$	(anhydrite)	2.23_6
$CaSO_4 \cdot 0.5H_2O$	(hemihydrate)	0.433
$CaSO_4 \cdot H_2O$	(gypsum)	1.70_0
KCl	(sylvite)	0.16_5
$MgCO_3$	(magnesite)	12.3
$NaCl$	(halite)	0.940_1

[a]Only minerals with a degree of saturation greater than 0.1 are presented in the table.

the experimental values calculated by Krumgalz and Millero (1983a) from the previously mentioned published experimental data, is presented in Figure 13-2. As can be seen from this figure, the theoretically calculated solubilities (solid line) are in fairly good agreement with all existing data obtained in laboratory conditions.

Since the degree of gypsum oversaturation in Dead Sea waters is not very large ($\Omega_{gypsum} < 1.7$), we expect that the mixing of Dead Sea water with other possible water sources can drastically change the gypsum precipitation-dissolution equilibrium in both directions. Consider the mixing of Dead Sea water with freshwater, which takes place almost every year when winter rainwater and water from freshwater springs reach the Dead Sea. As can be seen from Figure 13-3, the dilution of Dead Sea water by freshwater results in a drastic decrease in gypsum oversaturation. Mixtures of Dead Sea water with more than 10% wt. of freshwater became undersaturated with respect to gypsum.

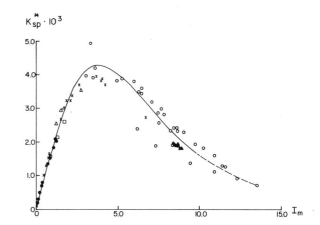

Figure 13-2 Stoichiometric solubility product of gypsum in different solutions at 298.15°K. The smooth curve is calculated from equation 4a. The experimental data have been taken from: filled triangle, Schnerb and Yaron (1952); open square, Madgin and Swales (1956); open triangle, Ostroff and Metler (1966) at 28°C; filled circle, Marshall and Slusher (1966); cross, Marshall and Slusher (1968; data recalculated from various sources at 30°C); open circle, Katz et al. (1981). (The figure was taken from Krumgalz and Millero, 1983a with some additions).

Table 13-6 Thermodynamic solubility products at 25°C of minerals formed from major ions of Dead Sea waters

Mineral (chemical name)	Formula	K_{sp}	References
Anhydrite	$CaSO_4$	$4.346 \cdot 10^{-5}$	Harvie et al., 1984
Antarcticite	$CaCl_2 \cdot 6H_2O$	$1.3917 \cdot 10^4$	Harvie et al., 1984
Aragonite	$CaCO_3$	$6.033 \cdot 10^{-9}$	Harvie et al., 1984
Arcanite	K_2SO_4	0.01674	Harvie et al., 1984
Bischofite	$MgCl_2 \cdot 6H_2O$	$2.8538 \cdot 10^4$	Harvie et al., 1984
Bloedite	$Na_2Mg(SO_4)_2 \cdot 4H_2O$	$4.501 \cdot 10^{-3}$	Harvie et al., 1984
Burkeite	$2Na_2SO_4 \cdot Na_2CO_3$	0.16905	Harvie et al., 1984
Calcite	$CaCO_3$	$3.925 \cdot 10^{-9}$	Harvie et al., 1984
Calcium chloride	$CaCl_2 \cdot 4H_2O$	$5.213 \cdot 10^5$	Harvie et al., 1984
Carnallite	$KMgCl_3 \cdot 6H_2O$	$2.140 \cdot 10^4$	Harvie et al., 1984
Chloromagnesite	$MgCl_2$	$9.813 \cdot 10^{21}$	Pabalan and Pitzer, 1987
Dolomite	$CaMg(CO_3)_2$	$8.269 \cdot 10^{-18}$	Harvie et al., 1984
Eitelite	$NaMg_{0.5}CO_3$	$1.050 \cdot 10^{-4}$	Konigsberg et al., 1992
Epsomite	$MgSO_4 \cdot 7H_2O$	0.013151	Harvie et al., 1984
Gaylussite	$CaCO_3 \cdot Na2CO_3 \cdot 5H_2O$	$3.794 \cdot 10^{-10}$	Harvie et al., 1984
Glaserite	$NaK_3(SO_4)_2$	$1.576 \cdot 10^{-2}$	Harvie et al., 1984
Glauberite	$Na_2Ca(SO_4)_2$	$5.697 \cdot 10^{-6}$	Harvie et al., 1984
Gypsum	$CaSO_4 \cdot 2H_2O$	$2.628 \cdot 10^{-5}$	Harvie et al., 1984
Halite	$NaCl$	37.19	Harvie et al., 1984
Hemihydrate	$CaSO_4 \cdot 0.5H_2O$	$1.846 \cdot 10^{-4}$	Marshall and Slusher, 1968
Hexahydrite	$MgSO_4 \cdot 6H_2O$	0.02917	Pabalan and Pitzer, 1987
Huntite	$CaMg_3(CO_3)_4$	$3.436 \cdot 10^{-31}$	Robie et al., 1978
Hydrophilite	$CaCl_2$	$8.247 \cdot 10^{11}$	Robie et al., 1978
Kainite	$KMgClSO_4 \cdot 3H_2O$	0.6419	Harvie et al., 1984
Kalicinite	$KHCO_3$	1.912	Harvie et al., 1984
Kieserite	$MgSO_4 \cdot H_2O$	0.7541	Harvie et al., 1984
Labile Salt	$Na_4Ca(SO_4)_3 \cdot 2H_2O$	$2.128 \cdot 10^{-6}$	Harvie et al., 1984
Leonhardtite	$MgSO_4 \cdot 4H_2O$	0.1298	Pabalan and Pitzer, 1987
Leonite	$K_2Mg(SO_4) \cdot 4H_2O$	$1.050 \cdot 10^{-4}$	Harvie et al., 1984
Magnesite	$MgCO_3$	$1.466 \cdot 10^{-8}$	Harvie et al., 1984
Magnesium bromide hexahydrate	$MgBr_2 \cdot 6H_2O$	$2.028 \cdot 10^5$	Balarew et al., 1984
Magnesium chloride hydrate	$MgCl_2 \cdot H_2O$	$1.716 \cdot 10^{16}$	Pabalan and Pitzer, 1987
Magnesium chloride dihydrate	$MgCl_2 \cdot 2H_2O$	$7.493 \cdot 10^{12}$	Pabalan and Pitzer, 1987
Magnesium chloride tetrahydrate	$MgCl_2 \cdot 4H_2O$	$2.675 \cdot 10^7$	Pabalan and Pitzer, 1987
Magnesium sulfate	$MgSO_4$	$1.085 \cdot 10^5$	Pabalan and Pitzer, 1987
Mirabilite	$Na_2SO_4 \cdot 10H_2O$	0.05920	Harvie et al., 1984
Monohydrocalcite	$CaCO_3 \cdot H_2O$	$2.869 \cdot 10^{-8}$	Robie et al., 1978
Nahcolite	$NaHCO_3$	0.3953	Harvie et al., 1984
Natron	$Na_2CO_3 \cdot 10H_2O$	0.1497	Harvie et al., 1984
Nesquehonite	$MgCO_3 \cdot 3H_2O$	$6.807 \cdot 10^{-6}$	Harvie et al., 1984
Northupite	$Na_2CO_3.MgCO_3 \cdot NaCl$	$1.585 \cdot 10^{-5}$	Vancina et al., 1986
Pentahydrite	$MgSO_4.5H_2O$	0.05190	Pabalan and Pitzer, 1987
Pirssonite	$CaCO_3.Na_2CO_3 \cdot 2H_2O$	$5.714 \cdot 10^{-10}$	Harvie et al., 1984
Polyhalite	$K_2MgCa_2(SO_4)4 \cdot 2H_2O$	$1.804 \cdot 10^{-14}$	Harvie et al., 1984
Potassium bromide	KBr	13.43	Robie et al., 1978
Potassium carbonate	$K_2CO_3 \cdot (3/2)H_2O$	$1.078 \cdot 10^3$	Harvie et al., 1984
Potassium magnesium bromide	$KBr.MgBr_2 \cdot 6H_2O$	$3.4801 \cdot 10^5$	Balarew et al., 1993
Potassium sesquicarbonate	$2K_2CO_3.4KHCO_3 \cdot 3H_2O$	$1.021 \cdot 10^7$	Harvie et al., 1984
Potassium sodium carbonate	$KNaCO_3 \cdot 6H_2O$	0.7657	Harvie et al., 1984
Potassium trona	$K_2CO_3.NaHCO_3 \cdot 2H_2O$	$17.236 \cdot 10^2$	Harvie et al., 1984
Schoenite (picromerite)	$K_2Mg(SO_4)_2 \cdot 6H_2O$	$4.704 \cdot 10^{-5}$	Harvie et al., 1984
Sodium carbonate heptahydrate	$Na_2CO_3 \cdot 7H_2O$	0.3466	Harvie et al., 1984
Sylvite	KCl	7.941	Harvie et al., 1984
Syngenite	$K_2Ca(SO_4)_2 \cdot H_2O$	$3.563 \cdot 10^{-8}$	Harvie et al., 1984
Tachyhydrite	$Mg_2CaCl_6 \cdot 12H_2O$	$2.421 \cdot 10^{17}$	Harvie et al., 1984
Thenardite	Na_2SO_4	0.5159	Harvie et al., 1984
Thermonatrite	$Na_2CO_3.H_2O$	3.033	Harvie et al., 1984
Trona	$Na_2CO_3 \cdot NaHCO_3 \cdot 2H_2O$	0.0903	Harvie et al., 1984
Wegscheider's salt	$Na_2CO_3 \cdot 3NaHCO_3$	0.0652	Vanderzee, 1982

Halite solubility

The hydrological balance of the Dead Sea has been disturbed because of an excess of evaporation over inflow of freshwater, resulting in a significant increase in salt concentration in the upper layer of Dead Sea water. Since 1978, the Dead Sea appears to have been saturated with respect to halite in some parts of the water body, in addition to oversaturation with respect to other minerals. The strong stratification of the Dead Sea was disturbed in 1979 during an overturn. After the overturn of 1979, the water column of the Dead Sea was not strongly stabilized, and several new overturns occurred again during the last years (Anati et al., 1987). The overturns took the lower, more concentrated water mass to the surface of the Dead Sea, where it was subjected to evaporation. As a result, more favorable conditions for halite deposition were created. The theoretical prediction of the conditions under which halite deposition will occur is very important for the explanation of the formation and development of the Dead Sea water masses. The solubility product of halite is defined by:

$$K_{sp, NaCl} = a_{Na^+} \cdot a_{Cl^-} = (m_{Na^+, sat} \cdot \gamma_{Na^+}) (m_{Cl^-, sat} \cdot \gamma_{Cl^-}) \quad (4b)$$

The degree of halite saturation is given

as: $$\Omega_{halite} = \frac{m_{Na^+, i} \cdot m_{Cl^-, i}}{m_{Na^+, sat} \cdot m_{Cl^-, sat}} = \frac{m_{Na^+, i} \cdot m_{Cl^-, i}}{\dfrac{K_{sp, NaCl}}{\gamma_{Na^+} \cdot \gamma_{Cl^-}}} \quad (6a)$$

Using equation 6a, halite saturation in Dead Sea waters sampled at various occasions between 1979 and 1983 was estimated. The sampling locations are specified in Figure 13-1. Unfortunately, only a few estimations could be done because only limited data with comprehensive chemical analysis of Dead Sea waters are available in the literature (Krumgalz and Millero, 1989). The results, summarized in Table 13-7, demonstrate that upper Dead Sea waters were saturated or close to saturation with respect to halite only in very rare cases, such as February 1979 and September 1982. Thus, these calculations give a theoretical explanation of the field observations conducted by Steinhorn (1983), who found no halite in the sediments of the deeper parts of the northern basin of the Dead Sea until February 1979. In addition, there was no sedimentological evidence for massive halite precipitation in the northern basin before 1979 (Garber, 1980).

However, for the first time, in February 1979, just after the overturn of the whole water column, Steinhorn et al. (1979) observed in situ the formation of halite crystals at the Dead Sea surface. Halite precipitation should indeed occur under these conditions, because my calculations, using Steinhorn's (1983) results of Dead Sea water chemical analysis, demonstrated that the degree of halite saturation was equal to 1.004 at 25°C. The in situ water temperature in the Dead Sea at that time was equal to 22°C. Taking into consideration the temperature effect on halite solubility, we will definitely receive an even larger degree of halite saturation at in situ temperature. Millero (1979) demonstrated that the temperature effect on the activity coefficients over a 5°C temperature range is negligible. Therefore, the temperature changes of degree of halite saturation will be conditioned only by the temperature effect on the product of solubility. The latter effect can be estimated using the temperature dependence of $K_{sp,NaCl}$ described by Wood (1976). My calculations, conducted on his data, showed that the degree of halite saturation increased by 0.005 units with a 1°C decrease in the temperature range 20-30°C. Therefore, the real degree of halite saturation in Dead Sea waters in February 1979 at an in situ temperature of 22°C was equal to about 1.02. Thus, this

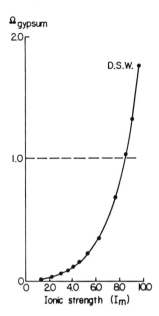

Figure 13-3 Gypsum saturation in mixtures of Dead Sea water (D.S.W.) and freshwater.

calculation explains the field observation of Steinhorn et al. (1979) concerning the halite precipitation at the Dead Sea surface in February 1979.

The calculations in Table 13-7 also show that, two and a half years later (September 1982, after the complete overturn), the upper waters of the Dead Sea were again very close to saturation conditions at in situ temperature. The degree of halite saturation ranged from 0.91 to 0.97. Under such conditions, two possible developments could occur with the coming winter season. If winter rains were to start earlier than normal, then the inflow of freshwater would dilute the upper layer of Dead Sea water before it reaches oversaturation with respect to halite, and halite deposition would not occur. The other development could occur if winter rains were to start later, and winter cooling of the upper layer of Dead Sea water would lead to its oversaturation with respect to halite. In such a case, halite deposition would start. Field observations in 1982 proved that the second development occurred under natural conditions. Halite deposition took place later in November and December 1982, when the in situ temperature of the upper layer dropped below 25°C as a result of seasonal cooling. Massive halite deposition along the coast and salt crystals floating on the Dead Sea surface were observed at that time by Levy (1985) and Anati et al. (1987). Halite deposition during that period was also documented by Gat el al. (1983), who collected halite in sediment traps deployed at different depths.

Heavy rains, occurring at the end of 1982 and the beginning of 1983, again led to the undersaturation of Dead Sea waters with respect to halite from the beginning of 1983, at low in situ temperature. Strong support for these findings is given by Levy (1988b), who demonstrated, by studying both sediment traps and the grab and core samples taken in the Dead Sea, that massive halite precipitation did not take place in the northern basin of the Dead Sea until the spring of 1983. However, according to Anati (1993, p. 2191), "the Dead Sea brine started to precipitate salts (~1983)." Since our data (Table 13-7) referred only to a specific period (January-April 1983) and place (River Mishmar and meteorological stations), they are not in contradiction with Anati's general statement.

During the last two decades, numerous studies related to a planned canal connecting the Dead Sea and the Mediterranean

Table 13-7 Degrees of halite saturation calculated at 25°C for Dead Sea waters sampled at various occasions during 1979-1983

Sampling location[a]	Date of sampling	Depth (m)	Temp. (°C) in situ	Ω_{halite} at 25°C	Ω_{halite},[b] at in situ temp.
Mitzpe Shalem	6 Feb 79	0	ca. 22	1.00$_4$	1.02
A-483	11 Nov 80	0	c	0.823	—
River Mishmar	21 Apr 82	0	25.48	0.861	0.86
	21 Apr 82	18	24.03	0.881	0.88
	21 Apr 82	36	21.61	0.940	0.96
	31 May 82	0	c	0.905	—
	31 May 82	5	25.80	0.906	0.90
	31 May 82	19	24.28	0.919	0.92
	31 May 82	34	22.80	0.979	0.99
	28 Jul 82	3.5	33.3	0.935	0.89
	28 Jul 82	19.5	32.7	0.929	0.89
	28 Jul 82	34	22.9	0.946	0.96
	7 Sep 82	0	33.5	1.00$_9$	0.97
	7 Sep 82	3	33.5	0.961	0.92
	7 Sep 82	10	32.1	0.947	0.91
	7 Sep 82	19	33.4	0.959	0.92
	7 Sep 82	27	23.1	0.91	0.92
River Mishmar and Meteorological Station	Jan–Apr 83	50–320	ca. 22[d]	0.93	0.95
River Mishmar	27 Dec 83	0	c	0.93	—

[a]Sampling locations are specified in Figure 13-1.
[b]The degrees of halite saturation at in situ temperature were calculated by the equation

$$\Omega_{halite}^{t,\,insitu} = \Omega_{halite}^{25} + 0.005\,(25 - t_{insitu})$$

[c]Temperature was not measured during the particular sampling.
[d]Stiller and Gat (pers. com., 1994).

Sea have been made, and possible effects of the canal on the ecology of the Dead Sea region have been discussed (Levy, 1982; Steinhorn and Gat, 1983). In the mid-1980s, the execution of this project was postponed indefinitely, because of financial problems in Israel. The signing of a peace treaty between Israel and the Kingdom of Jordan may revive this project by changing the canal route using a connection of the Dead Sea with the Gulf of Eilat (Aqaba) instead of with the Mediterranean Sea. One of the effects of this project would be the transformation once again of the Dead Sea into a stable meromictic lake for some considerable time. The fulfillment of this project would lead to mixing of the two different water masses. The mixtures of Dead Sea water and seawater either of the Mediterranean or of the Gulf of Eilat would be undersaturated with respect to halite. However, these mixtures would naturally be evaporated at a very high rate in this arid climate. To understand the chemical processes involving halite that would occur under such extreme natural conditions, Krumgalz and Millero (1989) conducted a mathematical simulation of the evaporation process of different Dead Sea—Mediterranean Sea water model mixtures to various degrees. The results of such modeling permitted the prediction of halite points (halite saturation) for either Dead Sea water or its mixtures with Mediterranean Sea water subjected to evaporation to various degrees.

Acknowledgments

This work was supported by the Ministry of Energy and Infrastructure (Jerusalem, Israel) and in some parts by the United States–Israel Binational Science Foundation (B.S.F.).

REFERENCES

Ananthaswamy, J., and Atkinson, G., 1982, Thermodynamics of concentrated electrolyte mixtures, 1. Activity coefficients in aqueous sodium chloride - calcium chloride at 25°C: *Journal of Solution Chemistry*, v. 11, p. 509–527.

Anati, D. A., 1993, How much salt precipitates from the brines of a hypersaline lake? The Dead Sea as a case study: *Geochimica et Cosmochimica Acta*, v. 57, p. 2,191–2,196.

Anati, D. A., Stiller, M., Shasha, S., and Gat, J. R., 1987, Changes in the thermo-haline structure of the Dead Sea, 1979–1984: *Earth and Planetary Science Letters*, v. 84, p. 109–121.

Atkinson, G., Raju, K., and Howell, R. D., 1991, The thermodynamics of scale prediction: SPE International Symposium on Oilfield Chemistry, Anaheim, Calif., Feb. 20–22, 1991, p. 209–215.

Balarew, C., Christov, C., Valyashko, V., and Petrenko, S., 1993, Thermodynamics of formation of carnallite type double salts:

Journal of Solution Chemistry, v. 22, p. 173–181.

Bloch, R., Littman, H. Z., and Elazari-Volcani, B., 1944, Occasional whiteness of the Dead Sea: *Nature*, v. 154, p. 402-403.

Downes, C. J., and Pitzer, K. S., 1976, Thermodynamics of electrolytes. Binary mixtures formed from aqueous NaCl, Na_2SO_4, $CuCl_2$, and $CuSO_4$ at 25°C: *Journal of Solution Chemistry*, v. 5, p. 389–398.

Esteso, M. A., Hernandez-Luis, F., Gonzalez-Diaz, O., Fernandez-Merida, L., Khoo, S. K., and Lim, T. K., 1991, Comparative analysis of the activity coefficients for the system NaBr-NaFormate + H_2O at 25°C by the methods of Scatchard, Pitzer and Lim: *Journal of Solution Chemistry*, v. 20, p. 417–429.

Friedman, G. M., 1965, On the origin of aragonite in the Dead Sea: *Israel Journal of Earth-Sciences*, v. 14, p. 79–85.

Friedman, G. F., 1993, Biochemical and ultrastructural evidence for the origin of whitings: A biologically induced calcium carbonate precipitation mechanism–Comment: *Geology*, v. 21, p. 287.

Garber, R., 1980, The sedimentology of the Dead Sea [Ph.D. thesis]: Troy, New York, Rensselaer Polytechnical Institute, 170 p.

Garrels, R. M., and Thompson, M. E., 1962, A chemical model for seawater at 25°C and one atmosphere total pressure: *American Journal of Science*, v. 260, p. 57–66.

Gat, J. R., Stiller, M., Levi, Y., Kaushansky, P., and Spencer, D. W., 1983, Chemical precipitation in the Dead Sea: U.S.–Israel BSF Report, project no. 2218/80, Rehovot, Israel, Weizmann Institute of Science, 35 p.

Greenberg, J. P., and Moller, N., 1989, The prediction of mineral solubilities in natural waters: A chemical equilibrium model for the Na-K-Ca-Cl-SO_4-H_2O system to high concentration from 0 to 250°C: *Geochimica et Cosmochimica Acta*, v. 53, p. 2,503–2,518.

Hara, R., Tanaka, Y., and Nakamura, K., 1934, Rep. Tohoku Imp. Univer. 11, 199. Cited from Marshall and Slusher (1968).

Harvie, C. E., Eugster, H. P., and Weare, J. H., 1982, Mineral equilibria in the six-component seawater system, Na-K-Mg-Ca-SO_4-Cl-H_2O at 25°C. II. Compositions of the saturated solutions: *Geochimica et Cosmochimica Acta*, v. 46, p. 1,603–1,618.

Harvie, C. E., Moller, N., and Weare, J. H., 1984, The prediction of mineral solubilities in natural waters: The Na-K-Mg-Ca-H-Cl-SO_4-OH-HCO_3-CO_3-CO_2-H_2O system to high ionic strengths at 25°C: *Geochimica et Cosmochimica Acta*, v. 48, p. 723–751.

Harvie, C. E., and Weare, J. H., 1980, The prediction of mineral solubilities in natural waters: The Na-K-Mg-Ca-Cl-SO_4-H_2O system from zero to high concentration at 25°C: *Geochimica et Cosmochimica Acta*, v. 44, p. 981–997.

Harvie, C. E., Weare, J. H., Hardie, L. A., and Eugster, H. P., 1980, Evaporation of seawater—Calculated mineral sequences: *Science*, v. 208, p. 398–500.

Helgeson, H. C., 1969, Thermodynamics of hydrothermal systems at elevated temperatures and pressures: *American Journal of Science*, v. 267, p. 729–804.

Holmes, H. F., Baes, C. F., and Mesmer, R. E., 1979, Isopiestic studies of aqueous solutions at elevated temperatures. I. KCl, $CaCl_2$ and $MgCl_2$: *Journal of Chemical Thermodynamics*, v. 10, p. 983–996.

Holmes, H. F., and Mesmer, R. E., 1986, Isopiestic studies of aqueous solutions at elevated temperatures. VIII. The alkalimetal sulfates: *Journal of Chemical Thermodynamics*, v. 18, p. 263–275.

Katz, A., Starinsky, A., Taitel-Goldman, N., and Beyth, M., 1981, Solubilities of gypsum and halite in the Dead Sea and in its

mixtures with seawater: *Limnology and Oceanography*, v. 26, p. 709–716.

Khoo, K. H., Lim, T. K., and Chan, C. Y., 1979, Activity coefficients for the system HBr + $CaBr_2$ + H_2O at 298.15°K: *Journal Chemical Society Faraday Transactions I*, v. 75, p. 1067–1072.

Kim, H.-T., and Frederick, W. J., Jr., 1988a, Evaluation of Pitzer ion interaction parameters of aqueous electrolytes at 25°C. 1. Single salt parameters: *Journal of Chemical Engineering Data*, v. 33, p. 177–184.

Kim, H.-T., and Frederick, W. J., Jr., 1988b, Evaluation of Pitzer ion interaction parameters of aqueous mixed electrolyte solutions at 25°C. 2. Ternary mixing parameters: *Journal of Chemical Engineering Data*, v. 33, p. 278–283.

Konigsberg, E., Schmidt, P., and Gamsjager, H., 1992, Solid-solute phase equilibria in aqueous solution. VI. Solubilities, complex formation, and ion-interaction parameters for the system Na^+-Mg^{2+}-ClO_4^--CO_2-H_2O at 25°C: *Journal of Solution Chemistry*, v. 21, p. 1,195–1,216.

Krumgalz, B. S., Fainshtein, G., Gorfunkel, L., and Nathan, Y., 1990, Fluorite in recent sediments as a trap of heavy metal contaminants in an estuarine environment: *Estuarine and Coastal Shelf Science*, v. 30, p. 1–15.

Krumgalz, B. S., and Gat, Y., 1984, Some thermodynamic properties of Dead Sea brines under conditions of halite saturation and possible precipitation: Israel Oceanogr Limnol Res, Haifa, Israel, IOLR Rep. H4/84, 83 p.

Krumgalz, B. S., Malister, A., Ostrich, I. J., and Millero, F., 1992, Heat capacity of concentrated multicomponent aqueous electrolyte solutions at various temperatures: *Journal of Solution Chemistry*, v. 21, p. 635–649.

Krumgalz, B. S., and Millero, F. J., 1982, Physicochemical study of the Dead Sea waters. I. Activity coefficients of major ions in Dead Sea water: *Marine Chemistry*, v. 11, p. 209–222.

Krumgalz, B. S., and Millero, F. J., 1983a, Physicochemical study of the Dead Sea waters. III. On gypsum saturation in Dead Sea waters and their mixtures with Mediterranean Sea water: *Marine Chemistry*, v. 13, p. 127–139.

Krumgalz, B. S., and Millero, F. J., 1983b, Physico-chemical study of the Dead Sea: Report to BSF. Israel Oceanogr Limnol Res, Haifa, 86 p.

Krumgalz, B. S., and Millero, F. J., 1989, Halite solubility in Dead Sea waters: *Marine Chemistry*, v. 27, p. 219–233.

Kurilenko, V. V., Filippov, V. K., Charykov, N. A., and Shwarz, A. A., 1990, The application of Pitzer's method for hydrogeochemical modeling of the process of modern evaporative basins' development: *DAN SSSR*, v. 311, p. 193–196.

Levy, Y., 1982, Calculations of chemical compositions of mixed layers of Mediterranean Sea and Dead Sea water: Jerusalem, Geological Survey of Israel, Rep. MGG/1/82, 36 p.

Levy, Y., 1985, Modern halite precipitation in the Dead Sea: Jerusalem, Geological Survey of Israel, Rep. GSI/7/85.

Levy, Y., 1988a, Sedimentary reflection of modern (1983–1985) halite precipitation from Dead Sea water: Jerusalem, Geological Survey of Israel, Jerusalem, Rep. GSI/12/88, 22 p.

Levy, Y., 1988b, Recent depositional environments in the Dead Sea: Jerusalem, Geological Survey of Israel, Mar. Geol Mapping Division, Rep. GSI/42/88, 21 p.

Leyendekkers, J. V., 1973, The chemical potentials of seawater components: *Marine Chemistry*, v. 41, p. 75–88.

Lima, M. C. P., de and Pitzer, K. S., 1983a, Thermodynamics of saturated aqueous solutions including mixtures of NaCl, KCl, and CsCl: *Journal of Solution Chemistry*, v. 12, p. 171–185.

Lima, M. C. P., de and Pitzer, K. S., 1983b, Thermodynamics of saturated electrolyte mixtures of NaCl with Na_2SO_4 and with $MgCl_2$: *Journal of Solution Chemistry*, v. 12, p. 187–199.

Linke, W. F., (ed.), 1958, *Solubilities, inorganic and metal-organic compounds. A compilation of solubility data from the periodical literature* (4th ed.): Washington, D.C., American Chemical Society, v. 1, p. 1487; v. 2, p. 1914.

Macaskill, J. B., White, D. R., Robinson, R. A., and Bates, R. G., 1978, Isopiestic measurements on aqueous mixtures of sodium chloride and strontium chloride: *Journal of Solution Chemistry*, v. 7, p. 339–347.

Madgin, W. M., and Swales, D. A., 1956, Solubilities in the system $CaSO_4$-$NaCl$-H_2O at 25° and 35°: *Journal of Applied Chemistry*, v. 6, p. 482–487.

Marshall, W. L., and Slusher, R., 1966, Thermodynamics of calcium sulfate dihydrate in aqueous sodium chloride solutions, 0–110°: *Journal of Physical Chemistry*, v. 70, p. 4,015–4,027.

Marshall, W. L., and Slusher, R., 1968, Aqueous systems at high temperature. Solubility to 200°C of calcium sulfate and its hydrates in sea water and saline water concentrates, and temperature-concentration limits: *Journal of Chemical Engineering Data*, v. 13, p. 83–93.

Mayer, J. E., 1950, The theory of ionic solutions: *Journal of Chemical Physics*, v. 18. p. 1426-1436.

Meissner, H. P., and Manning, M. P., 1983, Prediction of solubilities and activity coefficients in sodium-potassium-magnesium chloride brines, *in* Newman, S. A., ed., *Chemical engineering thermodynamics*: Ann Arbor Science, p. 339–348.

Millero, F. J., 1979, Effects of pressure and temperature on activity coefficients, *in* Pytkowicz, R. M., ed., *Activity coefficients in electrolyte solutions*, Vol. II: Boca Raton, FL, CRC Press, p. 63–151.

Millero, F. J., 1982, Use of models to determine ionic interactions in natural waters: *Thalassia Jugoslavia*, v. 18, p. 253–291.

Millero, F. J., and Thurmond, V., 1983, The ionization of carbonic acid in Na-Mg-Cl solutions at 25°C: *Journal of Solution Chemistry*, v. 12, p. 401–412.

Moller, N., 1988, The prediction of mineral solubilities in natural waters: A chemical equilibrium model for the Na-Ca-Cl-SO_4-H_2O system, to high temperature and concentration: *Geochimica et Cosmochimica Acta*, v. 52, p. 821–837.

Mucci, A., and Morse, J. W., 1985, Auger spectroscopy determination of the surface-most adsorbed layer composition on aragonite, calcite, dolomite, and magnesite in synthetic seawater: *American Journal of Science*, v. 285, p. 306–317.

Mucci, A., Morse, J. W., and Kaminsky, M. S., 1985, Auger spectroscopy analysis of magnesium calcite overgrowths precipitated from seawater and solutions of similar composition: *American Journal of Science*, v. 285, p. 289–305.

Neev, D., 1963, Recent precipitation of calcium salts in the Dead Sea: Israel Research Council Bulletin, v. 11G, p. 153–154.

Neev, D., 1964, The Dead Sea, recent sedimentary processes [Ph.D. thesis]: Jerusalem, The Hebrew University, 407p.

Neev, D., and Emery, K. O., 1967, The Dead Sea. Depositional processes and environments of evaporites: Jerusalem, Geological Survey of Israel Bulletin 41, 147 p.

Ostroff, A. G., and Metler, A. V., 1966, Solubility of calcium sulfate dihydrate in the system $NaCl$-$MgCl_2$-H_2O from 28° to 70°C: *Journal of Chemical Engineering Data*, v. 11, p. 346–350.

Pabalan, R. T., and Pitzer, K. S., 1987, Thermodynamics of concentrated electrolyte mixtures and the prediction of mineral solubilities to high temperatures for mixtures in the system Na-K-Mg-Cl-SO_4-OH-H_2O: *Geochimica et Cosmochimica Acta*, v. 51, p. 2,429–2,443.

Pabalan, R. T., and Pitzer, K. S., 1988a, Apparent molar heat capacity and other thermodynamic properties of aqueous KCl solutions to high temperatures and pressures: *Journal of Chemical Engineering Data*, v. 33, p. 354–362.

Pabalan, R. T., and Pitzer, K. S., 1988b, Heat capacity and other thermodynamic properties of Na_2SO_4 (aq) in hydrothermal solutions and the solubilities of sodium sulfate minerals in the system Na-Cl-SO_4-OH-H_2O at 300°C: *Geochimica et Cosmochimica Acta*, v. 52, p. 2,393–2,404.

Pabalan, R. T., and Pitzer, K. S., 1991, Mineral solubilities in electrolyte solutions, *in* Pitzer, K. S., ed., *Activity coefficients in electrolyte solutions* (2nd ed.): CRC Press, p. 435–490.

Peiper, J. C., and Pitzer, K. S., 1982, Thermodynamics of aqueous carbonate solutions including mixtures of sodium carbonate, bicarbonate and chloride: *Journal of Chemical Thermodynamics*, v. 14, p. 613–638.

Phutela, R. C., and Pitzer, K. S., 1983, Thermodynamics of aqueous calcium chloride: *Journal of Solution Chemistry*, v. 12, p. 201–207.

Phutela, R. C., and Pitzer, K. S., 1986, Heat capacity and other thermodynamic properties of aqueous magnesium sulfate to 473°K: *Journal of Physical Chemistry*, v. 90, p. 895–901.

Phutela, R. C., Pitzer, K. S., and Saluja, P. P. S., 1987, Thermodynamics of aqueous magnesium chloride, calcium chloride and strontium chloride at elevated temperatures: *Journal of Chemical Engineering Data*, v. 32, p. 76–80.

Pitzer, K. S., 1973, Thermodynamics of electrolytes. I. Theoretical basis and general equations: *Journal of Physical Chemistry*, v. 77, p. 268–277.

Pitzer, K. S., 1975, Thermodynamics of electrolytes. V. Effects of higher order electrostatic terms: *Journal of Solution Chemistry*, v. 4, p. 249–265.

Pitzer, K. S., 1979, Theory, ion interaction approach, *in* Pytkowicz, R. M., ed., *Activity coefficients in electrolyte solutions*, Vol. 1: CRC Press, p. 157–208.

Pitzer, K. S., 1981a, Characteristics of very concentrated aqueous solutions: *Physical Chemistry of the Earth (Chem. Geochem. Solutions High Temp. Pressures)*, v. 13–14, p. 249–272.

Pitzer, K. S., 1981b, The treatment of ionic solutions over the entire miscibility range: *Berichte der Bunsen-Gesellschaft für physikalische Chemie*, v. 85, p. 952–959.

Pitzer, K. S., 1983, Thermodynamics of electrolyte solutions over the entire miscibility range, *in* Newman, S. A., ed., *Chemical engineering thermodynamics*: Ann Arbor Science, p. 309–321.

Pitzer, K. S., 1986, Theoretical considerations of solubility with emphasis on mixed aqueous electrolytes: *Pure and Applied Chemistry*, v. 58, p. 1,599–1,610.

Pitzer, K. S., 1987, A thermodynamic model for aqueous solutions of liquid-like density, *in* Carmichael, I. S. E., and Eugster, H. P., eds., *Reviews in mineralogy*, Vol. 17: Mineral Society of America, p. 97–142.

Pitzer, K. S., 1991, Ion interaction approach: Theory and data correlation, *in* Pitzer, K. S., ed., *Activity coefficients in electrolyte solutions* (2nd Ed.): CRC Press, p. 75–153.

Pitzer, K. S., and Kim, J. J., 1974, Thermodynamics of electrolytes. IV. Activity and osmotic coefficients for mixed electrolytes: *Journal of the American Chemical Society*, v. 96, p. 5,701–5,707.

Pitzer, K. S., and Mayorga, G., 1973, Thermodynamics of electrolytes. II. Activity and osmotic coefficients for strong electrolytes with one or both ions univalent: *Journal of Physical Chemistry*, v. 77, p. 2,300–2,308.

Pitzer, K. S., and Mayorga, G., 1974, Thermodynamics of electrolytes. III. Activity and osmotic coefficients for 2-2 electrolytes: *Journal of Solution Chemistry*, v. 3, p. 539–546.

Pitzer, K. S., Olsen, J., Simonson, J., Roy, R., Gibbons, J. J., and Rowe, L., 1985, Thermodynamics of aqueous magnesium and calcium bicarbonates and mixtures with chloride: *Journal of Chemical Engineering Data*, v. 30, p. 14–17.

Pitzer, K. S., and Peiper, J. C., 1980, Activity coefficient of aqueous sodium bicarbonate: *Journal of Physical Chemistry*, v. 84, p. 2,396–2,398.

Pitzer, K. S., Peiper, J. C., and Busey, R. H., 1984, Thermodynamic properties of aqueous sodium chloride solutions: *Journal of Physical Chemistry Ref Data*, v. 13, p. 1–102.

Pitzer, K. S., Peterson, J. R., and Silvester, L. F., 1978, Thermodynamics of electrolytes. IX. Rare earth chlorides, nitrates, and perchlorates: *Journal of Solution Chemistry*, v. 7, p. 45–56.

Pitzer, K. S., Roy, R. N., and Silvester, L. F., 1977, Thermodynamics of electrolytes. 7. Sulfuric acid: *Journal of the American Chemical Society*, v. 99, p. 4,930–4,936.

Pitzer, K. S., and Silvester, L. F., 1976, Thermodynamics of electrolytes. VI. Weak electrolytes including H_3PO_4: *Journal of Solution Chemistry*, v. 5, p. 269–277.

Pitzer, K. S., and Silvester, L. F., 1978, Thermodynamics of electrolytes. 11. Properties of 3,2 4,2 and other high-valence types: *Journal of Physical Chemistry*, v. 82, p. 1239–1242.

Posnjak, E., 1940, Deposition of calcium sulfate from sea water: *American Journal of Science*, v. 238, p. 559-568.

Rao, N. K., and Ananthaswamy, Y., 1989, Thermodynamics of electrolyte solutions. Activity and osmotic coefficients of aqueous NaCl in the $NaCl-MgCl_2-H_2O$ system at different temperatures by the EME method: *Proceedings of the Indian Academy of Science*, v. 101, p. 433–437.

Rard, J. A., and Miller, D. G., 1981a, Isopiestic determination of the osmotic coefficients of aqueous Na_2SO_4, $MgSO_4$, and $Na_2SO_4-MgSO_4$ at 25°C: *Journal of Chemical Engineering Data*, v. 26, p. 33–38.

Rard, J. A., and Miller, D. G., 1981b, Isopiestic determination of the osmotic and activity coefficients of aqueous magnesium chloride solutions at 25°C: *Journal of Chemical Engineering Data*, v. 26, p. 38–43.

Rasumny, J., 1962, The solubility of Neocomian formations from Sdome region (Israel) in various aqueous solutions: *Congres National de Soc Des Savantes, Comp Rend*, v. 86, p. 387–388.

Reddy, D. C., Rao, N. K., and Ananthaswamy, J., 1988, Thermodynamics of electrolyte solutions:Electromotive force studies on aqueous solutions of KCl in KCl + $MgCl_2$ + H_2O system at 25°C: *Current Science*, v. 57, p. 287–290.

Robbins, L. L., and Blackwelder, P. L., 1992, Biochemical and ultrastructural evidence for the origin of whitings: A biologically induced calcium carbonate precipitation mechanism: *Geology*, v. 20, p. 464–468.

Robie, R. A., Hemingway, B. S., and Fisher, Y. L., 1978, Thermodynamic properties of minerals and related substances at 298.15 K and 1 Bar (10^5 Pascals) pressure and at higher temperatures: *Geological Survey Bulletin*, v. 1,452, 456 p.

Robinson, R. A., and Macaskill, J. B., 1979, Osmotic coefficients of aqueous sodium carbonate solutions at 25°C: *Journal of Solution Chemitry*, v. 8, p. 35–40.

Robinson, R. A., and Wood, R. H., 1972, Calculation of the osmotic and activity coefficients of seawater at 25°C: *Journal of Solution Chemitry*, v. 1, p. 481–488.

Rogers, P. S. Z., and Pitzer, K. S., 1981, High-temperature thermodynamic properties of aqueous sodium sulfate solutions: *Journal of Physical Chemistry*, v. 85, p. 2886-2895.

Rogers, P. S. Z., and Pitzer, K. S., 1982, Volumetric properties of aqueous sodium chloride solutions: *Journal of Physical Chemistry Reference Data*, v. 11, p. 15–81.

Roy, R. N., Gibbons, J. J., Wood, M. D., Williams, R. W., Peiper, J. C., and Pitzer, K. S., 1983, The first ionization of carbonic acid in aqueous solutions of potassium chloride including the activity coefficients of potassium bicarbonate: *Journal of Chemical Thermodynamics*, v. 15, p. 37–47.

Roy, R. N., Gibbons, J. J., Williams, R., Godwin, L., Baker, G., Simonson, J. M., and Pitzer, K. S., 1984, The thermodynamics of aqueous carbonate solutions. II. Mixtures of potassium carbonate, bicarbonate and chloride: *Journal of Chemical Ther-*

modynamics, v. 16, p. 303–315.

Saluja, P. P. S., Pitzer, K. S., and Phutela, R. C., 1986, High temperature thermodynamic properties of several 1:1 electrolytes: *Canadian Journal of Chemistry*, v. 64, p. 1328-1335.

Sarbar, M., Covington, A. K., Nuttall, R. L., and Goldberg, R. N., 1982a, Activity and osmotic coefficients of aqueous potassium carbonate: *Journal of Chemical Thermodynamics*, v. 14, p. 695–702.

Sarbar, M., Covington, A. K., Nuttall, R. L., and Goldberg, R. N., 1982b, The activity and osmotic coefficients of aqueous sodium bicarbonate solutions: *Journal of Chemical Thermodynamics*, v. 14, p. 967–976.

Schnerb, I., and Yaron, F., 1952, On the solubility of sodium chloride in Dead Sea brines: *Israel Research Council Bulletin*, v. 2, p. 197–198.

Scrivner, N. C., and Staples, B. R., 1983, Equilibria in aqueous solutions, industrial applications, *in* Newman, S. A., ed., *Chemical engineering thermodynamics*: Ann Arbor Science, p. 349–362.

Shaffer, L. H., 1967, Solubility of gypsum in sea water and sea water concentrates at temperatures from ambient to 65°C: *Journal of Chemical Engineering Data*, v. 12, p. 183–189.

Shalem, N., 1949, Whitings of the waters of the Dead Sea: *Nature*, v. 164, p. 72.

Silvester, L. F., and Pitzer, K. S., 1978, Thermodynamics of electrolytes. X. Enthalpy and the effect of temperature on the activity coefficients: *Journal of Solution Chemistry*, v. 7, p. 327–337.

Simonson, J. M., Roy, R. N., Connole, J., Roy, L. R., and Johnson, D. A., 1987, The thermodynamics of aqueous borate solutions. I. Mixtures of boric acid with sodium or potassium borate and chloride: *Journal of Solution Chemistry*, v. 17, p. 791–803.

Spencer, R. J., Moller, N., and Weare, J. H., 1990, The prediction of mineral solubilities in natural waters: A chemical equilibrium model for the Na-K-Ca-Mg-Cl-SO_4-H_2O system at temperatures below 25°C: *Geochimica et Cosmochimica Acta*, v. 54, p. 575–590.

Steinhorn, I., 1983, In situ salt precipitation at the Dead Sea: *Limnology and Oceanography*, v. 28, p. 580–583.

Steinhorn, I., and Assaf, G., 1980, The physical structure of the Dead Sea water column, 1975–1977, *in* Nissenbaum, A., ed., *Hypersaline brines and evaporitic environments*: Amsterdam, Elsevier, p. 145–153.

Steinhorn, I., Assaf, G., Gat, J. R., Nishry, A., Nissenbaum, A., Stiller, M., Beyth, M., Neev, D., Garber, R., Friedman, G. M., and Weiss, W., 1979, The Dead Sea—Deepening of the mixolimnion signifies the overture to overturn of the water column: *Science*, v. 206, p. 55–57.

Steinhorn, I., and Gat, J. R., 1983, The Dead Sea: *Scientific American*, v. 249, p. 102–109.

Stiller, M., Rounich, J. S., and Shasha, S., 1985, Extreme carbon-isotope enrichments in evaporating brines: *Nature*, v. 316, p. 434–435.

Thurmond, V., and Millero, F. J., 1982, Ionization of carbonic acid in sodium chloride solutions at 25°C: *Journal of Solution Chemistry*, v. 11, p. 447–456.

Tialowska-Mocharla, H., and Atkinson, G., 1985, Thermodynamics of concentrated electrolyte mixtures. 6. Activity coefficients of aqueous $CaCl_2-CaBr_2$ mixtures at 25°C: *Journal of Physical Chemistry*, v. 89, p. 4,884–4,887.

Tishchenko, P. Y., 1989, Activity coefficients of sodium chloride in the $NaCl-CaCl_2-H_2O$ system at different temperatures. Use of the Pitzer equation: *Russian Journal of Physical Chemistry*, v. 63, p. 1292–1296.

Tishchenko, P. Y., Bychkov, A. S., Hrabeczy-Pall, A., Toth, K., and Pungor, E., 1992, Activity coefficients for the system

NaCl+Na$_2$SO$_4$+H$_2$O at various temperatures. Application of Pitzer's equations: *Journal of Solubility Chemistry*, v. 21, p. 261–274.

Vancina, V., Plavsic, M., Bilinski, H., and Branica, M., 1986, Preparation and solubility of northupite from brine and its adsorption properties for Cu(II) and Cd(II) in seawater: *Geochimica et Cosmochimica Acta*, v. 50, p. 1,329–1,336.

Vanderzee, C. E., 1982, Thermodynamic properties of solutions of a hydrolyzing electrolyte: Relative partial molar enthalpies and heat capacities, solvent activities, osmotic coefficients, and solute activity coefficients of aqueous sodium carbonate: *Journal of Chemical Thermodynamics*, v. 14, p. 1,051–1,067.

Venkateswarlu, C., and Ananthaswamy, J., 1988, Thermodynamics of electrolyte solutions: Activity coefficients of NaCl in NaCl+MnCl$_2$+H$_2$O system at 25°C: *Indian Journal of Chemistry*, v. 27A, p. 768–771.

Whitfield, M., 1973, A chemical model for the major electrolyte component of seawater based on the Bronsted-Guggenheim hypothesis: *Marine Chemistry*, v. 1, p. 251–266.

Whitfield, M., 1974, A comprehensive specific interaction model for sea water. Calculation of the osmotic coefficient: *Deep-Sea Research*, v. 21, p. 57–67.

Whitfield, M., 1975a, An improved specific interaction model for seawater at 25°C and 1 atmosphere total pressure: *Marine Chemistry*, v. 3, p. 197–213.

Whitfield, M., 1975b, The extension of chemical models for sea water to include trace components at 25°C and 1 atm pressure: *Geochimica et Cosmochimica Acta*, v. 39, p. 1,545–1,557.

Whitfield, M., 1979, Activity coefficients in natural waters, *in* Pytkowicz, R. M., ed., *Activity coefficients in electrolyte solutions*, Vol. 2: CRC Press, p. 153–300.

Wood, J. R., 1976, Thermodynamics of brine-salt equilibria. II. The system NaCl-KCl-H$_2$O from 0 to 200°C: *Geochimica et Cosmochimica Acta*, v. 40, p. 1,211–1,220.

14. HALITE DEPOSITION FROM THE DEAD SEA: 1960 – 1993

ITTAI GAVRIELI

THE SEDIMENT RECORD OF THE DEAD SEA

Yam Hamelach, the Hebrew name for the Dead Sea, translates to "the Salt Sea," testifying to the lake's high salt content (340 g/l, Table 14-1). The lake is characterized by its ionic ratios which are typical of Ca-chloride solutions (i.e. $Ca/(SO_4 + HCO_3) > 1$). The Dead Sea developed through a complex path: Pliocene or Miocene seawater that once occupied the Dead Sea depression experienced evaporation, salt deposition, migration into the surrounding rocks, water-rock interactions, reemergence as surface brines, and further evaporation (Starinsky, 1974). Additional salts were introduced to this terminal lake via freshwater inflow. The unique composition and the high salinity of the Dead Sea results in it being close to saturation to oversaturated with respect to the major evaporitic minerals.

The sediments of the Dead Sea, as recorded from short cores, resemble that of its precursor, Lake Lisan (Begin et al., 1974; Katz and Kolodny, 1989) and consist of alternating varve-like white aragonitic and dark detritic clay laminae. Though crusts of gypsum were formed on exposed objects close to the shores and in the shallow coastal plains of the lake in the early 1960s (Neev and Emery, 1967), only minor traces of it were identified within the sediment (Levy, 1980a). Most of the gypsum that precipitated from the brine and reached the seafloor has redissolved: bacterial sulfate reduction within the sediments decreased the concentration of sulfate in the pore water, thereby promoting gypsum dissolution (Nissenbaum and Kaplan, 1976). Aragonite and gypsum have been directly observed to crystallize from the Dead Sea when sudden "whitening" of the surface took place (Bloch et al., 1944; Neev and Emery, 1967), thereby implying that the lake was saturated with respect to both minerals. In fact, thermodynamic calculations indicate that the lake is significantly oversaturated with respect to gypsum (Katz et al., 1981). However, because of the relative depletion of sulfate and bicarbonate in the Dead Sea (Table 14-1), massive crystallization of gypsum and aragonite from the brine require input of sulfate and bicarbonate from springs, rivers, and runoff waters. The limited inflow of runoff and river water over the last decades has thus resulted in a decrease in the rate of precipitation of these minerals.

Halite (NaCl) accounts for over 27% of the total weight of dissolved salts in the Dead Sea. Massive halite precipitation can, therefore, be initiated from this brine by its evaporation. This process is well exhibited in the evaporation ponds of the Dead Sea Works and the Arab Potash Co., the two potash industries south of the lake, where halite is the major salt to precipitate. The Dead Sea has experienced several periods of halite deposition as is evident by these observations: (1) The sediments of the southern Dead Sea basin are interbedded with several halite layers, particularly in its northern parts (Neev and Emery, 1967). (2) A 6.5-m-thick halite layer was penetrated in a drill hole near the Dead Sea coast, at Wadi Ze'elim, and dated to about 10,000 yr B.P. (Yechieli et al., 1993). (3) Four continuous reflections in the deep northern basin are interpreted as contacts between rock-salt and marl layers (Ben-Avraham et al., 1993). (4) An apparently continuous halite layer is present in the submarine section of the northern basin, at elevations below -430 to -440 m (Neev and Emery, 1967). At the present-day

Dead Sea surface level of -410 m, this continuous layer should be found in water depths of 20–30 m. This layer probably represents the latest episode, excluding the modern one, of halite deposition from the Dead Sea. Age estimates for the time of deposition of this halite bed range from 1,500 yr B.P. (Neev and Emery, 1967) to 500 ±500 yr B.P. based on ^{14}C analyses of carbonate mud deposited between halite crystals (Levy, 1984). The absence of rock salt above the 20 to 30 m depth contour could be the result of a later dissolution by the diluted upper water mass.

A range of morphologies of halite crystals was identified within the upper rock-salt layer of the northern basin in several cores that penetrated a maximum halite thickness of 20 cm (Levy, 1984). These included single euhedral 0.5 to 2.0 cm cubes, hoppers or chevrons formed from 0.5 to 1.0 cm euhedral cubes, aggregates of subhedral crystals, and massive aggregates of granular, rounded white or yellowish halite grains. No correlation between halite morphology and water depth or depth within the cores was found, and different morphologies were encountered in the same core.

Over the last decades, the Dead Sea has experienced a negative water budget, which has resulted in a salinity increase. This initiated halite precipitation, a process that began in 1983 (Levy, 1985, 1988; Stiller et al., chapter 15, this volume) and continued undisturbed until February–April, 1993. This chapter's compilation of the hydrographical, sedimentological, and chemical studies of the lake presents the major mechanisms by which halite precipitates from the lake and correlates them with the halite saturation index (Ω_{halite}).

LIMNOLOGICAL BACKGROUND

The overall negative water budget of the Dead Sea during this century caused a lake level drop of more than 19 m since 1912, from -389.0 m (Klein, 1986) to -408.4 m on October 1991 (Beyth et al., 1993). In 1976, the water level dropped to the level of the Lynch Straits, thereby disconnecting the southern basin from the northern basin, effectively drying out the former. Thus, since 1976, the Dead Sea has been confined to the larger and much deeper northern basin. The negative water budget also resulted in an increase in the salinity of the more diluted upper water mass (epilimnion) and a gradual disappearance of the stratification of the lake (Beyth, 1980). During the winter of 1978–1979, the lake overturned and the water column was homogenized (Steinhorn et al., 1979; Steinhorn and Gat, 1983). This process ended a period of nearly 300 years during which the lake remained stratified with a "fossil" and anoxic bottom water mass (Stiller and Chung, 1984).

The rainy winter of 1979–1980 resulted in a sharp sea level rise of about 2 m and the onset of a new stratified (meromictic) period, which lasted 3 years and ended in December 1982 when the water column overturned (Anati et al., 1987). Between 1981 and 1988, the lake level dropped at an average rate of 0.8 m/yr (Anati and Shasha, 1989) and continued to drop at about the same rate until 1990 (Anati, chapter 8, this volume). No long-lasting stratification of the water column occurred between 1983 and 1990. This holomictic period was characterized by the

Table 14-1 Average chemical compositions (g/l) of the Dead Sea between 1960 and 1993.

Year	Ω_{halite}	Density	TDS	K	Na	Mg	Ca	Cl	Br	SO$_4$	Alkalinity	Na/Cl	Mg/K
1960 UWM[a]	0.62	1.205	299.9	6.50	38.5	36.2	16.4	196.9	4.6	0.5	0.23	0.301	8.95
1960 LWM[a]	1.01	1.229	332.1	7.59	39.7	42.4	17.2	219.3	5.3	0.42	0.22	0.279	8.99
1960 average[a]			322.1	7.26	39.2	40.7	16.9	212.4	5.1	0.47	0.22	0.284	9.01
1965 UWM[b]	0.68	1.209	304.9	6.80	37.9	38.8	15.3	201.5	4.6			0.290	9.18
1965 LWM[b]	1.03	1.230	333.2	7.85	38.1	44.4	16.8	220.8	5.2			0.266	9.10
1976[b]	1.09	1.232	336.0	7.67	40.5	43.2	16.8	222.7	5.1			0.281	9.06
1977[c]	1.14	1.232	339.2	7.76	39.7	43.9	17.4	224.5	5.3	0.47	0.22	0.273	9.09
1979 surface[d]	1.11	1.233	337.4	7.67	40.1	43.4	17.4	223.7	5.2			0.276	9.10
1979[c]	1.18	1.232	340.9	7.85	40.0	44.5	16.9	225.7	5.3	0.44	0.27	0.273	9.11
1983[e]	1.07	1.232	336.7	7.66	38.2	44.1	17.8	223.3	5.3	0.44		0.264	9.25
1985[f]	1.12	1.234	339.3	7.88	38.4	45.0	17.5	224.6	5.2	0.47	0.26	0.263	9.18
1990[g]	1.12	1.235	340.0	7.89	36.7	46.4	17.1	225.6	5.5	0.45	0.25	0.251	9.47
1991[h]	1.10	1.235	339.6	7.92	37.2	45.8	17.0	225.6	5.4	0.48	0.24	0.254	9.31
1992[i]	1.18	1.234	343.5	7.77	36.4	46.6	17.6	228.9	5.5	0.49	0.26	0.245	9.65
1993[j]	1.08	1.234	338.7	7.81	36.4	45.9	17.3	225.0	5.6	0.41	0.26	0.249	9.45
end brines[k]		1.35	477.1	2.80	2.8	90.5	40.0	341.0				0.012	52.00

[a]Neev and Emery, 1967; UWM, upper water mass; LWM, lower water mass.
[b]Beyth, 1980; UWM, upper water mass; LWM, lower water mass.
[c]Calculated from the deep samples in Steinhorn (1980).
[d]Steinhorn (1983); surface brine in central Northern basin.
[e]Calculated from Beyth and Olshina (1983).
[f]Calculated from depth profiles in Gavrieli and Beyth (1986).
[g]Calculated from depth profiles in Beyth et al. (1990).
[h]Calculated from depth profiles in Beyth et al. (1991).
[i]Calculated from depth profiles in Gavrieli et al. (1993).
[j]Calculated from depth profiles in Gavrieli et al. (1994).
[k]Epstein et al. (1975).

formation in spring of a 15 to 30 m-thick upper warmer, sometimes diluted, Dead Sea brine (Anati and Stiller 1991; Anati, chapter 8, this volume). During the following summer months, this layer was heated to temperatures of 35–36°C, while evaporation increased its salinity to above that of the deep waters. The stability of the layering was thus maintained by the temperature difference between the two water bodies. Annual overturns occurred in December as a result of the continuous cooling of the upper water body in the autumn and early winter months. The holomictic period was characterized by a secular increase in the salinity of the deep brine. Superimposed on this increase was a smaller salinity decrease that occurred during the springs and early summers (Anati and Stiller, 1991). The holomictic period ended in the rainy winter of 1991–1992 which introduced 1.4×10^9 m^3 of water to the lake (Beyth et al., 1993), forming a diluted upper water body that still existed in the summer of 1994.

Massive spontaneous halite crystallization from the Dead Sea brine began following the December 1982 overturn. It continued throughout the above described holomictic period and ceased only in February-March 1993 (Anati, chapter 8, this volume), more than a year after the onset of the new meromictic phase. Insight into the mechanisms through which the modern halite precipitation took place may give us a clue about the conditions under which halite precipitated in the past. To understand these mechanisms, the halite saturation index (Ω_{halite}) of the Dead Sea brines must be determined. Following is a discussion of the thermodynamic calculations involved.

HALITE SATURATION INDEX

The saturation index of a solution with respect to halite, Ω_{halite}, is defined as:

$$\Omega_{halite} = \frac{a_{Na}a_{Cl}}{K_h} \tag{1}$$

$$= \frac{m_{Na}\gamma_{Na}m_{Cl}\gamma_{Cl}}{K_h}$$

$$= \frac{m_{Na}m_{Cl}\gamma_{\pm NaCl}^2}{K_h}$$

where a_i = activity of i; m_i = molality of i [moles/(kg H$_2$O)]; γ_i = ionic activity coefficient of i; $\gamma_{\pm NaCl}$ = mean activity coefficient of NaCl; and K_h = thermodynamic solubility constant of halite.

The increased solubility of halite with temperature is reflected in its thermodynamic solubility product, K_h (Gavrieli et al., 1989): at 25°C, $K_h = 38.3$; at 35°C, $K_h = 40.4$; and at 50°C, $K_h = 42.4$.

In multicomponent brines of high ionic strength (I), such as the Dead Sea ($I > 9$ m), the activities of ions or electrolytes can be calculated using either Pitzer's equations or Harned's rule, respectively. Traditionally, the latter model has been applied to calculate the saturation index of the Dead Sea (Lerman, 1967, 1970, Starinsky, 1974; Gavrieli et al., 1989), though it has also been calculated on the basis of Pitzer's equations (Krumgalz and Millero, 1982, 1989; Krumgalz, chapter 13, this volume). Following is a brief description of Harned's rule as it applies to the Dead Sea.

Harned's rule applied to the Dead Sea brine system

The mean activity coefficient of halite, $\gamma_{\pm NaCl}$, in brines of high ionic strength can be calculated following Harned's rule (Harned and Owen, 1964):

$$\log(\gamma_{\pm NaCl}) = \log(\gamma_{\pm NaCl(0)}) + \sum a_{NaCl-i} I_i \quad (2)$$

where $\gamma_{\pm NaCl(0)}$ is the NaCl mean activity coefficient at the ionic strength of the solution under study; α_{NaCl-i} is the interaction coefficient between NaCl and electrolyte i; and I_i is the ionic strength of electrolyte i in the solution under study;

$\gamma_{\pm NaCl(0)}$ is calculated as follows (Gavrieli et al., 1989):

$$25°C: \log(\gamma_{\pm NaCl(0)}) = -0.003200I^2 + 0.095612I - 0.485274 \quad (3)$$

$$35°C: \log(\gamma_{\pm NaCl(0)}) = -0.003310I^2 + 0.101182I - 0.499898 \quad (4)$$

$$50°C: \log(\gamma_{\pm NaCl(0)}) = -0.003470I^2 + 0.106009I - 0.522852 \quad (5)$$

The interaction coefficients in the Dead Sea brine system were shown to be a function of the temperature and the ionic strength of the specific brine (Gavrieli et al., 1989). Since NaCl + MgCl₂ + CaCl₂ compose more than 90% by weight of the salts in the Dead Sea, their interaction coefficients are the most important in determining the halite saturation index of the lake. In the range of ionic strength of 9–16 m, the interaction coefficients of NaCl with MgCl₂ and CaCl₂ were found to be:

$$25°C: \quad \alpha_{NaCl-MgCl_2} = 0.00261I - 0.00905 \quad (6)$$

$$\alpha_{NaCl-CaCl_2} = 0.00150I - 0.00650 \quad (7)$$

$$35°C: \quad \alpha_{NaCl-MgCl_2} = 0.00283I - 0.01972 \quad (8)$$

$$\alpha_{NaCl-CaCl_2} = 0.00207I - 0.01851 \quad (9)$$

$$50°C: \quad \alpha_{NaCl-MgCl_2} = 0.00226I - 0.01416 \quad (10)$$

$$\alpha_{NaCl-CaCl_2} = 0.00156I - 0.01315 \quad (11)$$

In somewhat more diluted brines (ionic strength < 9 m) of the Dead Sea system, the more applicable interaction coefficients are those given by Lerman (1967, 1970): that is, $\alpha_{NaCl-MgCl_2} = 0.018$ and $\alpha_{NaCl-CaCl_2} = 0.0095$.

The other interaction coefficients used for the calculation of Ω_{halite} in the Dead Sea brine system are (Lerman, 1967, 1970) $\alpha_{NaCl-KCl} = -0.0134$, $\alpha_{NaCl-Na_2SO_4} = -0.025$, and $\alpha_{NaCl-KCl} = 0.0268$. A detailed account of how these equations were derived is beyond the scope of this chapter and can be found in Gavrieli et al. (1989).

For historical reasons, most of the analyses of Dead Sea brines in the literature are presented in units of g/l and often do not include the density of the brine. To convert these analyses to units of molality, the density of the brine must be approximated. A linear correlation was found to exist between the total dissolved solids in g/l (TDS) and the density (Gavrieli et al., 1989), whereby

$$Density = 0.00074TDS + 0.98327 \quad (12)$$

The reliability of the saturation index calculations is a function of the reliability of the analyses of the chemical composition of the brine. A 1% error in the magnesium concentrations introduces a 2% error in the calculated Ω_{halite}, whereas a similar error in the analyses of chloride and sodium introduces a 1% error in the Ω_{halite}. Because of their relative lower concentrations in the Dead Sea brine, errors in the calcium, bromine, and potassium analyses introduce only minor errors in the calculated saturation index. In view of this, only the more reliable analyses could be considered in the following discussion.

HALITE SATURATION INDEX IN THE DEAD SEA: 1960–1993

Chemical compositions and the corresponding saturation indices of the Dead Sea brine during the period of 1960–1993 are presented in Table 14-1 and Figure 14-1. The table includes the frequently cited compositions between 1960 and 1976 (see Beyth, 1980), average compositions calculated for the deep waters presented by Steinhorn (1980) for 1977 and 1979, and average compositions calculated from depth profiles sampled at site En Gedi-320 (cord: 102 N, 197 E, water depth of 320 m) between 1983 and 1993 (see Table 14-1 for references). In view of the sensitivity of the saturation indices to analytical uncertainties, the calculated values only denote the general trend and cannot be regarded as exact figures that represent the state of the lake during a given year. Because new interaction coefficients were used, a slight difference exists between the previously published Ω_{halite} figures (Beyth, 1980) and those presented here for 1960 to 1977.

The halite saturation index of the Dead Sea was first calculated by Lerman (1967, 1970), who demonstrated that the hypolimnion (lower water mass) in 1959–1960 was close to saturation (Ω_{halite} = 1.01, Table 14-1, Fig. 14-1). Similar conclusions were drawn previously by Neev and Emery (1967), who based their conclusions on bottom sediments and the observation that halite precipitates from the brine upon slight evaporation. Dur-

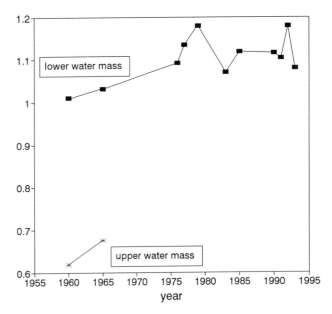

Figure 14-1 Calculated saturation indices (Ω_{halite}) at 25°C for average compositions of the Dead Sea between 1960 and 1993 (see Table 14-1).

ing that period (1959–1960), the less saline epilimnion (upper water mass) was undersaturated (Ω_{halite} = 0.62–0.68), as may be expected from such relatively diluted brine. During the late 1970s, as the lake was approaching its major 1978–1979 overturn, the salinity of the epilimnion increased, leading to an increase in its saturation index (Beyth, 1980). The rise in the saturation index of the lower water mass during this period to oversaturation values of 1.09 and 1.14 in 1976 and 1977, respectively (Table 14-1), is believed to be at least partly due to mixing with evaporated Dead Sea brines (end brines) returned to the lake by the potash industries (see hereafter). This would also account for the decrease and increase in the Na/Cl and Mg/K ratios, respectively, during this period (Fig. 14-2).

During February 1979, immediately after the historic 1978–1979 overturn, the halite saturation index of the deep water was calculated to have been 1.18. A certain increase in the saturation value relative to the 1977 value is to be expected because the overturn introduced the more saline epilimnion brines into the deep waters. However, the value of 1.18 appears to be too high: The salinity increase between the pre-1979 and the February 1979 brines is only by 1.5‰ (Anati et al., 1987; Anati and Stiller, 1991), which should correspond to an increase in saturation index of less than 0.01. Therefore, it is possible that the high oversaturation value obtained for February 1979 is a sampling or analytical artifact.

During the short meromictic period between December 1979 and December 1982, "practically no seasonal or any other change in the temperature or salinity of (the) deep water" was detected (Anati and Stiller, 1991, p. 346). However, the salinity of the deep water during this period was 1.2 g/kg greater than the pre-1979 value, a 4.5‰ salinity increase compared with the pre-1979 salinity. This change in salinity was calculated to lead to an increase of 0.03 in the saturation index relative to the pre-1979 value. Thus, even if the calculated saturation index of February 1979 is imprecise, it is evident that during the short meromixis period of December 1979 to December 1982, the deep waters were oversaturated with respect to halite. The oversaturation of the lake during this period is evident by the onset of halite crystallization on ropes suspended to below the pycnocline (Stiller et al., chapter 15, this volume).

The December 1982 overturn ended the short 1979–1982 meromixis and must have further increased the salinity of the deep water. No analysis of the brine immediately following the overturn is available. However, by the summer of 1983, the saturation index of the deep brine was 1.09—clearly lower than the saturation index during the short meromictic period. The decrease is believed to result from the onset of massive halite crystallization from the brine following the December 1982 overturn. During the summers of 1985, 1990, and 1991, saturation indices of 1.11–1.12 were maintained, reaching 1.18 during the summer of 1992. However, by the summer of 1993, the saturation index dropped to the lower oversaturation value of 1.08 (Table 14-1, Fig. 14-1).

MODERN HALITE SEDIMENTATION

Halite has been precipitating in the Lynch Straits and later in the southern part of the northern Dead Sea basin since at least the 1970s (Epstein et al., 1975; Druckman and Beyth, 1977). This precipitation is the result of mixing of the industrial end brines of the Israeli and Jordanian potash companies with the Dead Sea brine (Epstein et al., 1975). These industries evaporate the Dead Sea brine in evaporation ponds, located in the southern basin, to the point where carnallite ($KMgCl_3 \cdot 6H_2O$) precipitates. Carnallite is harvested for the production of potash, whereas the remaining brines, the end brines, which are highly concentrated solutions of magnesium and calcium chlorides,

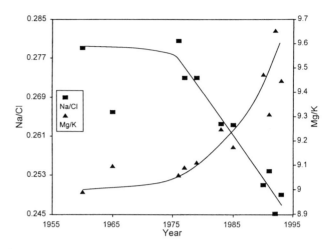

Figure 14-2 Mg/K and Na/Cl molar ratios of the average compositions of Dead Sea deep water between 1960 and 1993.

are allowed to flow back to the Dead Sea (Table 14-1, Epstein et al., 1975; Elata et al., 1977, Stiller et al., 1985). The mixing of the end brines with the Dead Sea brine causes outsalting—high halite oversaturation followed by halite precipitation (Epstein et al., 1975)—and has resulted in a build up of a salt delta in the Lynch Straits and later at the southern end of the northern basin (Druckman and Beyth, 1977; Gavrieli, 1987; Beyth et al., 1992). The continuous sea level drop over the past 10–15 years has exposed parts of this salt delta, but the inflow of fresh water during 1991–1992 has led to the dissolution of significant parts of it (Beyth et al., 1992).

Modern massive halite precipitation from the main Dead Sea body and the accumulation of halite sediments started in 1983 after the December 1982 overturn (Levy, 1985; Stiller et al., chapter 15, this volume). However, as argued previously, the deep waters attained oversaturation several years prior to that (Table 14-1) and not later than before the onset of the short 1979–1982 meromictic period. This is also evident from the precipitation of halite on ropes suspended to below the pycnocline in 1981–1983 (Stiller et al., this volume), a phenomena not known in the 1960s and 1970s. Halite continued to precipitate in the Dead Sea throughout the 1983–1991 holomictic period and ceased only during February–April 1993 (Anati, chapter 8, this volume). Nevertheless, the brine remained oversaturated (Ω_{halite} = 1.08, Table 14-1) as is evident by the halite crystallization on ropes suspended to below the pycnocline (A. Hecht, pers. com., 1994). This crystallization was continuing when the present publication was being prepared (summer 1994).

Studies based on cores retrieved in 1989 from the deeper parts of the lake estimated the halite accumulation rates between 1983 and 1989 to be 3 to 6 cm/yr (Levy, 1991, 1992). With these accumulation rates, the halite layer should be 30 to 60 cm thick by 1993. The newly deposited halite has been described by Levy (1991, 1992) to be composed of layers of unconsolidated crystals of two forms: (1) coarse crystals greater than 2 mm, which may reach a size of a few cm, and (2) fine grains, 100–400 microns in size. Both forms make up at least 20% by weight of each horizon, though often the distribution of the grain size is more uniform. The former generally form solid horizons within the halite layer, whereas the fine-grained halite forms a soft white sediment. Thin, millimeter-thick horizons of interlocked intermediate-sized halite crystals are sometimes encountered within the soft halite sediments. No correlation was found between depth and grain size, and

Figure 14-3 Cross cut of sediments from the Dead Sea collected by a grab sampler. Most of the sediment consists of fine-grained white halite layers interbedded with thin mud laminae and covered by a thin, brown mud layer. Transparent halite crystals, up to 0.5 cm grow within this layer.

Figure 14-4 The catcher of a gravity corer blocked by course, up to 1 cm, transparent halite crystals that form a consolidated layer within the sediment at a depth of 30 cm. Sampling was done few km east of Masada at water depth of about 70 m.

the two forms compose roughly similar portions of the newly formed halite.

Sampling with grab sampler and gravity corer by Gavrieli and Herut in 1993 (an ongoing study), a few km east of Masada at a water depth of about 70 m, has shown that the upper 20–30 cm of sediment are composed of fine-grained white halite layers interbedded with a few thin layers of brown clays (Fig. 14-3). Transparent crystals, up to 1 cm in size, formed a consolidated layer at a depth of about 30 cm which the gravity corer could not penetrate (Fig. 14-4). The top of the salt layer was covered by brown mud, a few mm thick (Fig. 14-3), which was probably deposited during the winter of 1992–1993. Individual transparent crystals of up to 0.5 cm were found to grow from within this mud layer.

The recent massive crystallization of halite from the Dead Sea is reflected in a decrease in the Na/Cl ratio of the deep water, from about 0.27–0.28 in the late 1970s, to less than 0.25 in 1992–1993 (Fig. 14-2). The 1983 composition of the Dead Sea, with its significantly lowered Na/Cl ratio, is the first to exhibit the massive halite precipitation. A contemporaneous increase in the Mg/K ratio of the deep brine (Fig. 14-2) suggests an efficient mixing of the end brines in the lake, a process that contributed to the increase in the saturation state and to the precipitation of halite from the Dead Sea.

MECHANISM OF HALITE CRYSTALLIZATION IN THE DEAD SEA

Several mechanisms of halite crystallization operate simultaneously in the lake, some of which, for example, outsalting as a result of the mixing of the end brines, can be directly observed, whereas others must be inferred indirectly. These include processes that operate on a local scale of time and space, such as evaporation from the surface brine, while others operate on the entire Dead Sea continuously. It is difficult to quantify each factor and determine its contribution to the total weight of halite precipitating from the Dead Sea. Following is a discussion of the processes that precipitate halite from the Dead Sea brine.

Cooling of the epilimnion

The annual overturn during the 1983–1991 holomictic period occurred in December as a result of the more saline epilimnion cooling from 33–35°C to about 24°C, that is, 1–2°C warmer than

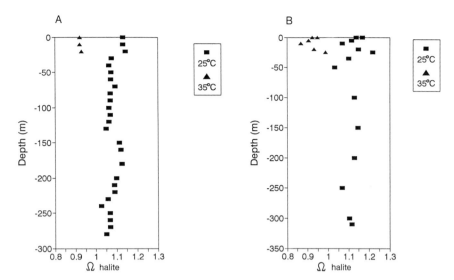

Figure 14-5   Calculated saturation indices (Ω_{halite}) of Dead Sea brines sampled at site En Gedi-320 on (A) September 1983, and (B) July 1990.

the hypolimnion (Anati and Stiller, 1991). Because of its high temperature during the summer months, the epilimnion remained undersaturated with respect to halite despite its higher salinity. Due to the autumn cooling, the epilimnion's saturation index increased to oversaturation, reaching values greater than those of the halite-precipitating hypolimnion (Fig. 14-5, Gavrieli et al., 1989). Under these conditions, halite is expected to spontaneously precipitate from the oversaturated brine. Though this process has never been observed in the Dead Sea, halite crystallization as a result of cooling was often observed when epilimnion brine samples were brought to the lab and allowed to cool.

Figure 14-5 presents two profiles of saturation indices calculated for brines sampled at site En Gedi-320 during September 1983 and July 1990. The saturation indices of the upper water body were calculated at ambient temperature of around 35°C, as well as at 25°C. The undersaturation state of the epilimnion at its ambient temperature is evident in both profiles (Ω_{halite} = 0.92–0.94). However, when calculated at 25°C, oversaturation values of 1.13–1.14 and 1.14–1.17 are attained for the 1983 and the 1990 upper water masses, respectively. These values are higher than the corresponding saturation indices of the lower waters, which were 1.05–1.09 and about 1.11, respectively. Moreover, additional evaporation and salinity increases in the upper water body were still to come in the late summer months, particularly for the July 1990 profile—a situation that would result in still higher saturation indices when the upper water cooled.

Halite precipitation due to cooling may, in principle, take place also on a diurnal basis, when the surface layer cools at night. This process is probably less significant because during the summer months, the surface brines are undersaturated and the cooling by 2–3°C (Anati, 1984) will only slightly increase their saturation index. During the winter, the surface brines are often diluted and are also far from saturation. Thus, if halite precipitation does occur as a result of diurnal temperature changes, it must occur during the autumn months and, as such, is part of the mechanism described herewith.

The annual overturns during the 1983–1991 holomictic period introduced a salinity increase to the hypolimnion that maintained its high oversaturation and promoted halite crystallization. The weight of halite that *could* have precipitated as a result of the annual cooling of the epilimnion should, therefore, give some independent estimate of the annual weight of halite

deposited by the Dead Sea as a whole. This calculation was carried out for the September 1983 epilimnion brines (Gavrieli et al., 1989) by incremental "precipitation" of NaCl from the brine until oversaturation of 1.08 at 25°C was attained. This value was chosen to represent the state at which the oversaturated Dead Sea brines no longer precipitate halite, as was the case in 1993. The calculation indicates that 3.1 g of halite can precipitate from 1 liter of epilimnion brine upon cooling from 35 to 25°C. With an epilimnion thickness of 25 m and a surface area of approximately 650 km² (Hall, Fig. 2-6, this volume), this would amount to about 5×10^7 tons of halite. If precipitation occurred homogeneously throughout the lake, a sedimentation rate of about 7.75 g/cm² would be obtained. At a density of 2 g/cm³ for the halite sediments, this would imply an accumulation rate of nearly 4 cm/yr. This value is within the range of the measured 3–6 cm/yr sedimentation rate deduced from cores sampled in 1989 at water depths of 60–100 m (Levy, 1991, 1992).

Halite crystallization from the surface brine

On February 6, 1979, halite crystals were observed on the surface of the central part of the Dead Sea (Steinhorn, 1983). Surface water temperature during that time of year was less than 20°C. On that day, calm weather conditions with no winds allowed surface evaporation to operate on a limited volume of the surface water, thereby increasing its salinity and thus its saturation index. After crystallization, the dendritic crystals were observed to sink. Chemical analyses of water samples taken with and without crystals were similar, within the accuracy of the measurements. The calculated saturation index for these brine at 25°C is 1.11 (Table 14-1), indicating oversaturation even at a temperature 5°C higher than the ambient temperature. It is reasonable to assume that the saturation index of the thinner interface layer from which the halite crystallized was higher, but the sampling introduced some less saline brine from below this interface layer. Thus, high oversaturation was required to initiate the crystallization observed on that particular occasion. Similar occurrences have since been described by numerous workers, generally under similar calm weather condition (e.g., Levy, 1985). Furthermore, some of these occurrences were during the summer months when the upper water mass was undersaturated because of its high temperature of 34–36°C.

The effect of evaporation and the consequent concentration increase on the saturation index of the Dead Sea surface brine was modeled. Calculations show that a 1% increase in concentration, corresponding to about 1% volume evaporation, increases the saturation index of the 1959–1960 lower water mass from 1.01 (25°C) to 1.06, whereas a 2% volume evaporation increases the index to 1.11. Likewise, a 1% and 2% evaporation of the 1979 surface brine increases its saturation index from 1.11 to 1.16 and 1.22, respectively. For the 35°C undersaturated 1990 surface brine ($\Omega_{halite} = 0.94$), evaporation of 3.3% was required to reach a saturation index of 1.11. Thus, for a significant change in the saturation index of a Dead Sea brine to occur under time and space limitations, a very thin surface layer must be evaporated. Such conditions can indeed be met only during calm weather conditions, when mixing of the surface water just below the water-air interface is limited. It is, therefore, to be expected that the relative mass of halite that precipitates from the Dead Sea by this mechanism is negligible.

Outsalting as a result of the mixing of end brines in the Dead Sea

The weight of halite deposited by mixing of the industrial end brines in the Dead Sea was estimated (Gavrieli, 1987). The calculations assumed end brine composition as given by Epstein et al. (1975; Table 14-1) and were carried out by calculating Ω_{halite} for a brine with a given mixing ratio. Halite was "deposited" from this brine until oversaturation of 1.05 was attained, after which the same brine was "remixed" with the Dead Sea brine and the procedure was repeated. This procedure was carried out for various mixing ratios; the results indicate that 50–100 g of halite are deposited by 1 liter of end brine before it totally mixes with the Dead Sea brine. Assuming that 0.2 km³ of end brines are dumped annually by the two industrial plants, the annual weight of halite deposited by mixing is 10–20 million tons (m.t.).

Bottom authogenic growth

The coarse halite crystals in the modern halite sediment have been previously attributed to slow crystallization from the oversaturated brines because of diurnal or seasonal cooling (Levy, 1991). However, it is suggested here that these large crystals are, in fact, the product of authogenic growth at the lake's bottom. This is particularly evident from the coarse-size halite crystals found within the mud that covered the modern halite layer in 1993 and which have developed through authogenic growth within the mud. Furthermore, no mm- to cm-size halite crystals, which are easily identified by the naked eye, have been encountered during sampling of the water column. The authogenic growth at the sediment surface or within the sediment can be the product of either recrystallization of finer grain halite or primary crystallization from the overlying water or from the interstitial brines. Indirect evidence for primary crystallization is found in the form of very coarse-size halite crystals that develop on ropes suspended to below the pycnocline. These cm-size primary crystals acquire their size within no more than a month (A. Hecht, pers. com., 1994), indicating that in the oversaturated Dead Sea brine, such halite crystals can grow once nucleation and growth sites are available. The fine halite crystals at the sea floor can serve as much better growth centers than the rope and therefore will promote the crystallization of the coarse-grained crystals at the lakes' bottom. This, however, does not imply that recrystallization of halite does not occur. Rather, evidence for halite recrystallization was found in the halite bed dated by Levy (1984) as 500 ±500 years old. There, aggregates of fine-grained halite crystals exhibit dissolution textures that were not found in the coarser

crystals (Levy, 1984), suggesting that at least some of the latter are later recrystallization products.

Spontaneous halite crystallization in the deep water

Spontaneous halite crystallization in the brine body followed by its deposition to the sea floor as fine-grained halite is believed to be the major mechanism by which halite is removed from the Dead Sea brine. Fine-grained halite crystals make up some 50% of the modern halite layer (Levy, 1991, 1992). Similar crystals have been observed to crystallize as a result of end brines outsalting and through surface evaporation, both of which are rapid crystallization processes. However, these latter processes cannot account for all the fine-grained halite in the modern halite sediments because, as discussed previously, the surface evaporation and crystallization is a rather limited process, whereas the mixing of the end brines and the consequent outsalting and halite precipitation are limited to the southern parts of the lake. Thus, spontaneous crystallization from the water body probably accounts for most of the fine-grained halite sediment.

Spontaneous halite crystallization is not expected to occur before high oversaturation develops: Crystallization *within* the water body (or any solution) requires higher oversaturation indices compared to heterogeneous crystallization on surfaces or around nucleation centers. As discussed previously, oversaturation characterized by $\Omega_{halite} \geq 1.1$ are reached during the mixing of the end brines and during the evaporation of the surface layer; in both cases, homogeneous crystallization is believed to occur. The fact that the massive halite crystallization from the Dead Sea did not begin before such high oversaturation was attained supports the model of spontaneous crystallization within the main Dead Sea water body. This process then accounts for the salinity decrease in the deep waters observed during the springs and early summers of the holomictic period (Anati and Stiller, 1991). Furthermore, winter deposition might also have taken place during that period through this mechanism: From December to February, the entire water column was well mixed and cooled by about 1°C, and in most years, salinity slightly increased (Anati and Stiller, 1991), probably because of negative water balance. These two factors must have further increased the oversaturation values of the water column and probably resulted in spontaneous halite crystallization. Direct evidence for the occurrence of spontaneous crystallization is still being sought.

Some of the fine-grained halite sediments probably serve as nucleation centers on which the coarse authogenic halite develop. Others apparently undergo recrystallization and are the intermediate-size halite crystals that form mm-size horizons within the soft halite sediment. However, most of the fine-grained halite crystals are still intact and form about 50% of the modern halite sediment.

MASS BALANCE CALCULATIONS

The total weight of halite deposited from the Dead Sea brine, including in the industrial evaporation ponds, over a given period of time can be calculated by means of mass balance. The following assumptions and calculations are in line with those presented by Levy (1980b), although different time periods are considered: (1) the decrease in the bulk weight of sodium in the lake over the years is due solely to halite precipitation, and (2) magnesium is a conservative element, whose concentration is affected only by dilution or evaporation; that is, there is no magnesium input to or output from the lake, and its bulk weight remains unchanged over the specified period. These assumption are valid in view of negligible sodium and magnesium input from runoff and spring water relative to their bulk content in the lake when time scales of 10–20 years are considered (Levy, 1980b). The effects of the industrial plants that harvest potash from the Dead Sea brine must also be considered. The pumping of brine from the Dead Sea, followed by evaporation, halite and carnallite ($KMgCl_3 \cdot 6H_2O$) deposition, and dumping of the remaining brine—the end brines—has its effect on the lake composition, as is evident by the increase in the Mg/K ratio over the years (Fig. 14-2). This change is mainly due to potassium removal from the brine because the difference between the mass of magnesium pumped from the Dead Sea into the evaporation ponds and that returned to the lake via the end brines is negligible compared to its total mass in the lake.

1960–1976

The enrichment factor of the average magnesium concentration (Table 14-1) in the Dead Sea between 1960 and 1976 is 1.063, implying, under the previous assumptions, that the volume of the Dead Sea has decreased from 150 km³ in 1960 (Hall and Neev, 1978) to about 141 km³ in 1976. Assuming these figures and the average sodium concentrations, the Dead Sea has lost about 170 million tons (m.t.) of sodium during this period. This is equivalent to 430 m.t. of halite, which implies a halite deposition rate of about 27 m.t./yr.

Most of the halite deposited during the 1960–1976 period, about 240 m.t., precipitated in the industrial evaporation ponds (Levy, 1980b), whereas most of the rest precipitated in the southern basin and in the Lynch Straits following the mixing of end brines with the Dead Sea brines. As discussed previously, the weight of halite deposited by "complete" mixing of 1 liter of end brine in the Dead Sea is 50–100 g (Gavrieli, 1987). The corresponding annual weight of halite thus deposited by the mixing process is 5–10 m.t. If we assume 1967 to be the year during which end brine dumping began (Levy, 1980b), the total weight of halite deposited by mixing over the time period of 1967–1976 amounts to 45–90 m.t. This is a maximum value because during this period, the southern basin and the epilimnion, in general, were less saline than the present day Dead Sea and, therefore, deposited less halite upon mixing. Between 1960 and 1976, no halite precipitation took place in the Dead Sea proper, excluding minor precipitation along the shores of the southern basin (Epstein et al., 1975). Thus, in terms of mass balance, about 100 m.t. of halite or about 25% of the total mass deposited between 1960 and 1976 are unaccounted for. This weight might have been removed as air-borne salts, but such a mechanism would also carry away some of the magnesium on which this calculation is based, thereby decreasing the enrichment factor and the weight of the "missing" halite. However, in view of the rough estimates made, the missing halite is probably within the uncertainty of the calculations.

1976–1991

A similar calculation for the 15 years between 1976 and 1991 emphasizes the increased precipitation rate of halite during this period and indicates that most of the additional halite has precipitated in the sea. A magnesium enrichment factor of 1.071 (based on average magnesium concentration from 1990, 1991, and 1992) suggests a Dead Sea volume decrease from 141 km³ to 132 km³. The corresponding mass balance calculation of sodium (based on average sodium concentration from 1990, 1991, and 1992) indicates that between 1976 and 1991, 850 m.t. of sodium were removed from the Dead Sea brine. This is equivalent to about 2,200 m.t. of halite and to a halite crystallization rate of about 150 m.t./yr.

Halite began to precipitate in the lake in 1983. Accumulation rates were estimated at 3–6 cm/yr (Levy, 1991, 1992), equivalent to 6–12 g/cm²yr (assuming a sediment density of 2 g/cm³). If we assume an area of about 650 km² (Hall, chapter 2, this volume), the annual weight of halite accumulating at the lake floor is 40–80 m.t. This should amount to 320–640 m.t. of halite deposited between 1983, when deposition began, and 1991. The weight of halite deposited in the industrial plants during this period was estimated based on an evaporation rate of 50% in the evaporation ponds, an average sodium concentration of 2.75 g/l in the end brines (Epstein et al., 1975), and an annual end brines discharge of 0.2 km³. The weight of halite deposited thus is 36 m.t. per year or 540 m.t. during 1976–1991. During this period, some 150–300 m.t. of halite were calculated to have been deposited by the mixing of the end brines in the Dead Sea and to have formed the salt delta. Thus, a maximum of 1,480 m.t. of halite are accounted for over the 1976–1991 time period, that is, about 33% short of the bulk 2,200 m.t. calculated by mass balance considerations. As for the 1960–1976 calculation, such uncertainty is considered reasonable in view of the rough estimates made.

This mass balance figure can be compared to a figure based on physical parameters. Anati (1993), based on the changes in the salinity-density curve, has argued that during the 1983–1991 holomictic period, the Dead Sea precipitated halite at a rate of 1.9×10^{-3} g/cm³/yr. On a basis of 135 km³, the average Dead Sea volume during this period, a halite precipitation rate of 256 m.t./yr was calculated, which is equivalent to about 2,050 m.t. over the 1983–1991 time period. During the preceding period of 1976–1982, halite was removed only in the evaporation ponds and due to end brines–Dead Sea mixing. Based on an annual precipitation of 36 m.t. in the evaporation ponds and 10 m.t. by the mixing process, this would amount to 276 m.t. precipitated during the 1976–1982 period. Thus, a calculated total of about 2,300 m.t. of halite was deposited from the Dead Sea during 1976–1991. This figure is very similar to the 2,200 m.t. deduced from mass balance considerations.

1991–1992

As discussed previously, despite the dilution of the upper water body by the 1991–1992 winter runoff, halite continued to precipitate from the main Dead Sea water body until April–March 1993 (Anati, chapter 8, this volume). The depth of the diluted layer was about 10 m, corresponding to less than 10% of the lake's volume. The change in the sodium concentration in the deep water between 1991 and 1992 was 0.8 g/l (Table 14-1). This value corresponds to about 100 m.t. of sodium, which is equivalent to about 250 m.t. of halite. This figure is probably an overestimate resulting from the relatively high 1991 sodium concentration. The latter is not in line with the general trend of decreasing sodium concentration with time and is probably due to some analytical or sampling artifact. If, as assumed previously for the 1976–1991 mass balance calculations, the average 1991–1992 sodium concentration is taken to represent the 1991 concentration, then the calculated weight of halite that was deposited between 1991 and 1992 is about 100 m.t.

SUMMARY

Massive halite precipitation from the Dead Sea began in 1983, coinciding with the onset of the holomictic period, and continued throughout this period (1983-1991). The extensive precipitation ceased in 1993, two years after the establishment of the modern meromictic period. The massive halite precipitation was initiated by the halite oversaturation index (Ω_{halite}) ≥ 1.1, and it ceased once oversaturation dropped to $\Omega_{halite} < 1.1$. The high oversaturation levels required for halite crystallization were maintained during the holomictic period by the negative water budget of the lake and through the annual December overturns which increased the salinity of the deep water.

The major mechanism by which halite was removed from the deep water was spontaneous crystallization and settling of the newly crystallized fine-grained halite to the lake's bottom. The coarse-grained halite crystals in the sediment are the result of authogenic growth and recrystallization. Annual cooling of the epilimnion and surface evaporation removed halite from the epilimnion. In the southern part of the northern Dead Sea basin, halite precipitated through mixing with the industrial end brines. However, this latter process took place prior to the onset of the massive halite crystallization in the Dead Sea and still prevails.

The calculated weight of halite deposited from the Dead Sea brine during the 1960–1976 and 1976–1991 periods is 430 and 2,200 million tons, respectively, that is, deposition rates of about 27 and 150 m.t./yr. Whereas, during 1960–1976, most of the halite precipitated in conjunction with the potash industries around the lake, during 1976–1991 most of the mass precipitated to form a new sedimentary halite layer in the northern basin, and which began to accumulate in 1983.

Acknowledgments

I wish to thank Mr. M. Gonen, the skipper of the *Tiulit* and his crew for their assistance during the sampling cruises. Long and fruitful discussions were held with Dr. D. Anati. The review and comments from Dr. M. Stiller significantly contributed to the manuscript.

REFERENCES

Anati, D. A., 1984, The top 50 metres of the Dead Sea: Solmat Publication No. 28, 11 p.
Anati, D. A., 1993, How much salt precipitates from the brines of a hypersaline lake? The Dead Sea as a case study: *Geochimica et Cosmochimica Acta*, v. 57, p. 2,191–2,196.
Anati, D. A., and Shasha S., 1989, Dead Sea surface-level changes: *Israel Journal of Earth Sciences*, v. 38, p. 29–32.
Anati, D. A., and Stiller, M., 1991, The post-1979 thermohaline structure of the Dead Sea and the role of double-diffusive mixing: *Limnology and Oceanography*, v. 36, p. 343–354.
Anati, D. A., Stiller, M., Shasha, S., and Gat, J. R., 1987, Changes in the thermohaline structure of the Dead Sea: 1979–1984: *Earth and Planetary Science Letters*, v. 84, p. 109–121.
Begin, Z. B., Ehrlich, A., and Nathan, Y., 1974, Lake Lisan, the Pleistocene precursor of the Dead Sea: Jerusalem, Geological Survey of Israel Bulletin 63, 30 p.
Ben-Avraham, Z., Niemi, T. M., Neev, D., Hall, J. K., and Levy, Y., 1993, Distribution of Holocene sediments and neotectonics in the deep north basin of the Dead Sea. *Marine Geology*, v. 113, p. 219–231.
Beyth, M., 1980, Recent evolution and present stage of the Dead Sea brines: *in* Nissenbaum, A., ed., *Hypersaline brines and evaporitic environments*: Amsterdam, Elsevier Scientific Publishing Company, p. 155–166.
Beyth, M., Gavrieli, I., and Baumann, N., 1990, Mixing of end brines in the Lynch Strait, Dead Sea (16–19 July, 1990): Jerusalem, Geological Survey of Israel Report GSI/48/90, 10 p.
Beyth, M., Gavrieli, I., Anati, D. A., and Katz, O., 1991, Mixing of end brines, Southern Dead Sea (5–8 August, 1991): Jerusalem, Geological Survey of Israel Report GSI/34/91, 16 p.
Beyth, M., Gavrieli, I., Anati, D. A., and Katz, O., 1993, Effects of the December 1991—May 1992 floods on the Dead sea verti-

cal structure: *Israel Journal of Earth Sciences*, v. 41, p. 45–48.
Beyth, M., Katz, O., and Gavrieli, I., 1992, Propagation and retrogression of the Salt Delta, southern Dead Sea (1985–1992): Jerusalem, Geological Survey of Israel Rept. GSI/23/92, 22 p.
Beyth M. and Olshina, A., 1983, Mixing of end brines in the Lynch Straits, Dead Sea (29 Aug.–1 Sept., 1983): Jerusalem, Geological Survey of Israel Report MGG/7/83/, 6 p.
Bloch, R., Littman, H. Z., and Elazari-Volcani, B., 1944, Occasional whiteness of the Dead Sea: *Nature*, v. 154, p. 402.
Druckman, Y., and Beyth, M., 1977, Salt reefs—A product of brine mixing, Lynch Strait, Dead Sea: Jerusalem, Geological Survey of Israel Report MG/7/77, 15 p.
Elata, C., Goldwasser, Y., and Sher, E., 1977, Mixing of end brine with the inflow of seawater to the evaporation ponds of the Dead Sea: International Association for Hydraulic Research, 7th congress, Baden Baden, August 1977, v. 1., p. 355–362.
Epstein, J. A., Zelvianski, B., and Ron, G., 1975, Manganese in sodium chloride precipitating from mixing Dead Sea brines: *Israel Journal Earth Sciences*, v. 24, p. 112–113.
Gavrieli, I., 1987, The source of the halite bodies in the Southern Dead Sea, [M.Sc. thesis]: Jerusalem, The Hebrew University, 123 p. (in Hebrew).
Gavrieli, I., and Beyth, M., 1986, Mixing of end brines in the Lynch Straits, Dead Sea (11–13 August, 1985): Jerusalem, Geological Survey of Israel Report GSI/16/86, 17 p.
Gavrieli, I., Beyth, M., Anati, D. A., and Katz, O., 1993, Traces of end brines in the Southern Dead Sea—An outcome of the 1991/92 floods: Jerusalem, Geological Survey of Israel Report GSI/8/93, 31 p.
Gavrieli, I., Katz, O., Weinstein, R., and Anati, D. A., 1994, A plume of end brines in the transition layer of the southern Dead Sea: Jerusalem, Geological Survey of Israel Report GSI/9/94, 38 p.
Gavrieli, I., Starinsky, A., and Bein, A., 1989, The solubility of halite as a function of temperature in the highly saline Dead Sea brine system: *Limnology and Oceanography*, v. 34, p. 1,224–1,234.
Hall, J. K., and Neev, D., 1978, The Dead Sea geophysical survey, 19 July–1 August, 1974: Jerusalem, Geological Survey of Israel Report MG/1/78, 28 p.
Harned, H. S., and Owen, O. B., 1964, *The physical chemistry of electrolytic solutions*, 3rd ed.: New York, Reinhold Publishing Corporation, 803 p.
Katz, A., and Kolodny, N., 1989, Hypersaline brine diagenesis and evolution in the Dead Sea–Lake Lisan system (Israel): *Geochimica et Cosmochimica Acta*, v. 53, p. 59–67.
Katz, A., Starinsky, A., Taitel-Goldman, N., and Beyth, M., 1981, Solubilities of gypsum and halite in the Dead Sea in its mixtures with seawater: *Limnology and Oceanography*, v. 26, p. 709–716.
Klein, C., 1986, Fluctuations of the level of the Dead Sea and climatic fluctuations in Erez-Israel during historical times [Ph.D. thesis]: Jerusalem, The Hebrew University, 208 p. (in Hebrew).
Krumgalz, B., and Millero, F. J., 1982, Physico-chemical study of the Dead Sea waters. 1. Activity coefficients of major ions in the Dead Sea water: *Marine Chemistry*, v. 11, p. 209–222.
Krumgalz, B., and Millero, F. J., 1989, Halite solubility in the Dead Sea waters: *Marine Chemistry*, v. 27, p. 219–233.
Lerman, A., 1967, Model of chemical evolution of a chloride lake—The Dead Sea: *Geochimica et Cosmochimica Acta*, v. 31, p. 2,309–2,330.
Lerman, A., 1970, Chemical equilibria and evolution of chloride brines: *Mineralogical Society of America Special Paper*, No. 3, p. 291–306.
Levy, Y., 1980a, Chemistry of bottom sediments and interstitial

water from the Dead Sea: Jerusalem, Geological Survey of Israel Report MG/8/80, 11 p.

Levy, Y., 1980b, A quantitative approach to the recent halite precipitation in the Dead Sea: Jerusalem, Geological Survey of Israel Report MG/4/80, 12 p.

Levy, Y., 1984, Halite from the bottom of the Dead Sea: Jerusalem, Geological Survey of Israel Report GSI/48/84, 13 p.

Levy, Y., 1985, Modern halite precipitation in the Dead Sea: Jerusalem, Geological Survey of Israel Rept. GSI/7/85, 19 p.

Levy, Y., 1988, Sedimentary reflections of modern (1983–1985) halite precipitation from Dead sea water: Jerusalem, Geological Survey of Israel Report GSI/12/88, 15 p.

Levy, Y., 1991, Modern sedimentation in the Dead Sea, across from En Gedi: Jerusalem, Geological Survey of Israel Report TR-GSI/2/91, 14 p.

Levy, Y., 1992, Modern sedimentation in the Dead Sea (1982–1989): Jerusalem, Geological Survey of Israel Report TR-GSI/7/92, 9 p.

Neev, D., and Emery, K. O., 1967, The Dead Sea: Depositional processes and environments of evaporites: Third edition-Geological Survey of Israel Bulletin 41, 147 p.

Nissenbaum, A., and Kaplan, I. R., 1976, Sulfur and carbon isotopic evidence from biogeochemical processes in the Dead Sea ecosystem, in Nriagu, J. O., ed., Environmental Biogeochemistry. Volume 1, Carbon, nitrogen, phosphorus, sulfur, and selenium cycles: Ann Arbor, Ann Arbor Science Publishers Inc., p.

309–325.

Starinsky, A., 1974. Relationship between Ca-Chloride brines and sedimentary rocks in Israel [Ph.D. thesis]: Jerusalem, The Hebrew University, 176 p. (in Hebrew).

Steinhorn, I., 1980, The density of the Dead Sea water as a function of temperature and salt concentration: Israel Journal of Earth Sciences, v. 29, p. 191–196.

Steinhorn, I., 1983, In situ salt precipitation at the Dead Sea: Limnology and Oceanography, v. 28, p. 580–583.

Steinhorn, I., Assaf, G., Gat, J. R., Nishri, A., Nissenbaum, A., Stiller, M., Beyth, M. Neev, D., Grader, R., Friedman, G. M., and Weiss, W., 1979, The Dead Sea—Deepening of the mixolimnion signifies the overturn of the water column: Science, v. 206, p. 55–57.

Steinhorn, I., and Gat, J. R., 1983, The Dead Sea: Scientific American, v. 249, p. 102–109.

Stiller, M., and Chung, Y., 1984, Radium in the Dead Sea—A possible tracer for the duration of meromixis: Limnology and Oceanography, v. 29, p. 574–586.

Stiller, M., Rounick, J. S., and Shasha, S., 1985, Extreme carbon-isotope enrichments in evaporating brines: Nature, v. 316, p.434–435.

Yechieli, Y., Magaritz, M., Levy, Y., Weber, U., Kafri, U., Woelfli W., and Bonani, G., 1993, Late Quaternary geological history of the Dead Sea area, Israel: Quaternary Research, v. 39, p. 59–67.

15. HALITE PRECIPITATION AND SEDIMENT DEPOSITION AS MEASURED IN SEDIMENT TRAPS DEPLOYED IN THE DEAD SEA: 1981–1983

Mariana Stiller, Joel R. Gat, and Perla Kaushansky

The Dead Sea is a hypersaline brine, made up of a mixture of Na, Ca, Mg, and K chlorides, with bromide as an important minor element. In terms of Eugster and Hardie's (1978) classification of saline lakes, the Dead Sea evolved when bicarbonate-poor waters were enriched with magnesium over calcium and gypsum precipitated to produce a Mg-Na-(Ca)-Cl brine (path III B1a). Starinsky (1974) described the various stages by which Pliocene seawater has supposedly evolved into Dead Sea brines.

Until the late 1970s, the Dead Sea was in a meromictic state (Neev and Emery, 1967), with the lower (fossil) layer apparently close to saturation with respect to NaCl (Lerman and Shatkay, 1968) and supersaturated with respect to gypsum (Krumgalz and Millero, 1983). The top layer (40 m) and the transition zone, down to about 100 m, were somewhat more dilute. Under these circumstances, the chemical precipitation from the lake was restricted to that of gypsum and an occasional deposition of endogenic carbonate, with the bulk of the sediments made up of detrital material (Garber, 1980).

As is well known, the Dead Sea has been experiencing a water deficiency since the middle of this century, resulting in the decline of its water level (Klein, 1981) and a commensurate increase in the surface water salinity. Finally, the density of the top layer was so close to that of the deeper water masses that a complete overturn occurred in February 1979 (Steinhorn et al., 1979). At that time, the brine had approached saturation with respect to halite (NaCl), and crystals were indeed observed from shipboard under calm conditions during daytime (Steinhorn, 1983). Further, during the summer of 1979, halite crystals often appeared, shortly after sampling, as sediment in the sampling bottles, as a consequence of their exposure to lower temperatures in the laboratory than at sea. This suggests that in the summer of 1979, some precipitation might have taken place in the top layer of the sea, as a result of the diurnal cooling cycle.

Because of the unusually wet season of 1979–1980, a dilute surface layer was reestablished for 3 years (Stiller et al., 1984). Conditions for halite precipitation did not occur in the surface layer until late in 1982 (Anati et al., 1987). The deeper layers, from 50 m downward, however, remained close to the saturation point. This is evident from the pattern of halite crystallization that was observed throughout 1981–1982 on cables suspended deep into the lake. Halite crystals were sometimes more abundant at certain depths than at others.

In December 1982, cubic crystal growth on equipment deployed in the Dead Sea augmented. Other observations included fine white crystals floating on the surface in the northern part of the lake close to the mouth of the Jordan River on November 16, 1982. These crystals covered about 50% of the observed area and were identified as fine cubic euhedral halite, ranging in size from 150 to 400 microns (Levy, 1985).

A similar phenomenon was also observed on a number of occasions during December 1982 and thereafter in the southern part of the Dead Sea, east of Masada. Crystals were seen only when the sea was absolutely calm (mirror-like surface), and the crystal cover then increased from noon through late afternoon. With the appearance of even the most minute waves, the crystal colonies appeared to tip over and were seen to settle through the water column. Momentary rates of settling at the time were about 3 cm/min (Levy, 1985).

To quantify the process of halite sedimentation, a set of sediment traps were deployed in the Dead Sea from May 1981 to December 1983. This setup also enabled the monitoring of non-halite sedimentation of allogenic (detrital) origin.

In this chapter, we use the nomenclature of Jones and Bowser (1978), as follows: (1) minerals that have been formed within the water column and are later found in the sediments are *endogenic*; (2) sedimentary material derived from outside the lake is *allogenic*; and (3) *authigenic* phases are those produced within the sediments by diagenetic processes that may affect both endogenic and allogenic phases.

In this chapter, we present the results of the sediment-trap collections. Because certain aspects of the data interpretation are related to the degree of solubility of halite in the Dead Sea brines, we first report on some laboratory determinations of halite solubility.

HALITE SOLUBILITY IN THE DEAD SEA

In the Dead Sea brine, which is a multicomponent solution, the halite solubility is not only a function of total salinity and of temperature, but also of the ion composition of the brine. The following processes lead to changes in ionic composition (and hence in solubility):

(1) input of dissolved salts by rivers and springs, some of them high in mineral content (Mazor et al., 1969);

(2) chemical precipitation from the lake, especially the precipitation of halite whenever it occurs, but also of calcium sulfate and carbonate; and

(3) industrial exploitation of the lake's mineral resources, especially of KCl and of bromides (which are accompanied by massive halite deposits in the evaporation ponds).

These processes have been visibly effective in changing halite solubility over recent years, as we discuss subsequently.

Theoretical calculations by Krumgalz (chapter 13, this volume), based on single ion activity ratios showed the brines of 1983 (which on the evidence presented herein were precipitating halite) to be slightly undersaturated, and only the brine of February 1979 appeared slightly supersaturated (2% supersaturation at 22°C). Gavrieli (chapter 14, this volume) using Harned's rule, claims a supersaturation of about 8%. However, these calculations have not been tested in the field or laboratory, and, in some instances, they appear to contradict field observations. We have thus taken the approach of attempting the experimental determination of halite solubility.

The direct experimental determination of halite saturation in a Dead Sea-type brine is difficult. For example, the saturation of an undersaturated Dead Sea brine with excess NaCl (the method employed by Schnerb and Yaron, 1952) causes changes in mineral composition of the brine and thereby invalidates the results. Our attempt to approach equilibrium more subtly by cooling a slightly undersaturated solution and determining the saturation point by observing the appearance of solid particles through light scattering techniques (20° angle scattering of green light in a Sophica scattering chamber) failed with the

Dead Sea brines (Fig. 15-1). Unlike the case of a pure NaCl solution, in which the temperature of the saturation point could be determined within 1°C of its correct value by cooling it in the scattering chamber, in the natural Dead Sea brine, only a slight increase in light scattering was observed near the (assumed) saturation point. Upon further cooling, the intensity of scattered light actually dropped below the initial value. These findings can be interpreted as resulting from a delay in the appearance of a crystalline phase because of a tendency to form supersaturated solutions or, alternatively, as a case where the suspended matter acts as nuclei for the growing salt crystals so that the number of scattering centers does not increase. Once these crystals are large enough (or alternatively, where the supersaturation exceeds a critical value), they settle out and evidently clear up the solution by carrying with them the suspended matter. This method, although not reliable for the purpose of constructing the saturation curve for Dead Sea brines, enabled us to appreciate the cleansing effect of halite formation.

An alternative procedure, a reverse saturometer technique was used to determine the saturation point at the temperature below which the salt concentration decreases. For this purpose, aliquot samples of the brine were thermostated at a temperature range that straddled the assumed saturation temperature by about 10°C above and below. The salt concentration of each of these aliquots was then determined: a clear supernatant solution was used in those cases where halite had precipitated out of the solution. With a brine as complex as that of the Dead Sea, we can determine the density (ρ) of the brine much more precisely than the concentration of individual ionic components. A density measurement using a temperature gradient column (Kaushansky and Starobinets, 1982) was thus adopted as the analytical tool.

Figure 15-2 shows a typical curve obtained for such a reverse saturometer run with a brine sample collected in the spring of 1979. The intercept supposedly defines the saturation point for the particular solution, whereas the descending limb of the curve obviously describes a saturation curve as a function of temperature, provided equilibrium is established at every datum point. The further the departure from the initial saturation point, the greater the change in the composition of the brine because of halite removal.

The saturation curves of Dead Sea brines sampled during 1979–1982 and having salinities (ρ_{25}) of 1.232–1.234 have been

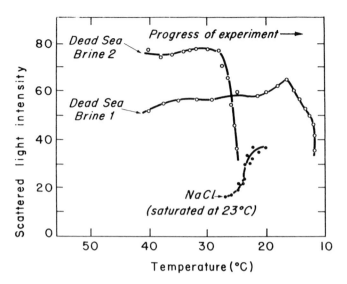

Figure 15-1 Intensity of scattered light from Dead Sea brines that are cooled below the NaCl saturation point. Data for a NaCl solution saturated at 23°C is given for comparison.

investigated. Because all of these brine samples were, in fact, different dilutions of the same ionic composition, the slopes of their saturation curves should be the same. The slope of the curves was found to be $(d\rho_{25,sat}/dT)_{DS} = (2.2 \pm 0.3) \times 10^{-4}$, whereas the slope for pure NaCl solutions (International Critical Tables) was $(d\rho_{25,sat}/dT)_{NaCl} = 1.27 \times 10^{-4}$. The temperature coefficient of the halite solubility in the Dead Sea brine is thus larger by a factor of about two than that of pure NaCl solutions. Translated into salinity, the temperature coefficient is equivalent to a change in solubility of 0.55 g NaCl/kg brine/°C, based on a measured change of ρ_{25} of 0.0004 g/cm^3 per gram of NaCl added to 1 kg of the test brine. This value can be compared to the calculated value of 0.7 g/kg brine/°C by Gavrieli et al. (1989).

We believe that this experimental procedure yielded a good estimate of the slope of the saturation curve. We chose not to

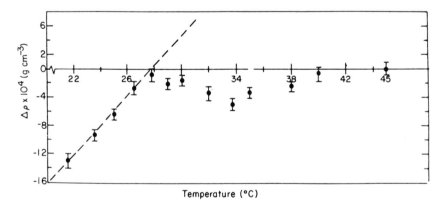

Figure 15-2 Reverse saturometer experiment with a Dead Sea brine from the spring of 1979. The horizontal axis represents the temperatures at which aliquots of brine have been equilibrated before the measurement of their density. The density measurements were recalculated as densities at 25°C (ρ_{25}) that serve as a measure of salinity (Steinhorn, 1980). The vertical axis ($\Delta\rho$) is the difference between the ρ_{25} values of the aliquot and that of the same brine, which has been equilibrated at 45°C. As long as the brine is undersaturated, $\Delta\rho$ is zero, that is, all the aliquots have the same salinity. When halite starts to settle out at lower temperatures, $\Delta\rho$ becomes negative, and the slope represents the saturation line of the brine. The intersection between the $\Delta\rho = 0$ line and the saturation line of the brine is the assumed saturation point of the respective brine.

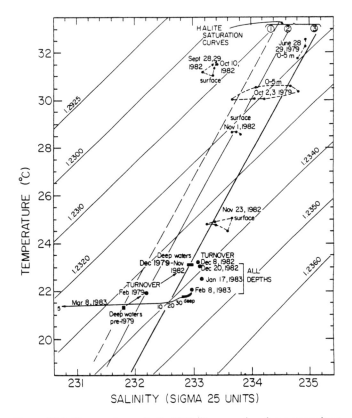

Figure 15-3 Temperature-salinity (T-S) diagram of surface waters from September 1982 through March 1983 and presumed halite saturation lines anchored on the following well-defined brines: (1) the deep waters, prior to 1979, (2) the overturned (mixed) water column in February 1979, and (3) deep waters for the period December 1979–November 1982. Diagonal lines represent isopycnals, that is, constant density lines. Sigma 25 units = (Density at 25°C - 1) x 1,000.

rely too much on the measured saturation points, but rather to fix the saturation curve on the temperature-salinity (T-S) diagram of the Dead Sea brines by relying on the field observations of the appearance of halite crystals in solutions with well-defined T-S properties, such as the deep waters during December 1979 through November 1982 (ρ_{25} = 1.23298; t = 23.15°C), assuming that the brines at that time were not in a highly supersaturated state (halite saturation line 3 in Fig. 15-3).

The effect of components other than NaCl, especially of the alkaline earth cations Mg and Ca, is to decrease the activity of Na and Cl ions, and thus the solubility of halite (a salting-out effect). Indeed, in the Dead Sea brine, the solubility of halite is only about one third of that in a pure NaCl solution. As long as the chemical changes in the Dead Sea brine are restricted to the dilution of the brine or its concentration by the addition or removal of water without any change in the chemical inventory, the solubility is well defined. But as soon as the loss of NaCl by precipitation is massive enough to change the ion composition appreciably, the solubility of NaCl decreases in the residual brine commensurate with the increase of total salinity (and density). For each specific ion composition, a different saturation line applies and the saturation line is shifted in the T-S space. In Figure 15-3, we have indicated two more saturation lines that are based on two occasions of supposedly saturated situations in the Dead Sea, namely, the deep waters preceding the 1979 overturn and the homogeneous water column of the Dead Sea at the time of the February 1979 turnover (lines 1 and 2, respectively in Fig. 15-3). A possible slight change in the slopes of the saturation lines under these circumstances was neglected in this reconstruction of the saturation lines.

SEDIMENT TRAPS

Description

The sediment trap design was based on the experience of scientists at Woods Hole Oceanographic Institute, as well as active cooperation with Dr. Derek Spencer. The setup consisted of a surface buoy, from which the sediment traps were suspended in the water column, and an anchoring system with a larger buoy (Fig. 15-4).

The sediment traps, shown in Figure 15-5, were built at the Weizmann Institute's workshop. They consisted of PVC cylinders (ID 18.8 cm), closed at one end and presenting an opening of 284 cm². They were made up of two parts held together by screws. The cylinders had a height-to-width ratio of three. A radial baffle array was mounted at the top of the cylinder to decrease the turbulence within the trap. The effective collecting area decreases to 240 cm² when the baffles are installed. Because halite crystals formed on the baffles, they were removed and not used after November 1981. The traps were mounted in pairs on stainless steel rods (Fig. 15-5). Usually three to five (sometimes up to six) pairs of traps were connected to each other by polyethylene ropes. The uppermost pair was connected to a surface buoy of 140-liter volume, and a weight of 25 kg was attached to the deepest pair.

The anchor setup consisted of a Danforth-type anchor, a 450-m polyethylene rope (16 mm), and a large buoy (200 liters). The buoys were prepared by filling large PVC containers with self-rising polyurethane. The smaller buoy of the sediment traps and the large buoy of the anchor were attached to each other by a 22-mm rope. The whole assembly of the traps (and the anchor) was placed into the water column and later retrieved on board with the help of a powerful winch (3-ton capacity) installed on the boat.

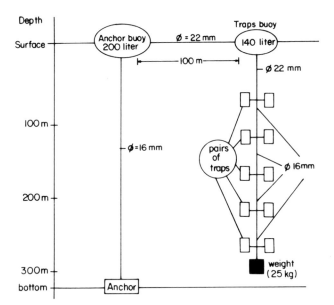

Figure 15-4 Sediment traps setup (schematic representation). The diameter (mm) of the ropes are designated on the figure. Ropes were wrapped in 4-cm diameter polyethylene sleeves. Sediment traps were usually deployed at depths of 70, 120, 170, 220, and 270 m.

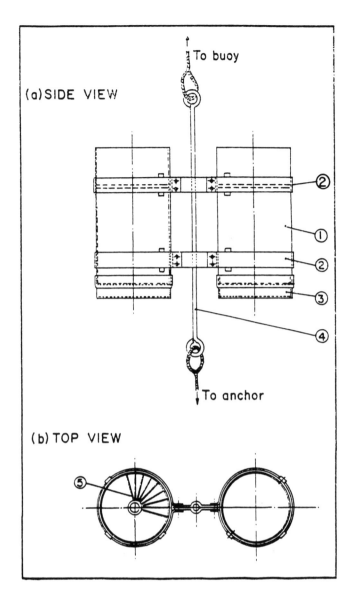

Figure 15-5 A sediment trap pair: (1) upper cylinder (PVC), (2) collar (SS), (3) base of lower cylinder (PVC), (4) holder (SS), (5) radial baffle array (perspex).

The deployment of the sediment traps in the Dead Sea presented severe technical problems related primarily to the encrustment of equipment by halite crystals. These problems are described under Remarks in Table 15-1.

Handling of the sediment traps and analytical procedures

Upon retrieving the sediment traps, the upper part of the cylinder was disconnected. Before that, the overlying water was allowed to discharge through the holes of the screws, which connect the two parts of the trap.

The water from within the lower cup was decanted down to a few centimeters above the sediment layer. Then the sediment slurry collected in the traps was transferred into plastic boxes, the cup was thoroughly rinsed with about 200 ml of the decanted Dead Sea water, and the "washings" were added to the sediment. The plastic boxes were not allowed to cool below 23°C (the temperature of the deep waters) to avoid halite formation as a result of cooling. The sediment was then centrifuged in the lab at 12,000 rpm and the supernatant Dead Sea

water was discarded. Repeated washings were performed with distilled water. The purpose of these washings was to dissolve halite crystals (if present) and to flush residual Dead Sea brine absorbed onto the detrital particles. The supernates of distilled water (the washings) were stored for chemical analysis; the weight of sediment (allogenic particles) collected by the trap was measured after drying in a vacuum at 40°C. The weight of settling particles collected by the pairs of traps agreed, on the average, within 2.6%.

The washings contained dissolved halite crystals and a small amount of remnant Dead Sea brine, which wet the centrifuged sediment. The washings were analyzed for sodium, chloride, and magnesium. Chloride was analyzed by titration (and included bromide), and sodium and magnesium by atomic absorption (AA). The amount of halite collected by the traps was calculated by two independent estimates, as described subsequently. The amount of chloride (moles) in the halite, Cl (halite), was calculated by subtracting the amounts of bromide and chloride originating from the DS brine: $A - 0.0377 \times Mg - 3.47 \times Mg = Cl$ (halite), where A is the number of moles of chloride and bromide in the washings determined by titration, Mg is the number of moles of magnesium determined by AA, and 0.0377 and 3.47 are average molar ratios of Br/Mg and Cl/Mg, respectively, in the DS brines of April–May 1982, as reported by Krumgalz and Millero (1989). The amount of sodium (moles) belonging to halite, Na (halite), was calculated by subtracting the sodium originating from DS brine: $B - 0.959 \times Mg = Na$ (halite), where B is the number of moles of sodium in the washings determined by AA and 0.959 is the average molar ratio Na/Mg in the previously mentioned DS brines (Krumgalz and Millero, 1989). The analytical uncertainty of the molar ratios used in these calculations is about 2%. Agreement between the two independent estimates was about 3%, on the average, suggesting that no other chloride mineral was present in addition to the halite.

When the precipitation of halite became extensive, starting during the sampling period of December 20, 1982 through January 15, 1983 and continuing thereafter, only the halite collected from one of each pair of traps was dissolved for quantitative analysis. The major amount of halite in these samplings (during 1983) was separated from the sediment particles with a 200-μm sieve (zooplankton nylon net). The halite fraction from one of the traps was then dissolved in distilled water and analyzed for Na, Cl, and Mg content as described previously. The procedure (centrifugation, etc.) was then applied to the fraction that passed through the 200-μm net. Halite deposition was calculated by summing halite content in both fractions. The major amount of halite collected by the pair-trap was stored for other studies.

Measurement of suspended matter in the Dead Sea brine

Suspended matter (SM) was measured by filtering Dead Sea water, sampled by Nansen bottles, through 0.45-μm Millipore membranes or 0.4-μm Nuclepore membranes. Although filtration through Millipore membranes is faster and more convenient, it is very difficult to wash the filter efficiently from Dead Sea salts. The weights of SM obtained with Nuclepore filters are thus more reliable. We later found that soaking the Millipore filter (already loaded with the suspended matter) in small volumes of distilled water provided an efficient wash and yielded comparable results to those from the Nuclepore filters. Therefore, we adopted this method in November 1982.

The deployment campaign of May 1981 – December 1983

The sediment trap setup was deployed at lat. 31°32'N, long. 35°27'E, near the center of the lake, where the water depth is

Table 15-1 Periods and depths of deployment of the sediment traps.

No.	Period	Days	Depth of deployment	Top of pycnocline[a] (m)	Remarks
	1981				
1	May 17–May 24	7	70,220,270	10	No halite on ropes.
2	May 24–July 20	57	70, 120, 170, 220, 270	10	Traps at 270 m lost bottoms. Small halite cubes on ropes.
3	July 20–Sept. 21	63	70, 170, 270	12.5	Larger halite cubes on rope 170–270 m.
4	Sept. 21–Nov. 23	63	70, 120, 170,220, 270	30	Polyethylene sleeve on ropes 0–70 m and 220–270 m. Large cubes on rope 170–220 m.
5	Nov. 23–Dec. 18		70, 120, 170,220, 270	20	Stormy weather prevented traps uplifting.
	1982				
6	Nov. 23, 1981– Jan. 28	66	Continuation of Nov. 23 deployment	25	Polyethylene sleeves as above. Larger halite cubes on rope 170–220 m. Collars removed from traps.
7	Jan. 28–Mar. 19	50	10, 70, 120, 170, 220, 270	10	All ropes (except 0–10 m) in polyethylene sleeve. Anchor rope torn, anchor lost. Trap assembly drifted till trap's weight touched bottom (~290 m).
8	Mar. 19–May 23	(65)	10, 70, 120, 170, 220, 270	15	Trap assembly lost. Only anchor with its rope floating were found.
9	May 23–July 25	(63)	not deployed	20	
10	July 25–Sept. 28	(65)	8, 16, 64, 164, 214, 264	25	All ropes wrapped from now on. Only anchor and its buoy found. Trap assembly lost.
11	Sept. 28–Oct. 10	12	10		Only one pair of reserve traps available for deployment.
12	Oct. 10–Nov. 1	22	70	25	As above.
13	Nov. 1–Nov. 23		70, 170, 270	40	Whole assembly lost, disappeared under water on Nov. 1 when boat disengaged from anchor buoy.
14	Nov. 23–Dec. 20	27	70, 120	M	Halite crystals visible in traps.
	1983				
15	Dec. 20, 1982–Jan. 16	27	70, 120	M	From now on, abundant halite always visible in traps.
16	Jan. 16–Feb. 8	23	70, 120, 170	M	Some halite on 0–70 m sleeves. Halite incrustation on trap walls and anchor rope.
17	Feb. 9–Mar. 8	27	70, 120, 170	5	Anchor rope in polyethylene sleeve. Large halite cubes on 0–70 m sleeve, halite incrustations on trap walls. Sediment and large halite crystals on trap weight.
18	Mar. 8–Apr. 12	35	50, 100, 150	10	Only few crystals on sleeve between 0–5 m.
19	Apr. 12–May 22	40	70, 120, 170	10	Less particulates and smaller halite crystals visible in traps. The 2 buoys disengaged from anchor, traveled southward, anchored on 170 m trap.
20	May 22–Jul. 6		no traps deployed	10	
21	Jul. 6–Aug. 25	50	70, 120, 170, 270	15	All ropes wrapped. Halite incrustations on trap walls, much smaller halite crystals in traps.
22	Aug. 25–Oct. 4	40	70, 120, 170, 270	15	All ropes wrapped. Halite incrustations on trap walls and much smaller halite crystals in traps.
23	Oct. 4–Nov. 10	37	70, 120, 170, 270	25	As above
24	Nov. 10–Dec. 14	34	70, 120, 170, 270	M	As above

[a]Depth at end of sampling period. M signifies mixed water column.

about 320 m. Table 15-1 details the periods and depths of deployment of the traps. In Table 15-1, we also note the location of the pycnocline (when present), as deduced form the hydrographic measurements. The Remarks column (Table 15-1) summarizes the modifications that were gradually introduced to adapt the sediment-trap system to the environmental conditions in the Dead Sea. Attempts to deploy traps above the pycnocline, at a 10-m depth, ended in damage of the assembly (January 28 to March 19, 1982) or its complete loss (March 19–May 23, 1982, and July 25–September 28, 1982). The only successful deployment at this depth was during a very short period, September 28 to October 10, 1982, with a single pair of traps at 10 m and with no other traps below the pycnocline. We concluded that the setup could not withstand the shear between the upper and lower water masses, and after October 1982, traps were no longer deployed at 10 m.

In 1983, after the onset of massive halite precipitation, it became quite difficult to fulfill buoyancy requirements because of ubiquitous halite incrustations, especially on the surface buoys. The period of deployment was, therefore, reduced from about 60 days in 1981 to 1982, to 27–35 days in the winter and 40–50 days in the summer of 1983.

Deposition of Sediments other than halite

The results of the collection of particulates other than halite in sediment traps are given in Table 15-2, expressed in terms of fluxes of sediment depositions. Table 15-3 summarizes the measurements of suspended matter (SM) in the water column.

The data in Table 15-2 and Figure 15-6 indicate that settling fluxes of about $0.6–0.8$ g/m^2/day prevailed during the summer and autumn of 1981. A marked increase during January–March 1982 to about $1.0–1.2$ g/m^2/day is probably due to an occasional flood (winter 1981–1982 was, however, relatively dry without major flooding events).

During the summer of 1982, the settling fluxes are believed to have been very small, less than 0.4 g/m^2/day, as suggested by the only two available data points: at a 10-m depth during September–October 1982 and a 70-m depth in October 1982. These data are in sharp contrast to the huge quantities collected afterward from November 1982 to March 1983, namely more than 10 g/m^2/day.

The floods of the end of February and beginning of March 1983 evidently contributed large amounts of suspended load to the surface layers of the Dead Sea. At a 70-m depth, the prominent flux of 17.4 g/m^2/day was measured for the period February 8–March 8, 1983. The SM concentration at 70 m was then also large: 5.7 mg/l. During the spring and summer of 1983, the settling fluxes decreased rapidly, attaining in October values that were as low as in the previous year, that is, less than 0.4 g/m^2/day (at 70 m).

Until early 1983, the amount of materials collected at all depths was rather similar. It is remarkable that from April 1983 onward, the amounts of sediment collected by the trap at 70 m were always the lowest and those collected by the deepest trap were always the highest (Fig. 15–6). This feature may reflect the gradual cleanup of SM from the water column and the absence of SM inputs from the lake surface. The measured SM concentrations do not indicate the cleanup trend as clearly as do the sediment traps data, possibly because of the larger experimental error in the SM data.

A renewed, but moderate, increase of the settling fluxes to about $1–2$ g/m^2/day was observed in November and December 1983. The fact that the uppermost traps, at 70 m, were still yielding the lowest deposition rates, and the deeper traps increasingly higher ones, suggests that there was no substantial input from the sea surface (no heavy floods) and much of the collected material may have originated from resuspended sediments.

Destratification as a result of autumn cooling and turnover mixing evidently plays an important role in producing resuspension as emphasized by the following observations. On November 23, 1982, the concentration of SM, shown in Table 15-3, was about 1 to 2 mg/l; even if we assume that halite formation had completely "cleaned up" the water column of its SM, then we should have measured a sedimentation rate of about 3.9 g/m^2/day (1.5 g/m^3 x 70 m/27 days) at a 70-m depth for November to December 1982. Actually, the measured value was much larger: 9.4 g/m^2/day. Moreover, instead of decreasing from November 23 to December 20, the SM concentrations increased to about 4.5 mg/l (Table 15-3). There were no floods during that period. One possible explanation is the resuspension of bottom sediments caused by the deepening of the pycnocline and the mixing currents associated with the December 1982 turnover.

The amount of resuspended sediments needed to increase the SM concentration in the whole lake from 1.5 mg/l to 4.5 mg/l is roughly 450,000 tons (3 g/m^3 x 145 x 10^9m^3). This represents a resuspended layer of less than 1 mm from the entire lake floor, which is about 800 km^2 (about 0.8 mm if the dry matter content of surface sediments is 0.7 g/cm^3 bulk sediment).

From December 20, 1982, to February 8, 1983, the concentrations of SM gradually decreased, indicating a rather rapid deposition of the resuspended sediments. The mean settling velocities for January 1983 are estimated at about 3 m/day [(10 g/m^2/day)/(3 g/m^3)], whereas for the same period of the previous year, the settling velocities were only about 0.5 m/day. A possible explanation is the entrainment of SM by larger halite particles.

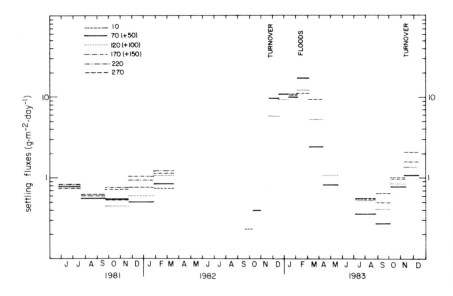

Figure 15-6 Settling fluxes of particulates other than halite in the Dead Sea, May 1981–December 1983. Note the different notations for the different depths of the traps. During March–April 1983, the traps were deployed at 50, 100, and 150 m instead of 70, 120, and 170 m.

Table 15-2 Fluxes of settling particulates in the Dead Sea, collected by sediment traps ($g/m^2 \cdot day$)

Collection Period		Depth of trap deployment (m)					
		10	70	120	170	220	270
1981	May 17–May 24	—	0.73	—	—	0.82	1.03
	May 24–Jul. 20	—	0.78	0.81	0.75[a]	0.84	[b]
	Jul. 20–Sept. 21	—	0.56	—	0.60	—	0.63
	Sept. 21–Nov. 23	—	0.55[a]	0.45	0.54[a]	0.76	0.72
1982	Nov. 23–Jan. 28	—	0.50	0.60	1.03[a]	0.94	0.76[a]
	Jan. 28–Mar. 18	0.74	0.85	1.07	1.11	1.22[a]	[c]
	Mar. 19–May 23	—	—	—	—	—	—
	May 23–Jul. 25	—	—	—	—	—	—
	Jul. 25–Sept. 28	—	—	—	—	—	—
	Sept. 28–Oct. 10	0.23	—	—	—	—	—
	Oct. 10–Nov. 1	—	0.39	—	—	—	—
	Nov. 1–Nov. 23	—	—	—	—	—	—
	Nov. 23–Dec. 20	—	9.43	5.68	—	—	—
1983	Dec. 20–Jan. 16	—	10.55	9.08	—	—	—
	Jan. 16–Feb. 8	—	9.89	10.66	10.45	—	—
	Feb. 9–Mar. 8	—	17.44	12.25	11.31	—	—
	Mar. 8–Apr. 12	—	2.35[d]	5.21[d]	9.26[d]	—	—
	Apr. 12–May 22	—	0.80	1.05	[e]	—	—
	May 22–Jul. 6	—	—	—	—	—	—
	Jul. 6–Aug. 25	—	0.35	0.51	0.55	—	0.55
	Aug. 25–Oct. 4	—	0.27	0.40	0.47	—	0.63
	Oct. 4–Nov. 10	—	0.78	0.83	0.94	—	0.98
	Nov. 10–Dec. 14	—	1.03	1.33	1.54	—	2.01

[a]Only one trap recovered; in duplicate trap, the bottom of cylinder was lost.
[b]Bottoms of cylinders lost in both traps.
[c]Large quantities of bottom material found in traps, as assembly "anchored" on weight and deepest pair of traps.
[d]Actual depths were 50, 100, and 150 m.
[e]Traps assembly "anchored" on the 170-m pair of traps, which "collected" bottom material and halite agglomerates.

Halite deposition

The results of the halite collections, presented in Table 15-4, can be subdivided into three periods.

May 24, 1981–January 28, 1982

The very small fluxes reported for this period might be, in part, artifacts of the collection method, because of halite crystals having formed on the ropes and fallen into the traps. One of the most striking observations was that this growth of halite did not occur at the same rate at all depths. At a depth of 170–270 m, the growth on the ropes was outstandingly large (see Table 15-1) in three out of four samplings. No significant differences in temperature or salinity were noted between waters at these depths and the other deep water layers. This phenomenon might be explained by the fact that very concentrated brine residues, the "end brines," are released from the evaporation ponds of the potash plants into the Dead Sea and spread within the lake waters (Beyth, 1978). This process may slightly increase the degree of supersaturation and thus trigger halite formation at the depth of its occurrence. It is suggested that this primarily affects the layer around 170-m deep (see also other evidence to this effect subsequently).

January 28, 1982–November 1, 1982

Limited data are available during this period because the collecting assembly was lost three times (see Table 15-1). The few existing data, however, are believed to be reliable because of the precautions taken with the deployments. Thus, the relatively large amount of halite ($3 \, g/m^2/day$) collected at 170 m in January–March 1982 is of special interest, as it is about two orders of magnitude larger than at the other depths. During this sampling period, the amount of SM at 170 m dropped from 1.8 mg/l on January 28 to 0.8 mg/l on March 19, 1982, suggesting that a process occurred that "cleaned" the Dead Sea water of suspended particles; perhaps this was the formation of halite crystals. The halite occurrence in the trap at 170 m during January–March 1982 thus does not appear to be a mere artifact, but an event probably triggered again by penetration of end brines into this layer.

At a 70-m depth, the values of $0.06 \, g/m^2/day$ in January through March 1982 and $0.07 \, g/m^2/day$ in October 1982

Table 15-3 Suspended matter (SM) in the Dead Sea (mg/liter).

Date		Filter	Depth (m)						
			0	70	120	170	220	270	310
1981	May 24	MP	—	9.9	—	—	2.8	7.4	—
	July 20	MP	—	25.4	2.4	6.7	10.7	—	—
		NP	—	3.2	1.9	1.2	1.6	—	—
	Sept. 21	NP[a]	—	4.7	—	2.2	—	1.7	—
	Nov. 23	NP	—	2.6	2.2	2.0	2.0	1.6	—
	Dec. 18	NP	—	1.4	—	—	—	—	—
1982	Jan. 28	NP	—	2.0	1.1	1.8	1.7	1.7	—
	Mar. 19	NP	2.4	2.8	1.0	0.8	1.2	—	—
	Nov. 1	MP[b]	—	1.4	—	—	—	—	—
	Nov. 23	MP	—	1.0	—	—	—	2.0	—
	Dec. 20	MP	—	4.4	4.6	—	—	—	—
1983	Jan. 16	MP	—	—	2.9[c]	—	—	—	—
	Feb. 8	MP	—	1.7	1.9	2.1	—	—	—
	Mar. 8	MP	—	5.7	2.5	2.7	—	—	—
	Apr. 12	MP	—	2.8	2.6	2.3	—	—	—
	May 22	MP	2.5	2.4	2.3	3.4	—	—	—
		NP	1.9	2.3	2.6	1.8	—	—	—
	July 6	MP	3.8	4.3	3.2	1.4	—	3.0	1.8
	Aug. 25	MP	1.5	1.3	1.8	0.9	1.4	2.4	1.6
	Oct. 4	MP	1.9	2.0	1.2	3.5	—	1.9	2.4
	Nov. 10	MP	1.1	1.9	2.0	1.3	0.5	2.6	0.7[d]
	Dec. 14	MP	2.9	2.3	1.8	1.7	—	2.2	2.9[d]

(MP) Millipore membrane, 0.45 µm type HA; (NP) Nuclepore membrane, 0.4 µm.

[a]Actual depths are 80, 150, and 250 m.

[b]Millipore data are reliable starting with this sampling and later on.

[c]Average of three measurements at a 100-m depth in center, northern ,and southern part of the lake (2.9, 2.7, and 3.1 mg/l, respectively).
[d]Actual depth is 305 m.

(marked < 0.1 in Table 15-4) can be regarded as background values at this depth, to be compared with the much larger amounts that were deposited during the following period.

November 23, 1982–December 14, 1983

This period is characterized by massive deposition of halite, always greater that 10 g/m²/day and in most cases greater than 100 g/m²/day. The highest measured value was 366 g/m²/day, collected between November 10 and December 14, 1983 (Table 15-4, Fig. 15-7) at the depth of 170 m. These fluxes are more than one order of magnitude larger than those of the allogenic sediments that were collected simultaneously (Table 15-2).

The halite deposition rate in the 70-m trap seemed to increase at the end of December 1982 and reached a peak value during January 1983 (240.9 g/m²/day). Then, a gradually decreasing rate was observed from February to May 1983. A second halite deposition peak appeared at this depth (70 m) in July–August 1983 and a third one, the highest, with 296.9 g/m²/day, in December 1983 (Fig. 15-7).

Until May 1983, the amounts of halite collected by the traps consistently decreased with depth (the only exception being the trap at 50 m during March–April 1983). Starting with July and August 1983, this pattern changed and from September 1983 to December 1983, the halite deposition measured in the 170-m trap became the most prominent one (Fig. 15-7).

We cannot exclude the possibility of some crystal growth within the traps; perhaps the small crystals that settled initially in the trap grew further within it. In fact, halite crystals grew everywhere on the trap exterior (on the edges of the cylinders, on the screws of the trap, and even on its outside walls) during the period December 1982–December 1983. However, the growth rate in the trap interior should be quite limited because the water within the trap is quiescent, whereas the amount of brine that comes in contact with the outside parts of the traps (and with the halite crystals growing on them) is by far larger. Thus, we should regard the measured quantities of halite as possibly a maximal estimate.

DISCUSSION

Yearly integrated sedimentation rates of particulates other than halite

The deployment of sediment traps from May 1981 to December 1983 enabled us to estimate rates of sedimentation at the lake's center both during meromictic conditions (multiannual stable stratification), which persisted until November 1982, and thereafter under holomictic conditions (complete mixing every winter).

In Table 15-5, annually integrated rates of sedimentation are given for May 1981 to May 1982 (meromictic regime) and for November 1982 to November 1983, which includes the December 1982 turnover.

The sedimentation rates estimated for 1981–1982 are about

Table 15-4 Fluxes of halite deposition in the Dead Sea, based on collection by sediment traps (g/m² day).

Collection Period		Depth of trap deployment (m)					
		10	70	120	170	220	270
1981	May 24–Jul. 20	—	<0.1	2.9	0.7	<0.1	—
	July 20–Sept. 21	—	0.47	—	0.21	—	0.65
	Sept. 21–Nov. 23	—	2.14	<0.1	0.18	1.58	3.60
1982							
	Nov. 23–Jan. 28	—	0.82	<0.1	0.17	2.11	5.93
	Jan. 28–Mar. 18	<0.1	<0.1	<0.1	3.18	<0.1	[a]
	Mar. 19–May 23	—	—	—	—	—	—
	May 23–July 25	—	—	—	—	—	—
	July 25–Sept. 28	—	—	—	—	—	—
	Sept. 28–Oct. 10	<0.1	—	—	—	—	—
	Oct. 10–Nov. 1	—	<0.1	—	—	—	—
	Nov. 1–Nov. 23	—	—	—	—	—	—
	Nov. 23–Dec. 20	—	20.91	10.19	—	—	—
1983	Dec. 20–Jan. 16	—	129.5	90.8	—	—	—
	Jan. 16–Feb. 8	—	240.9	157.5	85.5	—	—
	Feb. 9–Mar. 8	—	205.7	156.5	131.3	—	—
	Mar. 8–Apr. 12	—	99.4[b]	140.7[b]	108.0[b]	—	—
	Apr. 12–May 22	—	97.5	93.2	[c]	—	—
	May 22–July 6	—	—	—	—	—	—
	July 6–Aug. 25	—	223.9	115.6	146.6	—	112.2
	Aug. 25–Oct. 4	—	106.5	76.6	144.6	—	41.6
	Oct. 4–Nov. 10	—	147.6	188.2	306.8	—	127.0
	Nov. 10–Dec. 14	—	296.9	187.7	366.0	—	142.6

[a]Large quantities of bottom material found in traps, as assembly "anchored" on weight and deepest pair of traps.

[b]Actual depthes were 50, 100, 150 m.

[c]Traps assembly "anchored" on the 170-m pair of traps, which "collected" bottom material and halite agglomerates.

the same at all depths of the traps, on the average about 29 mg/cm²/yr. We assume that this value represents the sedimentation of allogenic particles during a relatively dry, meromictic year in the deep basin of the lake center. These values are in good agreement with the 30 mg/cm²/yr value measured with sediment traps deployed by Levy (1988) during the meromictic year 1980–1981. Also similar to our findings (see subsequent discussion), Levy (1988) claimed that only detrital minerals were precipitated during that period. The much larger sedimentation rates of about 143 mg/cm²/yr, calculated for 1982–1983, from the data of the 70-m and 120-m traps, include contributions from bottom sediments resuspended by the vigorous stirring during the December 1982 turnover, as well as the influx of allogenic material carried by the floods of the relatively wet year. As the overturn period is also the time when halite started to precipitate, we must also take into account the possible scavenging of SM from the water column by formation and settling of halite crystals, as discussed previously. By measuring halite deposition in bottom sediments from three locations for the holomictic years 1983–1985, Levy (1988) found detrital minerals associated with halite that settled at rates of 90 to 117 mg/cm²/yr. In a core representing the years 1983–1989, an average rate of 70 mg/cm²/yr was found (Levy, 1991).

During the entire period measurements were taken, there was no massive precipitation of aragonite or of gypsum (endogenic minerals) as evidenced by X-ray analysis of the non-halite fractions. The rather stable CO_2 percentage of the settling material indirectly confirms this statement: CO_2 was on the average about 18%, with minimal values of about 16% during the floods of February–March 1983 and maximal values of about 20% during September–October 1983 (Stiller, unpub. data).

Rates of halite formation

During most of the period covered by this investigation, the surface waters were rather diluted (Anati et al., 1987) and apparently not saturated with respect to halite. It was only in November 1982 that the salinity of the surface waters crossed over the halite saturation line (line 3, Fig. 15-3).

Starting with November 1982, the pattern of halite formation can be subdivided into several phases. In the first one, November 1982 to February 1983, halite formation at the lake surface was apparently triggered by evaporation at the air-sea interface. From December 1982 to February 1983 (the water column was well mixed at that time), the temperature of the water column dropped by 1°C and the average salinity decreased by about 0.14 g/kg (ρ_{25} decreased from 1.23311 to 1.23298; Anati et al., 1987; Stiller and Gat, 1983). This is equivalent to a precipitation rate of halite of 2.8×10^{-3} g/kg/day.

In the second phase, from March to May 1983, the upper 10 m became diluted by flood waters and were undersaturated with respect to halite; indeed, its formation in this layer ceased. Below 10 m, the formation of halite persisted, however, as is evident from the fact that the traps continued to collect halite, although at a gradually slower rate. Even if we assume a settling rate of only 4 m/day for the halite crystals (about 1/10 of that observed by Levy, 1985), the trap deployed in April–May 1983 at 70 m should have been empty of halite, but this was not the case. The average salinity of the deep waters (40–300 m) slightly decreased from February to April 1983, by about 0.11 g/kg (in April, ρ_{25} = 1.23288). This value gives a rough estimate for the rate of halite formation within the deep waters of 1.7 x 10^{-3} g/kg/day (0.11 g/kg/63 days). This value agrees well with the rate of halite formation of about 1.75 x 10^{-3}g/kg brine/day, inferred from the decrease in salinity of the isolated deep waters in 1991–1992 (Anati, Chapter 8, this volume).

If we assume that the precipitation rate of halite due to supersaturation within the water column was the same during the period of March to May and that of the previous phase, it follows that, in December 1982–February 1983, the contribution of halite formed at the surface by evaporative enhancement had been 1.1 x 10^{-3} g/kg/day (2.8 x 10^{-3}–1.7 x 10^{-3}). If all the NaCl from the daily evaporated film is being precipitated, then this evaporative contribution is equivalent to an evaporation rate of 0.23 cm/day (1.1 x 10^{-3} g/kg /day x 1.235 kg/l x 180 x 10^2 cm/ 105 g NaCl/l = 0.23 cm/day) or 11.5 cm for the entire period (the mean depth of the Dead Sea is 180 x 10^2 cm and its NaCl content is 105 g/l). This is in reasonable agreement with the estimated evaporation of 7.5 cm for the period December 20 to February 8 (monthly evaporation rates given by Stanhill (1994) for 1983–1987 have been used).

In the third phase, from July to the end of September 1983, the rate of halite collection rose (Fig. 15-7) especially at a 170-m depth. During that period, the salinity of the upper 10-m layer, which is slightly fresher, gradually increased; in August 1983, the salinity stratification became destabilizing, with the stability of the water column being maintained by temperature alone. However, the progressive deepening of the pycnocline took place under conditions that were not yet favorable for double-diffusive mixing and for the formation of salt fingers (Anati and Stiller, 1991), as is evident from the ratio $R_p = -\alpha\Delta T/\beta\Delta S$, which was larger ($R_p$ = 3.64) than the critical value of 1.7 ± 0.2. Therefore, the slight increases in average salinity and in temperature of the deep waters in August 1983 to 277.07 g/kg (ρ_{25} = 1.23297) and 21.93°C (from 21.90°C) are attributed to penetration of end brines, which are saltier and warmer than the deep waters. The increased halite yield at the trap deployed at 170 m, indicates that this layer was particularly affected. This is in accordance with the previously observed pattern at this depth.

During the last phase, from October to December 1983, several processes apparently combined to promote halite precipitation:

(1) Cooling of the upper layer resulted in bringing surface waters close to saturation with respect to halite, so that precipitation resulting from evaporation again became a possibility,
(2) Conditions became favorable for the formation of salt fingers and for double-diffusive mixing with R_ρ values of 1.76 and 1.52 in October and November, respectively. This process could then produce supersaturation below the pycnocline and within the deep waters (Anati and Stiller, 1991),
(3) Penetration and spreading of end brines probably persisted as evidenced by the ever-increasing collection rate of halite at 170 m water depth. Halite formation induced by human activities is thus superimposed on the natural process of halite formation in the Dead Sea.

The various processes responsible for halite formation in 1983, the presumed rates of formation, and the layers within which they took place are summarized in Table 15-6.

Salinity balance

It is interesting to compare the annual sedimentation rate of halite as inferred from the collection by the traps with the amount of salts missing from the lake as estimated from the salinities of the water masses in December 1982 and December 1983. The sediment trap setup collected halite from various water depths at one fixed station situated close to the lake center. This setup obviously does not provide information about the lateral homogeneity (or inhomogeneity) of halite formation and deposition in the Dead Sea.

The cumulative NaCl deposition as a function of the trap depth is shown in Figure 15-8. The most complete data sets are for 70 and 120 m, with only one collection period missing: May 22 to July 6, 1983. For this missing period, the data were evaluated by interpolation between the adjacent collection periods.

From December 1982 to December 1983, the largest NaCl accumulation was at a 70-m depth; 6.1 g/cm^2/yr (Fig. 15-8).

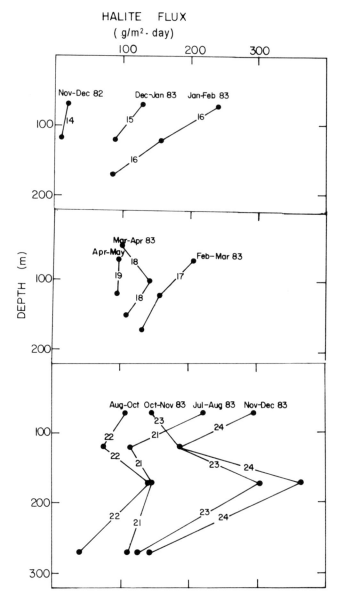

Figure 15-7 Depth profiles of the fluxes of halite collected by sediment traps (g/m^2·day) from November 1982 to December 1983. The numbers on the profiles represent the respective collection period (Table 15-1).

The enhanced collection rate of halite at 170 m probably started in August 1983 and obscures the previous pattern of decreasing cumulative halite with increasing depth. For 120, 170, and 270 m, the annual cumulative amounts are 4.5, 5.6, and 2.7 g/cm^2/yr, respectively. Levy (1991) measured the amount of halite accumulated at the lake floor from 1982 to 1989 in cores from the En Gedi area (close to the site where our traps were deployed). The average rates of sedimentation for 1982 to 1989 were 3 and 6 cm/yr at 270 and 240 m, respectively (or about 6 g/cm^2/yr to 12 g/cm^2/yr, assuming the halite density is 2 g/cm^3).

Between the two turnovers of December 1982 (salinity 277.22 g/kg brine, density 1.23384, and lake volume 144.7 km^3) and December 1983 (salinity 277.31 g/kg brine, density 1.23424, and lake volume 144.1 km^3; Anati et al., 1987), the lake level dropped by 0.72 m and the lake volume decreased by about 0.6 km^3. An estimate of the amount of salts lost from the Dead Sea during 1983, based on the salinities of the water masses, is about 170 x 10^6 tons salts/yr or 21.2 g/cm^2/yr or 0.953g/kg brine/yr.

About 30 x 10^6 tons NaCl are deposited annually in the Israeli and Jordanian evaporation ponds as part of the process of potash production. This accounts for only about 3.8 g/cm^2/yr of the missing salts. After adding this value to the largest number estimated by the traps, 6.1 g/cm^2/yr, we are still left with 11.2 g/cm^2/yr (or 90 x 10^6 tons) of missing salts that are unaccounted for. Perhaps this latter value represents the halite that is formed (and deposited) in the vicinity of the salt reefs at the southern edge of the northern basin, where the released end brines meet and mix with the Dead Sea water. In conclusion, less than half of the halite is deposited at the lake interior, whereas major deposition takes place at the lake's periphery in the reef area and in the evaporation ponds.

Our inventory of missing salts for 1982–1983 (170 x 10^6 tons/yr) is comparable with that found by other workers. Based on density and lake level measurements during 1983–1990 (Anati and Shasha, 1989; Anati, 1993), we can estimate that, during that period, the Dead Sea lost about 200 x 10^6 tons salts/yr (1.13 g salts/kg brine/yr). Most of this, obviously, is halite. Based on sodium measurements and respective mass balances, Gavrieli (chapter 14, this volume) estimates that, between 1976 and 1992, the Dead Sea has lost 2,400 x 10^6 tons of halite, averaging about 150 x 10^6 tons/yr (this includes the years 1976–1983 when halite deposition occurred only as a consequence of the potash production, but not within the lake).

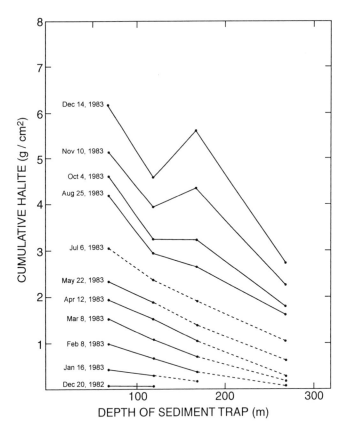

Figure 15-8 Cumulative deposition of halite at the various depths of the traps. Each line represents the cumulative amount from November 23, 1982, until the date marked near the line. The missing data for 170 and 270 m have been evaluated by extrapolation from the 70-m and 120-m data of the respective period to the respective depth and are designated by broken lines.

Comparison between the pattern of sedimentation of endogenic halite and of nonhalite particulates

The sedimentation flux of endogenic halite, as measured in the sediment traps in the center of the lake, appears to decrease

Table 15-5 Annual rates of sedimentation of particulates as measured by sediment traps deployed at the lake center (31°32'N, 35°27'E).

Depth of	Rate of sedimentation (mg/cm^2/yr)	
trap (m)	May 24, 1981–May 23,1982[a]	Nov 23, 1982–Nov 22, 1983[c]
70	24.1	149
120	26.9[b]	138
170	30.1	
220	33.6[b]	
270	31.7[b]	

[a]For the last two months of this year the values were estimated (at all depths) by interpolation between the settling rates of May 17–24, 1981, and those of the collection period January 28–March18, 1982.
[b]The missing data for one collection period (see Table 15-2) were estimated by interpolation between the data of adjacent depths for 120 and 220 m and between adjacent collection periods for 270 m.
[c]For the collection period May 21–July 6, the data were estimated by interpolation between adjacent collection periods, at respective depths.

with increasing depth. Until May 1983, this can be clearly seen in Figure 15-8. Later, this pattern is modified by the excess of halite collected at the depth of 170 m. We do not discuss this here; rather, we deal only with the deposition pattern of halite formed by natural processes.

In contrast, the annual (cumulative) sedimentation rates of nonhalite particles are quite similar at all the depths where traps were exposed (Table 15-5). These particles were almost exclusively of allogenic origin in 1981–1982 (no precipitation of aragonite or gypsum could be detected). During the 1982–1983 season, there was also a large contribution of resuspended material, which was mostly allogenic too.

Taken at its face value, the decreasing halite flux with increasing depth suggests that halite is redissolved while settling through the water column. However, this is not the case in the present situation. Data indicate that the lower water masses are saturated (or slightly supersaturated) with respect to halite. Thus, the lower masses can be expected to contribute additional halite to the settling crystals rather than redissolving it. An alternative mechanism that might result in the observed pattern is the preferential deposition of halite in the shallower parts of the lake.

Two factors contribute to the pattern of deposition of a sediment, namely, the location of the source and the particle size. Both of these factors differ in the case of the halite and the allogenic material. The source region for the halite, as discussed previously, is primarily at the surface, with a secondary site of formation within the water column. On the other hand, allogenic particles enter the Dead Sea at the surface as point sources at the outlets of rivers and wadis. Horizontal mixing in a lake, such as the Dead Sea, is much faster than vertical mixing (Mortimer, 1979), so that particles formed all over the surface of a lake have a chance to "touch" bottom at the rim of the lake.

The halite particles range from about 150 to 400 μm (Levy, 1985), whereas the allogenic particles are much smaller, from 2 to 4 μm (Neev and Emery, 1967; Garber, 1980). The removal process at the rim is expected to favor the larger particles, which sink faster and settle to the bottom, thus cleaning up a deeper segment of the water column before they are once more removed from the coast by horizontal currents. This fact, together with the lesser chance of resuspension of the larger particles, may explain the difference in the vertical accumulation pattern found in the sediment traps between the larger halite crystals and the much smaller allogenic particles. The net result of these processes would then be a focusing the finer sed-

iments in deeper parts of the lake and coarser sediments along its fringes. Thus, it is not so much the source region as it is the particle size that is the dominant factor; this is actually a well-recognized pattern in many lakes.

Resuspension of sediments occurs in the stratified Dead Sea from bottom layers just beneath the pycnocline when it deepens and from all over the lake floor at the time of overturn. Obviously, larger particles are much less affected by this process. The resuspended particles mix within the lake waters then start to resettle, mostly during the time when the lake is stratified. Resuspension causes, through homogenization in the water column, the relocation of the finer sediments.

POSTSCRIPT: THE GEOCHEMICAL EVOLUTION OF THE DEAD SEA

Whereas the overall water balance of the system (i.e, excess or deficiency of freshwater inflows above evaporation) is the primary factor in determining the level of the sea and its salinity, the related process of precipitation of minerals, especially halite, and its possible subsequent dissolution by rising water levels constitute the most important natural agent for geochemical change over the last centuries. The activities of the chemical factories in the region that exploit the mineral wealth of the Dead Sea brine (and, in particular, the potash companies) have become an additional factor in recent years, and their influence is now dominant.

A cycle of wet years following dry ones can reestablish lake levels to previous high-water stands. This cycle does not, however, necessarily restore the Dead Sea to its previous geochemical state. Two factors cause irreversibility of the evolution of the Dead Sea system. First, as discussed by Anati et al. (1987), the hydrographic evolution is such that each overturn leaves the lower water mass in a denser state than before. Since the lower water mass appears always close to saturation, there will be a loss of NaCl from the system at the end of each seasonal or meromictic cycle, as in 1979 and again in 1982–1983. To the extent that this precipitate is sedimented in the deeper parts of the lake, we can consider this loss to be irreversible, since their sediments are prevented, by the situation described previously, from coming in contact with any diluted (and undersaturated) brine. However, as our results indicate, an important fraction of the halite is deposited either in the shallower parts of the lake or in its periphery (e.g., the salt reefs near the Lisan Peninsula). These deposits can, in principle, be redissolved by rising

Table 15-6 Seasonality of processes causing halite supersaturation, and rates of halite formation, in 1983.

Time Period	Process	Layer	Rate of halite formation (g/kg brine/day)
Dec.[a]–Feb.	evaporation, cooling	entire water column	2.8×10^{-3}
March–May	remnant supersaturation	deep waters (below 20 m)	1.7×10^{-3}
July–Sept.	end-brine penetration	deep waters (max. at 170 m)	
Oct.–Dec.	evaporation cooling double diffusion end-brine penetration	surface upper layer (40 m) below pycnocline and in deep waters deep waters (max. at 170 m)	
whole year average[b]		for whole lake	2.6×10^{-3}

[a]December 1982.
[b]As estimated by the 1982–1983 salinity balance: 0.953 g/kg brine/yr (see text).

lake levels or by floods in the lake's coastal areas. The extent to which this halite is amenable to redissolution will determine the rate at which the geochemical evolution proceeds.

The release of the end brines from the potash companies also plays a significant role when the end brines penetrate into the deep water mass, augmenting the loss of halite. The consequence is a further change in the ionic composition of the deep water mass and hence a commensurate change in the solubility of halite.

It is also interesting to speculate on the possible effects of the proposed scheme to divert Mediterranean Sea (or Red Sea) waters to the Dead Sea to reestablish the high-water stand of decades ago (Steinhorn and Gat, 1983). Such an operation would introduce an additional irreversible geochemical factor, namely, the removal of calcium in the form of gypsum, further modifying the chemistry of the Dead Sea brine. Some gypsum and aragonite are also precipitated naturally, but this sedimentation does not seriously upset the geochemistry of the system because it compensates for the introduction of these chemicals by river and flood inflows.

Acknowledgments

We thank D. W. Spencer for his cooperation, advice, and interest in the sediment-trap program. Thanks are due to N. Bauman and S. Shasha for their skillful technical help and to M. Gonen, the skipper of the *Tiulit*. The sediment-trap program was supported by a United States–Israel Binational Science Foundation grant, no. 2218/80. Ship time was provided by grants from the Ministry of Energy and Infrastructure. The critical reading and useful comments of J. Kolodny and A. Starinsky are gratefully acknowledged.

REFERENCES

Anati, D. A., 1993, How much salt precipitates from the brines of a hypersaline lake? The Dead Sea as a case study: *Geochimica et Cosmochimica Acta*, v. 57, p. 2191–2196.

Anati, D. A., and Shasha, S., 1989, Dead Sea surface level changes: *Israel Journal of Earth Sciences*, v. 38, p. 29–32.

Anati, D. A., and Stiller, M., 1991, The post-1979 thermohaline structure of the Dead Sea and the role of double-diffusive mixing: *Limnology and Oceanography*, v. 36, p. 342–354.

Anati, D. A., Stiller, M., Shasha, S., and Gat, J. R., 1987, The thermohaline structure of the Dead Sea: 1979–1984: *Earth and Planetary Science Letters*, v. 84, p. 109–121.

Beyth, M., 1978, Mixing of end brines and state of "salt reefs" in the Lynch straits, Dead Sea, April–November 1978: Geological Survey of Israel Report MG/2/79, 7 p.

Eugster, H. P., and Hardie, L. A., 1978, Saline lakes, *in* Lerman, A., ed., *Lakes: Chemistry, geology, physics*; New York, NY., Springer, p. 237-293.

Garber, R. A., 1980, The sedimentology of the Dead Sea [Ph.D. thesis]: Troy, New York, Rensselaer Polytechnical Institute, 169 p.

Gavrieli, I., Starinsky, A., and Bein, A., 1989, The solubility of halite as a function of temperature in the highly saline Dead Sea brine system: *Limnology and Oceanography*, v. 34, p. 1224–1234.

Jones, B. J., and Bowser, C. J., 1978, The mineralogy and related chemistry of lake sediments, *in* Lerman, A. ed., *Lakes: Chemistry, geology, physics*: New York, NY, Springer, p. 179–235.

Kaushansky, P., and Starobinets, S., 1982, A temperature gradient column for the rapid measurement of the density of brines: *Marine Chemistry*, v. 11, p. 289–292.

Klein, C., 1981, The influence of rainfall over the catchment area on the fluctuations of the level of the Dead Sea since the 12th century: *Israel Meteorological Papers*, v. 3, p. 29–58.

Krumgalz, B. S., and Millero, F. J., 1983, Physico-chemical study of Dead Sea waters. III. On gypsum saturation in Dead Sea waters and their mixtures with Mediterranean Sea water: *Marine Chemistry*, v. 13, p. 127–139.

Krumgalz, B. S., and Millero, F. J., 1989, Halite solubility in Dead Sea waters: *Marine Chemistry*, v. 27, p. 219–233.

Lerman, A., and Shatkay, A., 1968, Dead Sea brines: Degree of halite saturation by electrode measurements: *Earth and Planetary Science Letters*, v. 5, p. 63–66.

Levy, Y., 1985, Modern halite precipitation in the Dead Sea: Jerusalem, Geological Survey of Israel Report GSI/7/85, 18 p.

Levy, Y., 1988, Sedimentary reflections of modern (1983–1985) halite precipitation from Dead Sea water: Jerusalem, Geological Survey of Israel Report GSI/12/88, 16 p.

Levy, Y., 1991, Modern sedimentation in the Dead Sea across from En Gedi (1982–1989): Jerusalem, Geological Survey of Israel Report TR-GSI/2/91, 14 p.

Mazor, E., Rosenthal, E., and Ekstein, J., 1969, Geochemical tracing of mineral water sources in the southwestern Dead Sea basin, Israel: *Journal of Hydrology*, v. 7, p. 246–275.

Mortimer, C. H., 1979, Some central questions of lake dynamics, *in Isotopes in lake studies*: Vienna, IAEA, p. 1–20.

Neev, D., and Emery, K. O., 1967, The Dead Sea, depositional processes and environments of evaporites: Jerusalem, Geological Survey of Israel Bulletin, 41, 147 p.

Schnerb, I., and Yaron, F., 1952, On the solubility of sodium chloride in Dead Sea brines: *Bulletin of the Research Council of Israel*, v. 2, p. 197–198.

Stanhill, G., 1994, Changes in the rate of evaporation from the Dead Sea: *International Journal of Climatology*, v. 14, p.465-471.

Starinsky, A., 1974, Relationship between Ca-chloride brines and sedimentary rocks in Israel [Ph.D. thesis]: Jerusalem, The Hebrew University, 176 p. (in Hebrew).

Steinhorn, I., 1980, The density of Dead Sea water as a function of temperature and salt concentration: *Israel Journal of Earth Science*, v. 29, p. 191–196.

Steinhorn, I., 1983, In-situ salt precipitation at the Dead Sea: *Limnology and Oceanography*, v. 28, p. 580–583.

Steinhorn, I., Assaf, G., Gat, J. R., Nishry, A., Nissenbaum, A., Stiller, M., Beyth, M., Neev, D., Garber, R., Friedman, G. M., and Weiss, W., 1979, The Dead Sea: Deepening of the mixolimnion signifies the overturn of the water column: *Science*, v. 206, p. 55–57.

Steinhorn, I., and Gat, J. R., 1983, The Dead Sea: *Scientific American*, v. 249, p. 102–109.

Stiller, M., and Gat, J. R., 1983, Hydrographic investigation and seasonal variations in the structure of the Dead Sea, September 1982–October 1983, Weizmann Institute Report No. 81-200-78 (in Hebrew).

Stiller, M., Gat, J. R., Bauman, N., and Shasha, S., 1984, A meromictic episode in the Dead Sea: 1979–1982: *Verhandlunden Internationale Vereinigung (für theoretische und angewandte) Limnologie*, v. 22, p. 132–135.

16. CARBON DYNAMICS IN THE DEAD SEA

Boaz Luz, Mariana Stiller, and A. Siep Talma

The Dead Sea is an ideal system for studying geochemical systems in the hypersaline environment. Although various important attributes of this lake have been extensively studied, the carbonate system and its isotopic composition have received less attention. A general overview of this system is given in the important monograph on the Dead Sea by Neev and Emery (1967), who noted (as did Bloch et al., 1944) whitenings and precipitation of aragonite crystals. Neev and Emery (1967) also were the first to report a few ^{13}C measurements of the total dissolved CO_2. Nissenbaum and Kaplan (1976) conducted the first extensive study of the isotopic aspects of the carbonate system. They report significant depletion of ^{13}C in the dissolved bicarbonate, and they attribute this depletion to incomplete isotopic equilibration of the dissolved CO_2 introduced by water influx of the Jordan River and winter floods. However, the reported isotopic values seem too "light" (depleted in the heavy isotope ^{13}C) when compared with the few measurements of Neev and Emery (1967), as well as those reported in this chapter (see subsequent discussion). Nissenbaum and Kaplan (1976) also suggested that organic-matter oxidation does not play an important role in the Dead Sea, but, as we discuss subsequently, this assumption may not be justified.

The concentration of dissolved bicarbonate reported in the previously mentioned studies is high, averaging about 3,000 μ mole/kg. A relatively low pH of about 6 is characteristic of the Dead Sea brines. The fact that aragonite occurs at such a low pH led Sass and Ben-Yaakov (1977) to conduct the careful alkalinity titrations that indicated similar high carbonate alkalinity values. However, Neev and Emery (1967) and Nissenbaum and Kaplan (1976) noted that the CO_2 yield obtained by acid extractions was always low and accounted for about one third of the expected amount based on alkalinity titrations. This discrepancy is resolved if the high borate alkalinity of the brine is taken into account (Schonfeld and Held, 1965; 45 mg/l boron).

Stiller et al. (1985) studied the behavior of carbon isotopes in the evaporation ponds of the Dead Sea Works Ltd., as well as in laboratory-controlled evaporation of Dead Sea brines. To avoid interference of the dissolved boron species, dissolved inorganic carbon was determined by acid extraction. Strong evaporation and the concomitant increase in salinity cause CO_2 to escape so the remaining dissolved bicarbonate fraction becomes highly enriched in ^{13}C. The possibility of such enrichment has been suggested by Katz et al. (1977), based on the equilibrium carbon isotopic fractionation between dissolved bicarbonate and atmospheric CO_2. However, the enrichment observed in the evaporating brines is greater and can only be accounted for by a kinetic isotope effect. This enrichment, as well as organic matter oxidation and biological production, are important mechanisms that affect the carbon isotopic composition of the Dead Sea. Carbon dynamics indicate that the Dead Sea is not as lifeless as its name implies.

THE DATABASE

The data set used in this chapter is based on analyses performed at the Weizmann Institute of Science (WI) for the period 1978 to 1985. Analyses of samples taken in 1992 were run at the Hebrew University of Jerusalem (HU, Oren et al., 1994). Samples taken before October 1983 were stored in glass bottles and processed in the laboratory in 1983. The rest of the analyses (with some exceptions, as indicated in Table 16-1) are based on shipboard sampling directly into preevacuated flasks containing a few ml of concentrated H_3PO_4 and a magnetic stirring bar. The flasks were weighed before and after they were filled with brine to determine the amount of sampled brine. In both laboratories, the concentration of the total dissolved inorganic carbon (TC) was determined by acid extraction of CO_2, cryogenic purification, and manometric determination of the gas yield. The experimental protocols used by the two laboratories differ in certain details. At WI, the method of Mook (1968, 1970) was followed. The reproducibility of the TC determinations was 4% (labeled WI-1 in Table 16-1), except for the October and November 1983 cruises, where the scatter between replicate analyses was 1.2% (WI-2 in Table 16-1). At HU (analyses labeled with HU in Table 16-1), the method was modified by the addition of a capillary between the sample flask and the first water trap on the vacuum extraction line. This minimized the amount of water vapor transfer from the sample to the trap without any adverse effect on CO_2 extraction. The details of this method will be published elsewhere. The precision based on replicate analyses is about 0.25%. The TC concentration is reported in μmole/kg.

Isotopic analyses of the extracted carbon dioxide were run at WI on a Finnigan MAT-250 isotope ratio mass spectrometer. Similar measurements were performed at HU on a modified VG-602 instrument. The carbon isotopic ratio of the TC is reported in the conventional delta notation:

$$\delta^{13}C = \frac{(^{13}C/^{12}C)_{samp} - (^{13}C/^{12}C)_{ref}}{(^{13}C/^{12}C)_{ref}} \cdot 1000 \qquad (1)$$

where samp is CO_2 extracted from a Dead Sea brine sample and ref is CO_2 derived from the PDB international standard. The calibration of both WI and HU with respect to PDB is based on the Cal Tech standard PDB-IV (S. Epstein). The precision is higher than 0.1 and 0.05‰ at WI and HU, respectively.

In Table 16-1, we list all the available data, but in the discussion that follows, we disregard a few outlier values. Most samples were obtained from a hydrographic station near the center of the lake (Fig. 16-1). In Table 16-1, we also list TC and ^{13}C values of several water samples taken along the lower Jordan River and samples from winter floods on the western side of the lake. The data of a few samples of pore waters extracted from the lake bottom sediments are also listed in Table 16-1.

LONG-TERM CHANGES IN CARBON INVENTORIES AND IN CARBON ISOTOPES

In 1978, the Dead Sea was approaching the end of a long meromictic period with a diminishing anoxic deep water mass below about 160 m. The meromictic pycnocline disappeared in early 1979, and the entire water column ventilated (Steinhorn et al, 1979). After the 1979 overturn, three distinct stages followed:

Table 16-1 Database for Dead Sea carbon studies (TC = total dissolved CO_2 μmole/kg; $\delta^{13}C$‰ versus PDB).

Date	Depth (m)	TC	$\delta^{13}C$
Jan. 18, 1963 (N&E)[a]	0		1.70
May 8, 1963 (N&E)	0		2.00
Jan. 6, 1978 (SB)[b]	0	884	0.38
	40		1.00
	100	992	0.27
	160	833	-1.29
	220	496	-2.00
	280	874	-3.14
Jan. 7, 1978 (SB)	180	788	-2.07
Jan. 8, 1978 (SB)	80	886	0.41
	220	639	-0.08
	315	767	-2.01
	325	1,187	-4.77
Aug. 15, 1978 (SB)	150	1,626	-14.56
	300	580	-2.48
Feb. 5, 1979 (SB)	5	866	-3.93
	10	628	-2.05
	25	587	0.37
	90	828	0.12
Feb. 6, 1979 (SB)	80	627	-1.47
	160	945	-2.83
	280	852	-4.88
Feb. 13, 1979 (SB)	175	588	2.02
June 27, 1979 (SB)	179	911	-2.40
Mar. 12, 1980 (SB)	0	983	-2.51
	10	876	-1.09
	20	799	-1.51
	50	842	-0.23
	100	731	-0.45
	200	805	0.70
	300	751	-4.84
Sept. 21, 1981 (SB)	0	687	4.18
	10	887	1.21
	80	715	1.40
Sept. 28, 1982 (SB)	300	927	-0.59
	300	965	-0.39
Nov. 1, 1982 (SB)	0	769	1.48
	25	817	0.64
	30	795	0.20
	50	817	-0.29
	100	748	0.28
	150	723	-0.30
	310	787	-0.66

Date	Depth (m)	TC	$\delta^{13}C$
Nov. 23, 1982 (SB)	0	764	0.91
	20	818	1.26
	50	769	-0.69
	60	743	-0.71
	120	708	0.31
	170	682	0.18
	220	826	-0.31
	270	765	0.25
	300	831	-0.01
Feb. 8, 1983 (SB)	0	784	0.50
	10	589	4.01
	20	819	0.39
	50	693	2.32
	100	1,099	-7.72
	200	799	0.18
	300	673	0.30
Oct. 3, 1983 (WI-2)[c]	0.5	903	1.32
	10	893	1.24
	15	893	1.40
	30	896	0.41
	50	906	0.33
	120	841	-0.10
	170	911	-0.27
	220	894	0.26
	310	890	0.20
Nov. 10, 1983 (WI-2)	5	899	0.89
	10	895	0.84
	20	882	1.00
	40	874	0.50
	70	885	0.39
	170	883	0.48
	305	907	0.45
Dec. 14, 1983 (WI-1)[d]	0.5	753	0.40
	168	874	2.33
	220	816	2.45
	330	791	0.62
Apr. 2, 1984 (WI-1)	0.5	814	0.54
	35	760	0.77
	310	1,096	0.72
May 16, 1984 (SB)	0.5	856	0.67
	50	817	0.64
	100	838	0.91
	170	783	0.44
	220	803	0.60
	310	804	0.63
July 3, 1984 (WI-1)	0.5	737	1.25
	10	761	1.23
	30	658	0.33
	50	713	0.51
	100	636	0.86
	170	746	0.68
	200	742	0.62
	220	739	0.70
	300	815	0.72

Table 16-1 (Continued) Database for Dead Sea carbon studies.

Date	Depth (m)	TC	$\delta^{13}C$	Date	Depth (m)	TC	$\delta^{13}C$
Sept. 5, 1984	0.5	719	1.33	Oct. 10, 1985 (con't)	150	680	0.93
(WI-1)	20	804	1.34	(SB)	200	610	0.62
	25	840	0.61		250	720	0.90
	50	774	0.49		270	660	0.89
	100	482	0.55		300	670	0.97
	170	663	0.71		310	730	0.96
	220	771	0.54				
	300	711	0.75	Dec. 17, 1985	0.5	620	1.25
Nov. 11, 1984	0	693	1.29	(SB)	10	550	1.29
(WI-1)					20	560	0.92
					30	700	0.63
Dec. 20, 1984	0	820	0.63		70	650	0.64
(SB)	50	601	0.75		150	590	0.92
	100	654	0.49		250	580	1.25
	200	687	0.65		300	650	1.13
	300	634	0.73	Mar. 18, 1992	300	846	1.65
Jan. 29, 1985	0.5	798	0.61	(HU)[e]			
(WI-1)	5	823	0.69	May 20, 1992	0		5.06
Mar. 17, 1985	0.5	700	0.98	(HU)	100		1.61
(WI-1)	20	756	0.95		310		1.64
	60	721	1.00	July 27, 1992	0	930	2.51
	100	768	1.03	(HU)	200	855	1.58
May 9, 1985	0.5	662	1.27	Sept. 22, 1992	2	1,030	2.39
(SB)	40	655	1.00	(HU)	4	1,025	2.48
	120	672	0.95		8.5	862	1.58
	220	666	0.98		14	862	1.40
					100	861	1.35
June 27, 1985	0	580	1.67		200	859	1.38
(SB)	10	640	1.04		310	859	1.43
	17.5	580	0.42				
	20	620	0.87	Dec. 11, 1992	100	868	1.21
	30	700	0.35	(HU)	200	860	1.24
	50	730	0.15		310	865	1.30
	250	640	0.38				
	300	590	0.68	*JORDAN RIVER* (WI-1)			
	310	700	0.71	Feb. 18, 1982	Degania	2,600	-4.80
				Feb. 18, 1982	Dalamya	4,700	-8.94
Aug. 5, 1985	0.5	610	1.46	Feb. 18, 1982	Yarmouk (Y)	3,700	-9.36
(WI-1)	15	660	1.59	Feb. 18, 1982	Jordan + Y	3,900	-8.78
	17.5	700	1.62	Feb. 18, 1982	Allenby b.	5,200	-8.64
	20	640	1.02	Feb. 18, 1982	Abdallah b.	4,800	-8.45
	50	580	0.45	Jan.14, 1993	Allenby b.	4,700	-8.94
	200	770	0.74		(HU)		
	250	790	0.71	*FLOODS*			
	300	640	0.62	Nov. 23, 1982	N. Zohar	3,500	-11.76
	310	730	0.67	Jan. 24, 1983	N. Hever	3,000	-8.93
				Jan. 24, 1983	N. Arugot	3,100	-8.98
Oct. 10, 1985	0	640	1.32	Feb. 8, 1983	N. Kidron	4,200	-7.55
(SB)	0.5	600	0.88	Nov. 26, 1984	N. David	3,090	-8.78
	20	530	0.90				
	25	730	0.85	*PORE WATER* (WI-1)			
	30	620	0.77	Aug 8, 1978	200m	516	-7.75
	50	676	0.93	May 28, 1980	4m	298	-6.44
	100	730	0.80	Oct. 15, 1984	318m	518	-3.97

TC = total dissolved CO_2 μmole/kg; $\delta^{13}C$‰ versus PDB, b. is bridge; N. is Nahal; [a]N&E, data from Neev and Emery (1967); [b]SB, analyses of brine stored in glass bottles; [c]WI-2, analyses made at WI (TC precision 1.2%); [d]WI-1, analyses made at WI (TC precision 4%); [e]HU, analyses made at HU (TC precision 0.25%).

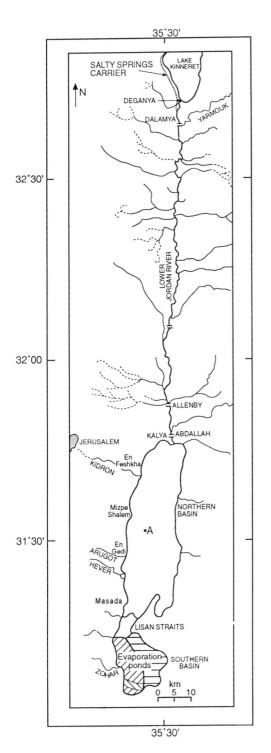

Figure 16-1 The Dead Sea and sampling station A.

tent to the surface (Table 16-1). The rise of deep water to the surface and incomplete homogenization in February 1979 are also indicated by a few relatively low values of dissolved ^{210}Pb, by low tritium concentrations in the upper water (Carmi et al., 1984; Stiller and Kaufman, 1984), and by the scattered pattern of the entire profile of these two isotopes.

The long-term changes that took place in the Dead Sea following its 1979 overturn are evident in Figures 16-2 and 16-3, which show the average TC and ^{13}C values below the pycnocline. The TC of the deep water mass in 1978 was about 850 μmole/kg. Following the 1979 overturn, it may have increased during the short meromictic stage of 1979–1982 to about 950 μmole/kg, because oxygen was then available and oxidation could take place in the deep waters and produce TC. Unfortunately, the data points available to substantiate this observation were too few. Concurrent with the TC change, ^{13}C sharply increased from -2.4‰ in 1978 to about 0‰ in November 1982. From the 1978 data shown in Figure 16-4, we can estimate an average value for the mixed water column of about -2‰, and, in fact, the mean of the scattered February 1979 profile is about -1.6 to -1.9‰. The low ^{13}C values before the overturn may indi-

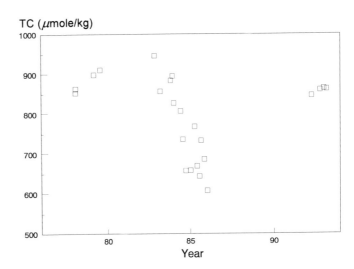

Figure 16-2 The long-term trends of average TC (total dissolved inorganic carbon) below the pycnocline.

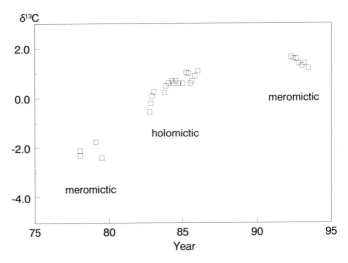

Figure 16-3 The long-term trends of average δ^{13}C of the TC below the pycnocline.

(1) a short meromictic episode that persisted until December 1982; (2) a holomictic stage, with regular yearly overturn in December, from 1983 to 1990; and (3) renewed meromixis since the winter of 1991–1992. It is evident from the ^{13}C data that in February 1979 the entire water column just started to mix and was not yet completely homogenous. The immediate effect of the overturn was to bring some deep water with low ^{13}C con-

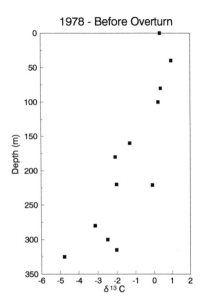

Figure 16-4 Composite depth profile of $\delta^{13}C$ of the TC in 1978.

cate decay of organic matter, probably by sulfate reduction in the anoxic water body. The higher values after the overturn result in part from homogenization of the water column. The rest of the rise observed during September–November 1982 may indicate penetration of surface waters that were enriched in ^{13}C (Table 16-1).

Between the end of 1982 and 1985, the TC values of the deep waters sharply decreased, whereas ^{13}C rose by about 1‰. The TC decrease occurred simultaneously with a massive precipitation of halite (Levy, 1988) from both the upper and the lower water masses (Anati and Stiller, 1991). It is possible that the halite crystals formed nucleation sites for aragonite precipitation from a brine supersaturated with aragonite. The observations of rapid aragonite precipitation during whitening events (Bloch et al., 1944; Neev and Emery, 1967) indicate supersaturation of the lake with respect to this mineral. A supply of bicarbonate to the lake can cause supersaturation to increase over several years. Supersaturation can then decrease when nucleation sites, such as halite crystals, facilitate the precipitation of aragonite. This hypothesis, of course, requires substantiation. But if it is valid, then aragonite precipitation should be accompanied by CO_2 escape to the atmosphere in the following reaction:

$$Ca^{+2} + 2HCO_3 \rightarrow CaCO_3 + CO_2 + H_2O \qquad (2)$$

Could this gas escape explain the observed ^{13}C enrichment in a Rayleigh distillation type process? This can be tested by assuming a kinetic isotope effect of about -20‰ (-19.4‰ at 45°C; Stiller et al., 1985) and a TC loss by gas escape of about 18% or remaining TC fraction (f) of 0.82. (This is about half of the change in TC during the period 1982–1985. The TC loss by aragonite formation will not significantly alter ^{13}C because of the small isotope effect involved in this process.) The Rayleigh equation can be expressed as follows:

$$\delta^{13}C = \delta^{13}C_0 + \varepsilon \ln f \qquad (3)$$

The calculated change ($^{13}C - ^{13}C_0$) is 3.8‰. It is about four times greater than the observed change and may indicate that

gas escape is not the only mechanism affecting ^{13}C in the Dead Sea or that the TC loss was not as large.

In March 1992, the TC of the deep waters rose almost to the October–November 1983 values and became more enriched in ^{13}C than previously known, indicating that the process of gas evasion from the Dead Sea had continued. It is unfortunate that there were no observations between 1985 and 1992, so the details of these changes cannot be studied. Nevertheless, it is of interest to compare the 230 μmole/kg increase in TC (from 620 μmole/kg in December 1985 to 850 μmole/kg in March 1992) with the rate of bicarbonate supply to the lake. Carmi et al. (1984) estimate the average yearly flow of freshwater to the Dead Sea at about 1.42 km³/yr. But substantial diversions of the Jordan and Yarmouk Rivers have gradually diminished the actual inflow during the 1980s to about 0.5 km³/yr (Anati and Shasha, 1989). The average TC of all the inputs (Jordan River and floods) is about 3,900 μmole/kg (Table 16-1), and the average inorganic carbon input is about 1.95×10^9 mole/yr. The volume of the Dead Sea in the late 1980s was about 142 km³ (Stiller et al., 1986; Gat and Anati, 1992). If we use 1.235 for the density of the Dead Sea brine, the added yearly TC is about 11 μmole/kg/yr, and the total supply between 1985 and 1992 is calculated as 70 μmole/kg. This figure does not compare well with the observed increase in TC, again casting doubt upon the reliability of the 1984–1985 TC data. However, other sources of TC, such as subsurface seepages and oxidation of suspended and dissolved organic matter, are possible, as we discuss subsequently.

SEASONAL CHANGES OF ^{13}C DURING HOLOMICTIC YEARS

During the period starting at the end of 1982 and ending at the beginning of 1992, the Dead Sea was in a holomictic state. It underwent regular cycles of overturn every winter when cooling rendered the water column unstable (Anati and Stiller, 1991). These seasonal cycles affected the vertical ^{13}C distribution (Table 16-1). The seasonal pattern of changes is shown in Figure 16-5. Mixing and a homogenous water column are evident in the ^{13}C profile of the 1984–1985 winter months. However, the consistent enrichment of ^{13}C throughout the entire water column in March 1985 compared to December 1984 indicates the loss of carbon dioxide to the atmosphere. By substitution of the ^{13}C change of 0.37‰ (0.37 = 1.02 - 0.67, the difference between the average values for this period), we obtain an estimate of the remaining TC fraction of 0.98. This means that about 14 μmole/kg were lost (assuming a TC of about 750 μmole/kg). For the entire lake volume, an amount of 2.4×10^9 mole was lost in 87 days. In other words, the winter evasion rate can be estimated at about 13 mole/m²/yr (assuming lake area of 770 km²). This is a minimum estimate because some oxidation probably takes place simultaneously with evasion, thus lowering the ^{13}C content.

Surface warming in the spring and summer, and the formation of a seasonal thermocline allow the upper waters and deep waters to attain different ^{13}C values. The upper waters become even more enriched in ^{13}C whereas the deep water mass is depleted in this heavy isotope, with minimal ^{13}C values just below the thermocline (about 30–70 m deep).

By similar calculation, we can estimate the summer evasion loss of the upper 20 m (about 15 km³) for the period May–September 1984 (112 days). The remaining TC fraction is calculated from the 0.65‰ ^{13}C enrichment (1.3 - 0.65 = 0.65‰) as 0.97. About 25 μmole/kg were lost (assuming TC of 810 μmole/kg) or 0.5×10^9 mole from the entire upper layer, yielding an evasion rate of about 2 mole/m²/yr. Thus the summer evasion rate is about six times less than the winter rate, suggesting the

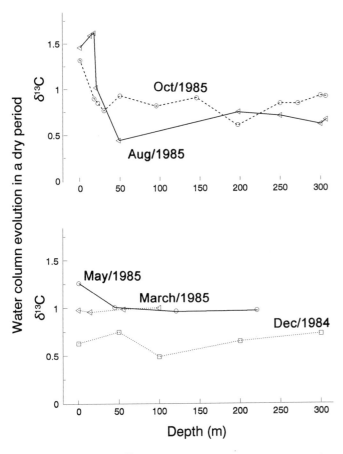

Figure 16-5 Changes in $\delta^{13}C$ of the TC from December 1984 to October 1985. The curve for June 1985 is shown in both plots. Note the decrease in values in the early summer and the increase in values in the later summer and early winter.

importance of the highly turbulent conditions that prevail during the winter.

After a thermocline is established, the deep water mass can, for a first approximation, be treated as a closed system. The depletion of ^{13}C in deep waters observed between March and August 1985 suggests the oxidation of organic matter depleted in ^{13}C. The ^{13}C of the organic carbon influx can be estimated as -23‰ (Stiller et al., 1988). From isotopic mass balance (eq. 4), we can calculate the change in ^{13}C by the addition of such organic carbon:

$$\delta^{13}C = x\delta^{13}C_{org} + (1-x)\delta^{13}C_0 \qquad (4)$$

where x is the added mole fraction of organic carbon ($^{13}C_{org}$), and $^{13}C_0$ is the isotopic composition before the addition. Oxidation of 12 μmole/kg of organic carbon can account for the observed ^{13}C decrease of about 0.38‰ between March 1985 (1.02‰) and August 1985 (0.64‰). The 12 μmole/kg change in TC is too small to be recorded, given the precision of the data for 1985. Oxidation also takes place in the upper waters, but its effects are masked by the enrichment caused by the CO_2 evasion. The minimum ^{13}C observed below the thermocline may suggest agglomeration of particles (hence more organic carbon) and further oxidation at this depth. The total amount of organic carbon oxidized in the entire lake (142 km³ or 175 × 10¹² kg) is large (2.10 × 10⁹ mole or 25,200 × 10⁶ gC).

There are five possible sources of organic carbon in the lake: (1) biogenic in situ production; (2) resuspended old bottom sediments containing organic carbon; (3) allogenic (Jordan River or winter floods) particulate organic carbon (POC); (4) allogenic dissolved organic carbon (DOC); and (5) the DOC reservoir of the lake itself. The first source is negligible because planktonic organisms were never observed during the period in question (Oren et al., 1994). Following the massive halite precipitation mentioned previously, a thick salt layer accumulated on the lake bottom, preventing the resuspension at overturn of sediments after 1983. The most plausible sources of organic carbon are allogenic POC, and DOC from seepages, from the Jordan River and floods, and from the lake storage.

The total input of allogenic particulates to the Dead Sea was calculated by Stiller et al. (1988) as 370,000 × 10⁶ g/yr, and 0.6% of this is POC (2,220 × 10⁶ g/yr). Because significant portions of the flow of the Jordan and Yarmouk Rivers were diverted, less than 10⁹ g/yr is probably a more realistic POC input. In any case, the estimated input is only a small fraction (about 4–9%) of the oxidized carbon in the lake. Vegetation is plentiful around the lakeshore where freshwater springs flow; this may be another source of POC, but it is hard to estimate the significance of this source.

There is only one measurement of DOC in the Jordan River, at its outflow from Lake Kinneret: 3.1 mg DOC/l (A. Spitzy, pers. com.). The present rate of the Jordan River flow is estimated as 0.5 km³/yr. We can thus calculate the DOC input as 1.5 × 10⁹ g/yr. Additional sources of DOC are the vegetated parts of the coastline and, possibly, subsurface seepages. The latter may be a very important source if the concentration of DOC in the seepage is similar to that of pore waters (about 100 mg/l, A. Spitzy pers. com.). With an estimated flow rate of 0.1 km³/yr (Stiller and Chung, 1984), the added DOC is calculated as 10.0 × 10⁹ g/yr, which is larger than the allogenic DOC input. Adding all the POC and DOC inputs yields a total of 12.5 × 10⁹ g/yr, or about half of the added TC to the deep water between March and August 1985 (25.2 × 10⁹ g), if we assume, probably unrealistically, that all the added organic carbon in its various forms becomes oxidized.

Endogenic DOC could also become a potential source of oxidizable carbon in the oxic, holomictic Dead Sea. DOC was measured in samples from various depths taken in the lake during the late 1970s and early 1980s. Concentrations varied from 3.4 to 4.8 mg DOC/l with no clear spatial or temporal trend (A. Spitzy, pers. com., 1986). We do not know how large the refractory fraction of this DOC is, but, in any case, it provides a large reservoir of organic carbon, about 580 × 10⁹ g. It could easily account for the missing carbon source in the previous calculations. Subsurface seepages (0.1 km³/yr) are expected to have higher TC than surface inflows and to be more depleted in ^{13}C than the latter. These seepages could thus provide a TC source, depleted in ^{13}C, without any oxidation having to take place.

Regardless of the source of the organic matter, if it is derived from oxidation, then we expect the concentration of dissolved oxygen to be lower. No measurements of oxygen exist for 1985, but Nishri and Ben-Yaakov (1990) report a 9-μmole/kg reduction between March 1980 and May 1980, and Levy (1980) reported a reduction of 6 μmole/kg for the same time span. Shatkay and Gat (pers. com., 1992) observed fluctuating oxygen levels between August 1987 and December 1989, with depletion below the seasonal thermocline reaching 6 μmole/kg. It seems that at least half of the added TC can be derived from organic matter oxidation in the depth of the Dead Sea. The rapid rate of change may indicate that this is a biochemical reaction. Obviously, more research is necessary to corroborate this hypothesis.

The upper water layer above the summer thermocline, as well as the entire water column in winter, are open to gas exchange with the atmosphere. In all circumstances, carbon dioxide is likely to escape from the Dead Sea. In the summer, it will do so because of its reduced solubility that results from the temperature and salinity rise (Stiller et al., 1985). In the winter in dry years, carbon dioxide may escape because of evaporation and ventilation which affects the entire water column. In rainy winters, CO_2 will escape because of supersaturation resulting from freshwater mixing with the brine (Nishri and Ben-Yaakov, 1990). Finally, supersaturation may result from the organic carbon oxidation of the previous summer. Because CO_2 escape results in a significant kinetic fractionation (of about -20‰), it is expected to enrich the TC in ^{13}C.

By using equation 3, we can estimate the "winter enrichment" caused by a 12 µmole/kg (or 1.6%) gas loss (similar to the estimated addition by oxidation). The calculated value is 0.32‰, which is in fair agreement with the enrichment of 0.37‰ of the entire water column observed between December 1984 and March 1985. Additional enrichment occurs above the pycnocline in the summer, when the change is greater (about 0.7‰) because the effect is concentrated on a much smaller fraction of the water column. It is not so obvious how this surface enrichment affects the deeper water when the pycnocline is strong. A mechanism by which water from above the pycnocline can be mixed to depth must be invoked to explain the enrichment that took place between August 1985 and October 1985 (Fig. 16-5). Such a mechanism has been proposed by Anati and Stiller (1991) in their attempt to explain the salinity increase and the temperature rise of the deep water that occurred at the same time. They have emphasized the role of double-diffusive mixing ("salt fingering") in the holomictic Dead Sea during late summer and fall. In the same way that the warmer and saltier brines penetrate the deep water as the result of this process, their higher ^{13}C content will enrich the ^{13}C of the deep-water mass.

Gat and Shatkay (1991) discuss the effect of the high salt content of the Dead Sea brine on the rate of gas exchange. Piston velocity is lower in the Dead Sea by up to one order of magnitude compared to normal seawater. Gas solubility in the brine is only one fourth of normal marine values. These two factors affect evasion and invasion rates in different ways. The invasion rate is lower in the Dead Sea brine by about one order of magnitude compared to seawater because reduced piston velocity and reduced solubility enhance each other. Under supersaturation, however, the effect of salt on these two factors works in opposite ways: It tends to reduce gas escape because of the lower piston velocity, but the reduced solubility enhances outgassing. Consequently, the evasion rate is greater than the invasion rate. We suggest that this situation will enhance the effect of Rayleigh distillation by allowing gas to leave the solution, thus causing fractionation. Once isotopic disequilibrium develops, the slow invasion will keep the system away from reequilibration.

As discussed in the previous section, the long-term (1982–1985) rise of ^{13}C is too small to be satisfactorily explained if a Rayleigh-type distillation alone is considered. From the ongoing discussion, it appears that this rise is the result of an interplay between inflow of TC and oxidation (both lowering ^{13}C) and CO_2 escape (increasing ^{13}C). Over the time span of the observations, gas escape, whatever its cause (reduced solubility from evaporation, supersaturation produced by mixing with freshwaters, by oxidation, and by aragonite formation), had a slightly greater enrichment effect than the depletion caused by oxidation.

CARBON ISOTOPES AND CARBON INVENTORIES IN 1992

Unusually heavy rainfall and large freshwater influx completely changed the hydrography of the Dead Sea in 1992. Gat and Anati (1992) report a highly stable water column with a strong halocline and pycnocline several meters from the surface. The reduced salinity and, perhaps, increased nutrient supply led to a massive algal bloom in May 1992. Following the crush of this bloom in June 1992, bacteria flourished in the upper waters (Oren et al., 1994). The ^{13}C and the carbon budget were greatly influenced by the hydrographic and biological changes.

Figure 16-6 shows the vertical profile of ^{13}C. There are no data for the winter of 1991–1992, but based on the hydrographic pattern from previous years, we can assume that the deep water data of March 1992 ($\delta^{13}C = 1.65‰$ and TC = 846 µmole/kg) approximately represent the surface waters before the beginning of the floods. In May during the algal bloom, the surface ^{13}C was 5.06‰, declining to about 2.5‰ in July. Between July and September, the surface ^{13}C did not change much.

Two ways to enrich the surface water in ^{13}C are possible: biological fixation of ^{13}C-depleted carbon and outgassing of CO_2 from a supersaturated solution. Supersaturation in carbon dioxide (and other gases) results from the mixing of fresh floodwaters with the Dead Sea brine (Nishri and Ben-Yaakov, 1990). Loss of carbon dioxide may be accompanied by precipitation of aragonite (see previous discussion). A crude mass balance of TC and carbon isotopes shows how this may work.

The level of the Dead Sea rose by about 2 m with the winter floods. The incoming flood water mixed with brine and formed a diluted surface water with a pycnocline at about 7 m in May. The average values of ^{13}C and TC in the floodwater can be taken as -8.9‰ and 3,900 µmole/kg, respectively (Table 16-1). The total added inorganic carbon is thus 7.8 mole/m². The total inorganic carbon of the brine that mixed to form the water mass above the pycnocline is about 5.2 mole/m² (5.2 ≈ 846 x 10⁻⁶ x 5,000 x 1.235, where 1.235 is the density of the brine before its mixing with freshwater) with $\delta^{13}C$ of 1.65‰. The calculated amount of inorganic carbon above the pycnocline is 13 mole/m², and its calculated ^{13}C is -4.7‰. Unfortunately, there are no data for TC in May, but if the July value (930 µmole/kg) is used in this calculation, then the actual amount above the pycnocline can be taken as 7.7 mole/m² (7.7 ≈ 930 x 10⁻⁶ x 7,000 x 1.177,

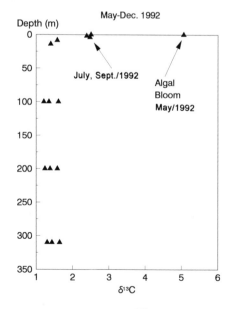

Figure 16-6 Composite depth profile of $\delta^{13}C$ of the TC in 1992.

where 1.177 is the density above the pycnocline in May). The difference between the actual inventory and the calculated value is 5.3 mole/m². A significant fraction (68%) of the carbon influx was lost by outgassing, biological uptake, and, possibly, aragonite precipitation.

Oren et al. (1994) estimated the total standing stock of particulate organic carbon in May as 2.2 mole/m². The ¹³C of this organic carbon is -13‰. The particulate fraction in May did not contain any aragonite detectable by x-ray diffractometer. Whether aragonite formed when floodwaters first entered the lake remains unknown. In any case, a significant portion (1.6 - 3.1 mole/m²) of the flood bicarbonate was lost by outgassing to the atmosphere. If aragonite did not precipitate, then the remaining fraction of inorganic carbon after the CO_2 escape is 0.75((13 - 3.1)/13 = 0.75). By substitution of this figure and a kinetic isotope effect (= -20 ‰) in equation 3, we calculate ¹³C of the remaining TC as 1.05‰. Biological uptake will further reduce the TC and will cause greater enrichment.

Following the crush of the algal bloom, ¹³C decreased from 5.06‰ to 2.51‰. As discussed previously, this change is not likely to result from equilibration with the atmosphere because of the slow gas exchange. It could have resulted from oxidation of part of the fixed carbon. However, there are no data to quantitatively test this possibility.

Below the pycnocline, the ¹³C changes systematically from high values in March through July to somewhat lower values in September and December (Fig. 16-7; Table 16-1). As Figure 16-8 and equation 5 show, TC and ¹³C values correlate well with a correlation coefficient of R = 0.891:

$$TC = 913 - 38.3137 \times {}^{13}C \qquad (5)$$

This indicates addition of TC by oxidation of organic carbon. The ¹³C of this added carbon is easily calculated from the slope of the regression equation as -21.1‰. Because the ¹³C of the algae and bacteria in the surface waters is -13‰, only a small fraction (about 20%) of the additional TC is derived from the plankton. A more likely source is allogenic organic matter with ¹³C of -23‰ (see previous discussion), which supplies most of the oxidizable organic carbon (about 80%).

Allogenic carbon is a major source destined for decay at the depth of the lake. About 18 µmole/kg of TC were added below the pycnocline between March and December 1992 (Table 16-1). The average thickness of the layer below the pycnocline is about 180 m and its density 1.235. Thus the total amount added below the pycnocline is calculated as 4.0 mole/m². Comparing this addition with the total particulate organic carbon in May (2.2 mole/m²) and noting that part of it must have been used for bacterial growth and oxidation at the surface (see the increase in the surface TC values in September 1992, Table 16-1) again indicates that particles derived from the algal bloom are not the only source of carbon added below the pycnocline. In fact, the amount of oxidized carbon in 1992 is in the same range as the estimated oxidation in 1985 (see previous discussion), which was based solely on allogenic carbon sources. If so, algal bloom and primary production are not a prerequisite for rapid oxidation in the Dead Sea. This rapid oxidation implies the existence of biochemical reactions—an intriguing topic for future research.

CONCLUDING REMARKS

From these data we can conclude that:
(1) The total dissolved carbon dioxide (TC) in the Dead Sea as determined by acid extraction is only about one third of that previously estimated by total alkalinity titrations. This discrepancy results from the high borate content of the Dead Sea brine.

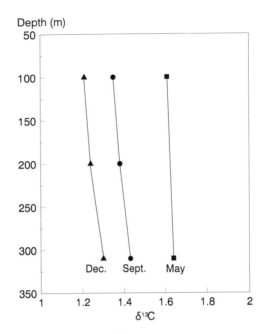

Figure 16-7 Depth profiles of $\delta^{13}C$ of the TC below the pycnocline from May 1992 to December 1992. Note the consistent and progressive depletion in ¹³C.

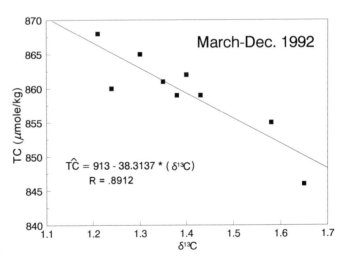

Figure 16-8 Correlation of TC with $\delta^{13}C$ of the TC for samples taken below the pycnocline in 1992.

(2) Large changes in carbon inventories and carbon isotopic composition have taken place in the Dead Sea since its historic overturn in 1979. As the deep water mass ventilated and changed from anaerobic to aerobic conditions, the TC became more and more enriched in ¹³C.
(3) Between 1982 and 1985, TC dramatically decreased by about 30%. This decrease probably resulted from coprecipitation of aragonite with the massive halite deposition that took place.
(4) Significant seasonal fluctuations in $\delta^{13}C$ are evident during the holomictic years. When the seasonal thermocline forms in the spring and early summer, the lowering of $\delta^{13}C$

below the thermocline indicates that oxidation of allogenic organic carbon is taking place. Later during the summer and fall, the increasing $\delta^{13}C$ of the deep waters suggests intrusion and downward mixing of ^{13}C-enriched surface waters. The enrichment at the surface results from CO_2 escaping to the atmosphere with a strong kinetic fractionation of the carbon isotopes.

(5) Massive rain, floods, and an algal bloom occurred in 1992. Large quantities of TC were introduced with the floodwater into the top layer of the Dead Sea. CO_2 became highly supersaturated and was partially removed by outgassing to the atmosphere, with the rest of the CO_2 taken up by the algal growth. Removal of CO_2 by both mechanisms, with strong kinetic and biological fractionations, led to a steep increase of $\delta^{13}C$ in the surface layer.

(6) A highly stable water column with a very strong pycnocline developed because of the floodwater dilution of the surface layer in 1992. Below the pycnocline, TC increased because of oxidation of organic matter. The amount of added TC and the TC versus $\delta^{13}C$ relation indicate that the major source of carbon is allogenic and that only a small portion is derived from the algal bloom.

Acknowledgments

We thank J. R. Gat and J. C. Vogel for their initiatives and for their input into the Dead Sea project. Thanks are due to D. A. Anati, A. Oren and E. Barkan for discussions and for sharing their data with us. We thank S. Shasha for technical assistance. This research has been supported by the ongoing Dead Sea project, Earth Science Administration, Ministry of Energy and Infrastructure, State of Israel. S. Talma's visit to Israel was supported by the South African CSIR/Israel NCRD exchange program.

REFERENCES

Anati, D. A., and Shasha, S., 1989, Dead Sea surface level changes: *Israel Journal of Earth Sciences*, v. 38, p. 29–32.

Anati, D. A., and Stiller, M., 1991, The post-1979 thermohaline structure of the Dead Sea and the role of double-diffusive mixing: *Limnology and Oceanography*, v. 36, p. 342–352.

Bloch, R., Littman, H. Z., Elazari–Volcani, B., 1944, Occasional whitenings of the Dead Sea: *Nature*, v. 154, p. 402.

Carmi, I., Gat, J. R., and Stiller, M., 1984, Tritium in the Dead Sea: *Earth and Planetary Science Letters*, v. 71, p. 377–389.

Gat, J. R., and Anati, D. A., 1992, Hydrography of the Dead Sea 1991–1992: Jerusalem, Israel Ministry of Energy and Infrastructure Report No. ES-37-92, 40 p.

Gat, J. R., and Shatkay, M. L., 1991, Gas exchange with saline waters: *Limnology and Oceanography*, v. 36, p. 988–997.

Katz, A., Kolodny, Y., and Nissenbaum, A., 1977, The geochemical evolution of the Pleistocene Lake Lisan-Dead Sea system: *Geochimica et Cosmochimica Acta*, v. 41, p. 1,609–1,626.

Levy, Y., 1980, Seasonal and long range change in oxygen and hydrogen sulfide concentrations in the Dead Sea: Jerusalem, Geological Survey of Israel, Report, MG/9/80, 11 p.

Levy, Y., 1988, Sedimentary reflections of modern halite precipitation from the Dead Sea water: Jerusalem, Geological Survey of Israel, Report, GSI/12/88, 16 p.

Mook, W. G., 1968, Geochemistry of the stable carbon and oxygen isotopes of natural waters in the Netherlands [Ph.D. thesis]: Groningen, Rijksuniversiteit, 137 p.

Mook, W. G., 1970, Stable carbon and oxygen isotopes of natural waters in the Netherlands: *Isotope Hydrology*, IAEA-SM-129/12, p. 163–190.

Neev, D., and Emery, K. O., 1967, The Dead Sea—Depositional processes and environments of evaporites: Geological Survey of Israel Bulletin, v. 41, 147 p.

Nishri, A., and Ben-Yaakov, S., 1990, Solubility of oxygen in the Dead Sea brine: *Hydrobiologia*, v. 197, p. 99–104.

Nissenbaum, A., and Kaplan, I. R., 1976, Sulfur and carbon isotopic evidence for biogeochemical processes in the Dead Sea ecosystem: *Environmental Biogeochemistry*, v. 1, p. 309–325.

Oren, A., Gurevich, P., Anati, D. A., Barkan, E., and Luz, B., 1993, A bloom of *Dunaliella parva* in the Dead Sea in 1992: Biological and biogeochemical aspect: *Hydrobiologia*, v. 297, p. 173-185.

Sass, E., and Ben-Yaakov, S., 1977, The carbonate system in hypersaline solutions—Dead Sea brines: *Marine Chemistry*, v. 5, p. 183–199.

Schonfeld, I., and Held, S., 1965, Spectrochemical methods for determining boron, barium, lithium and rubidium in Mediterranean and Dead Sea water: Tel Aviv, Israel Atomic Energy Commission, IA-1061.

Steinhorn, I., Assaf, G., Gat, J. R., Nishri, A., Nissenbaum, A., Weiss, W., Stiller, M., Beyth, M., Garber, R., and Friedman, G. M., 1979, The Dead Sea—Deepening of the mixolimnion signifies the overture to overturn of the water column: *Science*, v. 206, p. 55–57.

Stiller, M., and Chung, Y. C., 1984, Radium in the Dead Sea—A possible tracer for the duration of meromixis: *Limnology and Oceanography*, v. 29, p. 574–586.

Stiller, M., and Kaufman, A., 1984, ^{210}Pb and ^{210}Po during the destruction of stratification in the Dead Sea: *Earth and Planetary Science Letters*, v. 71, p. 390–404.

Stiller, M., Rounick, J. S., and Shasha, S., 1985, Extreme carbon-isotope enrichments in evaporating brines: *Nature*, v. 316, p. 434–435.

Stiller, M., Gat, J. R., and Shasha, S., 1986. Hydrographic survey annual report, November 1, 1984 to December 31, 1985: Jerusalem, Israeli Department of Energy, 33 p.

Stiller, M., Carmi, I., and Kaufman, A., 1988, Organic and inorganic ^{14}C concentrations in the sediments of Lake Kinneret and the Dead Sea (Israel) and the factors that control them: *Chemical Geology*, v. 73, p. 63–78.

17. THE RADIOCARBON CONTENT OF THE DEAD SEA

A. Siep Talma, John C. Vogel, and Mariana Stiller

Radiocarbon is a useful natural tracer with which to follow the path of CO_2 in the atmosphere and the air-sea system (Craig, 1957). Labelling atmospheric CO_2 with ^{14}C in the stratosphere provides the source, and the eventual decay of ^{14}C in the deep sea forms the sink of this tracer. Variations in the transfer rates and pool sizes along this pathway produce radiocarbon variations that can then be studied (Broecker and Peng, 1982).

The behavior of ^{14}C in lakes can be much more complex. Apart from the exchange of ^{14}C at the air-water interface, there are also other sources and sinks affecting the total inorganic carbon (TIC) pool of the water and its ^{14}C content, such as surface and underwater inflows and outflows, production and oxidation of organic matter, and solution and deposition of carbonate material. Isolation of parts of the lake from its sources can change the ^{14}C content by radioactive decay (half-life of 5,730 years). The decay of ^{14}C in the deep water of the Dead Sea, which was presumed to be isolated, was considered useful enough by Neev and Emery (1967) to attempt dating the hypolimnion of the Dead Sea water, and undertake the first ^{14}C sampling there.

The major feature of the atmospheric radiocarbon content during the past few decades has been the large pulse of radiocarbon produced by nuclear weapons which virtually doubled the atmospheric radiocarbon content in the northern hemisphere in the early 1960s (Levin et al., 1985). This peak was followed by a slow decrease to present-day atmospheric ^{14}C levels, which are still higher than at any time in history before 1955. The pattern of ingress of this ^{14}C pulse into water bodies is useful for studying the air-water exchange of CO_2 (Peng and Broecker, 1980).

The relative ^{14}C concentration (i.e., the $^{14}C/^{12}C$ ratio) of a sample is expressed as a percentage of a standard (0.95 NBS oxalic acid) that closely resembles the atmospheric ^{14}C content before nuclear weapons and industrial pollution effects manifested themselves (Stuiver and Polach, 1977); this concentration is named percent Modern Carbon units (pMC). Another way of expressing ^{14}C levels is the Lamont delta, $\Delta^{14}C$, which is the deviation, in permille, of the isotope ratio $^{14}C/^{12}C$ from the standard, after adjustment for isotopic fractionation based on measurement of ^{13}C (Broecker and Olson, 1961). In this chapter, we use percent Modern Carbon units because the ^{14}C variations in the Dead Sea are large enough for isotope fractionation effects to be ignored.

SAMPLING FOR RADIOCARBON IN THE DEAD SEA

Because of the large water quantity required for a ^{14}C determination by measuring its decay (typically 100 liters), some form of concentration of the total dissolved CO_2 from the water is necessary. This is done to reduce the sample volume sent to the radiocarbon laboratories, which are all remote from the study area in this project. Two methods are generally used: direct precipitation of the TIC as $BaCO_3$ from the water, and extraction of CO_2 gas from an acidified water sample and its subsequent absorption in hydroxide solution (Vogel, 1967). We used the latter method for this project. In the Dead Sea, the precipitation technique would have been impractical because of the coprecipitation of large amounts of sulphate and other salts at the high pH required for precipitation of barium carbonate.

The variable results obtained by the different laboratories working on the Dead Sea samples necessitate consideration of possible sample contamination during the entire handling procedure. During the CO_2 extraction process, leakage in some parts of the gas handling system can conceivably result in atmospheric air being introduced into the system. Atmospheric CO_2 (0.03%) can then be absorbed in the hydroxide solution and added to the sample CO_2. This would raise the ^{14}C levels of the sample extract. A 5% addition of atmospheric CO_2 to a typical Dead Sea sample, for instance, would increase its ^{14}C content by 4 pMC.

The first radiocarbon sample of Dead Sea water was collected by D. Neev in 1962 from a depth of 250 m (Neev and Emery, 1967; Vogel, 1980). CO_2 was extracted from this sample by Neev and was analyzed for ^{14}C by K. O. Munich in Heidelberg. At that time, the Dead Sea was stably stratified (i.e., meromictic). The upper mixed layer was 40 m thick, and the deep water mass below 100 m to the maximal depth of 325 m was anoxic and completely isolated from contact with the atmosphere. Between the two water masses, there was a transition layer from 40 to about 100 m.

A year later, Neev collected three additional samples (two surface and one deep), which were sent to Heidelberg where the CO_2 was extracted and the ^{14}C content measured. These three samples showed the expected lower ^{14}C content of deeper water (at 75 m) compared to that of surface water (Table 17-1). The surface water had a relatively low ^{14}C content, about 82.5 pMC, indicating a low uptake rate of atmospheric ^{14}C and the preponderance of low ^{14}C inputs into the Dead Sea. At that time, the atmospheric ^{14}C content had just increased from 98 to 160 pMC as a result of the atmospheric nuclear tests of the late 1950s (Levin et al., 1985), and the ^{14}C content of surface ocean water was slowly increasing from a base value of 95 pMC (Broecker and Peng, 1982). The much lower ^{14}C content of the Dead Sea surface water compared to that of the oceans and the realization that considerable "fossil" carbon input is likely prompted Neev and Emery (1967, p. 69) to state that "they (^{14}C determinations) cannot . . . be used for purposes of true age determination." The measurements and discussion presented in this chapter indicate that this remains a realistic statement.

The next samples specific for ^{14}C analysis were taken by one of the authors, M. Stiller, with J. Gat and Y. Levy in 1980. This was just after the major overturn of February 1979, which was the first time in recorded history that the entire Dead Sea water column mixed properly (Steinhorn, 1985). At the Pretoria radiocarbon laboratory, a minicounter for ^{14}C had just been commissioned (Vogel and Behrens, 1976). With this counter, small (50-mg carbon) samples could be counted for ^{14}C analysis (instead of the usual 1.5 g required for conventional ^{14}C counting), although there was some loss of precision. This meant that only 5 liters of Dead Sea water would be sufficient for a ^{14}C sample. This was well within the capabilities of the sampling program as it was conducted at that time. After samples were collected, gas extracts of the TIC were made at the Weizmann Institute in Rehovot and analyzed for ^{14}C in Pretoria

Table 17-1 Summary of radiocarbon measurements for Dead Sea water samples.

No.[a]	Site[b]	Depth (m)	Date	^{14}C (pMC)	δ^{14}C (‰)
H-1937	deep	250	Sept. 1962	81.1 ± 1.2	-229[c]
H-2018	surface	0	Jan. 1963	82.7 ± 0.5	-217[c]
H-2026	deep	75	Apr. 1963	75.5 ± 1.0	-280[c]
H-2040	surface	0	May 1963	82.2 ± 0.6	-222[c]
C-982	St 2	30	June 1976	92.6 ± 0.9	-122
C-981	surface	0	Dec. 1977	107.7 ± 1.6	+28[d]
Pta-3022	St 2	20-100	Jan. 1978	84.7 ± 2.5	-194
Pta-2754	various	180-320	Jan. 1978	71.3 ± 1.4	-314
C-959	St 2	237	Jan. 1978	74.1 ± 1.0	-287
C-954	St 2	280	Jan. 1978	77.6 ± 0.8	-261
C-980	St 1	310	Aug. 1978	80.7 ± 0.8	-226
Pta-3023	St 1	0-30	Oct. 1979	95.8 ± 3.1	-88
Pta-2758	St 1	0	Apr. 1980	86.9 ± 1.5	-170
Pta-2759	St 3	0	Apr. 1980	90.4 ± 1.7	-136
Pta-2756	St 1	250	Apr. 1980	84.6 ± 1.9	-195
Pta-2755	St 3	250	Apr. 1980	79.8 ± 1.5	-239
Pta-3024	St 2	300	Oct. 1980	89.1 ± 2.2	-150
Pta-3961	St 1	0.5	Oct. 1983	84.6 ± 2.2	-198
Pta-4248	St 1	5-15	Oct. 1983	85.4 ± 1.7	-189
Pta-4251	St 2	120-220	Oct. 1983	85.2 ± 1.6	-191
Pta-4250	St 1	305	Oct. 1983	85.4 ± 2.3	-188
Pta-3955	St 1	20	Nov. 1983	86.1 ± 2.1	-183
Pta-3965	St 1	40	Nov. 1983	84.9 ± 2.1	-193
Pta-3957	St 1	305	Nov. 1983	84.3 ± 2.1	-198
Pta-5548	St 2	320	Dec. 1990	83.0 ± 0.6	-212

[a]Analytical laboratories: H—Heidelberg; Pta —Pretoria; C—Bern/Zurich AMS.
[b]Sampling locations are also shown in Figure 17-3: St 1—Midway between Mizpe Shalem and En Gedi at 300-m depth; St 2—Opposite En Gedi (or 2–4 km north) at 300-m depth; St 3—Opposite inflow of Kidron River at 250-m depth.
[c]Originally described by Neev and Emery (1967); later updated by Vogel (1980).
[d]Possibly contaminated during AMS analysis from a previously analyzed sample (M. Andree, pers. com.).

(Table 17-1). Extractions were also made from samples that had been archived from earlier sampling cruises (1978 and 1979). The results varied somewhat (Fig. 17-1), but the lowest ^{14}C content obtained was 71.3 ±1.4 pMC for deep water prior to the overturn of 1979. This is lower than the 75.5 ±1 pMC found in 1963 for the water in the transition zone (at 75 m) of the old, stable Dead Sea.

Two complete profiles of ^{14}C samples were taken in October and November 1983 during A. S. Talma's stay at the Weizmann Institute to initiate the ^{13}C/TIC investigation in the Dead Sea (Luz et al., chapter 16, this volume). At this time, the Dead Sea showed annual overturns followed by efficient winter mixing (Anati and Stiller, 1991). The overturn preceding the 1983 sampling had occurred in December 1982, and efficient vertical mixing prevailed until mid-February 1983. This may explain the very constant 85.0 ±0.6 pMC ^{14}C found in both profiles, independent of depth (Fig. 17-2). This set of measurements, therefore, gives a good average ^{14}C value for the Dead Sea in 1983.

In 1982 and 1983, samples were also obtained from the lower Jordan River between Lake Kinneret and the Dead Sea at the outflow from Lake Kinneret (at Deganya) and at about 2 km downstream from the confluence with the Yarmouk River (2 km downstream of Dalamya; Fig. 17-3). A few of the floods entering the Dead Sea directly from the western side were also sampled. These various samples give a range of ^{14}C values entering the Dead Sea as surface water (Table 17-2).

In 1986, a further set of samples was prepared for analysis by accelerator mass spectrometry (AMS). This method has the capability to measure ^{14}C in very small samples (only about 1 mg carbon). This set of samples was taken from the archive collection of the pre-overturn period of 1976 to 1978 (Table 17-1). CO_2 extraction was done by M. Stiller in Rehovot, target preparation in Bern, and AMS analysis in Zurich by M. Andree. The results confirmed the lower ^{14}C content of the pre-overturn deep water, with varying higher values found closer to the surface. The 1978 values, in particular, confirmed the ingression of surface water into the deep water in August 1978 (sample C-

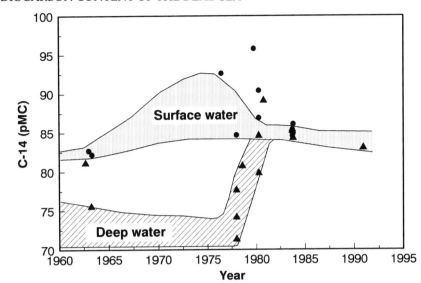

Figure 17-1 Plot of all ^{14}C analyses over time. Surface waters are marked by circles and deep waters by triangles. Boundaries of the presumed change in ^{14}C of surface and deep waters with time (based on pycnocline positions) are indicated. After 1980, surface and deep waters are within the same boundary.

980), just before the eventual overturn of February 1979 (Steinhorn, 1985), which was also indicated by other tracers, such as tritium and dissolved divalent iron (Carmi et al., 1984; Nishri and Stiller, 1984).

In 1990, a single deep water sample was collected by E. Visser and J. Kronfeld (Table 17-1) and analyzed in Pretoria (Pta-5548). In view of the regular annual overturns that had occurred since 1983, this sample can be considered to represent a good average of the entire water body. The ^{14}C decrease since 1983 (Table 17-1 and Fig. 17-1) suggests that the average ^{14}C content of the Dead Sea has marginally decreased (by 2 pMC) and that no further increases from the nuclear bomb ^{14}C pulse of the 1960s are likely.

GENERALIZED ^{14}C DISTRIBUTION OVER TIME

The general distribution of ^{14}C values with time is scattered (Fig. 17-1), even beyond the usual analytical error of 1–2 pMC. Before the 1979 overturn, the deep water was very stable and isolated (Neev and Emery, 1967), as confirmed by tritium measurements at that time (Carmi et al., 1984). As a first approximation, we can conclude that deeper waters have lower ^{14}C levels before the overturn of 1979. This is to be expected from a stable stratification that must have lasted for centuries (Stiller and Chung, 1984). Discrepancies beyond the analytical errors certainly exist; for example, the deep water of January 1978 (C-959 and C-954) shows a higher ^{14}C content for two samples at depths of 237 and 280 m (Table 17-1) than a single composite made up of eight samples between 180 and 320 m (Pta-2754).

To envisage a generalized ^{14}C pattern during this period, we can assume that the lowest ^{14}C value is probably the more reliable indicator of the average ^{14}C value of a water mass and that the system is fairly slow to respond, except at times of overturn. This does not preclude the fact that locally high ^{14}C patches may be found on the surface because of the influx of freshwater (Table 17-1 and 17-2). The highest ^{14}C value on the surface (108 pMC) was, in fact, sampled during the rainy season (December 1977) and could be due to localized freshwater inflow or to contamination (see footnote, Table 17-1).

Figure 17-1 shows an estimate of the progress of ^{14}C in deep and surface water of the Dead Sea with time. It is presumed that the lowest value of 72–74 pMC represents the deep water of the entire preoverturn period. The ^{14}C content of surface water increased between 1960 and 1976 because of the effect of freshwater inflow, bearing the nuclear bomb ^{14}C signal, into the Dead Sea. A ^{14}C peak may have been present in surface water

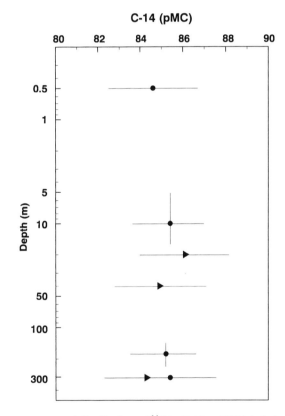

Figure 17-2 Vertical distributions of ^{14}C in October 1983 (circles) and in November 1983 (triangles). Samples that have been combined are indicated as vertical error bars.

sometime in the 1960s or early 1970s, but there are no measurements to support this. The deepening pycnocline after 1976 would effectively have introduced old, deep water into the surface layer, thereby lowering its ^{14}C content. The decrease in the upper mixed layer from 92.6 pMC in June 1976 to 84.7 pMC in January 1978 (Table 17-1, samples C-982 and Pta-3022) may be attributed to this process. The first overturn in February 1979 was rather weak, since it occurred late in the winter. By 1980,

Table 17-2 Summary of radiocarbon measurements for rivers flowing into the Dead Sea.

No.[a]	Site[b]	Date	^{14}C (pMC)	$\delta^{14}C$ (‰)
Pta-3932	Jordan at Deganya	Feb. 1982	107.8 ± 2.3	+28
Pta-3935	Jordan at Deganya	Oct. 1983	106.0 ± 1.6	+22
Pta-3930	Jordan + Yarmouk	Feb. 1982	86.3 ± 1.4	−169
Pta-3937	Jordan + Yarmouk	Oct. 1983	79.8 ± 1.6	−237
Pta-3741	Arugot	Jan. 1983	90.7 ± 1.1	−125
Pta-3748	Hever	Jan. 1983	113.1 ± 1.2	+94
Pta-3775	Kidron	Feb. 1983	97.7 ± 1.6	−61
Pta-3778	Zohar	Nov. 1982	100.0 ± 2.1	−27

[a]Analytical laboratory, Pta —Pretoria.
[b]Sampling locations are shown in Figure 17-3.

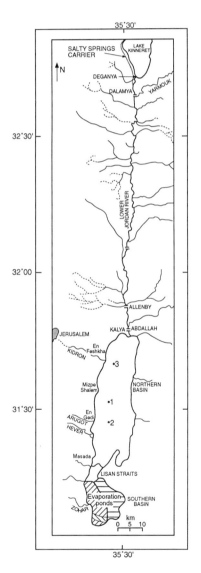

Figure 17-3 Map of the Dead Sea and part of its catchment area. The "Jordan + Yarmouk" samples (Table 17-2) were taken 2 km south of Dalamya.

however, the deep water had lost its separate character completely (Steinhorn, 1985), as shown by the ^{14}C content of deep water samples in April and October 1980 (Table 17-1).

In 1983, when the contamination possibility during CO_2 extraction was recognized and extra care was taken during this procedure, no discrepancies were observed and the entire profile yielded virtually the same ^{14}C content (Fig. 17-2): a well-mixed value of 85 pMC. As stated previously, the overturn preceding the 1983 sampling provided good mixing, and, therefore, vertical homogeneity could have been expected. By 1990, the ^{14}C content of the well-mixed Dead Sea had decreased by 2 pMC, even though the atmospheric ^{14}C content had been at least 40 pMC higher than that of the Dead Sea waters for the preceding three decades, indicating that the air-water exchange of CO_2 was quite small.

QUANTIFYING THE RADIOCARBON BALANCE IN THE DEAD SEA

Having established the essential values of the ^{14}C contents over time for both surface and deep water (Fig. 17-1), we seek a quantitative explanation of these numbers in terms of known properties of this lake. Since 1983, the Dead Sea has been represented as a well-mixed water body. Although it stratifies seasonally during spring, summer, and autumn, it completely mixes vertically every winter. The total radiocarbon contents of the Dead Sea can be estimated with reasonable accuracy for the well-mixed situations of 1983 and 1990. Some estimates can be made for 1980 from the range of values after the major overturn of 1979 and also for the stable situation prevailing in 1963 (and probably long before that time as well). The summary in Table 17-3 represents ^{14}C estimates of the entire water column for the periods when ^{14}C ranges seemed reliable and a representative average could be obtained.

Since 1960, the Dead Sea water volume has been declining (Anati and Shasha, 1989), as has the total inorganic carbon content (Luz et al., chapter 16, this volume). Substantial quantities of CO_2, $CaCO_3$, or both must have left the system. Luz et al. (chapter 16, this volume) show that this loss of carbon is essentially as CO_2 gas from surface water during the stratified periods. For chemical mass balance reasons, an equal amount of carbonate precipitation (probably as aragonite) should have occurred, although there is little observational evidence that this occurs on a continuous basis (Stiller et al., chapter 15, this volume). Whatever the process, the lost carbon will have the same ^{14}C content as the water (assuming good mixing through-

Table 17-3 Calculated ^{14}C budget of the Dead Sea.

Year	^{14}C content Avg. (pMC)	TIC[a] (mM/kg)	Total carbon (GM)[b]
<1963	75 ± 2[c]	0.9 (est)	166
1980	85 ± 2	0.91	164
1983	85.0 ± 0.6	0.88	156
1990	83 ± 1	0.83	144

[a]Refer to Luz et al. (chapter 16, this volume).
[b]1 GM = 1 gigamole = 1 km^3 · mMol/l based on water volumes (Steinhorn, 1981) and water levels and densities of Anati et al. (1987) and Anati and Shasha (1989).
[c]The average for 1963 has been calculated using Neev and Emery's (1967) water volumes, their ^{14}C data for the upper and transition water masses, and 72 pMC for deep water.

out the lake). The overall ^{14}C changes in the Dead Sea, therefore, suggest an above-average (>85 pMC) total ^{14}C input from 1963 to 1980 to raise the overall ^{14}C content of the Dead Sea water from 75 to 85 pMC. A below-average ^{14}C level for the input (<85 pMC) during the decade 1980 to 1990 must then be presumed. The ^{14}C level of the atmosphere was, at all times, considerably above the ^{14}C level of the Dead Sea. If any atmospheric ^{14}C exchanged with the Dead Sea water, there must have been an additional, low ^{14}C input to balance this high input.

Higher ^{14}C inputs are quite obvious: The ^{14}C content of the Jordan River was above 85 pMC during 1982 and 1983 at its outflow from Lake Kinneret (Deganya). Flood waters, at least on the west side of the Dead Sea, have ^{14}C contents greater than 85 pMC (Table 17-2). Any gaseous ^{14}C exchange between the water and the atmosphere will also have increased the ^{14}C content of the Dead Sea (Peng and Broecker, 1980; Carmi et al., 1985).

Possible low ^{14}C sources are "old" groundwaters flowing into the Dead Sea and dissolution of "old" carbonates. The only data available from aquifers in the vicinity of the Dead Sea are those of the En Feshkha springs (on the northwestern shore) with a ^{14}C content of 35 pMC (Mazor and Molcho, 1972). Carmi et al. (1984) estimated an annual inflow of less than 0.2 km^3/yr into the Dead Sea from this source; at most, this would reduce the ^{14}C content of the entire Dead Sea by only 0.1 pMC annually. Springs of unknown magnitude discharging into the lower part of the Jordan River could also contribute to a low ^{14}C input into the Dead Sea.

A detailed look at the ^{14}C measurements of the Jordan River system in 1982–1983 (Table 17-2) suggests that low ^{14}C inflows might be possible from that source, at least for some of the year. The Jordan River at Deganya represents the outflow of Lake Kinneret only and is small because most of the water of Lake Kinneret is presently fed into the National Water Carrier. The Salty Springs Carrier (SSC), shown schematically in Figure 17-3, is a water carrier (established in the mid 1960s) conveying high-salinity water from springs on the northwestern shore of Lake Kinneret toward the lower Jordan River and discharging at a point a few kilometers downstream of Deganya for disposal into the Dead Sea. The Yarmouk River joins the Jordan River flow just below this discharge point (Fig. 17-3). No isotopic data are available from the water of the SSC.

The maximum influence of this saline component on the Jordan River water is likely to be during the dry season, when minimum water amounts are released from Lake Kinneret and

when the Yarmouk water flow would also be low. The ^{14}C sample of the Jordan and Yarmouk Rivers in October 1983 (Pta-3937) represents such an event. This sample has a ^{14}C content significantly below that of the Dead Sea (79.8 pMC, Table 17-2). We can, therefore, postulate that the SSC could have contributed sufficient amounts of low ^{14}C water to the Jordan River to cause lowering of the Dead Sea ^{14}C level since 1983. With declining levels of freshwater contributions from Lake Kinneret and from the Yarmouk River during the past decade, the influence of the SSC water could well have become more significant. A detailed evaluation will require historical flow (and isotope) data from these various sources.

CONCLUSIONS

Radiocarbon measurements of the Dead Sea support the concept of isolation of the deep water mass before the major overturn of 1979. Although some inconsistent measurements have been made, the ^{14}C level of the Dead Sea has been remarkably constant over the past few decades, with very little influence from the atmospheric bomb ^{14}C pulse of the early 1960s. The slow ^{14}C changes appear to be caused by the various water sources of the Dead Sea. The recent slow decrease of ^{14}C in the entire sea is likely to have been caused by the low ^{14}C input from the waters of the Salty Springs Carrier. This input has become more important with the decline of water released from Lake Kinneret. CO_2 exchange with the atmosphere at the surface of the Dead Sea does not appear to have had much effect on the ^{14}C content of Dead Sea water.

Acknowledgments

We thank the various individuals who assisted with sampling and extraction of ^{14}C samples: Shlomo Shasha, Motti Gonen, Joel Kronfeld, and Ebbie Visser. We appreciate the initiatives of Joel Gat and Yitchak Levy to keep this aspect of the Dead Sea investigation going. We thank M. Andree for the AMS measurements of several Dead Sea samples and I. Carmi for critically reading the manuscript. This project was partially supported by the Israel Ministry of Energy and Infrastructure and the NCRD/CSIR scientific exchange program.

REFERENCES

Anati, D. A., and Shasha, S., 1989, Dead Sea surface–level changes: *Israel Journal of Earth Science*, v. 38, p. 29–32.
Anati, D. A., and Stiller, M., 1991, The post-1979 thermohaline

structure of the Dead Sea and the role of double-diffusive mixing: *Limnology and Oceanography*, v. 36, p. 342–354.

Anati, D. A., Stiller, M., Shasha, S., and Gat, J. R., 1987, Changes in the thermo-haline structure of the Dead Sea: 1979–1984: *Earth and Planetary Science Letters*, v. 84, p. 109–121.

Broecker, W. S., and Olson, I. A., 1961, Lamont radiocarbon measurements VIII: *Radiocarbon*, v. 3, p. 176–204.

Broecker, W. S., and Peng, T. H., 1982, *Tracers in the Sea*: New York, Lamont-Doherty Geological Observatory, 690 p.

Carmi, I., Gat, J. R., and Stiller, M., 1984, Tritium in the Dead Sea: *Earth and Planetary Science Letters*, v. 71, p. 377–389.

Carmi, I., Stiller, M., and Kaufman, A., 1985, The effect of atmospheric ^{14}C variations on the ^{14}C levels in the Jordan River system: *Radiocarbon*, v. 27(B), p. 305–313.

Craig, H., 1957, The natural distribution of radiocarbon and the exchange time of carbon dioxide between the atmosphere and sea: *Tellus*, v. 9, p. 1–17.

Levin, I., Kromer, B., Schoch-Fischer, H., Bruns, M., Münnich, M., Berdau, D., Vogel, J. C., and Münnich, K. O., 1985, 25 Years of tropospheric ^{14}C observations in Central Europe: *Radiocarbon*, v. 27, p. 1–19.

Mazor, E., and Molcho, M., 1972, Geochemical studies on the Feshkha Springs, Dead Sea basin: *Journal of Hydrology*, v. 15, p. 37–47.

Neev, D., and Emery, K. O., 1967, The Dead Sea—Depositional processes and environments of evaporites: Jerusalem, Geological Survey of Israel Bulletin 41, 147 p.

Nishry, A., and Stiller, M., 1984, Iron in the Dead Sea: *Earth and Planetary Science Letters*, v. 71, p. 405–414.

Peng, T. H., and Broecker, W. S., 1980, Gas exchange rates for three closed-basin lakes: *Limnology and Oceanography*, v. 25, p. 789–796.

Steinhorn, I., 1981, A hydrographical and physical study of the Dead Sea during the destruction of the long-term meromictic stratification [Ph.D. thesis]: Rehovot, Israel, Weizmann Institute of Science, 323 p.

Steinhorn, I., 1985, The disappearance of the long term meromictic stratification of the Dead Sea: *Limnology and Oceanography*, v. 30, p. 451–472.

Stiller, M., and Chung, Y. C., 1984, Radium in the Dead Sea—A possible tracer for the duration of meromixis: *Limnology and Oceanography*, v. 29, p. 574–586.

Stuiver, M., and Polach, H. A., 1977, Discussion—Reporting of ^{14}C data: *Radiocarbon*, v. 19, p. 355–363.

Vogel, J. C., 1967, Investigation of groundwater flow with radiocarbon, in *Isotopes in Hydrology*: Vienna, International Atomic Energy Agency, p. 355–369.

Vogel, J. C., 1980, Accuracy of the radiocarbon time scale beyond 15,000 BP: *Radiocarbon*, v. 22, p. 210–218.

Vogel, J. C., and Behrens, H., 1976, A mini counter for radiocarbon dating of small samples: *South African Journal of Science*, v. 72, p. 311.

18. IRON, MANGANESE, AND TRACE ELEMENTS IN THE DEAD SEA

Ami Nishri and Mariana Stiller

As a highly saline terminal lake that is characterized by relatively low pH values and restricted biological activity, the Dead Sea represents a unique environment for the study of the geochemical cycles of minor and trace elements. The geochemistry of these elements in the Dead Sea (DS) was first studied by Nissenbaum (1969, 1977) during the 1960s, and his work served as a basis for further studies. The period between 1976 and 1979 was characterized by a gradual deepening of the pycnocline, which had permanently separated the upper and lower water layers for about 260 years.

The disappearance of the pycnocline and complete turnover of the water masses in the winter of 1978–1979 allowed the introduction of oxygen into previously anoxic water layers. This event presented a unique opportunity to follow the geochemical cycles of various Eh-dependent minor elements, such as iron (Fe) and manganese (Mn). Following the overturn, the Dead Sea became saturated with respect to halite throughout the water column, and massive precipitation of this mineral started in 1983. This process affected the distribution of trace elements in the lake waters. In this chapter, we study the removal of Zn, Cu, Cd, and Pb from the water column and their coprecipitation with halite.

IRON IN THE DEAD SEA

The geochemistry of iron in stratified water bodies is strongly affected by the oxidation conditions prevailing along the water column. Iron in its soluble form is readily oxidized. Therefore, in oxic waters, there is practically no dissolved Fe^{+2}, and most of the iron appears as particulate trivalent Fe-oxyhydroxides. In anoxic environments, within the sediments or in the lower part of the water column, Fe-oxyhydroxides tend to be reduced, and the Fe^{+2}, that is released may then accumulate. In the presence of sufficient dissolved sulfide, insoluble FeS_x compounds may form from dissolved Fe^{+2}.

Iron during the deepening of the pycnocline and the turnover

Prior to 1979, the stratified DS contained two distinct water layers separated by a transition zone: an oxic, slightly less saline upper water mass (UWM), and a denser, anoxic and sulfide-containing lower water mass (LWM). During the winter of 1976–1977, the pycnocline separating these two layers was located between 80 m and 100 m. The continuous increase in salinity of the UWM narrowed the density difference between the layers, causing the pycnocline to deepen gradually, and during the winter of 1978–1979, a complete turnover of the water masses took place (Steinhorn et al., 1979) after 260 years of permanent stratification (Stiller and Chung, 1984). The changes in the redox conditions of the lower water stratum affected the spatial and temporal distribution of iron. The behavior of iron during stages of pycnocline deepening, at overturn, and shortly thereafter were studied by Nishri and Stiller (1984).

Between 1977 and the winter of 1979, during the stages of successive pycnocline deepenings, no Fe^{+2} was detected in the UWM, whereas the LWM contained about 340 µg/l. The disappearance of Fe^{+2} from the newly oxidized water strata at the zone of the pycnocline deepening was attributed to its oxidation. In addition to the more common phenomenon of Fe^{+2} disappearance from the uppermost layers of LWM, there was also a substantial decrease in Fe^{+2} levels in the lowest part of the LWM. This was attributed to oxidation of Fe^{+2} caused by sinking of highly evaporated, oxic surface waters that spread onto the bottom of the lake. The relatively higher activity of tritium measured in these bottom layers, compared with the rest of the LWM (Carmi et al., 1984), supported this assumption.

Oxidation of Fe^{+2} in the Dead Sea

As in other natural water systems where autoxidation may be ignored, the kinetics of Fe^{+2} oxidation in Dead Sea water (DSW) were assumed to be described by $-dFe^{+2}/dt = k\ pO_2\ a^2_{OH^-} Fe^{+2}$ (Nishri and Stiller, 1984), where k is the rate constant and pO_2 is the partial pressure of oxygen; the strong dependence upon pH is expressed by $a^2_{OH^-}$. Dead Sea waters are characterized by pH values of 6.0–6.35; hence, in comparison to sea water (pH 8.3), the rate of oxidation in DSW should be slower by about four orders of magnitude. The relatively long half-life of Fe^{+2} of about 820 min (Nishri and Stiller, 1984), determined experimentally for DSW, reflects the anticipated slow rate of oxidation.

The magnitude of k, 7.2×10^{-13} mole^{-2} atm^{-1} min^{-1}, is surprisingly similar to the k value measured by Singer and Stumm (1970) for HCO_3^- solutions, but it is about two orders of magnitude larger than the k values measured in natural waters (Kester et al., 1975; Murray and Gill, 1978).

Sedimentation of the particulate oxidation products from previously anoxic layers, and of particulate Fe in general, has been shown to be a relatively slow process (Nishri and Stiller, 1984). This was attributed to low settling velocities within the highly dense (1.23 g/cm^3) and viscous (about 3.1 centipoise at 22°C) DSW. This is also the reason that post-turnover DSW contains fivefold more particulate iron than any freshwater lake with comparable dimensions and input fluxes.

The iron budget

The iron budget in the Dead Sea was evaluated by a two-box model (Nishri and Stiller, 1984), taking into account fluxes between the UWM and the LWM. We assume that prior to 1976 a steady state inventory prevailed. The flux of terrigenous iron has been estimated to be less than 4,000 ton/yr (Nishri and Stiller, 1984). Given the morphometry of the DS, about 1,600 ton/yr of this terrigenous iron settles to the bed sediments underlying the UWM, and the rest, about 2,400 ton/yr, settles to the LWM. The residence time of particulate iron (presumably iron sulfide) in the LWM was assumed to be identical to that of ^{210}Pb, because ^{210}Pb is believed to settle by sorption to particulate FeS (Nishri and Stiller, 1984; Stiller and Kaufman, 1984). Thus, with a mean residence time for particulate iron in the LWM of $T = 4.4$ yr and a standing crop of about 48,000 tons, we estimate the overall sedimentation flux of particulate iron from the LWM to be about 10,800 ton/yr. Upward transport from the LWM to the UWM, by diffusion across the pycnocline,

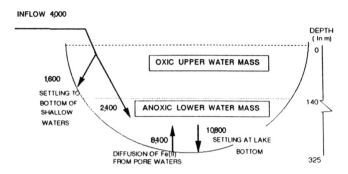

Figure 18-1 Schematic representation of major iron fluxes (ton/yr) in the Dead Sea, March 1977.

is relatively negligible (about 17 ton/yr), compared to fluxes of sedimentation from the UWM to the LWM and from the LWM to the bottom sediments (Fig. 18-1).

The steady state inventories of iron in the LWM were maintained by a large (8,400 ton/yr or 1 x 10⁻⁶ µM/cm²/sec) flux of Fe^{+2} diffusing from pore waters contained in bed sediments underlying the LWM. A similar flux (0.95 x 10⁻⁶ µM/cm²/sec) was evaluated independently by using Fick's first law and assuming a value for the molecular diffusion coefficient of 6.59 x 10⁻⁶ cm²/sec, identical to that of an infinitely diluted solution.

In summary, in the meromictic DS before the turnover, the major flux of iron into the DS was from diffusion of Fe^{+2} from the bed sediments into the LWM, where part of it formed FeS_x particulates. The major fluxes of iron removal were from sedimentation of FeS_x from the LWM and sedimentation of terrigenous iron all along the water column. In the holomictic DS as long as the LWM remains oxic, the cycling fluxes of iron and its standing crop will diminish substantially.

MANGANESE IN THE DEAD SEA

Chemistry of manganese in the Dead Sea

We expect the geochemistry of manganese, like that of iron, to be affected by the oxic conditions prevailing along the water column. Manganese carbonates are sometimes thought to be a major removal mechanism of manganese from natural waters. However, in contrast to iron, high concentrations of dissolved Mn in the oxic UWM of the DS were measured (Nissenbaum, 1977; Nishri, 1983, 1984). In the late 1970s, the average concentration of dissolved manganese in the Dead Sea was about 7.0 mg/l (measured by atomic absorption spectrometry (AAS); Nishri, 1983, 1984). Electron spin resonance (ESR) spectrometry of undiluted DSW spiked with $MnCl_2$ has shown (Nishri, 1984) that the peak height of the ESR signal is directly proportional to the dissolved manganese concentration and that the concentration measured by ESR is identical (±2%) to that measured by the AAS technique. Since ESR spectrometry measures only divalent Mn, it follows that most (>99%) of the manganese dissolved in DSW appears as Mn^{+2}. Particulate Mn (>0.45 µm, presumably Mn^{+4}) concentration measured in the DS (Nishri, 1984) is lower than 4 µg/l.

This very high concentration of dissolved Mn^{+2} raises the question of whether there is a solubility control for Mn^{+2} in DSW or whether it may further accumulate. Minerals that may precipitate Mn^{+2} are carbonates and oxides of manganese. Another possibility discussed in this chapter is the coprecipitation of Mn with aragonite, which is the Ca-carbonate mineral precipitating in this system.

The deepening of the pycnocline from 1976 to 1979, which was followed by introduction of oxygen into previously anoxic layers, did not cause the removal of Mn^{+2} from the water column (Fig. 18-2). The rate of oxidation of Mn^{+2} in DSW was determined experimentally at pH = 6.25, Mn^{+2} =7.0 mg/l, and T = 23°C. The rate of oxidation of Mn^{+2}, similar to that of Fe^{+2}, is described by first-order kinetics and is extremely slow, pK < -10.3 min⁻¹ (Nishri, 1984). According to this K value, at least 1,000 yr would be required (at pH 6.25) for detectable (-2%) oxidation of Mn^{+2} in the Dead Sea. However, as we show subsequently, in situ K values are somewhat larger.

Budget calculations

The Mn^{+2} flux from external sources (Jordan River) is estimated to be approx. 70 ton/yr (Nishri, 1984). A chemical analysis of the deep basin sediments shows that the average concentration of Mn is about 200 ppm (Nishri, 1983). If we assume that the rate of sedimentation in the Dead Sea is about 25 mg/cm²/yr (Stiller et al., chapter 15, this volume), it follows that the net Mn sedimentation in the DS amounts to about 40 ton/yr. This flux represents less than 0.004% of the present inventory of Mn^{+2} in the DS, of about 10⁶ tons, and implies that all forms of Mn removal to the sediments (in terms of yearly fluxes) are almost negligible. The net Mn input, that is, the surplus of input from external sources versus burial within the bed sediments, is similar to the net Mn sedimentation, about several tens of tons a year. If we assume steady state fluxes, this net Mn input suggests that Mn^{+2} accumulation within the DS took place over more than 10,000 years.

The effect of dilution on oxidation rate

Dilution of DSW by freshwater brings about an increase in pH, as well as a shift in the extent of ion complexation and ion pairing, which are well expressed by the calculated activity coefficients of the dissolved ions (Krumgalz and Millero, 1982). For instance, a 12.5-fold dilution increases the pH from 6.25 to 8.2 (Nishri, 1983). Dilution may affect the geochemistry of manganese not only through its effect on the activity of this ion but also because its rate of oxidation increases with pH. The effect of natural dilution processes on the in situ rate of Mn^{+2} oxidation is shown for the following case.

Stiller and Chung (1984) have estimated that, until the turnover of 1978–1979, the lake had been permanently stratified for about 260 years; in other words, the Jordan River inflow and occasional floods produced fluctuations of salinity only within the UWM. In the early 1960s most of the UWM (its upper part) was about 10% more diluted than the LWM, whereas the difference in Mn^{+2} concentration between the layers was about 60%: In the oxic UWM, the concentration of Mn^{+2} was approx. 4.5 mg/l, and in the anoxic, H_2S-containing LWM, it was 7.5 mg/l (Nissenbaum, 1969, 1977).

According to measurements of diluted DSW (Nishri, 1983), the average pH of the UWM in the 1960s must have been within the range 6.3–6.45, that is, 0.1–0.2 pH units higher than at the end of the 1970s. More than a decade later, shortly before the 1979 overturn, the salinity difference between the two water masses almost disappeared. At this time, the UWM contained about 6.0 mg Mn/l, compared to 7.5 mg/l in the slightly more saline LWM (Nishri, 1984). There was no detectable difference (±2%) in the overall inventories of Mn^{+2} in the Dead Sea between the 1960s and the 1970s. The increase in concentration observed in the UWM in the late 1970s was due to mixing of the UWM with layers from the LWM that were richer in Mn^{+2}. The redistribution of manganese along the water column during this decade was not affected by oxidation-reduction processes but was the result of destratification. Thus, over several years and at a pH of 6.15–6.3, it was not possible to detect oxidation followed by sedimentation of the Mn products to the bottom of the lake.

The dilution ratio observed in the UWM clearly shows that the vertical distribution of Mn^{+2} in the early 1960s was due not only to dilution but to additional geochemical processes. The 1960s vertical profile of Mn^{+2} thus seems to represent 260 years of meromixis during which redistribution of Mn along the water column took place. Slow oxidation that proceeded within the UWM gradually diminished its dissolved Mn^{+2} concentration. Redissolution of most of the tetravalent Mn settling products within the anoxic LWM gradually increased its Mn^{+2} concentration. If it is true, then substantial oxidation of Mn^{+2} because of the occasional freshening of the UWM can be detected only on a time scale of hundreds of years in the Dead Sea. The environmental K calculated accordingly is about an order of magnitude larger than the experimental one.

Other removal mechanisms

Over 75% (about 150 ppm) of the Mn buried in the deep basin bed sediments is associated with the carbonate phase (Nishri, 1983). This may suggest that Mn is removed from the water column either by a Mn-carbonate mineral or by coprecipitation with aragonite. One plausible mineral candidate for which the degree of saturation (DOS) in DSW was examined is $MnCO_3$, that is, rhodochrosite. Using Pitzer equations, Krumgalz (pers. com., 1982) calculated that the activity coefficient of Mn^{+2} in DSW is equal to 0.51 and that of Ca^{+2} is equal to 0.879. Because the concentration of Mn^{+2} in DSW is equal to 1.27×10^{-4} M, the activity of the free ion, $a_{Mn^{+2}}$, should be equal to $10^{-4.19}$ M. Similarly, because the concentration of Ca^{+2} is equal to 0.425 M, $a_{Ca^{+2}}$ is equal to 0.373 M.

Krumgalz also evaluated the activity coefficients of CO_3^{-2}; however, this value is still a matter of debate because the possible contribution of borate species to the titration alkalinity was ignored. Hence, as yet, we cannot directly estimate the degree of saturation (DOS) of DSW with respect to rhodochrosite or aragonite. However, because aragonite does precipitate in the DS (Neev and Emery, 1967), it seems reasonable to set a lower boundary for $a_{CO_3^{-2}}$ by assuming that DOS (aragonite) is equal to or greater than 1. Accordingly, considering that the equilibrium constant of aragonite is equal to $10^{-8.25}$, the lower boundary estimated for $a_{CO_3^{-2}}$ in DSW is $10^{-7.8}$ (Nishri, 1983). In this case, the ion activity product of $MnCO_3$ in DSW is equal to 10^{-12}. The solubility constant of rhodochrosite is equal to $10^{-10.4}$ (Li and Gregory, 1974), so it seems that the DS is about 40 times undersaturated in this mineral.

Because there is evidently Mn bound to carbonate minerals in the sedimentary record, coprecipitation of Mn with aragonite should also be considered as a removal mechanism. A corresponding sedimentation flux of about 10 ton/yr is obtained by multiplying the average concentration of Mn bound to sedimentary Ca-carbonate (about 150 ppm), by the net rate of sedimentation in the central part of the DS (25 $mg/cm^2/yr$; Stiller et al., chapter 15, this volume), by the average concentration of sedimentary aragonite (30%; Garber, 1980), and by the lake surface (800 km^2).

Sources of manganese in the Dead Sea

The average annual direct supply of Mn^{+2} to the DS via the Jordan River was estimated (Nishri, 1984) to be less than 1 ton/yr. The supply of Mn as solid particles from this source (ca. 70 ton/year) was evaluated by multiplying the average Mn concentration in suspended matter of the river (300 ppm) with the aver-

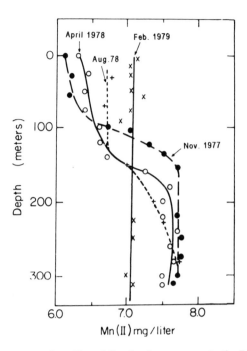

Figure 18-2 Vertical profiles of dissolved manganese in the Dead Sea during 1977–1979.

age input of allochthonous sediments (about 2.5×10^5 ton/yr). The Mn^{+2} contribution from saline water sources is unknown. The upward diffusional flux of Mn^{+2} from bottom sediments to the overlying lake water, $F(Mn^{+2})$, was evaluated (Nishri, 1983) by using Fick's first law, measured Mn^{+2} concentration gradients in pore waters, and a value of 1.3×10^{-7} cm^2/sec for the diffusion coefficient, $D(Mn^{+2})$ in this saline medium. A flux of 31 ton/yr was calculated for the deep basin sediments. Thus, the yearly Mn contribution to the DS from known sources is approx. 100 tons. Removal by coprecipitation (as discussed previously), is only about 10 ton/yr. This implies that the maximum net increase in inventory (assuming that all of the manganese associated with allochthonous material and incoming from the Jordan River dissolves within the lake waters) is on the order of 90 ton/yr—about 0.01% of the present-day inventory in the lake. In other words, under the present climatological and morphological conditions, the increase in the inventory of Mn^{+2} in the DS is extremely slow.

Manganese oxide occurrences along the shoreline

Manganese-containing deposits have been found in various localities along the western margin of the Dead Sea (Garber, 1980; Druckman, 1981). The drop in sea level during the last 35 years has exposed a pavement, parallel to the shoreline, composed of pebbles and boulders of detrital origin cemented by a finely varved aragonitic crust. In some places, the aragonite layers contain black laminae consisting of approximately 7% manganese by weight (Garber, 1980). The black surface crusts are often a manifestation of a layered structure with alternating white aragonite laminae and black Mn-enriched layers (Fig. 18-3). Infrared spectra of the black material (Nishri and Nissenbaum, 1993) suggests that it consists of MnO_2.

Druckman (1981) proposed that the black Mn-enriched material is formed by algae-mediated oxidation of Mn^{+2} dissolved in the DSW. However, Garber (1980) could not identify algal structures within these layers. Ehrlich and Zapkin (1981) could not grow Mn^{+2}-oxidizing bacteria on Dead Sea sediments in a

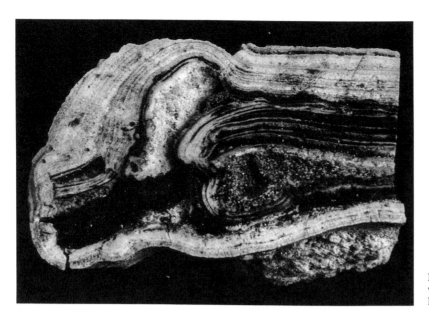

Figure 18-3 Layered structure within a boulder with alternating white aragonite laminae and black Mn-enriched layers

culture prepared in a synthetic Dead Sea water medium; they concluded that the salinity of the water was too high for these organisms to survive. Hence, there is no biologically mediated oxidation of Mn^{+2} in a Dead Sea water medium.

Ehrlich and Zapkin (1981) observed spore-forming oxidizing bacteria within the black material and proposed that, at least in part, the manganese-rich layers were formed by bacterial activity in a freshwater environment along the coast. But nonbiological oxidation of Mn^{+2} may also be favorable in this environment because of the relatively high pH of DSW and freshwater mixtures, as well as the relatively higher concentrations of Mn^{+2} prevailing in the coastal pore waters (Nishri and Nissenbaum, 1993). The rate of oxidation of Mn^{+2} increases with its concentration in DSW (Nishri, 1984). The following model for the formation of the black Mn-enriched layers was proposed (Nishri and Nissenbaum, 1993).

Mn-enriched carbonates are formed in nearshore sediments. The source of Mn is allochthonous Mn-oxides that either have been transported from the neighboring highlands or are erosional products of the coastal sediments. Once buried within the hostile (lower Eh-pH) environment of the DS sediments, these oxides are reduced, and the dissolved manganese goes into the pore water. This Mn^{+2} may diffuse up toward the sediment-water interface because of concentration gradients. If the concentration of Mn^{+2} exceeds about 270 mg/l, it may be in equilibrium with rhodochrosite (pH of pore water is $\approx$ 5.8), which could then be formed diagenetically within the sediments.

Along the coast near the outlets of wadis, there is often massive precipitation of aragonite that results from the introduction of HCO_3-bearing freshwater (floods and groundwater). Part of the Mn^{+2} that diffuses upward may coprecipitate with this aragonite. Both processes lead to the preferential accumulation of Mn-carbonates. The decline of the lake level between 1969 and 1991 caused a lakeward advance of relatively fresher groundwater. The Mn-containing DS sediments may therefore come in contact with a fresher media. Alteration through direct oxidation of Mn-carbonates to Mn-oxides, as suggested by Hager (1980), may facilitate the formation of the black layers enriched in Mn-oxide. A complementary pathway could be bacterial oxidation of manganese within the fresher media.

Further decline of the lake level provides a "fresh" source of Mn-oxides that may be subsequently eroded and transported to the "newly" formed nearshore bottom sediments. Thus, the receding lake level and the occurrence of aquiferrous ground- and floodwater at the outlets of wadis to the Dead Sea provide the appropriate environmental conditions for the formation of the Mn-oxide incrustations. This process cannot occur within the highly saline DSW.

THE HEAVY METALS: Zn, Cd, Pb, AND Cu

Published data on the vertical distribution of heavy metals (HM) prior to the Dead Sea's overturn in 1979 is scarce (Ariel and Eisner, 1963; Brooks et al., 1967; Nissenbaum, 1977; Stiller et al., 1984). In addition, agreement between the sets of data produced by various scientists is poor. Thus, in vertical profiles sampled in 1965, Cd and Pb concentrations seemed to be higher in the oxic UWM than in the anoxic LWM (Nissenbaum, 1977). In samples from the late 1970s (Stiller and Sigg, 1990), this differentiation in the Cd and Pb contents of the two water masses was not observed. Moreover, all concentrations were generally lower by an order of magnitude. Whether the decreases were real, related to scavenging by ferric oxyhydroxides formed during the destratification process, or whether the discrepancies were analytical artifacts remains the subject of debate.

Concentrations of HM in the homogeneous water column of December 1980 were in reasonable agreement with values estimated by "mixing" the measured content of HM of the stratified water column of August 1978 (Table 18-1; Stiller and Sigg, 1990). This fact indicates that the inventories of HM in the Dead Sea were almost unaffected by the turnover event in 1979. However, the inventory of Pb diminished by about 43% between 1976 and 1978, during the destratification process that preceded the turnover (Stiller and Sigg, 1990). Radioactive Pb exhibited behavior similar to that of stable Pb: Although the inventory of dissolved ^{210}Pb remained unchanged during the 1979 turnover, it decreased between 1977 and 1978 by about 36% (Stiller and Kaufman, 1984).

The data in Table 18-1 represent total HM concentrations. Because there was no systematic difference between data of

Table 18-1 Concentrations of heavy metals in the Dead Sea and in its end brine ($\mu g/l$).

Date	Depth (m)	Zn	Cd	Pb	Cu
Aug. 17, 1978	75 (UWM)	232	0.60	16.9	3.4[a]
Aug. 15, 1978	250 (LWM)	322	1.05	14.7	6.7[a]
Dec. 18, 1980	50 (M)[b]	363	0.84	17.8	6.3
May 9, 1985	80 (M)[b]	86	0.04	2.6	4.4
July 25, 1983	End brine	871	1.26	44.5	14.2

[a]At 30 m (UWM) and at 130 m (LWM) in 1976; lead concentrations at these depths in 1976 (see text) were 8.8 and 43.4 $\mu g/l$, respectively.
[b](M) = mixed, homogeneous water column.

acidified and nonacidified samples, we believe that the concentration of particulate HM was very small indeed. With regard to chemical speciation, anodic stripping voltammetry measurements indicate the presence of dissolved complexes $ZnCl_2^0$ and $CdCl_4^{-2}$, whereas for Pb and Cu, actual speciation is not well resolved (Stiller and Sigg, 1990).

Massive halite formation within the lake waters began in December 1982 and persisted during the holomictic period of 1983–1991. In contrast to the turnover that passed almost unnoticed, this abundant endogenic precipitation had a dramatic influence upon the HM concentrations. In May 1985 (Table 18-1), Cd practically vanished from the water column, Zn and Pb diminished markedly, and Cu was only slightly affected.

We examined coprecipitation of heavy metals with endogenic halite collected by sediment traps. Among the heavy metals, the Pb concentration in halite was the greatest, whereas Cd had the greatest apparent distribution coefficient, K_{app} (Table 18-2). With an estimated precipitation of about 330 x 10^6 tons of halite during 1983–1985, the disappearance of Cd from the water column is not surprising. The amount of Cd expected to coprecipitate with halite (about 127 tons) matches quite closely the Cd inventories of about 110 tons in 1976, 1978, and 1980. The diminished Pb inventory of 1985 can only partially be attributed to coprecipitation with halite, whereas the Zn and Cu losses from the water column are not at all accounted for by this process. To explain these losses, a different mechanism must be invoked. It is possible that the precipitation of halite "cleaned" the water column by disturbing the

stability of colloidal particles and triggering their settling as aggregates. If the HM were somehow attached to the surfaces of these colloidal particles, they might have been entrained and settled to the lake bottom.

The HM concentrations in the end brine (the residual solution after carnallite extraction from the Dead Sea brine) have also been measured (Table 18-1). Concentrations of Zn, Pb, and Cu are about 2.4 times greater than in the DS of 1980, whereas Cd is only 1.5 times greater. Each liter of end brine represents about 2.5 liters of DSW from which halite (250 g) and carnallite have been deposited by evaporation in solar ponds. Because of the very small distribution coefficients of Zn and Cu, the expected losses by coprecipitation with halite in the evaporation ponds are minimal and within the analytical error. It is remarkable that, as expected, the Zn and Cu enrichment factors coincide with the degree of evaporation in the end brine.

By the same reasoning, Cd and Pb, which have large distribution coefficients, should not be present at all in the end brine. But, similar to the behavior of Zn and Cu, Pb behaves conservatively, and only a partial loss is observed for Cd. The temperature dependency of K_{app} is not known. In the evaporation ponds, temperatures are usually higher (and the pH is lower) than in the Dead Sea. Perhaps under these conditions the K_{app} values for Pb and Cd are much smaller. Kinetic effects might also be involved. The evaporation rate in the solar ponds is lower than in the Dead Sea. If the rate of crystallization of halite in the ponds is also somewhat slower than in the DS, there should be less coprecipitation of HM (and smaller K_{app}). Confirmation of this hypothesis by adequate experiments is necessary.

Acknowledgments

We thank H. Foner for critical comments on the manuscript.

Table 18-2 Coprecipitation of heavy metals with halite

Metal	Halite[a] (ng/g)	K_{app}[b]
Zn	67	0.019
Cd	386	47.3
Pb	1,895	11
Cu	4.4	0.04

[a]Average of three measurements on halite collected by sediment traps in 1983 (Stiller and Sigg, 1990).
[b]Apparent distribution coefficient,
$K_{app} = [HM/Na]_{halite}/[HM/Na]_{DS}$

REFERENCES

Ariel, M., and Eisner, U., 1963, Trace analysis by anodic stripping voltammetry: 1. Trace metals in the Dead Sea brine—Zn and Cd: *Journal of Electroanalytical Chemistry*, v. 5, p. 362–374.

Brooks, R. R., Presley, J. B., and Kaplan, I. R., 1967, APDC–MIBK extraction system for the determination of trace metals in saline waters by atomic absorption spectrophotometry: *Talanta*, v. 14, p. 809–816.

Carmi. I., Gat, J. R., and Stiller, M., 1984, Tritium in the Dead Sea: *Earth and Planetary Science Letters*, v. 71, p. 377–389.

Druckman, Y., 1981, Subrecent manganese bearing stromatolites along the shorelines of the Dead Sea *in, Phanerozoic stromatolites*, Monthy, C., ed.: Berlin, Springer, p. 197–208.

Ehrlich, H. L., and Zapkin, M. A., 1981, Mn^{+2} oxidizing bacteria from the Dead Sea region of Israel (Abs.): Annual Meeting, American Society of Microbiologists, MGO, p. 183.

Garber, B., 1980, The sedimentology of the Dead Sea [Ph.D. thesis]: Troy, N.Y., Rensselaer Polytechnical Institute, 169 p.

Hager, L. H., 1980, Sorption of manganese and silica by clay and carbonate: *Marine Chemistry*, v. 9, p. 194–209.

Kester, D. R., Byrne, R. H., and Liang, Y. J., 1975, Redox reactions and solution complexes of iron in marine systems, *in* Church, T. M., ed., Marine chemistry in the Coastal Environment: American Chemical Society Symposium Series, v. 18. p. 23.

Krumgalz, B. S., and Millero, F. J., 1982, A physico chemical study of the Dead Sea waters: *Marine Chemistry*, v. 11, p. 209–222.

Li, Y. H., and Gregory, S., 1974, Diffusion of ions in sea water and in deep-sea sediments, *Geochimica et Cosmochimica Acta* v. 38, p. 703–714.

Murray, J. W., and Gill, G., 1978, The geochemistry of iron in Puget Sound: *Geochimica et Cosmochimica Acta*, v. 42, p. 9–19.

Nissenbaum, A., 1969, The geochemistry of the Jordan River–Dead Sea system [Ph.D. thesis]: Los Angeles, California, University of California, p. 288.

Nissenbaum, A., 1977, Minor and trace elements in Dead Sea water: *Chemical Geology*, v. 19, p. 99–111.

Nishri, A., 1983, The geochemistry of Manganese and Iron in the Dead Sea [Ph.D. thesis], Rehovot, Israel, The Weizmann Institute of Science, 149 p.

Nishri, A., 1984, The geochemistry of manganese in the Dead Sea: *Earth and Planetary Science Letters*, v. 71, p. 415–426.

Nishri, A., and Nissenbaum, A., 1993, Formation of manganese oxyhydroxides on the Dead Sea coast by alteration of Mn-enriched carbonates: *Hydrobiologia*, v. 267, p. 61–73.

Nishri, A., and Stiller, M., 1984, Iron in the Dead Sea: *Earth and Planetary Science Letters*, v. 71, p. 405–414.

Singer, P. C., and Stumm, W., 1970, Acidic mine drainage—The rate determining step, *Science*, v. 167, p. 1,121–1,123.

Steinhorn, I., Assaf, G., Gat, J. R., Nishri, A., Nissenbaum, A., Stiller, M., Beith, M., Neev, D., Garber, R., Friedman, G. M., and Weiss, W., 1979, The Dead Sea—Deepening of the mixolimnion signifies the overture to overturn of the water column: *Science*, v. 206, p. 55–57.

Stiller, M., and Chung, Y. C., 1984, Radium in the Dead Sea—A possible tracer for the duration of meromixis: *Limnology and Oceanography*, v. 29, p. 576–586.

Stiller, M., and Kaufman, A., 1984, ^{210}Pb and ^{210}Po during the destruction of stratification in the Dead Sea: *Earth and Planetary Science Letters*, v. 71, p. 390–404.

Stiller, M., Mantel, M., and Rapaport, M. S., 1984, The determination of trace elements (Co, Cu, and Hg) in the Dead Sea by neutron activation followed by x-ray spectrometry and magnetic deflection of beta-ray interference: *Journal of Radioanalytical Nuclear Chemistry*, v. 83, p. 345–352.

Stiller, M., and Sigg, L., 1990, Heavy metals in the Dead Sea and their coprecipitation with halite: *Hydrobiologia*, v. 197, p. 23–33.

19. MICROBIOLOGICAL STUDIES IN THE DEAD SEA: 1892-1992

Aharon Oren

LIFE IN THE DEAD SEA—HISTORICAL ASPECTS

Although it would be suitable to start a survey on the microbiology of the Dead Sea with a discussion of the pioneering work of Benjamin Elazari-Volcani (Wilkansky) in the late 1930s and early 1940s (Wilkansky, 1936; Elazari-Volcani, 1940; Volcani, 1944), the first report on the isolation of microorganisms from the Dead Sea was published more than a hundred years ago. In an investigation of the properties of Dead Sea sediments, with the intention of using these sediments as an aseptic substance, M. L. Lortet, a microbiologist working at the university of Lille (France), isolated endospore-forming anaerobic pathogenic bacteria of the genus *Clostridium* from sediment samples collected from the Dead Sea (Lortet, 1892; a French version of the paper was published in 1891). These bacterial isolates caused symptoms of tetanus and gas gangrene when the cultures were injected into experimental animals.

Though these *Clostridium* species do not show halophilic properties whatsoever and thus cannot be expected to multiply at the extremely high salinities characteristic of the Dead Sea water and sediment, their presence in the sediment is not at all surprising. Bacteria of the genera *Bacillus* and *Clostridium* are able to produce endospores that are extremely resistant to adverse conditions, such as high temperatures, prolonged desiccation, and exposure to high salt concentrations. Upon transfer to suitable, low-salt growth media, these endospores germinate and give rise to actively growing cultures. In a recent attempt to repeat the kind of experiments reported by Lortet, it was easy to isolate nonhalophilic anaerobic endospore-forming bacteria of the genus *Clostridium* from Dead Sea sediments (Oren, 1991a). However, there is no doubt that the viable endospores present in these sediments originated from bacteria that grew elsewhere, at low salt concentrations, and reached the Dead Sea at a later stage (e.g., with freshwater rain floods from the catchment area or from the air).

The existence of an endogenous community of microorganisms in the Dead Sea was first reported in 1936. Encouraged by the increasing amount of information on the existence of halophilic and halotolerant microorganisms inhabiting salt lakes all over the world, Elazari-Volcani set up a great number of enrichment cultures at high salt concentrations on a variety of growth substrates, using Dead Sea water or sediment as inoculum. In the course of his studies, a variety of microorganisms were isolated from the Dead Sea, comprising both bacteria (halophilic, halotolerant, and also nonhalotolerant types), eukaryotic algae (*Dunaliella* and possibly other types of unicellular photosynthetic eukaryotes), and even different types of protozoa (Wilkansky, 1936; Elazari-Volcani, 1940; Volcani, 1944).

The positive outcome of enrichment cultures, such as those of Elazari-Volcani's studies, implies that the inoculum used contained at least one viable microorganism that was able to multiply in the enrichment medium supplied. No further quantitative conclusions can be made about the abundance of the organisms thus obtained in the natural environment. The first quantitative measurements of community densities of microorganisms in the Dead Sea were taken only as recently as 1963–1964 (Kaplan and Friedmann, 1970). The unicellular green alga *Dunaliella* and one or a few types of mostly pleomor-

phic red bacteria were found to be the dominant types, and these were present in very high numbers: Populations of up to 4×10^4 *Dunaliella* cells were reported per ml of surface water (1964, sampling date not specified), whereas the bacterial community density in the surface water layer was between 2.3×10^6 and 8.9×10^6 cells/ml throughout the research period (December 1963-November 1964).

Very few additional quantitative measurements have been made on the biota of the Dead Sea lake prior to 1980. This lack of data is regrettable because the properties of the lake as an ecosystem have changed considerably during the years. The water balance has been negative since the beginning of this century, and this has caused extensive changes in salt concentrations, ionic composition, and the physical structure of the water column. All these parameters have a profound influence on the distribution of the microorganisms and their activities, as will be discussed subsequently. A systematic study, both qualitative and quantitative, of the microbiology of the Dead Sea has been conducted by the Division of Microbial and Molecular Ecology of the Hebrew University since the beginning of 1980.

This chapter presents an overview of the information that has accumulated on the biology of the Dead Sea in the hundred years that have passed since Lortet described the isolation of pathogenic clostridia from sediments collected from the lake. Additional data may be found in a number of review papers on the microbiology of the Dead Sea (Nissenbaum, 1975; Larsen, 1980; Oren, 1988, 1993).

THE MICROORGANISMS OF THE DEAD SEA

Among the microorganisms isolated from the Dead Sea, we find representatives of each of the three kingdoms in which living organisms are presently classified: the eukaryotes (eukarya), the archaeobacteria (archaea), and the eubacteria (bacteria).

The most important eukaryotic organism found in the Dead Sea is the unicellular flagellate green alga *Dunaliella parva* (Fig. 19-1A; also described as *Dunaliella viridis* by Kaplan and Friedmann, 1970). Representatives of the genus *Dunaliella* are found worldwide in hypersaline environments, and the presence of *Dunaliella* in the Dead Sea was reported in many qualitative and quantitative studies on the biology of the lake (Elazari-Volcani, 1940; Volcani, 1944; Kaplan and Friedmann, 1970; Oren and Shilo, 1982). The alga which is either the main or only primary producer in the Dead Sea, is often found in the upper water layers in very high numbers: Peak population densities were 4×10^4 cells/ml in 1964 (Kaplan and Friedmann, 1970), 8.8×10^3 cells/ml in 1980 (Oren and Shilo, 1982), and 1.5×10^4 cells/ml in 1992 (Oren, 1993b; Oren et al., 1995a). However, during certain periods (e.g., 1982-1991), no *Dunaliella* cells were observed in the many water samples examined. The reasons why the algae develop only in certain years are discussed subsequently.

In Volcani's enrichment culture studies (Elazari-Volcani, 1940; Volcani, 1944), other types of flagellate green algae also grew. These types were never encountered in massive numbers in the lake, and their quantitative importance to the biological

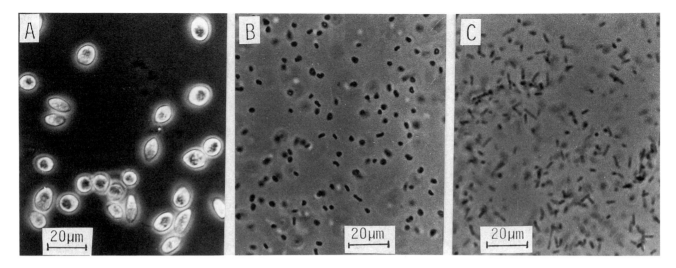

Figure 19-1 Dead Sea microorganisms: A, *Dunaliella parva*, B, *Haloferax volcanii*, and C, *Halobacterium sodomense*.

processes in the Dead Sea is probably negligible. The same is probably true for protozoa. Several types of halophilic ciliate and amoeboid protozoa were reported in enrichment cultures set up with water and sediment samples from the Dead Sea (Volcani, 1944). Direct examination of water samples has never yet shown such protozoa to be present in detectable numbers. Therefore, although the theoretical possibility exists that protozoa may be involved in controlling bacterial and algal community densities in the Dead Sea, no evidence for such a process has yet been obtained.

The prokaryotic community of the Dead Sea is dominated by extremely halophilic archaeobacteria of the family Halobacteriaceae. These bacteria are colored red because of a high content of C-50 carotenoid pigments (α-bacterioruberin derivatives) in their membranes. These pigments protect the cells against damage by high light intensities. Halophilic archaeobacteria are at times present in the Dead Sea in numbers high enough to impart a reddish color to the entire lake. Such a red coloration of the Dead Sea was reported in 1964 (Kaplan and Friedmann, 1970), in 1980 (Oren, 1983a), and in 1992 (Oren, 1993b).

Different types of halophilic red archaeobacteria isolated from the Dead Sea since Volcani's pioneering studies have been described as new species: *Halobacterium marismortui* (Volcani, 1944), *Halobacterium volcanii* (Mullakhanbhai and Larsen, 1975), and *Halobacterium sodomense* (Oren, 1983b). During recent years, the taxonomy of the Halobacteriaceae has undergone a thorough revision, based among other things on profound differences in the types of polar lipids present in the different species (Torreblanca *et al.*, 1986). Using the presently recognized classification, the previously mentioned *Halobacterium* species should be referred to as *Haloarcula marismortui*, *Haloferax volcanii*, and *Halobacterium sodomense*, respectively.

Haloarcula ("*Halobacterium*") *marismortui* was the first red archaeobacterium isolated from the Dead Sea (Volcani, 1944). The isolate was reported to grow at NaCl concentrations from 18% to saturation and was further characterized by a rapid reduction of nitrate. Unfortunately, Volcani's original culture has not been preserved. In the late 1960s, a red bacterium was isolated from the Dead Sea by Ginzburg et al. (1970). The properties of this organism closely fit the original species description of "*Halobacterium marismortui*" (Oren et al., 1988), enabling

a valid description of the species *Haloarcula marismortui* based on this new isolate (Oren et al., 1990).

Haloferax ("*Halobacterium*") *volcanii* (Fig. 19-1B) isolated from Dead Sea sediments by Mullakhanbhai and Larsen (1975), was named after Benjamin Elazari-Volcani, the pioneer of microbiological studies in the Dead Sea. This bacterium is extremely pleomorphic, and most cells are flat and irregularly shaped. It has a relatively low requirement for sodium (a property shared with the other representatives of the genus *Haloferax*), with an optimum NaCl concentration of 2–3 M. Moreover, the isolate proved extremely tolerant of high magnesium concentrations— relatively rapid growth was possible even at magnesium concentrations exceeding 1 M. This particular behavior toward divalent cations makes the strain especially suitable to life in Dead Sea brines, with their characteristic ionic composition of relatively low sodium concentrations (around 1.7–1.8 M) and very high divalent ion concentrations (about 1.8 M magnesium and 0.4 M calcium).

A third red archaeobacterial species, isolated from the Dead Sea and described as a new species, is *Halobacterium sodomense* (Oren, 1983b). Its successful isolation was enabled by supplementing growth media with starch, as suggested by Kritzman et al. (1973). *Halobacterium sodomense* is rod-shaped (Fig. 19-1C) and has the highest requirement for magnesium of all species known thus far; it grows optimally at a magnesium concentration of about 0.8 M. In addition, the isolate is even more tolerant of high magnesium concentrations than *Haloferax volcanii*. Under certain specific growth conditions, *Halobacterium sodomense* produces the purple pigment bacteriorhodopsin, a retinal-containing protein located in purple patches in the cell membrane. This pigment acts as a light-driven outward proton pump, enabling the cell to directly utilize energy of light of the proper wavelength (maximum absorption at 570 nm) for bioenergetic processes, such as the synthesis of adenosine triphosphate. Presence of bacteriorhodopsin in the bacterial community in the Dead Sea may be of great importance to enable bacterial survival during periods in which organic nutrients are in short supply, as will be discussed subsequently.

In addition to these three species of pleomorphic and rod-shaped red archaeobacteria, the coccoid species *Halococcus morrhuae* has been found in the Dead Sea (Elazari-Volcani, 1940).

Although from a quantitative point of view, the archaeobacteria are the dominant prokaryotes in the Dead Sea, it is relatively easy to isolate halophilic or halotolerant eubacteria from the lake. Enrichment cultures yielded a variety of such organisms, which for the most part have not yet been properly described (Volcani, 1944; Oren, 1981, 1988). Only in recent years have valid descriptions of some of these strains been published (Ventosa et al., 1989). In addition, it is possible to isolate from the Dead Sea nonhalophilic bacteria that are unable to grow at the high salt concentrations present in Dead Sea water. Such organisms may have developed elsewhere and reached the Dead Sea via the air or inflowing freshwater. The anaerobic, pathogenic, tetanus-causing and gas gangrene-causing *Clostridium* strains found by Lortet (1892) in Dead Sea sediments belong to this category. Aerobic endospore-forming bacteria of the genus *Bacillus* were isolated from the surface of the lake (Volcani, 1944). Even nonhalophilic bacteria that do not produce resistant endospores may survive in the hostile environment of the Dead Sea for a number of hours, as shown in a study in which the numbers of viable cells of *Escherichia coli* (a nonhalophilic bacterium) and *Vibrio harveyi* (a marine bacterium) were measured after different periods of suspension in Dead Sea water (Oren and Vlodavsky, 1985).

THE MICROBIOLOGY OF THE DEAD SEA— QUANTITATIVE ASPECTS

The *Dunaliella* population of the Dead Sea

As stated previously, very few quantitative data on the microbial communities in the Dead Sea were collected before 1980. Very little data are available on the size of the *Dunaliella* population before the 1979 overturn, during which the lake was "permanently" stratified. During the first quantitative study of life in the Dead Sea in the years 1963–1964, *Dunaliella* cells were enumerated only once (sampling date not specified), and numbers of 4×10^4 cells/ml were measured in surface water, decreasing by about 100-fold at a 50-m depth (Kaplan and Friedmann, 1970).

Extensive quantitative data are available on the microbial communities in the Dead Sea since 1980. A complete database exists on physical parameters, such as salinity and temperature, at different depths at the various sampling stations over time, enabling us to relate the biological phenomena to changes in the physical structure of the water column. At the beginning of 1979, the lake underwent a complete overturn, ending a long period of meromixis (Stiller et al., 1984). However, massive rain floods in the winter of 1979–1980 caused the formation of a new, relatively diluted surface water layer, which was separated from the more saline brines below by a pycnocline at a depth of 10 to 25 m (Anati and Shasha, 1989). Thus, another meromictic period was initiated, lasting until the end of 1982 (Stiller et al., 1984).

A mass development of the green unicellular alga *Dunaliella parva* (up to 8.8×10^3 cells/ml) was observed in the summer of 1980. This bloom was restricted to the upper water layer above the pycnocline. In this mixed layer, algal densities were quite uniform. The *Dunaliella* community declined rapidly at the end of the summer, and, by the end of the year, algal numbers had decreased to below 10^3 cells/ml (Oren, 1981, 1986a; Oren and Shilo, 1982).

Summer blooms of *Dunaliella*, such as those observed in 1980, are not an annually recurring phenomenon: no such blooms were observed in the years 1981–1991. Results of simulation experiments in the laboratory and in outdoor tanks indicate that two conditions are required for the development of

Dunaliella in Dead Sea water: (1) Phosphate must be available and (2) the salinity of the brine should yield specific gravity values below 1.21–1.22 g/ml (compared with a value of about 1.235 in undiluted Dead Sea waters; Oren and Shilo, 1985; Oren, 1986a). The large amounts of freshwater that entered the lake during the winter of 1979–1980 caused the formation of an upper layer, with a specific gravity of 1.20–1.22 g/ml, in which the *Dunaliella* population multiplied (Oren and Shilo, 1982).

During the years 1981–1991, the amounts of freshwater entering the Dead Sea during each winter was relatively small. The specific gravity of the upper water layer of the Dead Sea at the end of the winter rainy season never reached values below 1.22 g/ml, and therefore *Dunaliella* did not grow. As a result of the continuing negative water balance of the lake, a new overturn occurred at the end of 1982. This was followed by a period of nine years in which the Dead Sea was holomictic, with an annual overturn at the end of each summer (Anati and Shasha, 1989). Salinity values during this period were much higher than those conducive to the growth of *Dunaliella*.

In addition to supraoptimal salinities, lack of phosphate may also limit the development of *Dunaliella* in the Dead Sea. Laboratory simulation experiments showed that in Dead Sea water, diluted to yield sufficiently low salinity values, the yield of *Dunaliella* was proportional to the concentration of phosphate added (Oren and Shilo, 1985; Oren, 1986a). Data on the actual phosphate concentrations at different depths in the water column of the Dead Sea over time are lacking, but it is more than probable that the masses of freshwater entering during the winter from the Jordan River and as runoff from the catchment area, add substantial amounts of phosphate.

Unprecedentedly heavy rain floods during the winter of 1991–1992 caused a rise of almost 2 m in the water level of the Dead Sea. The lake became stratified again with a pycnocline at a depth of 4–5 m, isolating an upper layer of water with a specific gravity of 1.17–1.18 g/ml during the months April–June 1992. As expected from previous field observations and simulation experiments, *Dunaliella* developed again in the Dead Sea, with peak population densities even higher than those attained in 1980: Values of $1.2–1.5 \times 10^4$ cells/ml were found throughout the upper water layer in May 1992 (Oren, 1993a; Oren et al., 1995a). The 1992 *Dunaliella* bloom developed much earlier in the season than the 1980 bloom and reached higher peak values, but it also declined much faster. At the end of June 1992, less than 40 cells/ml were left in the upper water mass. During the decline of the bloom, thick-walled cells that morphologically resembled descriptions of zygotes of *Dunaliella* were frequently observed in the water samples (Oren et al., 1995a).

An unusual phenomenon, never reported before in the Dead Sea, was observed at the end of the summer of 1992. After the decline of the *Dunaliella* population in the surface layer above the pycnocline, renewed growth of *Dunaliella* occurred, but this time at a depth of 6–10 m at the lower end of the pycnocline. A maximal population density of 1,850 cells/ml was measured at an 8-m depth in September 1992 (Oren, 1993b; Oren et al., 1995a). The nature of this deep algal maximum is still far from understood. At these depths, the light intensity is very low because of the turbidity caused by the dense bacterial community present in the upper water layer. Measurements during the bacterial bloom in 1980 showed a tenfold decrease in light intensity for every 2 m of depth (Oren and Shilo, 1982). Moreover, simulation studies have shown that, at the salinities present (specific gravity 1.22-1.23 g/ml and higher), growth of *Dunaliella* is negligible. One hypothesis is that the development of the algae in the deeper layers may be related to the availability of nutrients, possibly phosphate. The possibility can not be excluded that the *Dunaliella* cells may migrate and may at times be found closer to the surface, where the salinity is lower and more light is available for photosynthesis. However, for technical reasons, sampling was done only around noon.

Reliable estimates of primary production in the Dead Sea are still lacking. This is due to logistic reasons but also to more fundamental questions about the methodology to be used. The commonly used methods to measure primary production are based on incorporation of $^{14}CO_2$, followed by filtration and determination of the amount of label retained in the particulate fraction. These methods may lead to a gross underestimation of primary production. *Dunaliella* cells are very fragile—they burst upon filtration and drying of the filters, thereby releasing the intracellular glycerol that may represent a substantial fraction of the newly produced organic carbon (Goldman and Dennett, 1985; Oren, 1993c). The few attempts that have been reported in the literature to measure primary production in other hypersaline lakes dominated by *Dunaliella* do not take this problem into account.

The bacterial community in the Dead Sea water column

As in the case of *Dunaliella*, the first quantitative estimates of bacterial community densities in the Dead Sea were made during the years 1963–1964. Between December 1963 and November 1964, bacteria were enumerated microscopically. Bacterial numbers at the surface of the lake increased from 2.3×10^6 cells/ml (March 1964) to a peak value of 8.9×10^6 cells/ml (November 1964). Bacterial densities declined with depth: At a depth of 50 m, bacterial counts were $2–3 \times 10^6$ cells/ml, whereas at 100 m, values of between $6–8 \times 10^5$ cells/ml were reported (Kaplan and Friedmann, 1970). Because no record exists on the physical structure of the water column during the period in which these measurements were performed, and because data on salinities and stratification patterns are lacking, it is difficult to draw any conclusions about the dynamics of the bacterial community during the research period. Kritzman and coworkers reported much lower bacterial numbers in 1973 (?) $1.5–6 \times 10^4$ cells/ml surface water, sampling date not specified); attempts to enumerate viable bacteria according to the number of colony-forming units on agar plates yielded somewhat lower numbers, between 1.7×10^4 and 4.5×10^4 (Kritzman et al., 1973).

A mass development of bacteria (up to 2×10^7 cells/ml, as enumerated microscopically) was observed during the summer of 1980. This bloom developed concomitantly with the *Dunaliella* bloom and was likewise restricted to the upper water layer above the pycnocline (Oren, 1983a). In this mixed layer, bacterial densities were quite uniform. The dense community of red archaeobacteria imparted a reddish color to the lake. The presence of this dense bacterial community was also reflected in high rates of incorporation of radioactively labeled simple organic compounds, such as glycerol and amino acids (Oren, 1992). As stated previously, the *Dunaliella* population declined rapidly at the end of the year. Bacterial numbers also decreased somewhat at the end of 1980, but a stable community of about 5×10^6 cells/ml remained present in the upper water layer as long as the lake was in a meromictic state (Stiller et al., 1984). At the end of 1982, we witnessed a sudden decline in numbers of microscopically recognizable bacteria in the upper water layers of the Dead Sea, from around $3–4 \times 10^6$ cells/ml to less than 10^6 cells/ml. This decline was a result of the final disappearance of the stratification of the lake that originated from the winter floods in 1979–1980 and the elimination of the separated shallow layer to which the bacterial community was formerly confined (Oren, 1985).

Because the growth of the heterotrophic bacterial community depends on organic matter produced by *Dunaliella* (see subsequent discussion), no new bacterial development was observed in the years 1983–1991. However, a small but viable bacterial community maintained itself throughout the water column to the bottom during these years, as indicated by low rates of incorporation of radioactively labeled organic substrates (Oren, 1990, 1991b).

As expected, the renewed mass development of *Dunaliella* in 1992 was accompanied by a new bacterial bloom. Red pleomorphic halophilic archaeobacteria multiplied once more, to reach peak values of about 3×10^7 cells/ml, again imparting a red coloration to the lake (Oren, 1993b).

From these observations, it is obvious that the development of bacteria in the Dead Sea is closely linked to the presence and activity of a community of algae. *Dunaliella* is the only primary producer in the Dead Sea, and the heterotrophic bacteria apparently develop at the expense of organic compounds released either by healthy or by degenerating *Dunaliella* cells. Members of the family Halobacteriaceae are chemoheterotrophs that grow on simple carbon compounds, such as amino acids, sugars, and related compounds. Though the metabolic potentials of the different isolates in culture have been well characterized, little is known about the nutritional mode of these bacteria in their natural habitat.

Glycerol is expected to be one of the principal nutrients used by natural communities of halophilic archaeobacteria. Glycerol is produced in large quantities by the unicellular green algae of the genus *Dunaliella*, which accumulate photosynthetically produced glycerol intracellularly as an osmotic solute, enabling the cell to withstand the high osmotic pressure of the surrounding medium (Ben-Amotz and Avron, 1973). Glycerol is indeed readily used by members of the Halobacteriaceae as carbon and as an energy source, either alone or in combination with other substrates.

More information on the possible role of glycerol as a nutrient for the bacterial community in the Dead Sea is provided in a study on the availability and turnover of glycerol in the Dead Sea in 1992 (Oren, 1993c). Using a newly developed sensitive colorimetric assay for glycerol and other polyols, Oren (1993c) measured apparent glycerol concentrations between 18 and 27 μM in the Dead Sea during the *Dunaliella* bloom (May 1992). This chemical assay probably overestimated the real glycerol concentrations present, as shown by labeled glycerol uptake experiments with the following results: values of $[K + S_n]$ (natural concentration + affinity constant) were in the range of 0.07–1.41 μM, V_{max} values were 160–426 nmol/l/h, and turnover times were 0.45–3.3 h. These results suggest that glycerol is rapidly taken up by the bacterial community present and is subject to rapid turnover. Below the pycnocline, no *Dunaliella* cells were observed, and apparent glycerol concentrations were below the detection limit. During the months of June and July 1992, the surface *Dunaliella* bloom declined rapidly. Simultaneously, the apparent glycerol concentrations measured decreased to undetectable levels.

The presence of a dense community of red archaeobacteria in the Dead Sea during bacterial blooms, such as those observed in 1980 and 1992, is revealed by the red color of the water, caused by the presence of the characteristic carotenoids (α-bacterioruberin and derivatives). However, the red color of the bacterial biomass does not imply that halophilic eubacteria do not play a significant role in the biological processes in the Dead Sea. As stated previously, halophilic and halotolerant eubacteria can easily be isolated from the Dead Sea by means of enrichment cultures. To evaluate the contribution of archaeobacteria and halophilic or halotolerant eubacteria to the bacterial activity in the Dead Sea, specific inhibitors have been used to inhibit the activity of either group. Bile salts (taurocholate, deoxycholate) in very low concentrations cause lysis of the members of the family Halobacteriaceae (with the exception of *Halococcus*), whereas eubacteria remain intact (Kamekura et al., 1988). Bile salts can thus be used as specific inhibitors of halobacterial activity in measurements of the heterotrophic activity of the community.

Amino acid incorporation in Dead Sea water samples (collected in 1989, 1990, 1991, and during the bloom of 1992) was completely abolished by either sodium deoxycholate (25 μg/ml) or sodium taurocholate (50 μg/ml), whereas pure cultures of representative halophilic eubacteria were not significantly inhibited under these conditions (Oren, 1989, 1991b). In addition, antibiotics that inhibit protein synthesis in archaeobacteria (anisomycin, which also affects eukaryotes), eubacteria (chloramphenicol), and eukaryotes (cycloheximide) were used to differentiate between the groups. Anisomycin completely inhibited amino acid incorporation in Dead Sea water samples. In contrast, chloramphenicol caused only a 30–40% inhibition, and cycloheximide did not affect uptake rates (Oren, 1990, 1991b). The partial inhibition by chloramphenicol may be due to a minor effect of this antibiotic on halophilic archaeobacteria. All these results suggest that the heterotrophic activity in the Dead Sea, both during the bacterial blooms and during the periods in between, is mainly due to archaeobacteria.

The Dead Sea has been an inoculum source for the isolation of representatives of all four recognized genera of neutrophilic aerobic halophilic archaeobacteria: *Halobacterium*, *Haloferax*, *Haloarcula*, and *Halococcus* (Oren, 1988). Until recently, very little was known about the contribution of the different genera and species to the bacterial community in the lake. Direct microscopic examination of Dead Sea water samples during the bacterial blooms of 1980 and 1992 showed a dominance of pleomorphic flat cells (characteristic of the genera *Haloferax* and *Haloarcula*). However, when grown under suboptimal conditions, *Halobacterium* isolates often lose their normal rod shape. Enumeration of colony-forming bacteria on plates, followed by characterization of the organisms developing, is of little use because plating efficiency is generally low—characteristically two or more orders of magnitude lower than the microscopically determined bacterial numbers. Such low efficiency is by no means limited to halophilic archaeobacterial blooms, but it is generally observed in attempts to quantify natural bacterial communities by means of viable counting.

To obtain more information about the nature of the dominant bacterium during the 1992 bloom, we exploited the differences in polar lipid composition, notably the glycolipid component, between the different genera of halophilic archaeobacteria (Torreblanca et al., 1986; Tindall, 1991). Analysis of the polar lipids in the Dead Sea biomass during the bloom by means of thin-layer chromatography showed the presence of one major glycolipid corresponding to the sulfated diglycosyl diether lipid characteristic of the genus *Haloferax* (Oren and Gurevich, 1993). Though a positive identification at the species level is not possible, the data prove that *Haloferax*-type organisms dominated during the 1992 bloom. Note added in proof: We recently isolated a new type of archaeobacterium from the Dead Sea with, while phylogenetically unrelated, has a lipid composition similar to that of the genus *Haloferax*. This organism, described as *Halobaculum gomorrense* (Oren et al., 1995b) may represent the dominant bacterium in the 1992 bloom.

There were no indications of significant amounts of specific glycolipids that indicate the presence of large numbers of Dead Sea archaeobacteria, such as *Halobacterium sodomense* or *Haloarcula marismortui*, or *Halobacterium* species, such as *H. halobium*, *H. salinarium*, and *H. saccharovorum*. This result does not imply that the dominant organism in the Dead Sea during the bloom

was identical to the species *Haloferax volcanii*, isolated in the past from the Dead Sea (Mullakhanbhai and Larsen, 1975). In view of the fact that *Haloferax volcanii* is a rapidly growing species, which also grows easily as colonies on plates, it is probable that other, not-yet-isolated species of the genus may be involved. We recently isolated a slow-growing *Haloferax* strain from the bacterial bloom in 1992 and are presently characterizing the isolate because it may represent the type of bacterium that dominated in the bloom.

SURVIVAL STRATEGIES OF MICROORGANISMS IN THE DEAD SEA

I do not intend to include in this chapter an overview of our understanding of the modes of adaptation of halophilic and halotolerant microorganisms to high salt concentrations. A wealth of information on the mechanisms enabling microorganisms to cope with high salt concentrations in their environment is available elsewhere. However, the Dead Sea presents specific, unusual problems to microorganisms that develop in the lake. The Dead Sea differs from most other hypersaline habitats by the extremely high concentrations of divalent cations (more than 1.8 M magnesium and more than 0.4 M calcium). As a result of the negative water balance in the Dead Sea during recent years and the consequent precipitation of halite, the ratio of divalent to monovalent cations has even increased (from 1.154 in 1977 to 1.276 in 1991; Oren, 1992). Halophilic microorganisms are well adapted to cope with high concentrations of sodium chloride, but molar concentrations of divalent cations are generally poorly tolerated. Extremely halophilic bacteria (*Halobacterium* and related organisms) that live in high NaCl environments are adapted to water activities of 0.75–0.88. Water activities in Dead Sea brines are much lower: The water activity calculated for Dead Sea water in 1979 was 0.6685 (Krumgalz and Millero, 1982)—very close to the lowest water activity ever shown to support life.

The metabolism of monovalent ions (Na^+, K^+) in both *Dunaliella* and halophilic archaeobacteria is well known. However, the molecular mechanisms that enable them to cope with extremely high concentrations of divalent cations, such as those found in the Dead Sea brines, have hardly been investigated (Oren, 1986b). Field observations and laboratory studies have demonstrated that the alga *Dunaliella* is somewhat less tolerant of the extremely high salinity (and, specifically, the high divalent cation concentrations) of the Dead Sea brines than the halophilic archaeobacteria found in the Dead Sea. The bacteria are dependent on substrates supplied by the algae for their growth, and this implies that the bacteria will develop only when the salinity is sufficiently low to enable algal growth. When the salinity becomes too high for the *Dunaliella* cells, they disappear from the water column. They may either die and sink to the bottom or form resting stages—cyst-like zygotes—such as those observed in the summer of 1993 (Oren et al., 1995a). Perhaps the algae survive as zygotes in the sediments until conditions become suitable for a new algal bloom. However, little experimental evidence exists to support this idea.

The halophilic archaeobacteria do not form resting stages to aid in their survival during unfavorable periods. Halophilic archaeobacteria have been isolated all over the world from hypersaline lakes with greatly varying ionic compositions. Some argue that the growth characteristics of the halophilic archaeobacteria with respect to their requirement for and tolerance toward divalent cations reflects the particular environment from which the organisms were isolated (Edgerton and Brimblecombe, 1981). In the case of Dead Sea archaeobacteria, this means a marked tolerance toward high magnesium con

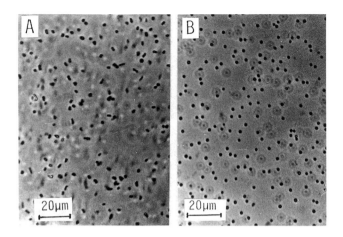

Figure 19-2 A, *Haloferax volcanii* cells; B, spherical *Haloferax volcanii* cells obtained after suspending them in medium lacking divalent cation

centrations, as shown for *Haloferax volcanii* (Mullakhanbhai and Larsen, 1975) and even more so for *Halobacterium sodomense* (Oren, 1983b). However, none of the isolates grows optimally at the high magnesium and calcium concentrations present in Dead Sea water, and all prefer much lower concentrations. Certain Dead Sea isolates of halophilic archaeobacteria show not only a high tolerance toward divalent cations, but also an unusually high requirement for magnesium for optimal growth. Organisms such as *Haloferax volcanii* not only require about 0.1–0.2 M magnesium for optimal growth, but, at lower magnesium concentrations, cells lose their native pleomorphic flat shape and form spheroplasts (Fig. 19-2), which rapidly lose their viability (Cohen et al., 1983; Oren, 1986b).

Certain halophilic archaeobacteria may survive under adverse conditions of nutrient starvation by using light energy absorbed by bacteriorhodopsin in the purple membrane. Though a wealth of literature has accumulated on the structure and function of bacteriorhodopsin, most strains of halophilic archaeobacteria are unable to produce the pigment, and those that do have the genes for bacteriorhodopsin make the purple pigment only under special conditions (low oxygen concentrations in the presence of light).

Of all the archaeobacterial isolates from the Dead Sea, only *Halobacterium sodomense* has been shown to be able to make bacteriorhodopsin (Oren, 1983b). Very few reports exist on the presence of bacteriorhodopsin in natural communities of halophilic archaeobacteria (Oren, 1992). The first such report came from the Dead Sea: The bacterial bloom present after the decline of the *Dunaliella* population at the end of 1980 had a very high bacteriorhodopsin content (Oren and Shilo, 1981). Light absorbed by bacteriorhodopsin may even drive a certain amount of CO_2 photoassimilation in the Dead Sea, as was shown in 1981 during a period in which no *Dunaliella* cells were observed in the lake (Oren, 1983c). Because the Halobacteriaceae are heterotrophs CO_2 cannot serve as the sole or main carbon source. However, light-dependent CO_2 fixation has been observed in cultures of bacteriorhodopsin-containing halobacteria, and the possible mechanisms include the carboxylation of propionyl-coenzyme A (Danon and Caplan, 1977) and the activity of glycine synthase (Javor, 1988). Recently, the presence of ribulose bisphosphate carboxylase was reported in a number of halophilic archaeobacteria (lacking bacteriorhodopsin),

including the Dead Sea isolates *Haloferax volcanii* and *Haloarcula marismortui* (Altekar and Rajagopalan, 1990), but its function in the cell is not clear. Thus, the contribution of the bacterial community to CO_2 assimilation in the Dead Sea may be greater than suspected in the past.

No bacteriorhodopsin was found in the bacterial bloom that developed in the Dead Sea in 1992—the bloom that, as discussed previously, probably consisted mainly of a not-yet-isolated type of *Haloferax*. Bacteriorhodopsin was never found to occur in any representative of the genus *Haloferax*; therefore, it is not clear whether the bacteriorhodopsin-containing bacterial bloom in 1980–1981 was dominated by the same species as the 1992 bloom.

During the period 1983–1991, the lake was holomictic, and overturns occurred each autumn. Not only did conditions never become favorable for the development of *Dunaliella* (and thus no new organic material became available from primary production processes), the conditions for the small bacterial community that remained present in the lake became even more extreme. As a result of the continuing drop in water level, halite precipitated from the lake, and the relative concentration of divalent cations steadily increased. In spite of these apparently highly unfavorable conditions, viable bacteria were present throughout this period in very low numbers, as was shown by measurements of incorporation of labeled organic compounds, such as glycerol and amino acids (Oren, 1989, 1991b, 1992). Inhibitor studies showed that this bacterial community consisted of archaeobacteria. Thus, for example, in December 1988, the number of particles microscopically resembling bacteria was about 8.4×10^5 cells/ml, probably an overestimate.

A low, but significant rate of incorporation of amino acids and other radioactively labeled substrates subsisted, which was abolished by low concentrations of bile salts (which cause lysis of halophilic archaeobacteria). Amino acid incorporation was also completely inhibited by the use of anisomycin, which inhibits halobacterial protein synthesis. Hardly anything is known about the survival mechanisms of this archaeobacterial community during years of apparent nutrient starvation. One possibility is the use of the light-driven proton pump bacteriorhodopsin as a means to derive energy from sunlight. *Haloarcula marismortui*, a Dead Sea isolate that was never shown to produce bacteriorhodopsin, was suggested to possess an inherently low need for maintenance energy, and its ion gradients across the membrane were reported to remain unchanged upon starvation (Ginzburg, 1978).

In summary, we still lack a basic understanding of the processes that enable the microbial communities in the Dead Sea to survive between the recently rare occasions that are conductive for algal and bacterial multiplication, which lead to dense blooms such as those observed in 1980 and 1992.

ANAEROBIC MICROBIAL PROCESSES IN THE DEAD SEA

Thus far in this chapter, all discussions centered around aerobic microbial processes that occur in the oxygenated water column. However, because the solubility of oxygen in hypersaline brines is low and the amount of organic matter available is often high, the anaerobic sediments of the Dead Sea may support a rich and varied community of anaerobic halophilic bacteria that are involved in the fermentative degradation of organic compounds. Anaerobic bacteria were found in Dead Sea sediments by Elazari-Volcani, who described lactose- and glucose-fermenting bacteria, as well as denitrifying bacteria, from sediments obtained from depths of 70–330 m (Elazari-Volcani, 1943; Volcani, 1944). Within the lake sediments, anaerobic conditions still exist today. During the past, reducing conditions also prevailed in part of the Dead Sea's water column. During the long meromictic period prior to the 1979 overturn, the lower water mass (below a depth of 40–60 m) lacked molecular oxygen, and low concentrations of sulfide were present.

The fact that sulfide was present in the lower water mass before the 1979 overturn is of special interest. Determinations of the stable sulfur isotope composition of the sulfide and sulfate in the anaerobic layers showed an enrichment in light sulfur isotopes in the sulfide characteristic of biological sulfate-reduction processes (Nissenbaum and Kaplan, 1976). The search for an extremely halophilic or halotolerant sulfate-reducing bacterium in the bottom sediments of the Dead Sea has as yet failed to yield an organism that may have been responsible for the accumulation of the sulfide (nor are halophilic sulfate-reducing bacteria from other habitats able to function at the salinity of the Dead Sea). However, attempts to isolate sulfate reducers from Dead Sea sediments have led to the isolation of a few novel and interesting types of halophilic anaerobic fermentative eubacteria.

The first such organism, isolated in 1980, was a long, slender rod-shaped bacterium described as a new genus and a new species: *Halobacteroides halobius* (Oren, et al. 1984b). This organism was quite abundant in Dead Sea sediments: 10^3–10^5 cells/ml were counted in Dead Sea bottom sediment collected from a water depth of 60 m. Similar organisms have since been isolated from other hypersaline environments too. *Halobacteroides halobius* is probably an endospore-forming bacterium, as it proved easy to grow this type of bacterium from hypersaline sediments after pasteurization for 10 min at 80–100°C (Oren, 1987).

Two additional endospore-forming fermentative anaerobic halophilic bacteria have been isolated from Dead Sea sediments. One of these produces gas vesicles at the time of the formation of the endospores, and these gas vesicles remain attached to the mature endospores after the degeneration of the vegetative cells. This isolate was originally described as *Clostridium lortetii* (Oren, 1983d), a new species of the anaerobic endospore-forming genus *Clostridium*, and named for M. L. Lortet, who isolated (nonhalophilic) clostridia from the Dead Sea a hundred years ago (Lortet, 1892). Later it became clear that the isolate is not related phylogenetically with the genus *Clostridium*, and therefore it was renamed *Sporohalobacter lortetii* (Oren et al., 1987). Another endospore-forming fermentative anaerobe, lacking gas vesicles, was isolated from Dead Sea sediment and described as *Sporohalobacter marismortui* (Oren et al., 1987).

Based on 16S rRNA sequence data, it was shown that these bacteria belong to the eubacterial kingdom and are related to each other, but no clear relationship could be demonstrated with any of the other major subgroups of the eubacterial kingdom. Therefore, they were classified in a new family, the Haloanaerobiaceae (Oren, et al. 1984a).

The mechanism of adaptation of anaerobic halophilic eubacteria to high salt concentrations resembles that of the halophilic archaeobacteria rather than that of the halophilic or halotolerant aerobic eubacteria. No organic osmotic solutes were detected in the cytoplasm, but high intracellular concentrations of sodium, potassium, and chloride ions were found in all cases examined, approximately balancing the osmotic value of the surrounding medium (Oren, 1986c).

Additional new species of anaerobic halophilic eubacteria that belong to this new interesting group of organisms, the first

representatives of which were isolated from the Dead Sea, have been isolated from other hypersaline environments all over the world. More information on the anaerobic halophiles is presented in two recent review articles (Oren, 1986d, 1991c).

EPILOGUE

The Dead Sea as a biotope is a highly dynamic system, with biological properties that vary greatly from year to year and from season to season. The drop in the water level of the lake in recent years, resulting in an increase in overall salinity and, more important, an increase in the relative abundance of divalent cations, has made the Dead Sea an even more hostile environment for life. Thus, during the years 1982–1991, no algae were observed in the lake, and the small archaeobacterial community, the remainder of a previous bloom, barely survived. The Dead Sea microorganisms seem to possess efficient mechanisms to overcome such long periods of adverse condition—mechanisms that for the most part are still unknown. The continuous presence of viable microorganisms suggests that abundant algal and bacterial life in the Dead Sea will be rapidly restored in those rare cases in which large amounts of rainwater cause a substantial dilution of the upper water layer. Such an event was witnessed only twice during the 13 years (from 1980 onward) in which systematic quantitative studies on the microbial ecology of the Dead Sea have been performed: once in 1980 (Oren and Shilo, 1982; Oren, 1988), and once in 1992 (Oren, 1993b).

Life in the Dead Sea in its present state thus seems to depend primarily on unusual "catastrophic" events of abundant rainfall in its catchment area. Although the behavior of a simple ecosystem like the Dead Sea, with only one primary producer and one or a few physiologically very similar heterotrophic bacteria, should be relatively easy to predict, a comparison of the two blooms of 1980 and 1992 shows that our real understanding of the factors underlying the development of the algal and bacterial communities is still extremely limited, in spite of the wealth of field and laboratory data that have been collected since indigenous life was first discovered in the Dead Sea in 1936.

Acknowledgments

These studies on the biology of the Dead Sea have been supported in the past by the Israeli Ministry of Energy and Infrastructure, and the Mediterranean-Dead Sea Co., Ltd. Recent work has been supported by the Israel Science Foundation administered by the Israel Academy of Sciences and Humanities, and by the Moshe Shilo Center for Biogeochemistry, BMFT-Minerva Gesellschaft für Forschung, München, Federal Republic of Germany.

REFERENCES

Alterkar, W., and Rajagopalan, R., 1990, Ribulose bisphosphate carboxylase activity in halophilic *Archaeobacteria*: *Archives of Microbiology*, v. 153, p. 169–174.

Anati, D. A., and Shasha, S., 1989, The stability of the Dead Sea stratification: *Israel Journal of Earth Sciences*, v. 38, p. 33–35.

Ben-Amotz, A., and Avron, M., 1973, The role of glycerol in the osmotic regulation of the halophilic alga *Dunaliella parva*: *Plant Physiology*, v. 51, p. 875–878.

Cohen, S., Oren, A., and Shilo, M., 1983, The divalent cation requirement of Dead Sea halobacteria: *Archives of Microbiology*, v. 136, p. 184–190.

Danon, A., and Caplan, S. R., 1977, CO_2 fixation by *Halobacte-*

rium halobium: *FEBS Letters*, v. 74, p. 255–258.

Edgerton, M. E., and Brimblecombe, P., 1981, Thermodynamics of halobacterial environments: *Canadian Journal of Microbiology*, v. 27, p. 899–909.

Elazari-Volcani, B., 1940, Studies on the microflora of the Dead Sea [Ph.D. thesis]: Jerusalem, The Hebrew University of Jerusalem, 119 p., (in Hebrew).

Elazari-Volcani, B., 1943, Bacteria in the bottom sediments of the Dead Sea: *Nature*, v. 152, p. 274–275.

Ginzburg, M., 1978, Ion metabolism in whole cells of *Halobacterium halobium* and *H. marismortui*, in Caplan, S. R., and Ginzburg, M., eds., *Energetics and structure of halophilic microorganisms*: Amsterdam, Elsevier/North Holland Biomedical Press, p. 561–577.

Ginzburg, M., Sachs, L., and Ginzburg, B. Z., 1970, Ion metabolism in a *Halobacterium*. I. Influence of age of culture on intracellular concentrations: *Journal of General Physiology*, v. 55, p. 187–207.

Goldman, J. C., and Dennett, M. R., 1985, Susceptibility of some marine phytoplankton species to cell breakage during filtration and post-filtration rinsing: *Journal of Experimental Marine Biology and Ecology*, v. 86, p. 47–58.

Javor, B. J., 1988, CO_2 fixation in halobacteria: *Archives of Microbiology*, v. 149, p. 433–440.

Kaplan, I. R., and Friedmann, A., 1970, Biological productivity in the Dead Sea. Part I. Microorganisms in the water column: *Israel Journal of Chemistry*, v. 8, p. 513–528.

Kamekura, M., Oesterhelt, D., Wallace, R., Anderson, P., and Kushner, D. J., 1988, Lysis of halobacteria in Bacto-Peptone by bile acids: *Applied and Environmental Microbiology*, v. 54, p. 990–995.

Kritzman, G., Keller, P., and Henis, Y., 1973, Ecological studies on the heterotrophic extreme halophilic bacteria of the Dead Sea [abs.]: 1st International Congress of Bacteriology, Vol. II, Jerusalem, p. 242.

Krumgalz, B. S., and Millero, F. J., 1982, Physico-chemical study of the Dead Sea waters. I. Activity coefficients of major ions in Dead Sea water: *Marine Chemistry*, v. 11, p. 209–222.

Larsen, H., 1980, Ecology of hypersaline environments, in Nissenbaum, A., ed., *Hypersaline brines and evaporitic environments*: Amsterdam, Elsevier, p. 23–29.

Lortet, M. L., 1892, Researches on the pathogenic microbes of the mud of the Dead Sea: *Palestine Exploration Fund*, v. 1892, p. 48–50.

Mullakhanbhai, M. F., and Larsen, H., 1975, *Halobacterium volcanii* spec. nov., a Dead Sea halobacterium with a moderate salt requirement: *Archives of Microbiology*, v. 104, p. 207–214.

Nissenbaum, A., 1975, The microbiology and biogeochemistry of the Dead Sea: *Microbial Ecology*, v. 2, p. 139–161.

Nissenbaum, A., and Kaplan, I. R., 1976, Sulfur and carbon isotopic evidence for biogeochemical processes in the Dead Sea ecosystem, in Nriagu, J. O., ed., *Environmental biochemistry*, Vol. 1. Ann Arbor, Michigan, Ann Arbor Science Publishers, Inc., p. 309-325.

Oren, A., 1981, Approaches to the microbial ecology of the Dead Sea: *Kieler Meeresforschungen*, Sonderheft v. 5, p. 416-424.

Oren, A., 1983a, Population dynamics of halobacteria in the Dead Sea water column: *Limnology and Oceanography*, v. 28, p. 1,094-1,103.

Oren A., 1983b, *Halobacterium sodomense* sp. nov., a Dead Sea halobacterium with extremely high magnesium requirement and tolerance: *International Journal of Systematic Bacteriology*, v. 33, p. 381-386.

Oren, A., 1983c, Bacteriorhodopsin-mediated CO_2 photoassimilation in the Dead Sea: *Limnology and Oceanography*, v. 28, p. 33-41.

Oren, A., 1983d, *Clostridium lortetii* sp. nov., a halophilic obligately anaerobic bacterium producing endospores with attached gas vacuoles: *Archives of Microbiology*, v. 136, p. 42-48.

Oren, A., 1985, The rise and decline of a bloom of halobacteria in the Dead Sea: *Limnology and Oceanography*, v. 30, p. 911-915.

Oren, A., 1986a, Dynamics of *Dunaliella* in the Dead Sea, in Dubinsky, Z., and Steinberger, Y., eds., Environmental quality and ecosystem stability, Vol. IIIA, Proceedings, Third International Conference of the Israel Society for Ecology & Environmental Quality Sciences: Jerusalem, June 1986, p. 351–359.

Oren, A., 1986b, Relationships of extremely halophilic bacteria towards divalent cations, in Megusar, F., and Gantar, M., eds., Perspectives in microbial ecology, Proceedings, Fourth International Symposium on Microbial Ecology, Ljubljana, Slovenia, August 1986, p. 52-58.

Oren, A., 1986c, Intracellular salt concentrations of the anaerobic halophilic eubacteria *Haloanaerobium praevalens* and *Halobacteroides halobius*:*Canadian Journal of Microbiology*, v.32, p.4-9.

Oren, A., 1986d, The ecology and taxonomy of anaerobic halophilic eubacteria: *FEMS Microbiology Reviews*, v. 39, p. 23-29.

Oren, A., 1987, A procedure for the selective enrichment of *Halobacteroides halobius* and related bacteria from anaerobic hypersaline sediments: *FEMS Microbiology Letters*, v. 42, p. 201-204.

Oren, A., 1988, The microbial ecology of the Dead Sea, in Marshall, K. C., ed., *Advances in microbial ecology*, Vol. 10: New York, Plenum Publishing Company, p. 193-229.

Oren, A., 1989, Halobacteria in the Dead Sea in 1988-1989: Novel approaches to the estimation of biomass and activity, in Spanier, E., Steinberger, Y., and Luria, M., eds., Environmental quality and ecosystem stability, vol. IV-B, Proceedings, Fourth International Conference of the Israel Society for Ecology & Environmental Quality Sciences, Jerusalem, June 1989, p. 247–255.

Oren, A., 1990, The use of protein synthesis inhibitors in the estimation of the contribution of halophilic archaebacteria to bacterial activity in hypersaline environments: *FEMS Microbiology Ecology*, v. 73, p. 187–192.

Oren, A., 1991a, Tetanus bacteria and other pathogens in the Dead Sea? *Salinet*, v. 6, p. 84–85.

Oren, A., 1991b, Estimation of the contribution of archaebacteria and eubacteria to the bacterial biomass and activity in hypersaline ecosystems: novel approaches, in Rodriguez-Valera, F., ed., *General and applied aspects of halophilic bacteria*: New York: Plenum Publishing Company, p. 25–31.

Oren, A., 1991c, The genera *Haloanaerobium*, *Halobacteroides*, and *Sporohalobacter*, in Balows, A., Trüper, H. G., Dworkin, M., Harder, W., and Schleifer, K. H., eds., *The prokaryotes: A handbook on the biology of bacteria—Ecophysiology, isolation, identification, applications*, 2nd. ed.: New York, Springer-Verlag, p. 1,893–1,900.

Oren, A., 1992, Bacterial activities in the Dead Sea, 1980–1991: Survival at the upper limit of salinity: *International Journal of Salt Lake Research*, v. 1, p. 7–20.

Oren, A., 1993a, Ecology of extremely halophilic microorganisms, in Vreeland, R. H., and Hochstein, L. I., eds., *The biology of halophilic bacteria*. Boca Raton, CRC Press, p. 25–53.

Oren, A., 1993b, The Dead Sea - Alive again: *Experientia*, v. 49, p. 518–522.

Oren, A., 1993c, Availability, uptake, and turnover of glycerol in hypersaline environments: *FEMS Microbiology Ecology*, v. 12, p. 15–23.

Oren, A., Ginzburg, M., Ginzburg, B. Z., Hochstein, L. I., and Volcani, B. E., 1990, *Haloarcula marismortui* (Volcani) sp. nov., nom. rev., an extremely halophilic bacterium from the Dead

III. QUATERNARY HISTORY OF THE LAKE AND ITS ENVIRONMENT

20. GEOMORPHOLOGY OF THE DEAD SEA WESTERN MARGIN

Dan Bowman

MORPHOLOGICAL SETTING

The Dead Sea is the terminal lake for drainage systems of the Kingdom of Jordan, the southern Negev, including parts of the Sinai, and wadis of the Judean and Hebron mountains in the west (Fig. 20-1). The Jordan River with its 2,730 km² basin, drains areas up to the Golan Heights and southern Lebanon.

The Dead Sea, 410 m below mean sea level in 1993, occupies the lowest point on the continents (see Hall, chapter 2, this volume). The Lisan Peninsula (Fig. 20-1) divides the shallow southern basin from the 330-m-deep northern one. The water level of the northern Dead Sea basin was probably about 40 m below its present level until 1,500 years ago, when it rose and transgressed into the southern basin, which is delimited by the Amazyahu fault (Neev and Emery, 1967).

The level of the Dead Sea during the first quarter of this century was about 392.5 m below mean sea level (MSL). The Dead Sea level dropped to -403.5 m between 1929 and 1979 (Klein, 1982), exposing the shallow southern basin, which is used by both Israel and Jordan as evaporation pans.

The fault escarpments of the rift rise steeply to the east and to the west of the lake, forming a well-defined graben that is morphologically, structurally, and bathymetrically asymmetric, with the major boundary fault in the east. Widowinski and Zilberman (1994) proposed that this asymmetry reflects a first-order half-graben structure. The western fault escarpment, up to 400 m high, is composed of resistant dolomite and limestone of Cenomanian and Turonian age (Bentor and Vroman, 1960). A sequence of normal faults form a stepped morphology with some antithetic faulting.

The Dead Sea area is extremely arid, with a mean annual precipitation of 50 mm. Summer temperatures are regularly in the 30–40°C range with a mean of 15°C in January. Surface soil temperatures may reach over 50°C in the summer. Annual potential evaporation is over 2 m, and the mean annual humidity is 45–50% (Survey of Israel, 1985). The watersheds of some western drainage systems are at an altitude of 800–1,000 m with mean annual precipitation exceeding 600 mm, causing intensive floods. The erosional recession of the gorges through the western rift escarpment is directly related to the surrogate measure of river discharge, that is, the drainage area. Knickpoint recession is greater as catchment size increases (Frostick and Reid, 1989).

The Dead Sea area attained its recent structural shape during the Late Pliocene to Early Pleistocene (Garfunkel, 1981). Morphologically, its age is mainly related to the Dead Sea stage, from Upper Pleistocene to recent. Being well preserved and dated, the margins of the Dead Sea are ideal for morphological research.

ALLUVIAL FAN MORPHOLOGY

The asymmetric structure of the rift made the western margin of the Dead Sea, mainly the Masada plain (Fig. 20-1) between the fault escarpment and the lakeshore, an ideal environment for the development of alluvial fans. The gorges and canyons reflect incision since the Early Pleistocene, when the last major tectonic pulse in the Dead Sea rift occurred. The canyons have been inundated by all deep-water bodies that emerged in the graben since then. The last one was Lake Lisan, which occupied the graben from 50,000 to 12,000 yr B.P. (Kaufman, 1971; Neev and Hall, 1977; Begin et al., 1985). Its sediments, known as the Lisan Formation, cover the entire area and are partly preserved within the canyons.

An array of well-defined alluvial fans mantle the western rift escarpment. These include (Fig. 20-1) the fans of Nahal (Wadi) Og, Nahal Kidron, Nahal Darga, Nahal Arugot, Nahal Hever, Nahal Mishmar, Nahal Rahaf, Nahal Mor, Nahal Yeelim, Nahal Boqeq, Nahal Hemar, Nahal Lot, and Nahal Hamarmar. These fans were initially deposited below the level of Lake Lisan as small sublacustrine fan deltas, each with an area of 0.5–5.5 km² and slope gradients of 0.5–3.5°, and nourished by drainage basins 17–235 km² in area. The sublacustrine fan deltas are partly composed of limnic Lisan sediments, originally deposited in the canyons from which they were subsequently washed out, and of coarse fluviatile load of dolomite, limestone and flint.

The fan deltas show a massive to crudely stratified and slightly cemented conglomeratic facies next to the rift escarpment. The size of their Cenomanian dolomite and Campanian flint components ranges from pebble-cobbles to giant boulders, some 4 to 5 m in diameter. The thickness of these fans at their apex reaches several tens of meters.

The fault escarpment is mantled by Pleistocene talus veneers, tens of meters high, that are among the largest in Israel. Farther eastward, the sediments are characterized by a sandstone facies with granules and mudballs.

Basinward, a rhythmic sandy-muddy facies occurs with quartz sands and ripple cross-laminations. Each facies component dominates a few hundreds of meters and then interfingers downstream with the next one (Sneh, 1979). No downstream decline in clast size, nor in other textural characteristics, has been observed, indicating the typical fluviatile sedimentary immaturity of the Dead Sea area. Cyclic changes between coarse offlapping facies and limnic fine-grained onlapping sediments are typical and often abrupt, indicating that hydrological fluctuations in both wadis and lake, not tectonic processes, control the sedimentation (Frostick and Reid, 1989).

High-magnitude floods, represented by boulder-size alluvium, are quite rare within the fan deltas (Reid and Frostick, 1993). Suspended-load detrital lamina represent smaller annual floods. Variations in the proportion of the detrital layers reflect changes in flood frequency, that is, wetter and drier years, and climatic cycles of different wavelengths (Reid and Frostick, 1993).

The Lisan fan deltas at the foot of the escarpment are poorly sorted and layered, dipping moderately 3–6° lakeward with intercalations of transgressive beach units. Overlying these basal fan units are steeply inclined (24–34°), planar, and well-sorted foresets of a Gilbert-type delta (Frostick and Reid, 1989; Bowman, 1990) with topset capping that demarcates the uppermost marginal facies of the Lisan Formation. The topsets indicate fluviatile underwater extension, whereas the steep foresets suggest bedload transportation and dumping in a basin of weak littoral and fluviatile dispersal energy. The uppermost foreset-topset architecture sharply contrasts with the underly-

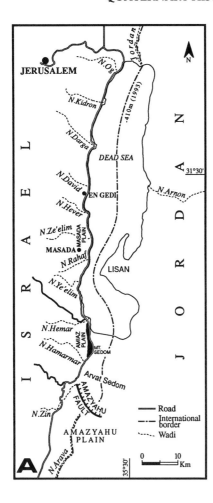

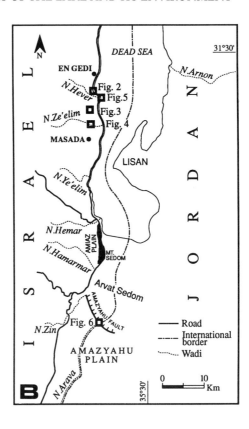

Figure 20-1 A, Location map. The outline of the Dead Sea shoreline is at the -395 m elevation; B, Location of the air photos.

ing sheet-like structure of the basal fan deltas. The Gilbert-type sublacustrine fan deltas were observed in the front of small catchment areas, such as Nahal Mor and Nahal Boqeq, and in front of medium catchment areas, such as Nahal Rahaf. No evidence of Gilbert-type sublacustrine fan deltas was found at the mouth of the largest drainage basins (Bowman, 1990).

Various data sources (Horowitz, 1979)—such as pollen, high fossil water table, widespread epipaleolithic sites in the Negev and the Sinai deserts, uppermost calcic horizons in the loess of southern Israel, and paleosols in northern Sinai—provide evidence of a wetter period at 18,000–12,000 yr B.P., which indicates the last highstand of Lake Lisan.

The typical fine-laminated aragonite, detritic, and gypsiferous Lisan Formation facies indicates the deeper lake environment. This facies is widely exposed in the Masada plain, the Amiaz plain west to Mount Sedom, and in the Amazyahu plain south to the Dead Sea. The softness of the sediments triggered the formation of a badland morphology with abundant piping. Lake Lisan also left its fingerprints in the form of shore terraces (Fig. 20-2) and regular, undissected surfaces, veneered with well-rounded beach sediments and coastal swash bars (Figs. 20-3B, 20-4B). Following the lowering of the lake level, these surfaces were disconnected from the hydrographic system and thus from most subaerial erosion, so they have been preserved.

Lisan and pre-Lisan surfaces are capped by beach sediments laid down during the last retreat of the lake. This morphological and sedimentary criterion provides a most useful field marker for distinguishing post-Lisan morphology from lacustrine Lisan deposits. Post-Lisan alluvial fans lack geomorphic beach features. A braided alluvial fan surface is characteristic of post-Lisan morphology (Figs. 20-3, 20-4). Since the drop of the Lake Lisan level, the major morphological trend has been entrenchment into the newly exposed sublacustrine fan deltas and into the lake sediments. The post-Lisan entrenchment provides spectacular exposures across the fan deltas. After the drop in Lake Lisan level, some wadis, such as Nahal Rahaf, were deflected around their exposed sublacustrine fan before resuming their eastward course (Nir, 1967).

North of the Dead Sea, Lisan sediments are entrenched by the 105-km long Jordan River. The Jordan River is a young Holocene river that meanders within a narrow alluvial plain and is flanked by Lisan badlands (Schattner, 1962).

After the lake level dropped, the newly exposed sublacustrine fan deltas became subareally entrenched. The eroded material was deposited in the front, forming a new fan veneer composed of coarse gravel with large boulders. Since the lake level dropped, fan head entrenchment has been ongoing with forward and lateral growth of the fan units forming a multicyclic fan complex. Each entrenchment was followed by basinward deposition of a younger fan, resulting in the formation of segmented telescopic alluvial fans (Fig. 20-3). The lowermost active toe components reach the lake shore, where the modern fan deltas are actively prograding into the Dead Sea as embryonic lobes and as Gilbert-type fan deltas interfingering with the lake sediments (Fig. 20-5). The main road along the lake is usually buried, but not entrenched, during winter floods, reflecting the migration of the depositional processes lakeward. Subsidence on some of the recent fans results from active piping and formation of "swallow holes" (Arkin, 1993).

The Ze'elim fan complex is an excellent example of a highly dissected, multilevel, telescopic fan with 15 well-preserved

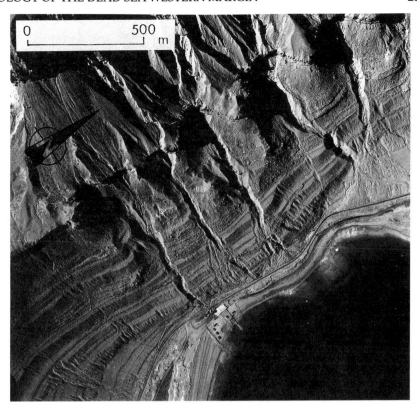

Figure 20-2 Well-preserved abandoned beaches of Lake Lisan along the rift scarp near En Gedi Spa. Note burial of the highest terraces under colluvial veneer, leaving no morphological marker for the highest stand of the lake. Abandoned shorelines of the Dead Sea are visible by the lakeshore.

alluvial fans that converge downward in the form of terraces (Fig. 20-4). The fan surfaces or fan terraces of Nahal Ze'elim and Nahal Lot show ferric oxide enrichment in the form of a varnish coating on the calcareous sand. Progressively higher and older fan surfaces disclose more stain (Bowman, 1982). Oxygenic conditions, basic for the development of varnish, were provided by the continuous entrenchment. The lowered water table permitted successive new, well-drained alluvial surfaces to form. The initial varnish stains on bars and on alluvial surfaces, only somewhat elevated above the active channel bed, indicate that in the extremely arid Dead Sea area, stain develops within a few tens of years.

The complex fan structure becomes evident also through the definition of intersection points, that is, sites where the channel emerges from an entrenched upper fan segment onto the active front fan, thereby indicating growth of the telescopic fan system. The intersection point is hydraulically important. It defines the change from confined to multiple channel flow conditions. Intersection points delimit the different fan surfaces within a telescopic system and provide a tool for deciphering the overall fan development and structure (Bowman, 1978). The advance of intersection points basinward is part of the overall downcutting trend and indicates a state of disequilibrium.

Because of cartographic generalization, the Israeli topographic sheets at a scale of 1:50,000 clearly demonstrate the structure of the segmented fans by contour offsets (Bowman, 1978). Lateral erosion during the post-Lisan entrenchment often continued asymmetrically southward or northward, thus destroying major parts of the original sublacustrine fan deltas and forming an asymmetric terraced cross section (Fig. 20-4).

Many of the Dead Sea alluvial fan complexes, even if reconstructed to their maximum initial dimensions before incision and avulsion, are underfit, that is, the fans show very low ratios of fan area to drainage area (Af/Ad = 0.009–0.055 (Bowman, 1974), which deviates markedly from the more common ratio of

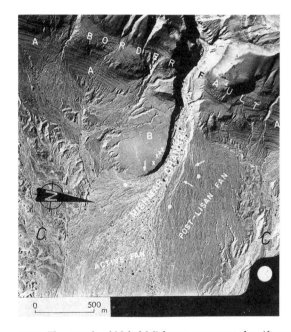

Figure 20-3 The mouth of Nahal Mishmar canyon at the rift escarpment showing a sublacustrine fan (B) with abandoned beaches (x) and a post-Lisan alluvial fan. Between them, the active fan is prograding eastward. Interfan areas (C), without a substantial capping veneer, are intensively eroded. (A) indicates the abandoned beaches of Lake Lisan along the escarpment of the rift.

0.2–0.3 (Denny, 1965). Such a discrepancy and the "young" appearance of the fan deltas, as single, non coalescing sedimentary bodies, reflect the complex paleomorphology of the Dead Sea area: Since the Lower Pleistocene, most of the sediments

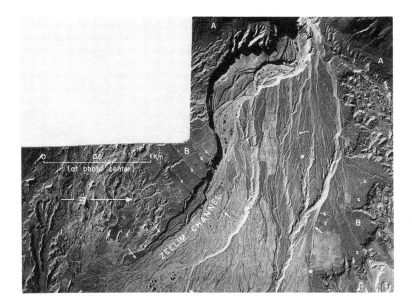

Figure 20-4 The mouth of the Nahal Ze'elim canyon at the rift escarpment showing a residual sublacustrine fan surfaces (B) with ancient coastal bars (x). At the center, post-Lisan fan terraces indicate a continuous fanhead entrenchment. Arrows indicate die-out zones and dots point to intersection points. (A) marks the raised beaches of Lake Lisan along the rift escarpment.

supplied to the rift were deposited as graben fill. The alluvial fans represent only the uppermost Pleistocene–Holocene stage, whereas the drainage basins supplied the rift with sediments at least since Pliocene–Pleistocene times. The absence of a sizable delta for Nahal Arnon, one of the largest rivers draining into the Dead Sea from Jordan (Fig. 20-1), is visible on the bathymetric map (Fig. 2-3c, this volume; Neev and Hall, 1979) and could well be the result of the same cause, although recent downfaulting should also be considered (Nir, 1967).

STEPPED BED MORPHOLOGY

The ephemeral channels which dissect the alluvial fans have a stepped bed morphology, composed of regular and rapid-like channel floor segments that are cyclically spaced. Stepping was studied along the Nahal Hemar, Nahal Ze'elim and Nahal Lot

channel beds (Bowman, 1977). Steep segments (2°19') are composed mainly of coarse gravel and boulders that cause a major drop in bed elevation. Water passes over the steep segment in the form of a rapid with supercritical velocity, causing a hydraulic jump. Regular segments, composed of well-sorted gravel, slope gently (0°25'), causing a subcritical flow. Average spacing between regular and rapid segments is 1.4 and 2.2 channel widths, respectively-distinctly smaller than the ratio range of 1:5 to 1:7 which is typical of perennial streams with episodic flooding.

Stepping is associated with a cyclic size variation within the gradual decrease of sediment size downstream. Stepping does not fade out toward the Dead Sea, thus indicating the prevalence of the immature fan conditions—steep slopes, coarseness, and poor sorting—down to the lakeshore. Megasteps, composed of huge boulders up to 5 m in diameter directly supplied

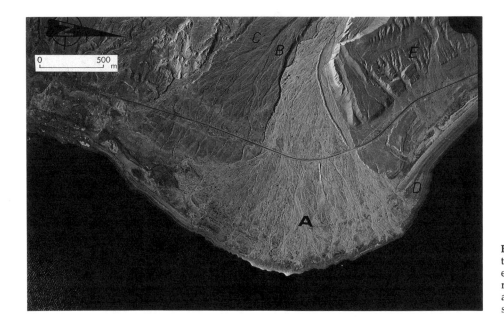

Figure 20-5 The active alluvial fan of the Nahal Hever canyon (A) entrenched in fossil fans (B, C). The raised shorelines of the Dead Sea (D) and of Lake Lisan (E) are well preserved.

from the canyons' walls, were observed in the canyons. Megasteps are not a cyclic phenomena. They create waterfalls and function as a local base level.

SEDIMENT TRANSPORT IN ERODIBLE MATERIAL

The caprock of Mt. Sedom, which consists of shales, marls, chalks, sandstones, gravel, and anhydrite, offers the possibility of evaluating the denudation rate and sediment transport in erodible material under extremely arid conditions (Gerson, 1972, 1977). Runoff in small (0.06–0.5 km²) watersheds was heavily laden with suspended sediment, 25% by weight, with many samples exceeding 50%. No correlation was found between such environmental factors as lithology, slope angle, catchment area, or rainfall intensity and the suspended sediment concentration or texture, which suggests that the availability of bed material is practically unlimited. The depth of scour per event and the movement of all sediment sizes including boulders is related to the high concentration of the suspension. The gradients of the channels, 0.01–0.05, typical of much larger desert drainage systems, are also related to the high competence and transportation potential, which cause a high average denudation rate of 1–2 mm/yr.

HOLOCENE REG EVOLUTION

The multilevel fan complexes of the Dead Sea area, mainly that of Nahal Ze'elim with its 15 fan terraces, provide a laboratory-like, generally-dated alluvial chronosequence for studying the effect of time on Holocene pedogenetic processes. The initial stages of the evolution of reg soil—an aridosol developed on a stable alluvial surface—were studied to assess the rates of development of the different properties over time and to evaluate their significance as relative time indicators (Amit and Gerson, 1986). The shattering processes of gravel by salt on post-Lisan alluvial sequences was the focus of another detailed study (Amit, 1990; Amit et al., 1993).

Maximum post-Lisan soil thickness was attained after several hundred to 2,000 years. After that period, there was no clear differentiation in fines and salts at various depths. The typical reg soils are of the AC and ABC types. Their maximum depth is 55 cm. The surface is characterized by bar and swell microtopography and by a moderately developed desert pavement—Ao horizon—covering 50% of the surface. The vesicular Av horizon is 0.5 cm thick and the thickness of the cambic B horizon is 5 cm. The gravelly C horizon composes most of the profile. Maximum salinity values are 16 mmho/cm. Desert pavement developed gradually with time. The reg soil did not indicate a clear Holocene climatic change in the Dead Sea area.

Gravel shattering by salt is dependent on the microenvironmental conditions of the soil profile, mainly on rapid and extreme variations of temperature and on the moisture conditions close to the soil surface (Amit, 1990; Amit et al., 1993). The maximum percentage of shattered gravel was 70%. The shattered gravel maintained its original shape. Although the rate of the shattering process is not linear, different shattering stages can be used as relative time indicators. The shattered gravel along the entire Ze'elim terrace sequence indicates continuous arid conditions in the Dead Sea area throughout the Holocene.

ABANDONED SHORELINES

The 17-m drop in the Dead Sea level during this century left a belt of well-preserved abandoned beaches along most of the Dead Sea periphery (Fig. 20-5,D). This untarnished belt, which is relatively light and patina-free, sharply contrasts with the higher Lisan levels. Klein (1961, 1982) measured the historical

high levels of the Dead Sea and mentioned their geomorphic and sedimentary markers: driftwood, loose gravelly swash bars forming ridge and runnel morphology often encrusted with alternating dark-gray calcite and white aragonite laminae (Neev and Emery, 1967), coated cliffs and building stones, beach rock, and abandoned coastal cliffs. These markers were identified even higher up (-369 m) on the lower fans, probably indicating the highest stand of the Dead Sea during Holocene. The Holocene history of the Dead Sea levels, based on the morphology of the Mt. Sedom karst system, is discussed by Frumkin (chapter 22, this volume).

At the fault escarpment, a second group of beach features is excellently preserved. Limnic sediments and colluvium exhibit shore features of Lake Lisan (Fig. 20-2). The stratigraphic evidence for fluctuations in the level of Lake Lisan are reviewed by Niemi (chapter 21, this volume). Huntington (1911), Blanckenhorn (1912), Picard (1943), Butzer (1958), and Neev and Emery (1967) mentioned the Lisan strandlines, measured some of their altitudes, and discussed their origin. Bowman (1971) defined, in three cross sections, 13 to 17 well-preserved and undisturbed lake terraces that exhibit a reasonable correlation along the western coast of Lake Lisan, composing, when integrated into one sequence, 28 terraces in the altitudinal range of 384–182 m below MSL.

The terraces consist of well-defined treads and risers (Fig. 20-2) that bear a definite geometric relationship to each other. The treads are 5–16° steep and relatively clean of gravel, which mainly covers the risers. This clear sedimentary zoning reflects the fossil subenvironments of the Lisan beaches. The treads functioned as the swash zone, from which the sediments were washed up and down the beach toward the steep (13–26°) risers.

The excellent preservation of the terraces is due to the aridity of the area and to their capping by colluvium and beach gravel, which are impregnated by aeolian silt. No significant enhancement of erosion took place after the impregnation. Deposited during the retreat of the Lisan, the capping veneer gradually thins out from the escarpment lakewards, making the lower areas vulnerable to erosion. Dissection gradually intensified eastward and badlands were formed. The degree of preservation, therefore, bears no relation to the relative age of the terraces.

The highest abandoned shorelines of Lake Lisan, identified along its southwestern corner in the northern Arava rift valley, are linear, level, and asymmetric swash bars. They exhibit a smooth surface of rounded pebbles and have a typical backshore structure. Their uppermost altitude, -150-m, is similar to the uppermost Lisan section observed by Yechieli (1987). The swash bars show only a very initial pedogenetic reg stage development with a low weight percentage of Fe^2O^3 coating the sand grains and a low salinity marked by the conductivity range of 10–50 mmho/cm. Such values typify Holocene alluvial surfaces (Dan, 1981; Gerson et al., 1985). These age indicators suggest that no older, pre-Lisan lake levels exist. The -150-m altitude is therefore proposed as the highest known stand of Lake Lisan (Bowman and Gross, 1992), replacing the traditional, frequently quoted level of -180 m.

BASE LEVEL EFFECT

The Dead Sea is an ultimate base level, that is, the lower boundary for incision by the surrounding fluvial systems. Because field data on the effect of lowering base levels is quite limited, the well-preserved river and beach terraces of Lake Lisan make the Dead Sea area a unique field model for studying the effect of base level lowering in an arid environment.

Fan-head entrenchment and knickpoints in the longitudinal

stream profiles in the Dead Sea area often have been hypothetically related to the effect of lowering the base level (Quennell, 1958; Willis, 1928; Nir, 1965; Sneh, 1979). Garfunkel and Freund (1981) also related the effect of a base level, toward which the fluviatile systems are graded, to the Dead Sea. Begin (1975) compared gradients along streams to apparent dips of rock strata in which the wadis are incised. Small-scale local structural and lithological variations explained most of the stream gradient variations. To Begin, the base level was less important than lithological constraints.

Fan terraces in the Dead Sea area converge downstream and die out toward intersection points in the post-Lisan fan complexes. The intersection points are never Lisan shorelines. Such field relations are of paramount importance for revealing the effect of the receding Lake Lisan by indicating disconnection between the fan terraces and the receding shorelines. The fan terraces do not converge toward abandoned beaches, and therefore their formation could not be controlled by a base level drop. The different number of Lisan raised beaches compared to the number of fan terraces suggests that two different systems were operating (Bowman, 1988).

The mean rate of post-Lisan entrenchment is in the range of 2.6–5.4 mm/yr (Bowman, 1974), which agrees with Gerson's 4.5 mm/yr erosional rate on Mt. Sedom (1972). Ages of Lisan deposits from elevations of -180 m and -370 m indicate a time interval of only 1,000 yr, suggesting a fast lowering of Lake Lisan (Neev and Emery, 1967). Such a rapid drop in lake level is related to lake floor subsidence or to an arid period and is one magnitude greater than the rate of entrenchment, implying that fan entrenchment lagged behind the rapid drop in lake level. The fall in base level was too rapid for headward erosion and therefore did not cause pronounced unconformities in the Lisan Formation. Base level drop allowed entrenchment but did not cause it. The receding lake level played only a passive role, neither controlling nor triggering entrenchment (Bowman, 1988).

Variations in the number of fan terraces in adjacent drainage basins rules out one overall regional triggering mechanism in the form of base level drop or tectonic subsidence. The formation of post-Lisan fan complexes is an inherent basin process that, since 12,000 yr B.P., has been controlled by climatic fluctuations (Neev and Hall, 1977) and by normal discontinuous downcutting episodes, which are an integral part of the fluvial system (Schumm, 1976).

The morphology of the Dead Sea area demonstrates a delayed transmission of the base level effect through the system. Such conditions indicate an out-of-grade state, which is also indicated by the overall recent erosional trend in the Dead Sea area. Parallelism between an active wadi bed and its terraces indicates equilibrium. The converging longitudinal profiles of Nahal Ze'elim and Nahal Lot suggest disequilibrium. Since the retreat of Lake Lisan, grade has not been yet reached.

RECENT MORPHOTECTONICS

Sublacustrine evidence of Dead Sea neotectonics, in the form of submarine faulting and slumps, is given by Niemi and Ben Avraham (chapter 6, this volume). Upper Pleistocene to historic surface faulting episodes along the Dead Sea margin have left conspicuous morphotectonic expressions. The neotectonic features in the Lisan Formation include well-defined fault scarps with vertical displacements of 0.5–5 m and lengths over 1 km. These scarps are discontinuous, trending north-south subparallel to the main border fault. Some of the fault scarps developed into wash slopes or debris slopes. The steepest, with 40–80° slope gradients, comprise free slopes, which indicate recent faulting, that is, a few thousand years or less in age. Small grabens in the form of 15- to 40-m-wide depressions and

faulted post-Lisan alluvial fan veneers have also been observed. The pre-Lisan Feshha Conglomerate is also faulted (Mor, 1987). All this normal faulting, which took place over the last 50,000 years, suggests extension in the east-southeast and west-northwest direction, and migration of the active tectonics from the western escarpment eastward into the rift (Gardosh et al., 1990).

Trenches east of Jericho indicate deformation of the Lisan and post-Lisan sediments (Reches and Hoexter, 1981; Gardosh et al., 1990). Folding, normal faulting, push-up swells and reverse faulting were also exposed here. Pottery artifacts younger than 3,000 yrs and amino acid racemization of terrestrial snails in the deformed sequence yielded an age estimate of 3,000–4,000 yr B.P.

The lower parts of Nahal Yeelim and Nahal Mor are linear, oversized canyons, 0.5 km wide, with subhorizontal slickenside lineations on their walls, suggesting a major pre-Lisan shearing stage (Gilat, 1991). Faulted Lisan shore terraces at the exit of Nahal Yeelim, with displacements of about 7 m, indicate post-Lisan neotectonics (Agnon, 1982, 1993).

South of the Dead Sea, the linear Amazyahu scarp, accompanied by a line of springs, stands out as a transverse fault within the graben fill (Fig. 20- 6). Some of the wadis to the south of the Amazyahu fault are still "hanging" above Arvat Sedom. Tensional cracks along the Amazyahu scarp indicate its recency and its ungraded stage. Reactivation of the Amazyahu fault is clearly indicated where its trace crosses Nahal Arava by downfaulting its terrace (Fig. 20-6, C). Additional evidence for reactivation of the Amazyahu fault are the linear and sharp scarps along its front, which form a terrace morphology without the typical gravel capping of the raised shorelines, thereby suggesting neotectonic origin (Bowman and Gross, 1989). Linear scarps in the Lisan sediments west to Arvat Sedom were defined as Upper Pleistocene faults by Neev and Emery (1967).

Farther to the south, where Nahal Zin crosses the main road (Fig. 20-1), there is a very recent fault line in a post-Lisan alluvial fan veneer. The fault forms a linear scarp, perpendicular to the drainage network, and is also expressed as a stratigraphic displacement in the Lisan and post-Lisan units. Facets of the free face and of the crest are still preserved along the fault scarp, suggesting an initial stage in the degradation process of a very young fault, a few hundreds or thousands of years in age. This historic to modern age is corroborated by morphological dating techniques (Bowman and Gross, 1989). The young tectonic age is also deduced from the initial pedogenetic reg stage development, which lacks profile differentiation and shows only initial development of the vesicular A and the cambic B horizons.

The longitudinal profiles of the main wadis draining the northern Arava rift toward the Dead Sea—Nahal Zin, Nahal Amazyahu and Nahal Arava—all show undisturbed planar profiles without any trend toward concavity, which would be expected as an indication of grade to the recent Dead Sea level. These planar profiles suggest an overall young disequilibrium stage.

Nahal Zin and Nahal Amazyahu south to the Amazyahu escarpment exhibit local meandering and thalweg sinuosity, that is, a meandering channel within a straight valley with intensive lateral undercutting. Such local meandering, superimposed on planar longitudinal profiles, is unexpected in graded conditions. It may be the response to and a fingerprint of local neotectonic uplift and steepening south of the Amazyahu line.

FUTURE RESEARCH

The drainage basins to the west of the rift drain via bedrock canyons toward the Dead Sea. The climate in these basins ranges from extremely arid to semiarid or Mediterranean.

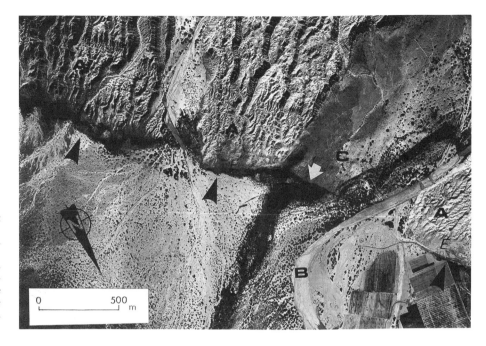

Figure 20-6 The Amazyahu fault (black arrows), at the southern Dead Sea basin, delimits the badlands formed in the lacustrine deposits (A). A terrace (C) of Nahal Arava (B) is cut (white arrow), suggesting neotectonic reactivation along the Amazyahu fault line. Compare the sharp fault trace on the terrace with the degraded Amazyahu scarp (E).

Because their size ranges from a few tens to a few hundreds km², these drainage basins promise interesting variability in size for bedload movement research, including topics such as residence time and rates of sediment yield. Refilling large pits, dug by the Dead Sea Works, by diverting a wadi into them, grants an opportunity to study sedimentary fill processes, including growth of fans and fan deltas, under quasi-laboratory conditions.

Magnitude and frequency of bedload transport events, controlled by bedrock geometry, are relevant subjects not yet studied in the canyons that drain toward the Dead Sea. The canyons also provide excellent sites for studying the rate and locus of vertical and lateral accretion and incision, lateral persistence modes of deposition, and aggradational-degradational process-structure relations in bedrock-controlled systems. Reconstruction of past floods and efforts to retrieve data about paleohydrological fluctuations, based on texture, structure, stratigraphy, and macroforms in fill units, are promising research topics in the Dead Sea area.

Well-developed sequences of pools in the bedrock canyons provide the opportunity to study bedload movement under various flow conditions, including storage and evacuation rates in potholes and in the alluvial reaches between pools, their textural characteristics, and spatial extent.

Boulders 4–5 m in diameter on some of the fan terraces and in the canyons, highlight the competence dilemma. Evaluating the transport and the deposition of boulders on the fans and under hydraulic jump conditions in the canyons, while taking into consideration the hydraulic effects of hiding and protrusion, is an additional research challenge.

The detailed textural and structural composition of the alluvial fans, as a potential response to tectonic episodes, has not yet been analyzed in detail. Earthquake deformation in the Lisan deposits has been suggested by El-Isa and Mustafa (1986). However, earthquake fingerprints in the youngest sedimentary record, that is, post-Lisan to recent alluvial sequences, have not been studied. At the front of the active alluvial fans along the Dead Sea shore, distinct liquefaction layers, which consist of load structures, can be observed. Their possible earthquake-induced origin is understudy by examining whether the criteria for correlating deformational structures with seismic events is satisfied.

The largest occurrence of talus in Israel is along the fault escarpment of the Dead Sea. These huge colluvium features may bear evidence of a number of faulting events or of climatic changes. Examination of different colluvial components, separated by soils of different catenary stages, may contribute to our paleoclimatic and paleotectonic knowledge.

The Dead Sea area exhibits a very rich inventory of environments, that is, active gravel beaches, abandoned beaches, raised shore terraces, active and abandoned coastal cliffs, active alluvial fans, fan terraces, fan deltas, talus veneers, neotectonic fault scarps, neotectonically deformed alluvium, and badlands in lacustrine deposits. However, the Dead Sea area has not yet been morphologically mapped, nor have any terrain evaluation methods been applied here. Scattering characteristics of the different roughness surfaces, texture contrasts, and reflectance studies may prove useful in future relief pattern classification.

Acknowledgments

Thanks to R. Amit, Y. Enzel, A. Schick, and E. Ziberman for constructive review comments and helpful suggestions.

REFERENCES

Agnon, A., 1982, Post-Lisan faulting at the Dead Sea graben border [abs.]: Israel Geological Society, Annual Meeting, p. 1.

Agnon, A., 1993, The fault escarpment south-west to the Dead Sea—From stratigraphy to historical morphotectonics: Israel Geological Society, Annual Meeting, Arad, Field trips guidebook, p. 81–97 (in Hebrew).

Amit, R., 1990, Shattered gravel in desert Reg soil—The effect of salts on the nature and rate of the weathering process [Ph.D. thesis]: Jerusalem, The Hebrew University, 238 p. (in Hebrew with English summary).

Amit, R., and Gerson, R., 1986, The evolution of Holocene reg (gravelly) soils in deserts—An example from the Dead Sea region: *Catena*, v. 13, p. 59–79.

Amit, R., Gerson, R., and Yaalon, D. H., 1993, Stages and rate of gravel shattering process by salt in desert Reg soils: *Geoderma*, v. 57, p. 295–324.

Arkin, J., 1993, "Karstic" sinkholes in alluvial fans: Israel Geological Society, Annual Meeting, Arad, Field trips guidebook, p. 71– 80, (in Hebrew).

Begin, Z. B., 1975, Structural and lithological constraints on stream profiles in the Dead Sea region: *Journal of Geology*, v. 83, p. 97–111.

Begin, Z. B., Broecker, W., Buchbinder, B., Druckman, Y., Kaufman, A., Magaritz, M., and Neev, D., 1985, Dead Sea and Lake Lisan levels in the last 30,000 years: Jerusalem, Geological Survey of Israel, Preliminary report GSI/29/85, 18 p.

Bentor, Y., and Vroman, A., 1960, The geological map of Israel, sheet 16, Mt. Sedom: Jerusalem, Geological Survey of Israel, 117 p.

Blanckenhorn, M., 1912, *Naturwissenschaftliche Studien am Toten Mer und in Jordantal*: Berlin, Friedlander, 478 p.

Bowman, D., 1971, Geomorphology of the shore terraces of the Late Pleistocene Lisan lake (Israel): *Palaeogeography, Palaeoclimatology, Palaeoecology*, v. 9, p. 183–209.

Bowman, D., 1974, River Terraces in the Dead Sea area—Morphology and genesis [Ph.D. thesis]: Jerusalem, The Hebrew University, 184 p. (in Hebrew with English summary).

Bowman, D., 1977, Stepped-bed morphology in arid gravelly channels: *Bulletin of the Geological Society of America*: v. 88, p. 291–298.

Bowman, D., 1978, Determination of intersection points within a telescopic alluvial fan complex: *Earth Surface Processes*, v. 3, p. 265–276.

Bowman, D., 1982, Iron coating in recent terrace sequences under extremely arid conditions: *Catena*, v. 9, p. 353–359.

Bowman, D., 1988, The declining but non-rejuvenating base level—The Lisan Lake, the Dead Sea area, Israel: *Earth Surface Processes*, v.13, p. 239–249.

Bowman, D., 1990, Climatically triggered Gilbert-type lacustrine deltas, the Dead Sea area, Israel: International Association of Sedimentologists Special Publication no. 10, p. 273–280.

Bowman, D., and Gross, T., 1989, Neotectonics in the Northern Arava Rift: Earth Science Section, Ministry of Energy and Infrastructure, Final research report, 53 p. (in Hebrew).

Bowman, D., and Gross, T., 1992, The highest stand of Lake Lisan: 150 meter below MSL: *Israel Journal of Earth Science*, v. 41, p. 233–237.

Butzer, K. W., 1958, *Quaternary stratigraphy and climate in the Near East*: Bonn, Bonner Geographische Abhandlungen, v. 24, 157 p.

Dan, J., 1981, Soil formation in arid regions of Israel: *in* Dan, J., Gerson, R., Koyumdjisky, H., and Yaalon, D. H., eds., *Aridic soils of Israel*: Beth Dagan, Volcani Center, Special publication 190, p. 17–50.

Denny, C. S., 1965, *Alluvial fans in the Death valley region, California and Nevada*: U.S. Geological Survey Professional Paper 466, 62 p.

El-Isa, Z.H., and Mustafa, H., 1986, Earthquake deformations in the Lisan deposits and seismotectonic implications: *Geophysical Journal of the Royal Astronomy Society*, v. 86, p. 413–424.

Frostick, L. E., and Reid, I., 1989, Climatic versus tectonic control of fan sequences—Lessons from the Dead Sea, Israel: *Journal of the Geological Society London*, v. 146, p. 527–538.

Gardosh, M., Reches, Z., and Garfunkel, Z., 1990, Holocene tectonic deformation along the western margins of the Dead Sea: *Tectonophysics*, v. 180, p. 123–137.

Garfunkel, Z., 1981, Internal structure of the Dead Sea Leaky transform (rift) in relation to plate kinematics: *Tectonophysics*, v. 80, p. 81–108.

Garfunkel, Z., and Freund, R., 1981, Active faulting in the Dead Sea Rift: *Tectonophysics*, v. 80, p. 1–26.

Gerson, R., 1972, Geomorphic processes of Mount Sodom [Ph.D. thesis]: Jerusalem, The Hebrew University, 234 p. (in Hebrew with English summary).

Gerson, R., 1977, Sediment transportation for desert watersheds in erodible material: *Earth Surface Processes*, v. 2, p. 343–361.

Gerson, R., Amit, R., and Grossman, S., 1985, Dust availibility in desert terrain—A study in the deserts of Israel and the Sinai: Report for the U.S. Army Research, Development and Standardization Group, U.K. Institute of Earth Sciences, The Hebrew University of Jerusalem, 190 p.

Gilat, A., 1991, Oversized U-shaped canyon development at the edge of rotating blocks due to wrench faulting on margins of the Dead Sea graben: *Terra Nova*, v. 3, p. 638–647.

Horowitz, A., 1979, *The Quaternary of Israel*. London, Academic Press, 394 p.

Huntington, E., 1911, *Palestine and its transformation*: Boston, Houghton Mifflin, 443 p.

Kaufman, A., 1971, U-series dating of Dead Sea basin carbonates: *Geochimica et Cosmochimica Acta* , v. 35, p. 1,269–1,281.

Klein, C., 1961, On the fluctuations of the level of the Dead Sea since the beginning of the 19th Century: Hydrological paper no. 7, Hydrological service of Israel, Ministry of Agriculture, Jerusalem, 83 p.

Klein, C., 1982, Morphological evidence of lake level changes, western shore of the Dead Sea: *Israel Journal of Earth Sciences*, v. 31, p. 67–94.

Mor, U., 1987, The Geology of the Judean Desert in the area of Nahal Daraja [M.Sc. thesis]: Jerusalem, The Hebrew University, 112 p. (in Hebrew with English Abstract).

Neev, D., and Emery, K. O., 1967, The Dead Sea—Depositional processes and environments of evaporites: Jerusalem, Geological Survey of Israel Bulletin 41, 147 p.

Neev, D., and Hall, J. K., 1977, Climatic fluctuations during the Holocene as reflected by the Dead Sea levels, *in* Greer, D. C., ed., *Terminal lakes*: Proceedings from International conference on Desertic Terminal Lakes, Ogden, Utah, p. 53–60.

Neev, D., and Hall, J. K., 1979, Geophysical investigations in the Dead Sea: *Sedimentary Geology*, v. 23, p. 209–238.

Nir, D., 1965, A geomorphological map of the Judean desert 1:100.000: *Scripta Hierosolymitana*, [Publications of the Hebrew University, Jerusalem], v. 15, p. 5–29.

Nir, Y., 1967, Some observations on the morphology of the Dead Sea Wadis: *Israel Journal of Earth Sciences*, v. 16, p. 97–103.

Picard, L., 1943, *Structure and evolution of Palestine*: Jerusalem, Department of Geology, The Hebrew University, 188 p.

Quennell, M. A., 1958, The structural and geomorphic evolution of the Dead Sea rift: *Journal of the Geological Society of London Quarterly*, no. 453, p. 1–24.

Reches, Z., and Hoexter, D. F., 1981, Holocene seismic and tectonic activity in the Dead Sea Area: *Tectonophysics*, v. 80, p. 235–254.

Reid, I., and Frostick, L., 1993, Late Pleistocene Rhythmite Sedimentation at the margins of the Dead Sea Trough—A Guide to palaeoflood frequency, *in* MacManus, J., and Duck, R. W., eds., *Geomorphology and sedimentology of lakes and reservoirs*: New York, John Wiley & Sons, p. 259–273.

Schattner, I., 1962, *The lower Jordan Valley—A study in the fluviomorphology of an arid region*: Scripta Hierosolymitana, [Publications of the Hebrew University, Jerusalem], v. 11, 123 p.

Schumm, S. A., 1976, Episodic erosion: A modification of the geomorphic cycle, *in* Melhorn, W. N., and Flemal, R. C., eds. *Theories of landform development*: Binghamton, New York, Publication in Geomorphology, State University of New York, p. 69–85.

Sneh, A., 1979, Late Pleistocene fan-deltas along the Dead Sea Rift: *Journal of Sedimentary Petrology*, v. 49, p. 541–552.

Survey of Israel, 1985, *Atlas of Israel*,:Tel Aviv, Survey of Israel.

Widowinski, S., and Zilberman, E., 1994, Large scale asymme-

tries across the Dead Sea Rift—A half-graben model for the formation of the Arava Valley and Dead Sea basin [abs.]: Israel Geological Society, Annual Meeting, p. 115.

Willis, B., 1928, Dead Sea problems—Rift valley or ramp valley: *Bulletin of the Geological Society of America*, v. 39, p. 490–542.

Yechieli, Y., 1987, The geology of the Northern Arava rift and the Machmal anticline, Hazeva area: Jerusalem, Geological Survey of Israel Report GSI/30/87, 94 p.

21. FLUCTUATIONS OF LATE PLEISTOCENE LAKE LISAN IN THE DEAD SEA RIFT

Tina M. Niemi

Lake Lisan was a large, saline lake that filled the Dead Sea rift in the late Pleistocene. At its peak level, Lake Lisan filled the fault valley for 220 km from the Sea of Galilee (Lake Kinneret) in the north to 30 km south of the Dead Sea (Fig. 21-1). Lake Lisan is just one in a series of inland lakes that filled the depression along the Dead Sea–Jordan transform. During its 45,000-year duration (approximately 60–13 ka), the level of Lake Lisan did not remain static. Fluctuations in lake level are recorded in the deposits of Lake Lisan, known as the Lisan Formation. Bentor and Vroman (1960, p. 68–69) wrote: "The repeated alternations of finely varved chalky and gypseous (sic.) lake sediments with horizons of coarse clastics in many sections near the mountain border, certainly reflect repeated changes of the Lisan lake level." The fluctuating levels were caused by late Quaternary climatic changes and the active tectonic setting of the region. These climatic and tectonic changes are not yet fully understood. In this chapter, I review the stratigraphy, radiometric age, and paleoshoreline indicators of the Lisan Formation and discuss the roles of tectonics and climate in controlling the level of Lake Lisan.

STRATIGRAPHY

Lartet (1869) first used the term *Lisan deposits* to describe ancient sediments of the Dead Sea in the area of the Lisan Peninsula (Fig. 21-1). Other early publications also mention these unique, laminated marl and aragonite deposits (e.g., Hull, 1886; Blanckenhorn, 1914, Picard, 1943). Bentor and Vroman (1960) first formally used the name Lisan Formation for a 40- to 80-m-thick Pleistocene marl that rests unconformably on either older lacustrine sediments or bedrock of various ages.

The maximum thickness of the Lisan Formation is uncertain. Thick stratigraphic sections are interpreted from seismic reflection data in the northern Dead Sea basin (Neev and Hall, 1979) and from borehole data mostly from the southern Dead Sea basin (Neev and Emery, 1967; Horowitz, 1979). Within the Dead Sea depression, the Lisan Formation overlies the Samra and Amora Formations, which have been identified in oil wells (see Gardosh et al., chapter 5, this volume). However, no clear stratigraphic boundaries are visible between the Lisan and underlying formations. Along the basin margin, the Samra Formation consists of lacustrine and fluvial sediments that have U-Th dates of 350 to 116 ka (Kaufman et al., 1992).

There remains some variation in the stratigraphic nomenclature of the Lisan Formation (Figure 21-2). Several members of the formation have been defined, based on lithologic changes and erosional unconformities. Langozky (1961) divided the formation into the lower clastic Hamarmar member and the upper evaporitic Amiaz member. Begin et al., (1974), using outcrops along the entire length of western exposures, considered only the evaporitic section as the Lisan Formation. These authors divided the Amiaz member of Langozky (1961) into two lithostratigraphic sections: a lower Laminated member and an upper White Cliff member. Horowitz (1979) included both the lower Hamarmar member (Langozky, 1961) and the upper "unnamed clastic unit" of Begin et al. (1974) in the Lisan Formation. Horowitz (1979) named the unnamed clastic unit

the Fatzael member. This unit is only found overlying the Lisan Formation north of Masada. A more recent 117-ka date for the Hamarmar section indicates that it belongs to the Samra Formation (Kaufman et al., 1992). The interbedded gravel, sand, silt, and clay of the Fatzael member occasionally contain freshwater *Melanopsis* gastropods. In Wadi Fatzael, epipaleolithic stone tools found within the section indicate an age of approximately 19–10 ka (Bar-Yosef et al., 1974). The Fatzael member, or unnamed clastic unit, is not commonly included in the Lisan Formation.

The Lisan Formation has been dated using $^{230}Th/^{234}U$ and ^{14}C dating methods. The U–Th dating method on the Lisan Formation yielded ages of 60–40 ka for the lower part and ages of 40–17 ka for the upper portion (Kaufman, 1971). The top of the Lisan Formation on Mt. Sedom has yielded slightly younger U-Th ages of 13 ±4 ka (Kaufman, 1971). Results of radiocarbon analyses suggest that the upper White cliff member accumulated between 33–15 ka (Kaufman, 1971) or between 35–15 ka (Vogel and Waterbolk, 1972). Comparison of the U-series and ^{14}C dating of the Lisan shows that the radiocarbon-dated carbonate gives slightly younger ages for the same stratigraphic position (Kaufman, 1971; Vogel, 1979, 1980). Recent reanalyses of Lisan Formation aragonite layers in the Dead Sea region (Kaufman et al., 1992) show that Lake Lisan began forming 63 ±7 ka and the upper White cliff member, which probably marks the maximum areal extent of the lake, began accumulating ca. 36 ±5 ka. High-precision $^{230}Th/^{234}U$ dates of aragonite layers from the Laminated member by Stein et al. (1992) yield similar ages for the lower section between 68 and 40 ka. Druckman et al. (1987) published a ^{14}C age of 14,600 yr B.P. for an organic layer beneath 3–4 m of oolitic lacustrine sediments at an elevation of about -190 m; they attribute this age to the highest highstand of Lake Lisan. This radiocarbon age may be too young by about 3,000 years, given published corrections of ^{14}C dates to the U/Th series (Bard et al., 1990).

The basin of Lake Lisan is divided into a narrow, 7-km-wide northern basin between the Sea of Galilee and Marma Feiyad, and a 15-km-wide southern basin from Marma Feiyad to 30 km south of Mt. Sedom (Fig. 21-1). The lithology of the Lisan Formation varies along the length and width of the lake (Fig. 21-3). Begin et al. (1974) divided the Lisan Formation into three facies—gypsum, aragonite, and diatomite—each with a specific regional distribution. The gypsum facies of the formation, located west of the Dead Sea, is marked by abundant gypsum beds. The Lisan Formation south and north of the Dead Sea is called the aragonite facies because of the abundance of aragonite and an increased detrital component. North of Wadi Malih in the Marma Feiyad area, the Lisan Formation is predominantly clastic but also contains diatomite layers. It is referred to as the diatomite facies.

Two broad depositional facies of the Lisan Formation are illustrated in Figure 21-4 and described subsequently. The finely laminated marls (often thought of as "typical" Lisan Formation) are lake-bottom sediments found at elevations below the maximum elevation of the lake shoreline. These varved sediments comprise the lacustrine facies. Toward the basin margin, clastic sediments predominate and form the fan-delta facies.

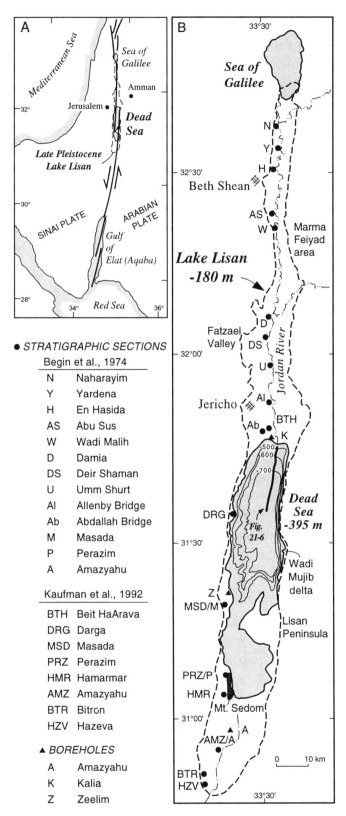

● STRATIGRAPHIC SECTIONS

Begin et al., 1974

N	Naharayim
Y	Yardena
H	En Hasida
AS	Abu Sus
W	Wadi Malih
D	Damia
DS	Deir Shaman
U	Umm Shurt
Al	Allenby Bridge
Ab	Abdallah Bridge
M	Masada
P	Perazim
A	Amazyahu

Kaufman et al., 1992

BTH	Beit HaArava
DRG	Darga
MSD	Masada
PRZ	Perazim
HMR	Hamarmar
AMZ	Amazyahu
BTR	Bitron
HZV	Hazeva

▲ BOREHOLES

A	Amazyahu
K	Kalia
Z	Zeelim

Figure 21-1 Location of Lake Lisan. A, Lake Lisan formed along the Dead Sea transform. B, Location of stratigraphic sections, boreholes, and other sites (modified after Begin et al., 1974).

Lacustrine facies

The lacustrine facies of varved sediments consists of very fine lamina of white aragonite alternating with dark gray detrital

Langozky (1961)		Begin et al. (1974)		Horowitz (1979)	Age (ka)
Lisan Formation		Lisan Formation	Unnamed clastic member	Lisan Formation	Fatzael member	
						~17-13
	Amiaz member		White Cliff member		Amiaz member (evaporitic)	36 ±5
			Laminated member			
						63 ±7
	Hamarmar member (formation)		Samra Formation		Hamarmar member (clastic)	

Figure 21-2 Stratigraphic nomenclature and U-Th age of members of the Lisan Formation.

sediment. The detrital layers are composed mainly of calcite and clay minerals. In addition to thick beds of gypsum, aragonite, and diatomite, the White Cliff member contains a higher percentage of aragonite than the Laminated member. Bentor and Vroman (1960) suggested that the varves were seasonal with the detrital layers deposited in the winter and the chemical layers deposited in the summer.

Both freshwater and brackish-water diatom species occur within the detrital, aragonite, and diatomite layers of the Lisan Formation. Begin et al. (1974) defined three diatom zones based on the restricted abundance of euryhaline forms. These zones are (1) *Nitzschia vitrea* for the base of the Laminated member, (2) *Nitzschia lembiformis* for the top of the Laminated member, and (3) *Rhopalodia gibberula* for the White Cliff member north of the Dead Sea. North of Beth Shean, the uppermost section of the formation contains the freshwater species *Gomphonema* sp. A southward decrease in the number of diatoms preserved in the formation suggests a salinity gradient which increased toward the south in Lake Lisan. Because Lake Lisan's drainage basin was the same as the Dead Sea's (see Fig. 1-2, this volume), freshwater influx was greatest in the northern Jordan Valley, and lowered the salinity of the water in this region. In addition, the northern water mass may not have mixed with the water to the south because of a sill that restricted circulation. The southernmost portion of the lake near the Amazyahu section, was slightly less saline than the Dead Sea section to the north due to freshwater input from the south (Begin et al., 1980).

Begin et al. (1974) also interpreted the decrease in diatoms vertically in the Lisan stratigraphic section as a result of an increase in salinity of the lake over its lifespan. Corroborating evidence for an increase in salinity with time was presented by Gardosh (1987) based on the trend in ionic ratios of detrital lamina. In contrast, geochemical analyses of water-soluble salts in aragonite layers of the Lisan Formation by Katz et al. (1977) and Katz and Kolodny (1989) showed an initial high salinity of the Lake Lisan brine (similar to the present Dead Sea), which decreased two- or threefold in the upper Lisan sediments. These authors suggested that Lake Lisan evolved from a "small, yet deep, hypersaline Dead Sea-like, water body" with an estimated water depth of 400–600 m (Katz et al., 1977, p. 1609).

Although only euryhaline forms are found in the aragonite lamina, both euryhaline and freshwater diatoms are present in the detrital lamina. Over 150 freshwater taxa have been identified (Begin et al., 1974). Because there are no deep-water lakes with salinities comparable to those of Lake Lisan, there are no modern environmental analogs for the lake and its diatom population. Two interpretations of the deposition of the diatom-rich sediments are possible: either the diatoms are planktonic forms that lived in a stratified water body, or the diatoms are littoral, benthic forms, like modern shallow-water species. Ehrlich and Noël (1988) suggested that the diatomite layers were deposited in the littoral zone as a result of desiccation. Diatoms within the dark detrital layers of the laminated sediment were transported from the littoral and surface freshwater zone and were deposited basinward with other detritus.

Fan-delta facies

Lake Lisan filled a pre-existing depression. The Lisan Formation was deposited in the arms of the lake, that extended into canyons that drained the flanks of the Dead Sea rift. Where the streams emerged from the rift escarpment, a series of fan deltas developed. With the final lowering of Lake Lisan, the streams draining the rift margins incised across their older fan deltas and built younger deltas into the Dead Sea. The fan-delta sediments exposed in the cliffs of the incised drainage consist of clast-supported, subhorizontally to cross-stratified gravels and sand of fluvial, beach, and deltaic origin.

Sneh (1979) summarized the fan-delta sequences along the major tributaries of Lake Lisan along the western margin of the Dead Sea. Fining-upward sequences indicating rising lake levels during deposition are overlain by subaerial alluvial-fan sediments. Clastic sediments at the top of the Lisan Formation at the mouth of relatively small drainage basins exhibit a distinct Gilbert-style (bottomset, foreset, and topset) deltaic sequences (Bowman, 1990). The change in hydrologic regime reflected in the sedimentary sequence led Bowman (1990) to infer a wetter climatic phase for the Lisan Formation between 18 and 14 ka.

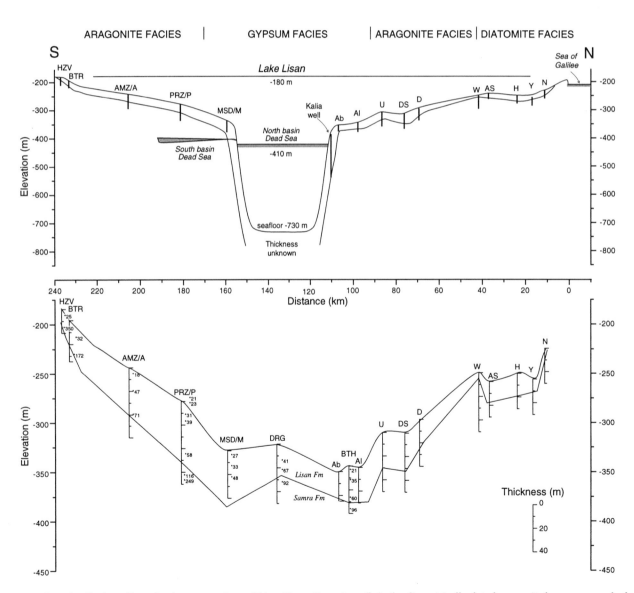

Figure 21-3 Longitudinal profile and columnar sections of Lisan Formation. Ages (ka) of radiometrically dated aragonite layers are marked. Figure modified after Begin et al. (1974) and Kaufman et al. (1992).

Figure 21-4 Photographs of the Lisan Formation in Nahal Mor (located 9.5 km south of Masada): A, lacustrine facies with algal stromatolites; B, fan-delta facies interbedded lacustrinal marls.

PALEOSHORELINE INDICATORS

The maximal shoreline elevation of Lake Lisan is often cited at -180 m based on the top elevation of the Lisan marls (e.g., Neev and Emery, 1967; Begin et al., 1974). Clark (1988) observed that the maximum elevation of Lisan marls in the Wadi el-Hasa was -160 m in the southeastern margin of the basin. He attributed the higher elevation to post-Lisan tectonic uplift. Further evidence for a higher maximal extent of Lake Lisan is present southwest of the Dead Sea: lacustrine deposits (Yechieli, 1987) and beach bars (Bowman and Gross, 1992) in the Hazeva area at an elevation of -150 m suggest a possible upward revision of the maximal Lake Lisan highstand.

The final retreat of Lake Lisan is marked by lake terraces (-182 to -384 m in elevation) along the western border escarpment (Bowman, 1971; also chapter 20, this volume). Only the last regressional phase is preserved at the surface; terraces developed during low levels of the lake were either buried or eroded. Terraces below -270 m may represent fluctuations of the Dead Sea in the Holocene (see Frumkin, chapter 22, this volume).

Cycles of lake lowering and rising are recorded in the lithology of the Lisan Formation, where laminated aragonitic lacustrine deposits are interbedded with alluvial-fan and deltaic sediments. Additional evidence that sets limits on the elevation of paleoshorelines include algal stromatolites, archaeological sites, and subsurface data from seismic reflection records and sediment cores.

Sedimentary cycles

A detailed study of seven regressive-transgressive depositional cycles in a 70-m section of Lisan Formation exposed in the Amazyahu canyon was presented by Sneh (1982). Each cycle is composed of a basal cross-bedded, pebble gravel that grades upsection to ripple-bedded sand. The sand beds are overlain by oolitic mudstones and lacustrine chalks. The elevation of the lowest aragonite in the Amazyahu section suggests that the early lake level phase (60 ka) may not have reached above -280 m. U-Th dating of the aragonitic layers higher in this section indicate that Lake Lisan rose above -260 m only about 47 ka, when the lake began to fluctuate frequently between lacustrine and clastic facies. The lake continued to rise to -240 m by 18 ka (Kaufman et al., 1992).

Cycles of clastic and laminated lacustrine facies were attributed to climatic fluctuations by Manspeizer (1985) and Frostick and Reid (1989). Manspeizer (1985) investigated the facies distribution of fan-delta sequences along the western margin of the Dead Sea and focused on the Nahal Darga exposure. Because of the absence of syndepositional faulting, Manspeizer (1985) concluded that facies in the Lisan Formation reflect climatic rather than tectonic cycles. This conclusion was also reached by Frostick and Reid (1989). From stratigraphic section measurements of exposures along the banks of Nahal Rahaf (3.5 km south of Masada), Frostick and Reid (1989) identified seven clastic intervals of braided-stream and alluvial-fan sediments interbedded with laminated lacustrine and beach-facies sediments. Given the abrupt boundaries between lake sediments and alluvium, Frostick and Reid (1989) concluded that the clastic cycles denoted a change in climatic conditions with high-magnitude floods depositing the horizontal beds of fluvial sediments. They suggested that pulses of tectonic uplift or subsidence would be manifested as sequences that coarsen upwards. Such sequences are absent in the stratigraphic section.

Algal stromatolites

Blanckenhorn (1912) first described tufa deposits along the eastern margin of the Dead Sea. Tufa deposits located at elevations -180 to -370 m have also been described along the western shore of the Dead Sea (Bentor and Vroman, 1960; Neev and Langozky, 1960; Neev and Emery, 1967; Buchbinder et al., 1974; Begin, 1975; Buchbinder, 1981). Neev and Emery (1967) suggested that the tufa was formed from algal mats. The microscopic fabric of the deposits was investigated by Buchbinder et al. (1974) and Buchbinder (1981). The microfabric consists of radially rimmed spheroids of aragonite—a structure formed by coccoid blue-green algae (a non-filamentous unicellular algae)—and smooth laminations indicative of filamentous algae.

Algal stromatolite structures are found at various elevations within the Lisan Formation. A series of bowl-shaped algal stromatolites at an elevation of -240 m occur 5 km north of Jericho. Similar algal structures are found within the Lisan Formation at Nahal Mor, 9.5 km south of Masada at an elevations of -340 to -370 m (Fig. 21-4). The stromatolites were formed in the photic zone in shallow, highly saline water. An increased salinity would correspond to the shrinking of Lake Lisan to lower elevations where stromatolites were formed (-240 and -370 m), and indicates steep drops in lake level over relatively short time intervals. A curve of levels of Lake Lisan was compiled primarily from the elevation and radiocarbon age data of organic carbon from algal stromatolites by Begin et al. (1985). This curve (Fig. 21-5A) shows three Lake Lisan lowstands, with a minimum elevation of -370 m, at approximately 35–28 ka, 24–22 ka, and 17 ka. The lake level fluctuation curve is compared to curves based on estimates of runoff to evaporation ratios for the Lisan Formation (Neev and Emery, 1967; Begin et al., 1974; Neev and Hall, 1977).

Archaeology

Geoarchaeological investigations provide information on paleoshorelines. An in situ prehistoric site with Mousterian artifacts (150–40 ka) found within the lower Lisan Formation, 15 km north of Jericho at an elevation of -222 m (Bar-Yosef et al., 1974; Bar-Yosef, 1987), indicates low levels of Lake Lisan during deposition of a portion of the Laminated member. Tools of possible epipaleolithic age (19–10 ka) eroding from a Lisan Forma-

tion outcrop ca. 2 km south of Wadi Fatzael may indicate a Lake Lisan level lower than -215 m during that period (Schuldenrein and Goldberg, 1981). Late Natufian cultural remains overlying the Lisan Formation at the Salibiya site (9 km north of Jericho) clearly demonstrate that the level of Lake Lisan was below about -230 m by 10–12 ka (Schuldenrein and Goldberg, 1981). Paleolithic artifacts have also been found in the Lisan Formation in Jordan (Vita-Finzi, 1966).

On the southeast margin of the Dead Sea in the Wadi Hasa drainage, artifacts of Kebaran affinity (ca. 18-14 ka) are found in the upper Lisan Formation (Copeland and Vita-Finzi, 1978). This deposit has been eroded and filled with an inset terrace that contains lithics dated to the Natufian period. This buttress unconformity indicates a lowstand followed by a highstand of Lake Lisan between the Kebaran and Natufian cultural periods.

The distribution of surface archaeological sites southeast of the Dead Sea clearly indicates Neolithic sites at -300 m (Neeley, 1992), but lack Upper Palaeolithic and epipaleolithic artifacts. The paucity of prehistoric lithics at lower elevations is not conclusive evidence that the lake maintained a high level. Lowstand sites have a very low preservation potential because of erosion and burial by subsequent highstand lake deposits.

Recent survey and excavation of an in situ archaeological site, Ohalo II, located near the southern outlet of the Sea of Galilee provide conclusive evidence that by 19 ka, saline Lake Lisan had retreated to the south (Nadel and Nir, 1994). Fish bones associated with the site indicate the presence of a freshwater body, possibly the beginning of the Sea of Galilee. Koucky and Smith (1986) used the distribution of archaeological sites, combined with geomorphic and geologic evidence, to suggest that a freshwater lake (Lake Beisan) with a maximal shoreline elevation of -100 m was ponded north of the Marma Feiyad region between 12–5 ka. It is unclear whether this separation was produced by a fall in the level of Lake Lisan or by tectonic uplift or subsidence of a structural block.

Seismic stratigraphy

The morphology of the Dead Sea floor was first studied from echo sounder records from the submerged portion of the Lisan Peninsula by Emery (1963) and from records of the southwest sector of the Dead Sea by Neev and Emery (1967). Neev and Emery (1967) observed a dendritic drainage pattern that they interpreted as incision caused by subaerial erosion during a near complete desiccation of the basin. The submerged canyon heads lie between -425 and -645 m in elevation and extend to the seafloor (-725 m) where they appear as flat-bottomed channels that suggest subsequent sediment infilling. Based on the interpretation of single-channel seismic reflection data from the Dead Sea, Neev and Hall (1976, 1977, 1979) attributed this desiccation to the final lowering of Lake Lisan at the Pleistocene–Holocene boundary.

Seismic reflection data (Neev and Hall,1976, 1977, 1979) also show that the delta of the Jordan River ranges in thickness from 300 to 100 m and contains 16 km^3 of sediment. Since there is no direct age control, the timing of the initiation of the delta is uncertain. Neev and Hall (1976, 1977, 1979) concluded that the delta began forming only after the final lowering of Lake Lisan, about 17–13 ka. It seems more likely that the delta started to form during lowstands (-370 m) of Lake Lisan between 35–13 ka (see Fig. 21-5A; Begin et al., 1985). During these lowstand phases, the basin north of the Marma Feiyad area would have been disconnected from the southern Lake Lisan basin and may have held freshwater. There is some evidence that this northern subbasin may have a different stratigraphic history (Macumber and Head, 1991). The northern intermediate lake would have fed water to an ancestral Jordan River, much as the Sea of Galilee flows into the Jordan River today.

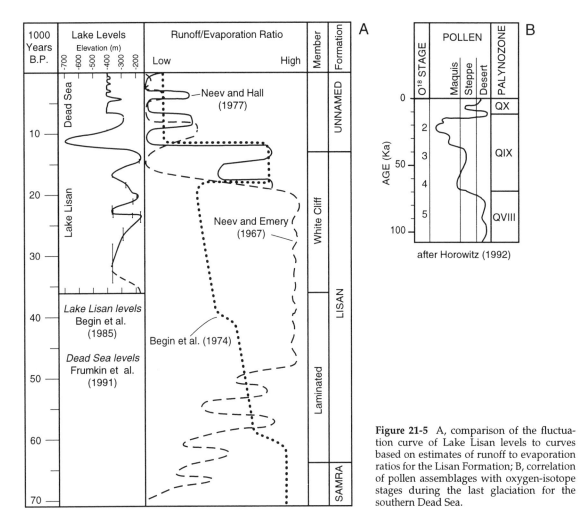

Figure 21-5 A, comparison of the fluctuation curve of Lake Lisan levels to curves based on estimates of runoff to evaporation ratios for the Lisan Formation; B, correlation of pollen assemblages with oxygen-isotope stages during the last glaciation for the southern Dead Sea.

The single-channel seismic reflection profiles of the submarine Jordan River delta (Neev and Hall, 1976, 1979) were reexamined using the techniques of seismic sequence stratigraphy (Niemi and Ben-Avraham, 1994). The relative level of the lake can be reconstructed by examining the detailed geometry of the reflectors within the deltaic sedimentary wedge. The base of each sequence is defined by erosional truncation of underlying reflectors and onlapping reflectors that are interpreted as lowstand deposits. The maximum flooding surface is identified as the point where onlapping reflectors change to downlapping reflectors and marks a change from transgression to regression during the lake's highstand. On the seismic line shown in Figure 21-6, which extends from the mouth of the Jordan River across the submerged Jordan delta in the Dead Sea, five cycles of lowstand and highstand sequences were identified. These cycles correlate well with the lake level curve (Fig. 21-5A) of Begin et al. (1985). A major erosional event is marked by the distinct truncation at the base of sequence 2. This erosional event is correlated to incision of the channels along the western flank of the Dead Sea (Neev and Hall, 1979), which occurred at the final retreat of Lake Lisan around 13–12 ka.

Sediment Cores

Very little is known about the subsurface lithology of the Lisan sediments in the northern Dead Sea basin because of the lack of long sediment cores. The 1934 Kalia water well (Picard, 1943),

at a surface elevation -400 m near the north shore of the Dead Sea (Fig. 21-3), contains 118 m of interbedded marl and clastic sediments resting on bedrock of probable Eocene age (Rotstein et al., 1991). All or part of these sediments may represent Lisan Formation and Holocene deposits.

A research borehole at the distal section of the Zeelim alluvial fan (Fig. 21-1) at a surface elevation of -394 m provides the only radiometrically dated section in the northern Dead Sea basin. The 34.5-m stratigraphic section consists of a basal layer of laminated aragonite overlain by a clay layer with halite crystals, a 6.5-m halite layer, and 24 m of interbedded fine and coarse clastic beds (Yechieli et al., 1993). U-Th dating of the basal aragonite yielded an age of 21 ka, whereas ^{14}C dating of wood in the overlying clay yielded an age of 11 ka. The hiatus in deposition or the erosion between 21 and 11 ka correlates with the extreme drop of the lake level described previously. [14]C dating of wood overlying the 6.5-m-thick halite indicates that the halite was deposited during arid conditions sometime between 11 and 8.5 ka.

Palynological analyses of the Lisan Formation from well chippings from oil exploration wells mostly in the southern basin of the Dead Sea was presented by Horowitz (1987, 1992). Ten palynozones were defined using the Amazyahu 1 borehole (Fig. 21-1) as a type section (Horowitz, 1987, 1992). Zones X and IX, defined as the top 100 to 200 m of section, were correlated to the Holocene Dead Sea and late Pleistocene Lake Lisan, respectively. Zone IX is very high in arboreal pollen, indicating

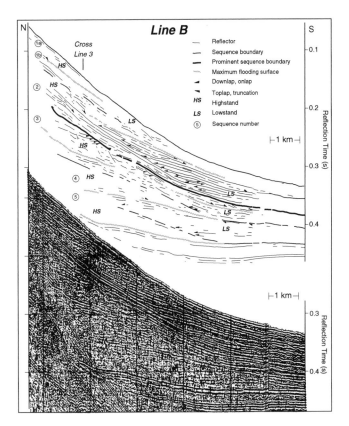

Figure 21-6 Seismic sequence stratigraphy of the Lisan Formation. Five cycles of lowstand and highstand sequences correlate well with the lake level curve (see Fig. 21-5). A major erosional event is marked by the dramatic truncation at the base of sequence 2. This erosional event is correlated to incision of the channels along the western flank of the Dead Sea, which occurred at the final retreat of Lake Lisan around 13–12 ka.

a wet climate during the deposition of Lake Lisan. Horowitz correlated the Lake Lisan pluvial to the Würm (Wisconsin) glacial period based on the similar shape of the pollen diagrams to oxygen isotope curves (Fig. 21-5B). Interstadial periods and a shift to drier climatic conditions within the Lisan section are represented in the pollen spectra by a decrease in arboreal pollen and an increase in steppe vegetation.

TECTONIC VERSUS CLIMATIC CONTROLS OF LAKE LISAN LEVELS

The level of a closed basin lake, such as Lake Lisan or the Dead Sea, is a result the amount of runoff from precipitation in the watershed that reaches the lake less the amount of water lost from the surface of the lake by evaporation. However, in an actively subsiding basin like the Dead Sea, tectonic motion may also cause oscillations in the lake level directly, or by structurally blocking inflow to the lake. The Lisan Formation crosses the escarpment of the rift valley, where it was deposited within the wadi canyons with no appreciable displacement, implying quiescence along the marginal fault system. Langozky (1963, p. 23) stated that "along the margins of the whole Rift Valley there is clear evidence that the Lisan Formation is undisturbed." A recent detailed study of the Lisan Formation near Masada (Marco et al., 1996) documented some small scale syndepositional faults. Seismic reflection data show that active faults lie at the floor of the Dead Sea (Ben-Avraham et al., 1993). But did pulses of basin subsidence away from the main escarpment

contribute to or control lake fluctuations? Or are changes in lake level mainly a result of shifts in climatic conditions?

Tectonism

Spectacular deformation horizons are found between undeformed beds within the Lisan Formation. Bentor and Vroman (1960) suggested that the slumped layers interbedded between horizontal layers represented soft sediment deformation features caused by turbidity currents during flash floods. El-Isa and Mustafa (1986) studied the intraformation deformation within the Lisan Formation on the Lisan Peninsula on the eastern margin of the Dead Sea. Folded and faulted layers and layers in which the laminae were destroyed were attributed to earthquake-induced deformation and mixing of the water-saturated bottom sediments of Lake Lisan. The number and thickness of deformed layers observed in a 3.8-m thick section were used to derive the frequency and magnitude of paleoearthquakes. Intraformational faulting within the Lisan Formation exposed along the western margin of the Dead Sea has also been attributed to seismogenic processes (Agnon and Marco, 1994; Marco et al., 1996).

Tectonic deformation in the Dead Sea-Jordan transform basin has varied over time, has migrated, and in some cases has even been inverted. For example, once a site of subsidence, the Lower Pleistocene lake sediment beds of the famous Ubeidiya Formation, which contain hominid stone tools, have been compressed into steeply dipping beds. Compressional folding of the Samra Formation below the Lisan Formation south of the Dead Sea (Sneh, 1982; Yechieli, 1987; Kaufman et al., 1992) indicates a change in tectonic regime prior to Lisan deposition (60 ka). Folding as young as 4 ka was documented along the transform north of the Dead Sea (Gardosh et al., 1990). The southern basin of the Dead Sea is marked by the scarp of Amazyahu fault, which has as much as 150 m of post-Lisan vertical offset (Neev and Emery, 1967).

The Lisan Formation varies considerably north and south of the Beth Shean basin. The findings of Macumber and Head (1991) suggest that the northern basin may have acted independently. Lake Lisan deposits do not extend beneath the Sea of Galilee, indicating that this basin subsided some time after 19 ka when Lake Lisan retreated from this region. Koucky and Smith (1986) interpreted a highstand (-100 m) of freshwater Lake Beisan at 12–11 ka. The southern outlet of this intermediate lake may have been structurally blocked by uplift in the Marma Feiyad area. This may have cut the flow of water to the Dead Sea region, causing a major fall in lake level to the south. Neev and Emery (1967) suggested that subsidence of the northern and southern basins of the Dead Sea occurred at the end of the Lisan deposition leaving the Lisan Peninsula as a horst block. This subsidence, controlled both by faulting and diapiric rise of salt, ". . .may even have been responsible for the rapid recession of the Lisan Lake" (Neev and Emery, 1967, p. 26).

Horowitz (1979) followed Neev and Emery (1967) by suggesting that the demise of Lake Lisan about 18 ka was caused by tectonic subsidence of the north basin of the Dead Sea. This opinion was also held by Begin et al. (1974), who described the maximum water depth of Lake Lisan as 200 m (between -180 and -380 m). The absence of a Lisan or present-day delta at the mouth of Wadi Mujib led Nir (1967) to suggest post-Lisan downfaulting of several hundred meters. Given the present elevation of the seafloor of the Dead Sea at -730 m, a subsidence rate of 19 to 29 mm/yr is required for a 350-m drop in the basin floor over 18,000 to 12,000 years.

This rate is an order of magnitude higher than other estimates for rates of subsidence of basins along the Dead Sea–Jordan transform. Most researchers estimate the Dead Sea subsidence rate to be approximately 1 mm/yr, or about 10% of

Table 21-1 Subsidence rates of basins along the Dead Sea–Jordan Transform.

Basin	Period	Subsidence Rate (mm/yr)	Reference
Hula	Since 400 ka	0.52	Kafri et al. (1983)
Hula	Since 4 Ma	0.5–0.75	Heimann (1990)
Sea of Galilee	4.4 Ma–600 ka	0.21	Heimann (1990)
Sea of Galilee	Pleistocene	1.56–2.35	Braun et al. (1991)
Dead Sea	Since 1.6 Ma	3.75	ten Brink and Ben-Avraham (1989)
North of Dead Sea	Holocene	0.83	Gardosh et al. (1990)

the long-term average geologic slip rate of 6 to 10 mm/yr (Zak and Freund, 1966; Freund et al., 1970; Garfunkel et al., 1981; Joffe and Garfunkel, 1987). Table 21-1 summarizes some of the published rates of subsidence for the Dead Sea rift basins (S. Marco, pers. com., 1995).

Seismic reflection profiles (Neev and Hall, 1976) of the 300-m-deep northern basin of the Dead Sea do not support the idea that subsidence of the deep northern Dead Sea basin began only in post-Lisan times. Eastward thickening of a wedge of sediment greater than 500 m thick from the western margin to the center of the basin is visible on the seismic profiles, suggesting that basin subsidence has been an ongoing process that predates 18 ka. This agrees with the geochemical data of Katz et al. (1977) that indicate that Lake Lisan evolved from a deep-water brine. Continued subsidence enhanced low levels of Lake Lisan. Once the lake was lowered below the sill at Marma Feiyad, the freshwater influx was apparently captured and dammed in a northern intermediate basin, thereby exacerbating lowering in the southern portion of Lake Lisan. Evidence from high resolution seismic reflection profiles indicates that the northern Dead Sea basin is still actively subsiding (Ben-Avraham et al., 1993; Niemi and Ben-Avraham, chapter 6, this volume).

Climate

The Lisan Formation was deposited during the Würm (Wisconsin) glacial period, which, on a global scale, is generally marked by climatic conditions colder than the present. Changes in the hydrologic regime of the catchment area of Lake Lisan must have affected the level of Lake Lisan. The main question is whether there is a direct correlation between climatic cycles during this ice age and fluctuations of the lake.

The pollen assemblages of Israel have been correlated to the European glacial-interglacial cycles and oceanic oxygen-isotope stages (Fig. 21-5B; Horowitz, 1992). Based on these data, Horowtiz (1992) interpreted glacial periods as pluvial or wet periods that were highly vegetated, with a southern shift of the Mediterranean biozone and widespread soil formation, whereas interglacial periods correspond to the pervasive desert conditions during interpluvials. Small excursions (or interstadials) within glacial or interglacial cycles represent climatic conditions similar to the present, with wet winters and dry summers.

Travertine deposits, with a radiocarbon age of 41–22 ka, associated with springs near Beth Shean are the age equivalent of the White Cliff member of the Lisan Formation. The travertine deposits may represent wetter climatic conditions associated with the maximal highstand of the lake (Kronfeld et al., 1988). Kronfeld et al. (1988) further argue that the presence of travertine suggests increased rainfall (rather than cooler temperatures

and decreased evaporation). Pollen spectra show wetter and cooler conditions (Horowitz, 1992).

Evidence of increased rainfall in the Negev was presented by Goodfriend and Magaritz (1988). Radiocarbon dating of carbonate nodules in paleosols in the Negev desert shows approximate periods of pedogenesis at 39–34 ka, 31–23 ka, and 15–11 ka separated by drier climatic periods characterized by erosion. Goodfriend and Magaritz (1988) suggested that the paleosols correlate with globally warmer periods or interstadials within the last glacial (Würm) and at the beginning of deglaciation (Allerod). Contrary evidence was presented by Horowitz (1987), who correlated wet intervals with glacials rather than interstadials. Given the resolution of the radiometric dating of the carbonate nodules, the same data can just as easily be interpreted as correlating with the two-stage late Würm glacial maxima at 20–23 ka and 14–15 ka (Broecker and Denton, 1989).

There is no global synchroneity in lake level fluctuations (Street and Grove, 1979). Lower solar radiation during glacial periods would decrease rainfall, causing aridity in the equatorial zone (Street and Grove, 1979). Thus, although glacials at midlatitude are marked by pluvial periods and high lake levels as in the Dead Sea region, they are marked by low lake levels in the tropics. Radiocarbon-dated marls, shells, diatomite, and algal stromatolites from the Arabian desert (lat. 18°–23°N) show highstand lakes about 36–17 ka and 9–6 ka separated by an interval of intermediate lake level (McClure, 1976). These data correlate well with lake level data from the Dead Sea region.

Much confusion has undoubtedly been introduced by the use of European glacial chronological terminology to describe the Dead Sea region paleoclimate and paleoenvironments. However, these climatic cycles, as described by the pollen diagrams (Horowitz, 1992) seem to explain the changing levels of Lake Lisan. The interstadial between Early and Middle Würm glacial periods (~40–30 ka) is marked by an unconformity and major lithologic break in the lacustrine facies of the Lisan Formation between the Laminated and White Cliff members. Middle Würm (32–22 ka) represents the maximum areal extent of Lake Lisan. By the mid-Late Würm interstadial (22–18 ka), the lake had shrunk to -240 m elevation into a sabkha plain as seen in the algal stromatolites. It fell as low as -370 m. Evidence of Upper Paleolithic culture is scarce. The northern part of Lake Lisan may have retained a freshwater to brackish-water lake after the saline waters drained around 19 ka.

The Late Würm is marked by an explosion in epipaleolithic culture and an arboreal pollen peak at 13–11 ka. These data indicate a brief return of wet climatic conditions, which may be the local manifestation of the Younger Dryas event. There is growing evidence that the Younger Dryas event, a major episode of climatic cooling that occurred during the last deglaciation around 11–10 ka, was a global event (Kudrass et al., 1991; Alley et al., 1993). Ice core data from Greenland show that the

Younger Dryas lasted for 1,300 ±70 years and that onset and termination of glacial conditions occurred very rapidly, within one to two decades (Mayewski et al., 1993). Lake Lisan had probably become a series of intermediate lakes during this event. The lake in the Dead Sea region was at a very low level, perhaps because of obstructions of inflow to the lake.

SUMMARY

The stratigraphic section exposed in deep wadi incisions shows distinct cycles of sedimentation during deposition of the Lisan Formation. These cycles are also visible in seismic reflection profiles of the ancestral and modern Jordan River delta and in trends of pollen type analyzed from borehole samples. Evidence from archaeological sites and shallow-water algal stromatolites provides limiting elevations for the paleoshoreline of Lake Lisan. Together, these data suggest that the fluctuations of Lake Lisan levels have apparently been driven by variations in precipitation and evaporation. Tectonic basin subsidence no doubt has also influenced lake level elevation throughout the Pleistocene and has played a role in the final desiccation of the lake.

The sediments of Lake Lisan known as the Lisan Formation were deposited during the Würm glacial period. A residual brine, perhaps of Lake Samra, was diluted to form Lake Lisan when the onset of the glaciation brought a wet Mediterranean climate to the Dead Sea rift around 63 ka. The lake rose to at least -280 m but then apparently receded. The lake rose again around 48 ka to at least -260 m. A lithologic change occurred at 36 ka when the salinity of the lake rose and chemical precipitation increased. Lake Lisan continued a cycle of rise and fall until it reached a maximum elevation of -180 m by 17–15 ka. Major changes occurred within the lake basin after 15 ka. In the Dead Sea region, the level of Lake Lisan apparently fell to -700 m. The lowering of Lake Lisan was facilitated by continued subsidence of the Dead Sea basin and possibly by structural blocking of the inflow from the north.

There remain many unresolved questions about the cycles represented within the Lisan Formation. Clastic sections that may be equivalent in age to the Lisan Formation have not been adequately correlated. Because the top of the "classic" lacustrine facies of the Lisan Formation is weathered, the final stage of Lake Lisan is not well dated. Nor are the paleogeographic relation of the clastic and freshwater deposits overlying the Lisan Formation clearly defined. Our understanding of the fluctuations of Lake Lisan would benefit from high-resolution dating of sediment cores and pollen sequences. More radiometric dating of aragonite layers within the clastic sections is also needed to define the timing of lake level highstands and to bracket the age of erosional cycles.

Acknowledgments

I thank Zvi Ben-Avraham for sponsoring this research and Karen Grove, Shmuel Marco, and Yossi Yechieli for very helpful reviews.

REFERENCES

Agnon, A., and Marco, S., 1994, Fault controlled slumping in Lake Lisan sediments, Massada Plain: Israel Geological Society, Annual Meeting, Nof Ginosar, February 29–March 2, 1994, p. 1.

Alley, R. B., Meese, D. A., Shuman, C. A., Gow, A. J., Taylor, K. C., Grootes, P. M., White, J. W. C., Ram, A., Waddington, E. D., Mayewski, P. A., and Zielinski, G. A., 1993, Abrupt increase in Greenland snow accumulation at the end of the Younger Dryas event: *Nature*, v. 362, p. 527–529.

Bard, E., Hamelin, B., Fairbanks, R. G., and Zindler, A., 1990, Calibration of the ^{14}C timescale of the past 30,000 years using the mass spectrometric U-Th ages from Barbados corals: *Nature*, v. 345, p. 405–410.

Bar-Yosef, O., 1987, Prehistory of the Jordan rift: *Israel Journal of Earth Sciences*, v. 36, p. 107–119.

Bar-Yosef, O., Goldberg, P., and Leveson, T., 1974, Late Quaternary stratigraphy and prehistory in Wadi Fazael, Jordan Valley—A preliminary report: *Paléorient*, v. 2, p. 415–428.

Begin, Z. B., 1975, The geology of Jericho sheet (1:50,000): Geological Survey of Israel Bulletin 67, 35 p.

Begin, Z. B., Broecker, W., Buchbinder, B., Druckman, Y., Kaufman, A., Magaritz, M., and Neev, D., 1985, Dead Sea and Lake Lisan levels in the last 30,000 years: Geological Survey of Israel, Preliminary report GSI/29/85, 18 p.

Begin, Z. B., Ehrlich, A., and Nathan, Y., 1974, Lake Lisan, the Pleistocene precursor of the Dead Sea: Geological Society of Israel Bulletin 63, 30 p.

Begin, Z. B., Ehrlich, A., and Nathan, Y., 1980, Stratigraphy and facies distribution in the Lisan Formation—New evidence from the area south of the Dead Sea, Israel: *Israel Journal of Earth Sciences*, v. 29, p. 182–189.

Ben-Avraham, Z., Niemi, T. M., Neev, D., Hall, J. K., and Levy, Y., 1993, Distribution of Holocene sediments and neotectonics in the deep north basin of the Dead Sea: Marine Geology, v. 113, p. 219–231.

Bentor, Y. K., and Vroman, A. J., 1960, The geological map of Israel, 1:100,000, Sheet 16, Mt. Sedom: Jerusalem, Geological Survey of Israel, 117 p.

Blanckenhorn, M., 1912, *Naturwissenschaftliche Studien am Toten Meer und im Jordanthal*: Berlin, Friedländer and Sohn, 478 p.

Blanckenhorn, M., 1914, *Syrien, Arabien und Mesopotamien*: Heidelberg, Handbuck des Regione Geologie.

Bowman, D., 1971, Geomorphology of the shore terraces of the late Pleistocene Lake Lisan (Israel): *Palaeogeography, Palaeoclimatology, Palaeoecology*, v. 9, p. 183–209.

Bowman, D., 1990, Climatically triggered Gilbert-type lacustrine fan deltas, the Dead Sea area, Israel: International Association of Sedimentologists Special Publications 10, p. 273–280.

Bowman, D., and Gross, T., 1992, The highest stand of Lake Lisan: ~150 meter below MSL: *Israel Journal of Earth Sciences*, v. 41, p. 233–237.

Braun, D., Ron, H., and Marco, S., 1991, Magnetostratigraphy of the hominid tool bearing Erk el Ahmar Formation in the northern Dead Sea rift: Israel Journal of Earth Science, v. 40, p. 191–197.

Broecker, W. S., and Denton, G. H., 1989, The role of ocean-atmosphere reorganizations in glacial cycles: *Geochimica et Cosmochimica Acta*, v. 53, p. 2465–2501.

Buchbinder, B., 1981, Morphology, microfabric and origin of stromatolites of the Pleistocene precursor of the Dead Sea, Israel, *in* Monty, C., ed., *Phanerozoic stromatolites*: Springer-Verlag, Berlin, p. 181–196.

Buchbinder, B., Begin, Z. B., and Friedman, G. M., 1974, Pleistocene algal tufa of Lake Lisan, Dead Sea area, Israel: *Israel Journal of Earth Sciences*, v. 23, p. 131–138.

Clark, G. A., 1988, Some thoughts on the southern extent of the Lisan Lake as seen from the Jordan side: *Bulletin of the American School of Oriental Research*, v. 272, p. 42–43.

Copeland, L., and Vita-Finzi, C., 1978, Archaeologcial dating of geological deposits in Jordan: *Levant*, v. 10, p. 10–25.

Druckman, Y., Magaritz, M., and Sneh, A., 1987, The shrinking of Lake Lisan, as reflected by the diagenesis of its marginal oolitic deposits: *Israel Journal of Earth Sciences*, v. 36, p. 101–106.

Ehrlich, A., and Noël, D., 1988, Sedimentation patterns in the

Pleistocene Lake Lisan, Precursor of the Dead Sea—New data from nannofacies analysis of the diatom floras: *Cahiers de Micropaléontologie*, v. 3, p. 5–20.

El-Isa, Z. H., and Mustafa, H., 1986, Earthquake deformations in the Lisan deposits and seismotectonic implications: *Geophysical Journal of the Royal Astronomical Society*, v. 86, p. 413–424.

Emery, K. O., 1963, A reconnaissance study of the floor of the Dead Sea, *in* Barney, R. E., *Search for Sodom and Gomorrah*: Kansas City, Missouri, CAM Press, p. 305–320.

Freund, R., Garfunkel, Z., Zak, I., Goldberg, M., Weissbrod, T., and Derin, B., 1970, The shear along the Dead Sea rift: *Royal Society of London Philosophical Transactions*, v. A267, p.107–130.

Frostick, L. E., and Reid, I., 1989, Climatic versus tectonic controls of fan sequences—Lessons from the Dead Sea, Israel: *Journal of the Geological Society of London*, v. 146, p. 527–538.

Frumkin, A., Magaritz, M., Carmi, I., and Zak, I., 1991, The Holocene climatic record of the salt caves of Mount Sedom, Israel: *The Holocene*, v. 1, p. 191–200.

Gardosh, M., 1987, Water composition of the late Quaternary lakes in the Dead Sea rift: *Israel Journal of Earth Sciences*, v. 36, p. 83–89.

Gardosh, M., Reches, Z., and Garfunkel, Z., 1990, Holocene tectonic deformation along the western margins of the Dead Sea: *Tectonophysics*, v. 180, p. 123–137.

Garfunkel, Z., Zak, I., and Freund, R., 1981, Active faulting in the Dead Sea rift: *Tectonophysics*, v. 80, p. 1–26.

Goodfriend, G. A., and Magaritz, M., 1988, Palaeosols and late Pleistocene rainfall fluctuations in the Negev Desert: *Nature*, v. 332, p. 144–146.

Heimann, A., 1990, The development of the Dead Sea rift and its margins in northern Israel during the Pliocene and the Pleistocene: Golan Research Institute, and Geological Survey of Israel Report GSI/28/90, 90 p.

Horowitz, A., 1979, *The Quaternary of Israel*: New York, Academic Press, 394 p.

Horowitz, A., 1987, Subsurface palynostratigraphy and paleoclimates of the Quaternary Jordan Rift Valley fill, Israel: *Israel Journal of Earth Sciences*, v. 36, p. 31–44.

Horowitz, A., 1992, *Palynology of arid lands*: Amsterdam, Elsevier, 546 p.

Hull, W., 1886, *The survey of Western Palestine, memoir on the geology and geography of Arabia Petraea, Palestine and adjoining districts, with special reference to the mode of formation of the Jordan—Arabah Depression and the Dead Sea*: London, Bentley and Son.

Joffe, S., and Garfunkel, Z., 1987, Plate kinematics of the circum Red Sea—A reevaluation: *Tectonophysics*, v. 141, p. 5-22.

Kafri, U., Kaufman, A., and Magaritz, M., 1983, The rate of Pleistocene subsidence and sedimentation as compared with those of other time spans in other Israeli tectonic regions: *Earth and Planetary Science Letters*, v. 65, p. 126–132.

Katz, A., and Kolodny, N., 1989, Hypersaline brine diagenesis and evolution in the Dead Sea–Lake Lisan system (Israel): *Geochimica et Cosmochimica Acta*, v. 53, p. 59–67.

Katz, A., Kolodny, N., and Nissenbaum, A., 1977, The geochemical evolution of the Pleistocene Lake Lisan–Dead Sea system: *Geochimica et Cosmochimica Acta*, v. 41, p. 1609–1626.

Kaufman, A., 1971, U-series dating of Dead Sea basin carbonates: *Geochimica et Cosmochimica Acta*, v. 35, p. 1269–1281.

Kaufman, A., Yechieli, Y., and Gardosh, M., 1992, Reevaluation of the lake-sediment chronology in the Dead Sea basin, Israel, based on new ^{230}Th/U dates: *Quaternary Research*, v. 38, p. 292–304.

Koucky, F. L., and Smith, R. H., 1986, Lake Beisan and the prehistoric settlement of the northern Jordan Valley: *Paléorient*, v. 12/2, p. 27–36.

Kronfeld, J., Vogel, J. C., Rosenthal, E., and Weinstein-Evron, M., 1988, Age and paleoclimatic implications of the Bet Shean travertines: *Quaternary Research*, v. 30, p. 298–303.

Kudrass, H. R., Erlenkeuser, H., Vollbrecht, R., and Weiss, W., 1991, Global nature of the Younger Dryas cooling event inferred from oxygen isotope data from Sulu Sea cores: *Nature*, v. 349, p. 406–409.

Langozky, Y., 1961, Remarks on the petrography and geochemistry of the Lisan Marl Formation [M.S. thesis]: Jerusalem, The Hebrew University (in Hebrew).

Langozky, Y., 1963, High level lacustrine sediments in the rift valledy at Sdom: The Bulletin of the Research Council of Israel, v. 12, p. 17–25.

Lartet, L., 1869, Essai sur la geologie de la Palestine et des cotrees avoisinantes, telles que l'Egypte et l'Arabie, Comprenant les observations reueillies dans le cours de l'expedition du Duc de Luynes a lar Mer Morte: These, Victor Masson et fils, Paris, 109 p.

Macumber, P. G., and Head, M. J., 1991, Implications of the Wadi al-Hammeh sequences for the terminal drying of Lake Lisan, Jordan: *Palaeography, Palaeoclimatology, Palaececology*, v. 84, p. 163–173.

Manspeizer, W., 1985, The Dead Sea rift: Impact of climate and tectonism on Pleistocene and Holocene sedimentation, *in* Biddle, K. T., and Christie-Blick, N., eds., *Strike-slip deformation, basin formation, and sedimentation*: S.E.P.M. Special Publications No. 37, p. 143–158.

Marco, S., Stein, M., Agnon, A., and Ron, H., 1996, Long-term earthquake clustering: A 50,000-year paleoseismic record in the Dead Sea graben: Journal of Geophysical Research, v. 101, p. 6,179-6191.

Mayewski, P. A., Meeker, L. D., Whitlow, S., Twickler, M. S., Morrison, M. C., Alley, R. B., Bloomfield, P., and Taylor, K., 1993, The atmosphere during the Younger Dryas: *Science*, v. 261, p. 195–197.

McClure, H. A., 1976, Radiocarbon chronology of late Quaternary lakes in the Arabian Desert: *Nature*, v. 263, p. 755–756.

Nadel, D., and Nir, Y., 1994, Water levels of Lake Kinneret—New archaeological evidence: Israel Geological Society, Annual Meeting, Nof Ginosar, Feb. 28–March 2, 1994, p. 73.

Neeley, M. P., 1992, Lithic period sites, *in* MacDonald, B., *The Southern Ghors and Northeast 'Araba archaeological survey*: Sheffield, U.K., Sheffield Archaeological Monographs 5, J. R. Collis Publications, University of Sheffield, p. 23–51.

Neev, D., and Emery, K. O., 1967, The Dead Sea: Depositional processes and environments of evaporites: Jerusalem, Geological Survey of Israel Bulletin no. 41, 147 p.

Neev, D., and Hall, J. K., 1976, The Dead Sea geophysical survey, 19 July–1 August 1974, Final Report No. 2: Jerusalem, Geological Survey of Israel, Marine Geology Division, Report No. 6/76, 21 p.

Neev, D., and Hall, J. K., 1977, Climatic fluctuations during the Holocene as reflected by the Dead Sea level: International Conference on Terminal Lakes, Weber State College, Ogden, Utah.

Neev, D., and Hall, J. K., 1979, Geophysical investigations in the Dead Sea: *Sedimentary Geology*, v. 23, p. 209–238.

Neev, D., and Langozky, Y., 1960, Tufa deposits (algal bioherms?) of the Lisan Lake: Symposium on the Pleistocene in Israel, Association for the Advancement of Science in Israel, p. 10–11 (in Hebrew).

Niemi, T. M., and Ben-Avraham, Z., 1994, Lake Lisan and Dead Sea levels from seismic reflection profiled in the north Dead Sea basin: Israel Geological Society, Annual Meeting, Nof Ginosar, Febuary 28–March 2, 1994, p. 76.

Nir, Y., 1967, Some observations on the morphology of the Dead Sea wadis: *Israel Journal of Earth Sciences*, v. 16, p. 97–103.

Picard, L., 1943, Structure and evolution of Palestine: Jerusalem, Bulletin of the Geology Department, The Hebrew University, v. 4, no. 2–4, 187 p.

Rotstein, Y., Bartov, Y., and Hofstetter, A., 1991, Active compressional tectonics in the Jericho area, Dead Sea rift: *Tectonophysics*, v. 90, p. 239–259.

Schuldenrein, J., and Goldberg, P., 1981, Late Quaternary paleoenvironments and prehistoric site distributions in the lower Jordan Valley–Preliminary report: *Paléorient*, v. 7, p. 57–71.

Sneh, A., 1979, Late Pleistocene fan-deltas along the Dead Sea rift: *Journal of Sedimentary Petrology*, v. 49, p. 541–552.

Sneh, A., 1982, Quaternary of the Northwestern 'Avara, Israel: *Israel Journal of Earth Sciences*, v. 31, p. 9–16.

Stein, M., Goldstein, S. L., Ron, H., and Marko, S., 1992, Precise TIMS ^{230}Th–^{234}U ages and magnetostratigraphy of Lake Lisan sediments (Paleo-Dead Sea): *EOS*, Transactions, American Geophysical Union, v. 73, p. 155.

Street, F. A., and Grove, A. T., 1979, Global maps of lake-level fluctuation since 30,000 yr B.P.: *Quaternary Research*, v. 12, p. 83–118.

ten Brink, U., and Ben-Avraham, Z., 1989, The anatomy of a pull-apart basin: Reflection observation of the Dead Sea basin: *Tectonics*, v. 8, p. 330–350.

Vita-Finzi, C., 1964, Observation on the late Quaternary of Jordan: *Palestine Exploration Quarterly*, v. January-June, p. 19–33.

Vogel, J. C., 1979, Dating the Lisan Formation, *in* Horowitz, A., *The Quaternary of Israel*: NY, Academic Press, p. 151–153.

Vogel, J. C., 1980, Accuracy of the radiocarbon time scale beyond 15,000 B.P.: *Radiocarbon*, v. 22, p. 210–218.

Vogel, J. C., and Waterbolk, H. T., 1972, Groningen radiocarbon dates—Geological samples Dead Sea Series (Lisan): *Radiocarbon*, v. 14, p. 46–47.

Yechieli, Y., 1987, The geology of the northern Arava rift and the Mahmal Anticline, Hazeva area: Israel Geological Survey Report GSI/30/87 (in Hebrew).

Yechieli, Y., Magaritz, M., Levy, Y., Weber, U., Kafri, U., Woelfli, W., and Bonani, G., 1993, Late Quaternary geological history of the Dead Sea area, Israel: *Quaternary Research*, v. 39, p. 59–67.

Zak, I., and Freund, R., 1966, Recent strike-slip movements along the Dead Sea rift: *Israel Journal of Earth Sciences*, v. 15, p. 33-37.

22. THE HOLOCENE HISTORY OF DEAD SEA LEVELS

Amos Frumkin

Lake level changes are widely used as paleoclimatic indicators that allow regional and global climatic reconstructions (Street and Grove, 1979). Climatic inference is especially valid for terminal lakes like the Dead Sea, when compared to lakes with outlets. The main climatic factors that control lake levels in arid, terminal settings are precipitation and evaporation within the lake and its catchment area. Such a lake is an excellent indicator of variations in water balance on a timescale of 100–1,000 years (Street-Perrott and Perrott, 1990).

Three approaches are commonly used to reconstruct lake level history. The sedimentologic approach uses lake deposits and investigates their lithology, mineralogy, paleontology, and palynology. This approach may yield continuous records of lake water chemistry and relative level changes. The second approach relies on geomorphic evidence, such as terraces. This approach allows a better determination of lake levels but long-term records are rare, especially where there have been several lake level fluctuations. The third approach uses historical records and the location of archaeological sites to set limits on lake level elevations. Better results and higher temporal and spatial resolutions may be obtained by combining results from these three approaches. The purpose of this chapter is to present the Holocene history of Dead Sea levels using the three approaches to lake level reconstruction, based on evidence from Mount Sedom caves and other studies.

PREVIOUS WORK

The sedimentologic approach was first used by Neev (1964), who studied cores drilled in the southern basin of the Dead Sea. Dating of the section was based on two ^{14}C dates of disseminated organic material: 4,410 ±320 ^{14}C yr from a depth of 3.7 m and 9,850 ±150 yr B.P. from a depth of 10.5 m. These dates are not beyond doubt because older organic matter could have been mixed in the samples. The dated material was found in a core consisting mainly of marls (laminated shales and aragonite) and coarse alluvium, at the southern part of the southern basin, east of Mount Sedom. This core was correlated with others drilled farther north at the center of the basin, where several salt layers are fully developed within the marls. Dead Sea levels and climatic implications based on these finds were further discussed by Neev and Emery (1967) and Neev and Hall (1977), and by Begin et al. (1985) who presented a lake level curve. The salt layers were attributed to an arid climate with low lake levels, permitting the southern basin of the Dead Sea to dry out, whereas marls were attributed to moister climates when the Dead Sea level rose, depositing suspended load carried by floods. Begin et al. (1985) suggested that the Dead Sea level oscillated between -370 and -700 m during the Holocene. However, the previous interpretations of the timing of the southern basin stratigraphy is under reevaluation (Neev and Emery, 1995).

Yechieli et al. (1993) studied sediments of the southwestern part of the Dead Sea's northern basin in three boreholes in the alluvial fan of Nahal Ze'elim. One borehole penetrated the entire Holocene sequence, and four pieces of wood were ^{14}C dated to 8,255, 8,390, 8,440, and 11,315 yr B.P. Salt beds deposited after 11,315 yr B.P. and before 8,440 yr B.P. were attributed to a low water level at the transition between Lake Lisan and the present Dead Sea. Aragonite within a clay layer of 8,440–8,255 yr B.P. indicated a relatively high lake level. Clay with gravel and some aragonite deposited after 8,255 yr B.P. were attributed to somewhat lower Dead Sea level. Later changes from fine to coarse alluvial sediments were interpreted as relating to the proximity of the Dead Sea shore to the drill sites (i.e., where the shore was relative to the drill sites during deposition). The presence of aragonite associated with fine sediments suggested several periods of lake level greater than -400 m, flooding the study sites during the mid-Upper Holocene. However, because of the absence of ^{14}C dates after 8,255 yr B.P., the timing of these events is not clear.

Horowitz (1992, p. 327–339) studied the palynology of several boreholes around Mount Sedom. The resolution of the Holocene pollen diagrams presented for this region is not high enough to allow comparison with the present study. However, arboreal pollen factors from archeological sites in Israel (Horowitz 1992, p. 418) allow such a comparison (see subsequent discussion).

High-resolution seismic profiles of the Dead Sea's northern basin are reported by Ben-Avraham et al. (1993). Five strong reflectors were identified in a sediment sequence representing an estimated period of between 8,000 and 2,500 years, based on sedimentation rates. The four subbottom reflectors were presumed to represent contacts between salt and marl. Niemi and Ben-Avraham (chapter 6, this volume) correlate this evidence with four lowstands of Dead Sea level during the past 5,000 years inferred from Mount Sedom caves (Frumkin et al., 1991). Seismic reflection data from the northern basin of the Dead Sea reveal relict canyons and buried channels of the Jordan River, suggesting that the Dead Sea level fell below -400 m in the Holocene (Niemi and Ben-Avraham, 1993), and possibly lower than -486 m (Niemi and Ben-Avraham, chapter 6, this volume).

Goodfriend et al. (1986) studied salt-covered shells of land snails found under a boulder near Qumran. They hypothesized that the salt cover originated in the splash zone of the Dead Sea, suggesting a Dead Sea level of -280 m between 7,100 and 6,400 ^{14}C yr B.P.

Klein's studies of Dead Sea levels during historical periods have been based on a wide variety of historical, archaeological, morphological, and climatic lake level indicators (Klein, 1965, 1982, 1986). Such diverse sources have varying credibility and should be evaluated carefully. Naturally, more evidence is available for recent periods, so the lake level curve becomes more accurate. Klein has suggested large fluctuations in the Dead Sea level during the last few thousand years to -330 m in the first century B.C.E. (Before the Common Era). Lake level changes between 1979 and 1988 were discussed by Anati and Shasha (1989) and by Anati (chapter 8, this volume).

EVIDENCE FROM THE MOUNT SEDOM CAVES

Frumkin et al. (1991) used a morphological approach in the Mount Sedom karst system to construct a curve of the Dead Sea levels. This curve has been modified to take into account new data.

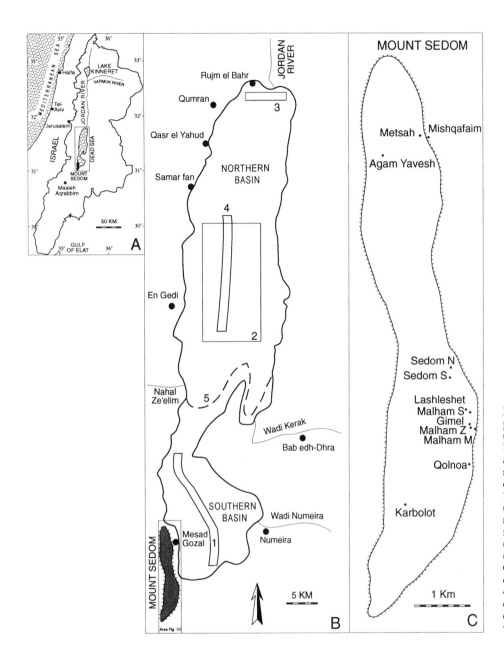

Figure 22-1 Sites with evidence pertaining to the Dead Sea level (shown at a level of -396 m). A, Dead Sea catchment area; B, approximate locations of relevant studies within the Dead Sea, (1) cores (Neev, 1964; Neev and Emery, 1967), (2) 3.5-kHz seismic profiles (Ben-Avraham et al., 1993), (3) seismic reflection profiles revealing relict canyons of the Jordan River (Niemi and Ben-Avraham, 1993), (4) profiles of [226]Ra (Stiller and Chung, 1984), and (5) south shore of the Dead Sea today (1993, level -410 m); C, detail of Mount Sedom on the southwestern shore of southern basin with the names of caves where samples were ^{14}C dated.

Mount Sedom is an exposed salt diapir along the western shore of the southern Dead Sea basin (Fig. 22-1). The mountain is 11 by 1.5 km, rising to -160 m, about 250 m above the 1992 level of the Dead Sea. It consists of Neogene (?) rock salt of marine origin that pierces through younger lake evaporites and clastics commonly assigned a Neogene-Pleistocene age (Zak and Bentor, 1968; Horowitz, 1992, p. 332; Stein et al., 1994). The salt is covered by a less soluble cap rock that has favored the development of allogenic karst.

Runoff collected from the small catchments is captured in fissures, forming underground dissolution channels within the salt diapir. Most caves have a low active flood channel, as well as high and dry levels that were active in the past (Fig. 22-2). Water flowing through these vadose cave passages to the Dead Sea deposits wood fragments embedded in alluvial cave sediments. Thirty-three wood samples from 12 inactive cave passages have been ^{14}C dated and range in age from 6,000 B.C.E. to 1,800 C.E. For a full list of samples, locations, and dates, see Frumkin et al. (1991) and Frumkin et al. (1994). [14]C dates were

calibrated with CalibETH (Niklaus, 1991) to allow comparison with historical evidence. In this chapter, calendrical ages are used, unless mentioned otherwise. C.E. and B.C.E. represent Common Era and Before the Common Era, respectively. For conversion graphs between conventional ^{14}C ages in yr B.P. and calendrical ages (C.E. and B.C.E.), see Pearson et al. (1986) and Stuiver and Pearson (1986).

Active cave passage profiles tend to be subhorizontal, sloping asymptotically toward base level. It is hypothesized that a lowering of the Dead Sea level is followed by rapid downcutting of caves draining to the southern basin of the Dead Sea. On the other hand, rising of the lake level is assumed to be followed by rapid alluvial aggradation and upward dissolution in the caves. If the base level change lasted for hundreds of years or more, these processes would take place until a new cave profile is achieved, equilibrated with respect to the new Dead Sea level. These assumptions are supported by downcutting rates measured in Mount Sedom caves, showing that they are dynamic and highly responsive to base level changes (Frumkin

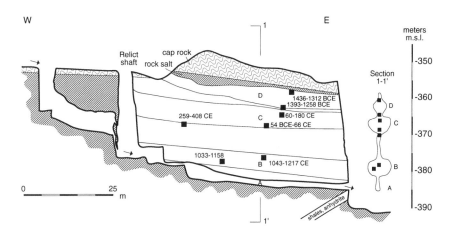

Figure 22-2 Schematic profile of Mishqafaim Cave: A, modern flood channel hanging above base level because of an insoluble layer; B, C, D, relict passages. These wide passages are usually connected to one another by narrower entrenchments (section 1-1'). Arrows show the present flood route through the cave. Black squares locate ^{14}C-dated pieces of wood, with one standard deviation range of each date given in calendrical years. Dead Sea level and diapir rising are assumed to have affected cave levels.

and Ford, 1995). If a cave profile does not have enough time to equilibrate with a rapidly lowering Dead Sea level or if a resistant rock layer is encountered, the cave passage may be hanging above a falling base level or submerged under a rising lake level. Three out of 28 active cave outlets in the eastern escarpment are hanging above base level today, all of them attributed to insoluble layers inhibiting downcutting at the outlet (Fig. 22-2). The other 25 caves slope asymptotically toward base level, at least in the downstream part of the channels. This data can justify using a salt cave as an indicator of lake level changes lasting for at least hundreds of years.

The caves are most significant as lake level indicators for those periods when the Dead Sea waters reached the Mount Sedom escarpment, whose base is about -390 m (Fig. 22-3). Lower lake levels exposing the foot of Mount Sedom need not change caves level considerably (Bowman, 1988). Nevertheless, because the caves are vadose, they can always be used as indicators for the maximum possible lake level.

Mount Sedom is an actively rising diapir (Zak and Freund, 1980). The rising rate must be taken into account when reconstructing lake levels (Frumkin et al., 1991). Determining lake levels from a rising diapir is highly dependent on the accuracy to which diapir rising rate can be determined. This rate is presumably not constant. Because the available evidence is limited, I do not use one single diapir rising rate in this chapter. To demonstrate the range of possibilities, I use four average rising rates of 0, 3, 6, and 9 mm/yr to construct four possible Dead Sea level curves (Fig. 22-4).

The first curve assumes no net diapir rise with respect to mean sea level (M.S.L.). This is similar to the assumption underlying other Dead Sea level studies—an assumption that is justified in areas relatively stable during the Holocene. In Mount Sedom, it would be justified if the rate of diapir rise with respect to its surrounding was similar to the subsidence rate of the entire region (southern basin of the Dead Sea and Mount Sedom together). The southern Dead Sea basin has been subsiding at a rate of 1 m/ka (Neev and Emery, 1967) to 10 m/ka (Horowitz, 1992, Fig. 8.5.4) during the Holocene along active normal faults bordering it on all sides (Ben-Avraham et al., 1990; Gardosh et al., 1990; Garfunkel et al., 1981). This is indicated also by eastward tilting of Mount Sedom during the Holocene (Frumkin, 1992).

The second lake level curve is based on a diapir uplift rate of 3 mm/yr, comparable to the value of 3.5 mm/yr calculated by

Figure 22-3 The eastern escarpment of Mount Sedom, with three outlets of Malham Cave. The lowest outlet drains an active channel sloping asymptotically toward the present base level. The Dead Sea reached the base of the escarpment in 1911 (lake level -391 m, Klein, 1965). Since then, the foot of the mountain at the foreground dried as the water level fell. Lake levels higher than -390 m had an effect on cave levels. The highest outlet seen drained a major channel dated to Stage 3 (4,200–2,300 B.C.E.), when Dead Sea level rose against the escarpment.

Zak (1967) for southwestern Mount Sedom and adopted in a preliminary report of this study (Frumkin and Zak, 1991). But this is only one possibility, not necessarily pertinent for all parts of Mount Sedom.

The third lake level curve is based on a linear correlation found between the age of subhorizontal cave passages and the

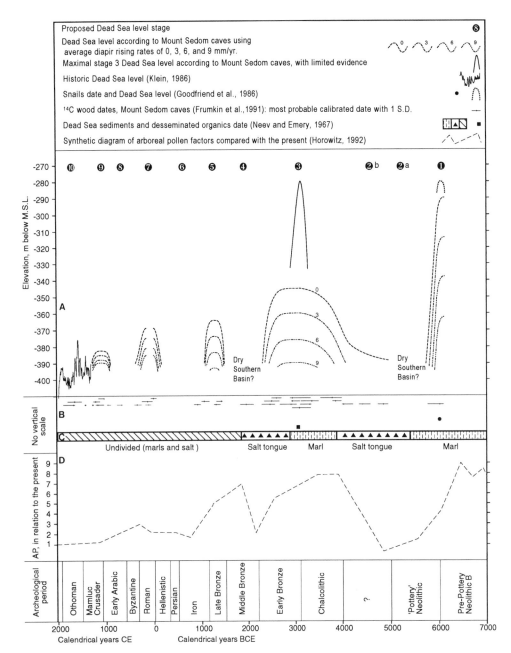

Figure 22-4 A, Dead Sea level curves derived from Mount Sedom caves and other sources; B, ^{14}C dates of wood from Mount Sedom caves (33 dates), organics in Dead Sea sediments (1 date, after Neev and Emery, 1967), and snails (1 date, after Goodfriend et al., 1986); C, sediment types of the southern basin, after Neev and Emery (1967); and D, synthetic diagram of arboreal pollen factors from archeological sites in Israel, as compared with the present day, after Horowitz (1992).

height above the currently active channel, indicating a mean downcutting rate of about 6 mm/yr. Assuming that the Dead Sea level fluctuated around an average level that remained fairly constant throughout the last 8,000 years, the observed average entrenchment would be the caves' reaction to a diapir rising at a rate averaging 6 mm/yr (Frumkin, 1996).

The fourth curve is based on a direct measurement of the diapir rising taken between 1990 and 1992 (Frumkin, 1992). In the Sedom Cave, 9 mm/yr of bedrock was displaced with respect to the foot of the eastern escarpment. However, this figure may vary temporally and spatially.

The rising rates of 6 and 9 mm/yr are actually measured with respect to the plain immediately outside of the Mount Sedom boundary faults (Fig. 22-5). Applying these rates to construct Dead Sea level curves with respect to M.S.L. would be justified

if the plain at the foot of Mount Sedom is stable with respect to M.S.L., that is, if the rising rate of the foot of the mountain caused by its proximity to the diapir is balanced by the subsidence of the entire region (southern basin of the Dead Sea plus Mount Sedom).

Fig. 22-4 also shows the correlation among the Dead Sea level curve, the Dead Sea sediments (Neev and Emery, 1967), and arboreal pollen factors from archeological sites in Israel (Horowitz 1992, Fig. 10.3.4). For most of the period the correlation is good because both lake level and arboreal pollen integrate the precipitation-evaporation ratio over a large area. The correlation is not good for the last 2 ka, when climatic fluctuations decrease and human impact increases (Baruch, 1990).

Neev and Emery (1967) and later Begin et al. (1985) suggest that the Dead Sea levels did not rise above -390 m during most

Figure 22-5 Southeastern face of Mount Sedom. The large relict outlets in the center, at modern elevations of -362 m to -353 m, are dated to Stage 3. Modern outlets are small, at the level of the foreground plain, outside of the Mount Sedom boundary faults. The plain, formerly part of the southern basin of the Dead Sea, is occupied today by potash evaporation ponds.

of the Holocene. This is possible if the Mount Sedom diapir has been rising at a mean rate of 9 mm/yr throughout the last 8,000 years.

HISTORY OF THE DEAD SEA LEVELS

Lake Lisan, the precursor of the Dead Sea, covered the Mount Sedom diapir during the Late Pleistocene and deposited the Lisan Marls on top of it (Zak, 1967; Begin et al., 1980). During the rising of the diapir, the top of the salt was dissolved, creating a residual cap rock up to 50 m thick.

Druckman et al. (1987) proposed rapid fall of Lake Lisan after 14,000 yr B.P. based on ^{14}C dating of detrital plant remains in oolitic sediments south of the Dead Sea. Based on contours of submarine canyons on the seafloor of the northern basin of the Dead Sea, which could not have developed underwater, Neev and Emery (1967) suggested an almost complete desiccation of the lake, thus ending the Lisan phase. This is supported by massive salt beds deposited in the southern basin prior to 9,850 ^{14}C yr B.P. (Neev, 1964) and in the southern part of the northern basin, between 11,300 and 8,400 ^{14}C yr B.P. (Yechieli et al., 1993). Later Holocene sediments encountered in boreholes within the northern basin do not include massive salt beds (Stiller et al., 1988; Yechieli et al., 1993). Based on these dates, the extremely dry event marking the end of Lake Lisan took place between 11,000 and 10,000 ^{14}C yr B.P.

The ensuing Dead Sea phases recorded in the Mount Sedom caves are divided into 10 stages, based on the dating of cave levels and the Dead Sea level changes inferred from them. Most of the evidence considered herein represents a rough average of lake levels during certain periods of time. However, short-term fluctuations may have a large amplitude, as seen in the measured record of the 20th century (Fig. 22-4).

Stage 1: 10,000–5,800 B.C.E.

An age of 9,850 ^{14}C yr B.P. obtained from disseminated organics within a marl sequence in the southern basin indicated that the Dead Sea refilled by this time to a level higher than -400 m (Neev, 1964). Neev and Emery (1967) suggested that the marl was deposited during a moist period that ended at about 6,000 B.C.E.

More recent boreholes from the Dead Sea shore have yielded ages of 8,440, 8,390, 8,255, and 8,385 ^{14}C yr B.P. from wood within clays overlying the halite sequence of the transition between Lake Lisan and the present Dead Sea (Yechieli et al., 1993). The clay sequence is attributed to the refilling of the basin to form the Dead Sea. A large influx of detritus by the Nahal Ze'elim was inferred from a high sedimentation rate about 8,300 ^{14}C yr B.P.

No caves predating about 7,100 ^{14}C yr B.P. (the earliest date from Lashleshet Cave) have been found in Mount Sedom. This may indicate Early Holocene submergence of the Sedom rock salt under a high Dead Sea level. Submergence of the salt could also occur under low Dead Sea levels, if the diapir had not yet risen to expose the salt.

Goodfriend et al. (1986) presented evidence for a -280-m Dead Sea level sometime between 7,100 and 6,400 ^{14}C yr B.P. This could indicate the end of a high stage submerging Mount Sedom before 6,000 B.C.E. (about 7,000 ^{14}C yr B.P.). The Dead Sea must have been below the -210-m elevation of Jericho during the Prepottery Neolithic occupation of the site, which took place approximately from 10,300 to 8,000 ^{14}C yr B.P. (Burleigh, 1984; Bar-Yosef, 1987). Kadan et al. (1995) report a transgression ~9000-8000 yr B.P.

The oldest caves in Mount Sedom are concentrated in a small area in the southeastern part of the mountain. The oldest wood (from Lashleshet Cave) is dated to about 6,000 B.C.E., indicating that the first emergence of rock salt above Dead Sea level took place not long before 6,000 B.C.E. The cave, at a modern elevation of -290 m, lacks an open outlet toward base level, so it could have formed at any height above the contemporaneous Dead Sea. The maximum possible lake level during formation of the Lashleshet Cave was -362 to -290 m, depending on which diapir rising rate is assumed. The first exposure of salt to erosional processes indicates a gradual fall of lake level at the end of Stage 1, after a long period of high levels.

Other studies (e.g., Goldberg and Bar-Yosef, 1982; Magaritz and Goodfriend, 1987) suggest a moist climate in Israel during the Early Holocene. The available evidence thus suggests a high Dead Sea level during Stage 1. However, more evidence is needed for detailed reconstruction of the lake level.

Stage 2: 5,800–4,200 B.C.E.

No wood samples from Mount Sedom caves have been dated during the interval 5,800–5,000 yr B.C.E. (Stage 2a), although the rock salt was subjected to dissolution. Lashleshet Cave was entrenching rapidly beneath its large Stage 1 channel, forming a narrow canyon.

Malham Cave was the first to have an open outlet passage to the Dead Sea; this passage is dated to 5,000-4,200 yr B.C.E. (Stage 2b) by two wood samples. Its development followed the fall of the Dead Sea level during Stage 2a, when rock salt first became exposed at the base of the southeastern escarpment of Mount Sedom.

In the southern Dead Sea basin, a salt tongue extending farther south than any other Holocene salt unit is attributed to Stage 2. This is the only salt tongue represented in borehole CA-27, whose organics were used to date the Holocene deposits of the entire basin (Neev, 1964). Interpolation between the dates suggested that the salt tongue was deposited at about 6,500–5,500 ^{14}C yr B.P., and it was postulated that this was the driest period during the last 10,000 yr (Neev and Emery, 1967).

A gradual desiccation of the Dead Sea after 8,200 ^{14}C yr B.P. is indicated by increasing gravel and decreasing aragonite in the sediments from boreholes at the Nahal Ze'elim fan (Yechieli et al., 1993). At a level of -404 m, no aragonite was deposited in the gravel and sand sequence, indicating that the area became subaerially exposed.

The combined evidence suggests that the Dead Sea level dropped dramatically, to a level lower than -404 m, at the end of Stage 1 and during Stage 2, causing the southern basin to become dry. The high Early Holocene level was never to be reached again.

Stage 3: 4,200–2,300 B.C.E.

Twelve wood samples from seven Mount Sedom caves are dated to this period. These were found in large passages, typically three times wider than presently active channels in the same caves. Most of these high and dry passages have large outlets, at modern elevations of -362 to -353 m (Fig. 22-5). Dead Sea dissolution features are found even higher, up to -345 m on the wall of a Stage 3 passage in Malham Cave.

The ceilings of Stage 3 passages are usually flat and near-horizontal (if not disturbed by breakdown), and the floors are covered by alluvial deposits that become fine grained in the upper layers. These features indicate upward solution of the ceiling by water filling the passage completely. Aragonite sediments were deposited up to the ceilings of three cave passages near the outlets, to a modern level of -359 m (Fig. 22-6). This type of deposit has been found only in Stage 3 passages, no more than 100 m upstream from the outlets. The deposits were analyzed for ^{14}C but contained none; this suggests redeposition of older sediment, probably from the Late Pleistocene Lisan Formation that covers the mountain. This sediment is believed to have been carried in floodwaters through the caves as suspended load and to have been deposited near the outlets when these were submerged by a rising Dead Sea.

Morphological evidence indicates that passages became completely filled with water and sediment some time during Stage 3. The Dead Sea level reached -389 to -345 m, depending on which diapir uplift rate is assumed (Fig. 22-4). High levels were maintained long enough for passages to become wide and filled with sediment.

Most plant fossils in the cave deposits are similar to those found today in small catchments on Mount Sedom or nearby. Exceptions are four oak samples (*Quercus calliprinos* Webb.) from three different caves dated to a short time interval—3,500–2,700 yr B.C.E. Oak does not grow today in the Dead Sea region because it needs more than 400 mm of annual precipitation, eight times more than the present amount. The soil profiles on Mount Sedom do not support such a dramatic change of conditions during the Holocene. Therefore, the oak must have been transported by some agent from a nearby Mediterranean environment. The agents could have been humans, but no archaeological remains are found on Mount Sedom to support this. In a preliminary report, Frumkin et al. (1991) suggested that the Dead Sea itself could be the transporting agent, implying that its level rose to -324 to -280 m, submerging the whole cave system. This is feasible because there are major dissolution and collapse features associated with the higher parts of the Stage 3 caves. Some portions of the caves were completely destroyed, leaving collapse dolines at the surface. Such a peak level could not have lasted many years, however, because no large subhorizontal cave passages have been found at -324 to -280 m. The

maximum level suggested by the oak twigs (-280 m) is shown on Figure 22-4, above the three curves indicating the highest Dead Sea dissolution features.

The timing of the Stage 3 climax is possibly recorded by the highest temporal concentration of wood twigs between 3,300 and 2,900 yr B.C.E. and the oak twigs dating to 3,500–2,700 yr B.C.E.

Sedom Cave N passage started to develop at about 2,700 yr B.C.E. and does not contain features typical of older Stage 3 caves, such as detrital aragonite, high–level Dead Sea dissolution features, and major collapse structures. This suggests that the Sedom Cave N passage postdates the highest Dead Sea level. Three wood samples from this passage date the last part of Stage 3 to 2,700–2,300 yr B.C.E. A much narrower canyon that is entrenched at the bottom of this passage marks the onset of Stage 4.

Neev (1964) reported a date of 4,410 ±310 ^{14}C yr B.P. for disseminated organic material within a laminated clay unit in core CA-27 from a -403-m level, 3.4–4.0 m beneath the lake bottom, in the southern Dead Sea basin. Neev and Emery (1967) suggested that a moist period occurred approximately between 5,500 and 4,300 ^{14}C yr B.P., when the laminated clay was deposited. This period coincides with Stage 3 of the Mount Sedom cave record.

In the Nahal Ze'elim boreholes, Stage 3 (inferred from the age-depth ratio) is represented by clay and sand, with aragonite at -403 m, indicating submergence by the Dead Sea (Yechieli et al., 1993).

Some morphological evidence from the western shore of the Dead Sea suggests a stable lake level between -359 and -355 m (Klein, 1982). These features include a bar of rounded boulders and a shingle beach, a notch and an abrasion cave at the northern corner of the Samar delta, a cliff with caves in the talus north of En Gedi, and the base of bluffs cutting Lisan outcrops. A wooden coffin of the first century B.C.E. found in one of the caves near En Gedi shows that the cliff and caves are older. The Mount Sedom evidence demonstrates that Stages 4–10 levels were all lower than -359 m, suggesting that the morphological features at -359 to -355 m may represent Stage 3.

Archaeological evidence from the eastern side of the Dead Sea valley provides further indications. The lowest level Early Bronze site (probably a farmstead or a working area) is BDS-18 at -351 m in the alluvial fan of Wadi Kerak (McConaughy, 1981). This suggests a maximum Dead Sea level below -351 m during the Early Bronze period, although it does not exclude a short phase of higher level.

Two Early Bronze cities, Bab edh-Dhra and Numeira, were located on higher parts of the eastern shore alluvial fans. Organic material excavated in the cities yielded ^{14}C dates (Rast and Schaub, 1980; Weinstein, 1984) comparable to those from Stage 3 in the Mount Sedom caves. The cities border Wadi Kerak and Wadi Numeira. Both these valleys had gentle U-shaped cross sections during the occupation of the sites, but below this level they became narrow, deeply cut (28-50 m) canyons (Donahue, 1985). The downcutting was attributed to a lowering of the base level that increased the stream gradient and initiated rapid incision after the Early Bronze occupation. Donahue (1985, p. 136) suggested that "Although no specific data have yet come to light documenting fault movement, the change from an aggradational to a degradational or erosional regime was most likely caused by a fault movement and uplift of the area." However, the good morphological and dating correlation between the two wadis and the Mount Sedom caves suggests that this was probably due to Dead Sea level changes, rather than to local tectonics. The wide cross sections of the channels during the Early Bronze age may correspond to a higher Dead Sea level, and the rapid incision after it probably was an adjustment to a falling lake during Stage 4. A high groundwater table in Bab edh-Dhra during Stage 3 (Donahue, 1985) was probably associated with the high Dead Sea level.

A wide range of other paleoclimatic evidence indicates that Stage 3 was moist in Israel (Frumkin et al., 1994) and in other areas of the Levant (Issar, 1990; Weiss et al., 1993). Mount Sedom evidence indicates that, during Stage 3, the Dead Sea reached its highest level in the last 7,000 years.

Stage 4: 2,300–1,500 B.C.E.

After the climax of Stage 3 and during Stage 4, cave passages associated with deep and rapid incision became narrower throughout Mount Sedom. Cave gradients increased, permitting less sedimentation of alluvium and wood fragments. One large tree stump, jammed in the narrow canyon of the Sedom N passage is dated to 2,100–1,800 yr B.C.E., fixing the time of incision. The downstream part of this relict passage stands at about -380 m today. Thus, the Dead Sea probably fell to this level or lower after Stage 3.

Rapid canyon downcutting also occurred in surface channels along the eastern Dead Sea shore (Donahue, 1985), indicating that a common base level control was causing incision on both sides of the lake.

Three salt tongues are reported from a depth of 1.7–4.56 m in core CA-10 in the southern Dead Sea basin (Fig. 21 in Neev, 1964). These do not appear in the nearby CA-27 core, where organic matter from a depth of 3.4–4 m is dated to 4,410 ±310 ^{14}C yr B.P. Neev (1964) estimated an age of 4,300 to 3,500 ^{14}C yr B.P. For the salt tongues, based on correlation and age-depth ratio. A sequence of clay and silt overlies the salt tongues in core CA-10, while farther north, in the center of the southern basin, a salt layer several meters thick was deposited continuously during Stage 4 and later. This indicates that the Dead Sea level was falling during Stage 4, and the south basin may have dried out. One of the reflectors observed in seismic profiles of the northern basin may also be attributed to a salt layer deposited during this stage (Ben-Avraham et al., 1993; Niemi and Ben-Avraham, 1993).

An interesting historic account of Stage 4 is found in Genesis 14:3, as follows: "All these joined forces in the Vale of Siddim, that is, the Salt Sea." This has been interpreted during the last 2,000 years as describing an inundation of the previously dry valley by the Dead Sea (e.g., Midrash Tanchuma, Lech-Lecha 8; Flavius, trans. 1961, 1,9,1). Possibly, the southern basin of the Dead Sea (Vale of Siddim) had been dry during Stage 4—the period described in Genesis 14:3—and became submerged by the rising Dead Sea during the moister Stage 5, when the book of Genesis is believed to have been written.

A rapid fall of the Dead Sea level during Stage 4 is clearly indicated by Mount Sedom caves and Dead Sea sediments. The southern basin of the Dead Sea probably dried out completely—the result of climatic deterioration proved by a wide spectrum of evidence (Frumkin et al., 1994).

Stage 5: 1,500–1,200 B.C.E.

Passage D in Mishqafaim Cave, at an elevation of -363 m (Fig. 22-2), is dated to this stage by two wood samples. It suggests that the Dead Sea rose to about -393 to -363 m, depending on which rate of diapir rising is assumed (Fig. 22-4).

Two historical descriptions suggest a high Dead Sea level during this stage: Genesis 14:3 (mentioned previously under Stage 4) and Joshua 15:2–5, which states, "And their south border was from the shore of the Salt Sea, from the bay that looketh southward ...and the east border was the Salt Sea, even unto the end of Jordan, and their border in the north quarter was from

Figure 22-6 The remains of a flat, near-horizontal ceiling of Stage 3 passage in Malham Cave, 10 m from the outlet. Detrital aragonite sediment filled the passage reaching up to the ceiling at -359 m, indicating a rising Dead Sea level at 4,200–2,300 B.C.E. Vertical rock salt layers at the top were disrupted by collapse.

the bay of the sea at the uttermost part of Jordan." According to these verses (as well as Joshua 18, 19), the Dead Sea ended with a bay, both to the south and to the north. A bay or an estuary forms at the northern end when the water level rises to invade the Jordan River canyon (first suggested by Clermont-Ganneau, 1902). Stage 5 transgression was observed also by Kadan et al. (1995).

Undated morphological features along the western shore of the lake, indicating a stable Dead Sea level at -375 m, led Klein (1982, 1986) to postulate that this level was attained during Stage 5. This is in agreement with the evidence presented herein.

The Dead Sea rose during Stage 5, submerging the southern basin. However, this stage was shorter and less significant than preceding high-level stages.

Stage 6: 1,200–100 B.C.E.

In Mishqafaim Cave, a high-gradient narrow passage formed that is attributed to Stage 6 by interpolation between dated twigs found in passages D and C (Fig. 22-2). The downcutting indicates a fall of the Dead Sea level. Two wood samples from Agam Yavesh Cave are also dated to this stage (Frumkin et al., 1991).

Archaeological remains along the western Dead Sea shore may help reconstruct Stage 6 lake levels (Klein, 1986). Mesad Gozal, at -392 m below the eastern side of Mount Sedom, is dated by potsherds to the 11th and beginning of the 10th century B.C.E. (Aharoni, 1964). Finds from two structures that served as boat docks on the northwestern shore are dated to the eighth and seventh centuries B.C.E., plus a second phase of use during Stage 7 (Bar Adon, 1989). Both structures indicate a Dead Sea level of -400 to -395 m during the eighth and seventh centuries B.C.E. (Klein, 1986).

Rock anchors dated to the third century B.C.E. were found at -405 m near En Gedi (Nissenbaum and Hadas, 1990; Hadas, 1993). The anchors were probably thrown from a ship, indicating a Dead Sea level higher than -405 m.

The available evidence suggests that the Stage 6 Dead Sea was typically around -400 to -395 m, and probably did not rise much above this level. The southern basin of the Dead Sea was covered by shallow water.

Stage 7: 100 B.C.E. –400 C.E.

Cave passages of this stage are wider than than those of the previous and subsequent stages. A wide level in Malham Cave M passage was dated to Stage 7 by one piece of wood. Passage C in Mishqafaim Cave, dated to this stage by three wood samples, is now at -368 m (Fig. 22-2). This indicates that the Dead Sea level could have risen to between -385 and -368 m, depending on which rising rate is assumed (Fig. 22-4). Halite was deposited within the center of the southern Dead Sea basin during the last 3,500 years, except for three thin marl layers of 10–30 cm each (Neev, 1964 and pers. com., 1994). This sequence suggests a generally dry climate with three episodes of higher lake levels. The lower or middle marl layer may be attributed to Stage 7.

Klein (1986) describes aragonite crusts on the northern wall of a building in the Samar fan, south of Qumran. One layer reportedly covers the plaster matrix of the wall at -378 m. The structure was built during the eighth and seventh centuries B.C.E., and it was deserted in the first century C.E. (Bar Adon, 1989). If this aragonite is accepted as evidence for submergence by the Dead Sea, as suggested by Klein, the -378-m level could have occurred during Stage 7. Morphological evidence along the western Dead Sea shore indicates a stable paleolake level at -378 to -375 m (Klein, 1982). This may also be attributed to Stage 7. A high Dead Sea level is reported also by Kadan et al. (1995).

A time limit for this Dead Sea level is established by archaeological remains at lower levels. Coins from several sites at -386 to -393 m indicate that the Dead Sea level was lower at least during the first century and early second century C.E. (Klein, 1986). Rujm el Bahr and Qasr el Yahud were used as boat docks during that time (Bar Adon, 1989), indicating a lake level of -400 to -395 m.

Consequently, the Dead Sea could have risen to about -375 m either during the first century B.C.E. or during the latter part of the second to third century C.E. By the fourth century C.E., lake level seems to have dropped again to about -390 m, as indicated by Byzantine archaeological remains (Klein, 1986). Other paleoclimatic evidence from Israel indicates that Stage 7 was slightly moister than preceding and subsequent periods (Issar and Tsoar, 1987).

Klein (1986) argues for a rise of the Dead Sea up to -330 m during the first century B.C.E. There is no supporting evidence in Mount Sedom, where such a high level would have submerged most of the caves, such as Mishqafaim (Fig. 22-2). This apparent disagreement led me to reexamine the evidence presented by Klein for a -330-m level:

(1) An aragonite line on a wall in Qumran, referred to by Klein (1986), has not survived to 1990. Only faint remnants of it appear in Klein's photograph (1986, plate 31). An earlier photograph taken during excavation of the site (de Vaux, 1973, plate IX) shows a light-colored line that may originate from the stratigraphy existing in any archaeological site. Aragonite is naturally a component, as Qumran is built on a Lisan Formation terrace consisting mainly of aragonite. However, the aragonite was not necessarily deposited by the Dead Sea.

(2) Klein (1986) refers to a description of a sediment from Qumran locus 130: "Near the northwest corner of the secondary building, it reaches a thickness of 75 cm. As it extends towards the east it grows progressively thinner" (de Vaux, 1973, p. 23). The layer was interpreted by de Vaux's explanation to be flood water sediment from water diverted into Qumran by

a well preserved and intact aqueduct. A short-term inundation by Dead Sea water could hardly produce a local sediment 75 cm thick.

Average Dead Sea level during Stage 7 was probably higher than in Stages 6 and 8, except during the first century C.E. and the beginning of the second century C.E. when it dropped to about -400 m. Peak levels were probably below -368 m and did not reach the previous Holocene peaks.

Stage 8: 400–900 C.E.

At Mount Sedom, a narrow canyon was entrenched 10 m from passage C down to passage B in Mishqafaim Cave (Fig. 22-2). Downcutting is attributed to Stage 8 by interpolation between dated twigs found in passages C and B. Wood dated to this stage was also found in Agam Yavesh Cave. Downcutting was probably in response to lowering of the base level to the present Dead Sea level. Klein (1986) suggested a Dead Sea level ranging from -400 to -390 m, based on historic evidence. Neev and Emery (1967) suggested the Dead Sea level was as low as -436 m during Stage 8, based on the morphology of the northern basin of the Dead Sea. Other evidence in Israel suggests increased aridity at the seventh century C.E. (Issar et al., 1989).

Dead Sea level during Stage 8 was lower than during Stages 7 and 9, possibly varying between -436 and -390 m.

Stage 9: 900–1300 C.E.

Cave passages widened in Mount Sedom, and twigs became abundant. Five wood samples from three caves were dated to this stage. The elevation of passage B in Mishqafaim Cave (Fig. 22-2), dated to this stage, is -382 m. It indicates a Dead Sea level of -389 to -382 m, depending on the diapir uplift rate. A 10- to 30-cm-thick marl layer overlying the rock salt layer that plasters the flanks of the northern basin from its trough up to the level of -436 m (Neev, 1964 and pers. com., 1994) may have been deposited during this stage.

An age of 930 ^{14}C yr for plant material at a depth of 29 m in the Lisan Strait, just above a pair of halite and gypsum laminae, indicates that Dead Sea level began rising more than 1,000 years ago (Neev and Emery, 1967), that is, at the beginning of Stage 9.

Klein (1986) suggests high Dead Sea levels (up to -350 and -370 m) during the tenth, eleventh, and thirteenth centuries C.E., respectively, and lower levels (-396 to -390 m) during most of the twelfth century. The relatively high levels are in general agreement with the Mount Sedom evidence. However, a peak level of -350 m suggested by Klein during the tenth and eleventh centuries C.E. would have submerged many caves, such as Mishqafaim, but the cave morphology does not support this. An alternative possibility is that the eastern part of the diapir has subsided some 30 m during the last millennium, but, once again, this is unlikely.

To clarify the issue, I reexamined Klein's (1986) evidence for a -350 m level, as follows:

(1) A major point was cited in an eleventh-century document supposed to have been written on "Zoar, the island near the Salt Sea" (Goitein, 1975). However, Gill (1983, p. 168, footnote 311) has shown that this was not the case. Zoar is not mentioned in the document, which was probably written in Egypt, and the "Salt Sea" is the Mediterranean, not the Dead Sea.

(2) A marl sediment was found covering scroll fragments in Qumran cave 4 (at -350 m). "The fragments from cave 5 were lying under more than one metre of natural deposit. The many fragments ... from cave 4 ... were coated with a marl sediment which had accumulated and solidified over a long

period " (de Vaux, 1973, p. 100). Klein (1986) suggests that the marl sediment was deposited by the Dead Sea during Stage 9.

A geological survey was conducted in Qumran caves 4 and 5 for the present study. The caves are artificial excavations in a terrace of littoral facies of the Lisan Formation, including marl and unconsolidated fan deposits. Bedrock properties cause gradual disintegration of the cave ceiling, covering the floor with a marly deposit. This was probably the marl sediment that sealed off the scroll fragments. Submergence of the caves in the Dead Sea would probably result in destruction of the scrolls rather than in sealing them.

Other caves in Qumran terrace were partly or totally destroyed by disintegration and collapse: "The fragments from caves 7 to 10 had evidently been deposited before the erosion or the collapses which carried away the greater part of these caves" (de Vaux, 1973, p. 100). The proposed -350 m level is therefore not supported by sufficient evidence.

Evidence from both Klein (1986) and the present study agrees about a rise in Dead Sea level at the beginning of Stage 9, although exact levels are not agreed upon. There seems to be little evidence for levels higher than -382 m.

Stage 10: 1300 C.E.—present

Three samples of wood from two Mount Sedom caves are dated to this stage. Cave passages began downcutting after the end of Stage 9, indicating a falling lake level. Later Dead Sea levels seem to fluctuate with a smaller amplitude.

Historic evidence presented by Klein (1986) increases in reliability during Stage 10, until the actual lake levels were measured beginning in the early 20th century (Masterman, 1913). Klein (1986) presents a continuous lake level curve for the last 1,000 years that is in general agreement with the Mount Sedom cave evidence, which has a lower resolution. According to Klein's curve, the lake level dropped 25 m around 1300 C.E.—the beginning of Stage 10. Later lake levels were typically around -390 to -400 m until 1970, with an episode of -375-m level. ^{14}C dates of fossil trees from the recently exposed shore (Raz, 1993, p. 75, 80) indicate a lake level lower than -400 m during the 14th century C.E., followed by a rise in lake level that submerged the trees, covering them with aragonite crusts. At the beginning of the 19th century, the lake level dropped to -402 m and rose to -391 m at the end of the 19th century (Klein, 1986).

Neev and Emery (1967) suggested a gradual rise in the Dead Sea from about -436 m more than 1,000 years ago to about -400 m 300 years ago, with a maximum of -391 m at the end of the 19th century.

Radium balance was used by Stiller and Chung (1984) to estimate the age of the meromictic structure of the Dead Sea. They suggested that meromixis stabilized after an episode of low water about 1720 C.E. This is supported by halite crystals of similar age (inferred from age-depth ratio) in a borehole from the northern basin of the Dead Sea.

Seismic reflection profiles from the northern end of the Dead Sea show a relict Jordan River canyon at -486 m (Niemi and Ben-Avraham, 1993 and pers. com., 1993). The canyon is not filled by sediments, indicating a very low Dead Sea level sometime during Stage 10, which would have allowed downcutting by the Jordan River.

The 20th century Dead Sea levels were studied by Masterman (1913), Ashbel (1965), Klein (1965), Anati and Shasha (1989), Raz (1993), Anati (chapter 8, this volume). The modern fluctuations, affected by increasing human interference, are beyond the scope of this chapter.

A Dead Sea level of -405 to -390 m was common during the later part of Stage 10, with shallow water covering the southern basin. A lower-level event appears to have occurred earlier during this stage, probably before the 18th century C.E. The present low level is induced by human activity.

GLOBAL CORRELATION

The extremely low-level event about 11,000–10,000 ^{14}C yr B.P. that marks the end of Lake Lisan and the onset of the Holocene Dead Sea seems to correlate with the low levels of eastern and northern African lakes during the cool Younger Dryas event (Gasse et al., 1980; Servant and Servant-Vildary, 1980; Roberts et al., 1993; Williamson et al., 1993).

Lakes in eastern and northern Africa reached their highest levels during the Early Holocene, 10,000–6,000 ^{14}C yr B.P. (COHMAP Members, 1988). This was explained by Milankovich forcing: Changing orbital parameters increased seasonality in the northern hemisphere, causing high-intensity monsoons to penetrate farther north than they do today (Kutzbach and Street-Perrott, 1985). The high lake level event reached southern Arabia, where lakes persisted from 9,000 to 6,000 ^{14}C yr B.P. in regions that are presently arid (McClure, 1976; Whitney, 1982). It was formerly suggested that this event did not extend beyond 28° latitude (Roberts, 1982); while farther north, lakes had lower levels. The high Dead Sea level suggested here for Stage 1 may correlate with the African–Arabian high levels event.

A short, low lake level event took place in eastern and northern Africa between 6,200 and 5,800 ^{14}C yr B.P. (Street-Perrott and Harrison, 1984). This event seems to correlate with the low-level Stage 2a of the Dead Sea, although the Dead Sea dry event lasted longer.

The belt of high lake levels in Africa and Arabia appears to have migrated northward from 10,000 to 4,500 yr B.P. (Street and Grove, 1979). Lakes shrank rapidly after 4,000 ^{14}C yr B.P. to present levels (Gasse, 1980). The shrinkage to lower levels at around 4,000 ^{14}C yr B.P. was the most important Holocene change in the lakes located directly across the African tropics (Hamilton, 1982). This correlates well with the rapid fall of the Dead Sea level at the end of Stage 3.

The African lake levels have remained low since 4,000 ^{14}C yr B.P., with fluctuations of small amplitude (Street-Perrott and Harrison, 1984). This is generally the case for the Dead Sea also, although available evidence does not allow precise correlation of brief events (Stages 5–10).

The correlation between the Dead Sea and African lakes suggests that the Dead Sea level is affected by climate more than by local tectonic subsidence. The Dead Sea area may have received summer monsoons during the first half of the Holocene, as did the Sahara and Arabia. A rising Dead Sea level could result from increased summer cloudiness, even without a change in precipitation (Goldberg and Rosen, 1987). More evenly distributed precipitation throughout the year could have supported the early agricultural communities developing in the Levant during the first half of the Holocene.

CONCLUSION

Holocene Dead Sea levels are lower than the levels reached by its predecessor, Late Pleistocene Lake Lisan (Begin et al., 1985). Furthermore, Dead Sea fluctuations are small compared with Lake Lisan fluctuations. The amplitude of fluctuations has decreased even more during the Late Holocene.

For most of the Holocene, a fair qualitative agreement exists between the Dead Sea levels suggested by the sedimentologic approach, the geomorphic approach, and the historical approach. Quantitative disagreements are attributed to the limitations of accuracy inherent in each approach.

The Dead Sea level was high during the Early Holocene. There were large amplitude fluctuations in mid-Holocene: In Stages 2a and 4, the southern basin probably completely dried out, whereas in Stage 3, the Dead Sea rose to its highest level in the last 7,000 years. This rise is clearly indicated by Mount Sedom caves and Dead Sea sediments.

After Stage 3, the lake level fell considerably. The Late Holocene Dead Sea seems to have stabilized at an average low level, with three short episodes of higher levels (Stages 5, 7, and 9) every 1,000 years. These episodes did not reach the high levels of earlier Stages 1 and 3. The time period represented by later stages is shorter than earlier stages, possibly because more evidence is available.

A comparison between the high levels achieved by Lake Lisan and the lower Dead Sea levels shows a general drying trend in the climate from the Upper Pleistocene to the present. Holocene Dead Sea fluctuations may be attributed to subtle climatic changes, which, unlike the larger climatic oscillations of the Late Pleistocene, are sometimes difficult to detect using other climatic indicators. The climatically induced lake level fluctuations are superimposed on tectonically controlled changes, which probably include a gradual drop in level as a result of subsidence of the Dead Sea basin. However, the gradual desiccation of the Dead Sea during the Holocene, as well as the first four Dead Sea stages, seem to correlate well with stages proposed for the tropical and north African lakes. Any model invoked to explain the African Holocene climatic changes (e.g., COHMAP Members, 1988; Street-Perrott and Perrott, 1993) should also consider similar changes in the Dead Sea area.

Acknowledgments

This study was based partly on a project prepared for the Dead Sea Works. It was financed by a grant from the research fund of the Society for the Protection of Nature in Israel and a grant from the Natural Sciences and Engineering Research Council of Canada. I am grateful to Mr. I. Carmi and the late Prof. M. Magaritz of the Weizmann Institute of Science for ^{14}C dating and participation in a preliminary report; Prof. I. Zak of the Hebrew University for supervision and participation in a preliminary report; Mr. J. Charrach of the Dead Sea Works for his enthusiastic help and comments on the manuscript; Prof. D. C. Ford and Prof. H. P. Schwarcz of McMaster University for reviewing the manuscript; Dr. D. Neev for useful discussion; Dr. Y. Elitzur for discussing geographical history matters; and the Israel Cave Research Center team for field assistance.

REFERENCES

Aharoni, Y., 1964, Mesad Gozal: *Israel Exploration Journal*, v. 14, p. 112–113.
Anati, D. A., and Shasha, S., 1989, Dead Sea surface-level changes: *Israel Journal of Earth Sciences*, v. 38, p. 29–32.
Ashbel, D., 1965, The rising and falling of the Dead Sea level: *Mada*, v. 9, no. 5, p. 255–260 (in Hebrew).
Atkinson, T. C., Briffa, K. R., and Coope, G. R., 1987, Seasonal temperatures in Britain during the past 22,000 years, reconstructed using beetle remains: *Nature*, v. 325, p. 587–592.
Bar Adon, P., 1989, *Excavations in the Judean Desert* : Jerusalem, The Department of Antiquities and Museums, 91 p. (in Hebrew).
Bar-Yosef, O., 1987, Prehistory of the Jordan Rift: *Israel Journal of Earth Sciences*, v. 36, p. 107–119.
Baruch, U., 1990, Palynological evidence of human impact on the vegetation as recorded in Late Holocene lake sediments in Israel, *in* Bottema, S., Entjes-Nieborg, G., and Van-Zeist, W., eds., *Man's role in shaping of the Eastern Mediterranean landscape*: Rotterdam, Brookfield, p. 283–293.
Begin, Z. B., Broecker, W., Buchbinder, B., Druckman, Y., Kaufman, A., Magaritz, M., and Neev, D., 1985, Dead Sea and Lake Lisan levels in the last 30,000 years, a preliminary report: Jerusalem, Geological Survey of Israel, report 29/85, 18 p.
Begin, Z. B., Nathan, Y., and Ehrlich, A., 1980, Stratigraphy and facies distribution in the Lisan Formation—New evidence from the area south of the Dead Sea, Israel: *Israel Journal of Earth Sciences*, v. 29, p. 182–189.
Ben-Avraham, Z., Niemi, T. M., Neev, D., Hall, J. K., and Levy, Y., 1993, Distribution of Holocene sediments and neotectonics in the deep north basin of the Dead Sea: *Marine Geology*, v. 113, p. 219–231.
Ben-Avraham, Z., ten-Brink, U., and Charrach, J., 1990, Transverse faults at the northern end of the southern basin of the Dead Sea graben: *Tectonophysics*, v. 180, p. 37–47.
Bowman, D., 1988, The declining but non-rejuvenating base level—The Lisan Lake, the Dead Sea area, Israel: *Earth Surface Processes and Landforms*, v. 13, p. 239–249.
Burleigh, R., 1984, Additional radiocarbon dates for Jericho (with an assessment of all the dates obtained), *in* Kenyon, K. M., and Holland, T. A., eds., *Excavations at Jericho*: London, British School of Archaeology in Jerusalem, v. 5, p. 760–765.
Clermont-Ganneau, C., 1902, Où était l'embouchure du Jourdain à l'époque de Josué?: *Recueil d'archéologie Orientale*, v. 5, p. 267–280.
COHMAP Members, 1988, Climatic changes of the last 18,000 years—Observations and model simulations: *Science*, v. 241, p. 1043–1052.
de Vaux, R., 1973, *Archaeology and the Dead Sea scrolls*: Oxford, Oxford University Press, 142 p.
Donahue, J., 1985, Hydrologic and topographic change during and after Early Bronze occupation at Bab edh-Dhra and Numeira, *in* Hadidi, A., ed., *Studies in the History and Archaeology of Jordan*: Amman, Department of Antiquities, v. II, p. 131–140.
Druckman, Y., Magaritz, M., and Sneh, A., 1987, The shrinking of Lake Lisan, as reflected by the diagenesis of its marginal oolitic deposits: *Israel Journal of Earth Sciences*, v. 36, p. 101–106.
Frumkin, A., 1992, The karst system of the Mount Sedom salt diapir [Ph.D. thesis]: Jerusalem, The Hebrew University of Jerusalem, 208 p. (in Hebrew, English abstract).
Frumkin, A., 1996, Uplift rate relative to base level of a salt diapir (Dead Sea, Israel), as indicated by cave levels, in Alsop, I., Blundell, D. and Davison, I., eds., Salt Tectonics, Special Publication no. 100: London, Geological Society, p. 41–47.
Frumkin, A., Carmi, I., Zak, I., and Magaritz, M., 1994, Middle Holocene environmental change determined from the salt caves of Mount Sedom, Israel, *in* Bar-Yosef, O., and Kra, R., eds., *Late Quaternary chronology and paleoclimates of the eastern Mediterranean*: Tucson, The University of Arizona, p. 315–332.
Frumkin, A., and Ford, D. C., 1995, Rapid entrenchment of stream profiles in the salt caves of Mount Sedom, Israel: *Earth Surface Processes and Landforms*, v. 20, p. 139–152.
Frumkin, A., Magaritz, M., Carmi, I., and Zak, I., 1991, The Holocene climatic record of the salt caves of Mount Sedom, Israel: *The Holocene*, v. 1, no. 3, p. 191–200.
Frumkin, A. and Zak, I., 1991, Holocene evolution of Mount Sedom Diapir based on karst evidence (abstract), in Weinberger, G.,eds., Annual meeting, Akko, Israel Geological Society, p. 35.
Gardosh, M., Reches, Z., and Garfunkel, Z., 1990, Holocene tec-

tonic deformation along the western margins of the Dead Sea: *Tectonophysics*, v. 180, no. 1, p. 123–137.

Garfunkel, Z., Zak, I., and Freund, R., 1981, Active faulting in the Dead Sea rift: *Tectonophysics*, v. 80, no. 1, p. 81–108.

Gasse, F., 1980, Late Quaternary changes in lake-levels and diatom assemblages on the southeastern margin of the Sahara: *Palaeoecology of Africa*, v. 12, p. 333–350.

Gasse, F., Rognon, P., and Street, F. A., 1980, Quaternary history of the Afar and Ethiopian Rift lakes, *in* Williams, M. A., and Faure, H., eds., *The Sahara and the Nile*: Rotterdam, A. A. Balkema, p. 361–400.

Gill, M., 1983, *Israel in the first moslem period*: Tel Aviv, Tel Aviv University, (in Hebrew), 688 p.

Goitein, S. D., 1975, A court record from Zoar on the Salt Sea: *Eretz-Israel*, v. 12, p. 200–202 (in Hebrew).

Goldberg, P., and Bar-Yosef, O., 1982, Environmental and archaeological evidence for climatic change in the Southern Levant, *in* Bintliff, J. L., and Van Zeist, W., eds., *Palaeoclimates, palaeoenvironments and human communities in the eastern Mediterranean region in later prehistory*: Oxford, BAR International Series, v. I33(i), p. 399–407.

Goldberg, P., and Rosen, A. M., 1987, Early Holocene palaeoenvironments of Israel, *in* Levy, T. E., ed., *Shiqmim I—Studies concerning Chalcolithic societies in the northern Negev Desert, Israel (1982–1984)*: BAR International Series, v. 356(i), p. 23–33.

Goodfriend, G. A., Magaritz, M., and Carmi, I., 1986, A high stand of the Dead Sea at the end of the Neolithic period—Paleoclimatic and archeological implications: *Climatic Changes*, v. 9, p. 349–356.

Hadas, G., 1993, Where was the harbour of 'En-Gedi situated?: *Israel Exploration Journal*, v. 43 no. 1, p. 45–49.

Hamilton, A. C., 1982, *Environmental history of East Africa, a study of the Quaternary*: London, Academic Press, 328 p.

Horowitz, A., 1992, *Palynology of arid lands*: Amsterdam, Elsevier, 546 p.

Issar, A. S., 1990, *Water from the rock*: Heidelberg, Springer Verlag, 213 p.

Issar, A. S., and Tsoar, H., 1987, Who is to blame for the desertification of the Negev, Israel?, The influence of climate change and climatic variability on the hydrologic regime and water resources, v. 168: Vancouver, The International Association of Hydrological Sciences, p. 577–583.

Issar, A. S., Tsoar, H., and Levin, D., 1989, Climatic changes in Israel during historical times and their impact on hydrological, pedological and socio-economic systems, *in* Leinen, M., and Sarnthein, M., eds., *Paleoclimatology and paleometeorology—Modern and past patterns of global atmospheric transport*: Dordrecht, Netherlands, Kluwer Academic Publishers, p. 525–541.

Kadan, G., Eyal, Y., Enzel, Y., 1995, Dead-Sea fluctuations and tectonic events in the Holocene fan-delta of Nahal Darga, in Arkin,Y. and Avigad, D., eds., Annual meeting, Zikhron Ya'aqov, Israel Geological Society, p. 52.

Klein, C., 1965, On the fluctuations of the level of the Dead Sea since the beginning of the 19th Century: Jerusalem, Israel Hydrological Service, Hydrological paper no. 7, revised edition, 83 p.

Klein, C., 1982, Morphological evidence of lake level changes, western shore of the Dead Sea: *Israel Journal of Earth Sciences*, v. 31, p. 67–94.

Klein, C., 1986, Fluctuations of the level of the Dead Sea and climatic fluctuations in Israel during historical times [Ph.D. thesis]: Jerusalem, The Hebrew University of Jerusalem, 208 p. (in Hebrew, English abstract).

Kutzbach, J. E., and Street-Perrott, F. A., 1985, Milankovitch forcing of fluctuations in the level of tropical lakes from 18 to

0 kyr B.P.: *Nature*, v. 317, p. 130–134.

Magaritz, M., and Goodfriend, G. A., 1987, Movement of the desert boundary in the levant from latest Pleistocene to early Holocene, *in Abrupt climate change — Evidence and implications*: Netherland, Reidel, p. 173–183.

Magaritz, M., Rahner, S., Yechieli, Y., and Krishnamurthy, R. V., 1991, $^{13}C/^{12}C$ ratio in organic matter from the Dead Sea area: Paleoclimatic interpretation: *Naturwissenschaften*, v. 78, p. 453–455.

Masterman, E. W. G., 1913, Summary of observations on the rise and fall of the Dead Sea, 1900–1913: *Palestine Exploration Fund Quarterly Statement*, v. 1913, p. 192–197.

McClure, H. A., 1976, Radiocarbon chronology of Late Quaternary lakes in the Arabian Desert: *Nature*, v. 263, p. 755–756.

McConaughy, M. A., 1981, A Preliminary report on the Bab edh-Dhra site survey: *Annual of the American Schools of Oriental Research*, v. 46, p. 187–190.

Neev, D., 1964, The Dead Sea: Jerusalem, Geological Survey of Israel, report Q/2/64, 407 p.

Neev, D., and Emery, K. O., 1995, *The destruction of Sodom, Gomorrah, and Jericho*: New York, Oxford University Press, 175 p.

Neev, D., and Emery, K. O., 1967, The Dead Sea—Depositional processes and environments of evaporites: Geological Survey of Israel Bulletin 41, p. 1–147.

Neev, D. and Emery, K.O., 1995, The destruction of Sedom, Gomorrah, and Jericho: Oxford University Press, 175 p.

Neev, D., and Hall, J. K., 1977, Climatic fluctuations during the Holocene as reflected by the Dead Sea levels, *in* Greer, D., ed., *Desertic terminal lakes*, Logan, Utah State University, p. 53–60.

Niemi, T. M., and Ben-Avraham, Z., 1993, Neotectonics and Late Quaternary paleoclimate of the Dead Sea: Geological Society of America, Abstracts with programs, v. 25, p. 391.

Niklaus, T. R., 1991, *CalibETH user's manual*: Zürich, ETH, 151 p.

Nissenbaum, A., and Hadas, G., 1990, Dating of ancient anchors from the Dead Sea: *Naturwissenschaften*, v. 77, p. 228–229.

Pearson, G. W., Pilcher, J. R., Baillie, M. G. L., Corbett, D. M., and Qua, F., 1986, High-precision ^{14}C measurement of Irish Oaks to show the natural ^{14}C variations from AD 1840 to 5210 BC: *Radiocarbon*, v. 28, 2B, p. 911–934.

Rast, W. E., and Schaub, R. T., 1980, Preliminary report of the 1979 expedition to the Dead Sea plain, Jordan: *Bulletin of the American Schools of Oriental Research*, v. 240, p. 21–61.

Raz, E., 1993, *Dead Sea Book*: Jerusalem, Nature Reserves Authority, 231 p. (in Hebrew).

Roberts, N., 1982, Lake levels as an indicator of Near Eastern palaeo-climates: a preliminary appraisal, *in* Bintliff, J. L., and Van Zeist, W., eds., *Palaeoclimates, palaeoenvironments and human communities in the eastern Mediterranean region in later prehistory*: Oxford, BAR International Series, v. I33(i), p. 235–271.

Roberts, N., Taieb, M., Barker, P., Damnati, B., Icole, M., and Williamson, D., 1993, Timing of the Younger Dryas event in East Africa from lake-level changes: *Nature*, v. 366, p. 146–148.

Stein, M., Agnonn, A., Starinsky, A., Raab, M., Katz, A., and Zak, I., 1994, What is the "age" of the Sedom Formation? Annual meeting, Israel Geological Society, Nof Ginosar, February 28-March 2m, 1994, p. 108.

Servant, M., and Servant-Vildary, S., 1980, L'environnement quaternaire du bassin du Tchad, *in* Williams, M., and Faure, H., eds., *The Sahara and the Nile*: Rotterdam, Netherlands, Balkema, p. 133–162.

Stiller, M., Carmi, I., and Kaufman, A., 1988, Organic and inorganic ^{14}C concentrations in the sediments of Lake Kinneret and the Dead Sea (Israel) and the factors which control them: *Chemical Geology*, v. 73, p. 63–78.

Stiller, M., and Chung, Y. C., 1984, Radium in the Dead Sea—A possible tracer for the duration of meromixis: *Limnology and Oceanography,* v. 29, no. 3, p. 574–586.

Street, F. A., and Grove, A. T., 1979, Global maps of lake-level fluctuations since 30,000 yr B.P.: *Quaternary Research,* v. 12, p. 83–118.

Street-Perrott, F. A., and Harrison, S. P., 1984, Temporal variations in lake levels since 30,000 yr B.P. ——An index of the global hydrological cycle, *in* Hansen, J. E., and Takahashi, T., eds., *Climate processes and climate sensitivity:* Washington, D.C., American Geophysical Union, p. 118–129.

Street-Perrott, F. A., and Perrott, R. A., 1990, Abrupt climate fluctuations in the tropics—The influence of Atlantic Ocean circulation: *Nature,* v. 343, p. 607–612.

Stuiver, M., and Pearson, G. W., 1986, High-precision calibration of the radiocarbon time scale, AD 1950–500 BC: *Radiocarbon,* v. 28, no. 2B, p. 805–838.

Weinstein, J. M., 1984, Radiocarbon dating in the Southern Levant: *Radiocarbon,* v. 26, no. 3, p. 297–366.

Weiss, H., Courty, M., Wetterstorm, W., Guichard, F., Senior, L., Meadow, A., and Curnow, A., 1993, The genesis and collapse of third millennium north Mesopotamian civilization: *Science,* v. 261, p. 995–1004.

Whitney, J. W., 1982, Geologic evidence of late Quaternary climate change in western Saudi Arabia, *in* Bintliff, J. L., and Van Zeist, W., eds., *Palaeoclimates, palaeoenvironments and human communities in the eastern Mediterranean region in later prehistory:* Oxford, BAR International Series, v. I33(i), p. 277–321.

Williamson, D., Taieb, M., Damnati, B., Icole, M., and Thouveny, N., 1993, Equatorial extension of the Younger Dryas event—Rock magnetic evidence from Lake Magadi (Kenya): *Global and Planetary Change,* v. 7, p. 235–242.

Yechieli, Y., Magaritz, M., Levy, Y., Weber, U., Kafri, U., Woelfli, W., and Bonani, G., 1993, Late Quaternary geological history of the Dead Sea area, Israel: *Quaternary Research,* v. 39, p. 59–67.

Zak, I., 1967, The geology of Mount Sedom [Ph.D. thesis]: Jerusalem, The Hebrew University, 208 p. (in Hebrew, English abstract).

Zak, I., and Bentor, Y. K., 1968, Some new data on the salt deposits of the Dead Sea area, Israel, Symposium on the Geology of Saline Deposits, Hannover, UNESCO, p. 137–146.

Zak, I., and Freund, R., 1980, Strain measurements in eastern marginal shear zone of Mount Sedom salt diapir, Israel: *American Association of Petroleum Geologists Bulletin,* v. 64, p. 568–581.

23. THE DEAD SEA REGION: AN ARCHAEOLOGICAL PERSPECTIVE

Itzhaq Beit-Arieh

From the perspective of archaeological settlements, the desert environment around the Dead Sea differs radically from all other desert regions. Whereas the predominant characteristic of deserts generally is their low settlement and population density, the region around the Dead Sea boasts a large number of ancient sites that extend from prehistoric to later periods. Many of the sites have been described in the New Encyclopedia of Archaeological excavations in the Holy Land (Stern, 1993). In this chapter, I briefly review the principal ancient sites in the Dead Sea region (Fig. 23-1), and discuss the reasons for the concentration of ancient sites in this area. The chronology of the archaeological periods is given in Table 23-1.

PREHISTORIC PERIODS

The earliest occupation of the Dead Sea region dates from the epipaleolithic period, 10,000 to 12,000 years ago evidenced by remnants of the Natufian culture excavated at Tel el-Sultan (Jericho). A relatively advanced level of construction was found at the same site from the Neolithic period (7,000 to 11,000 yr ago) (Kenyon, 1957; Stern, 1993). Although Jericho is located about 10 km northwest of the Dead Sea at the mouth of a perennial freshwater spring, its founding and development over consecutive periods can be attributed to a large extent, to its proximity to the Dead Sea with its natural mineral resources. An additional factor is the relatively comfortable, dry tropical climate of the Jericho depression, which is well-suited to plantation culture and the growing of special field crops.

A diversity of remains from the Neolithic period were excavated from a cave dwelling located in Nahal Hemar—an ephemeral wadi that flows from the western bank into the southern part of the Dead Sea located, 5 km south of En Boqeq (Fig. 23-1; Bar-Yosef and Alon, 1988). Among the excavated artifacts from the cave dwelling were wooden implements and fruit items. These exceptional finds, which could only have been preserved because of the arid climate, have appreciably increased our knowledge of this early period.

The Chalcolithic culture, flourishing between the end of the Neolithic ceramic period and the beginning of the Early Bronze Age, was of utmost importance in the formation of the historical era. Artifacts and structure from this period have been found in cave dwellings and other sites. Several phases within the long span of its existence can be distinguished. This last phase of the Chalcolithic period was the first to be discovered, at Tulelat el-Ghassul (Fig. 23-1, Mallon et al., 1940). Since its discovery, this culture has been named the Ghassulian culture after the name of this site.

During the late Chalcolithic period, in the fourth millennium B.C.E., the region was inhabited by a comparatively large population. Numerous settlements were founded, especially in the northern and western Negev. An outstanding feature of the period is the exploitation of the copper ore deposits in the vicinity of Feinan, Jordan. The period is also notable for the skill of its contemporary craftsmen in the production of basalt implements and for the high technological level displayed in its art objects.

The major site of this period in the area of the Dead Sea is Tuleilat el-Ghasul, about 5 km northeast of the sea and extend-

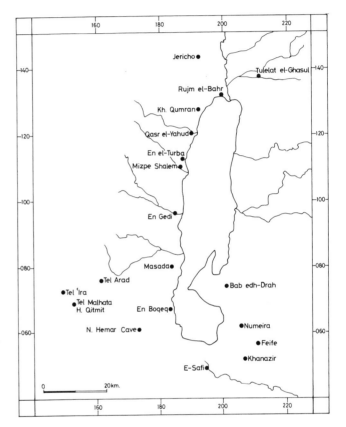

Figure 23-1 Archaeological sites around the Dead Sea.

ing over an area of 200 dunams (Levy, 1986; Stern, 1993). Another Chalcolithic site is the shrine built above the En Gedi springs on the western side of the Dead Sea (Ussishkin, 1980). This was designed to serve the cultic needs of the population that dwelt, for the most part, apparently, in caves in the cliffs of the wadis that discharge into the Dead Sea from the west.

The people of this particular area are thought to have used and even to have produced the rich and unique hoard of copper implements discovered in the 1960s in a cave at Nahal Mishmar, 8 km south of En Gedi, called "the cave of the treasure" (Bar-Adon, 1980). The hoard, one of the most spectacular finds in the ancient Middle East, contained over 400 artifacts, most of them made of copper. Included among these mostly ritual objects are 138 standards, 10 "crowns," and 240 maceheads. Perhaps the hoard was associated with ritual activities carried out at the nearby Chalcolithic shrine at En Gedi.

HISTORIC PERIODS

In the next period, the Early Bronze Age (3,200–2,200 B.C.E.), the region witnessed a flowering of settlement. In the western hinterland, the Canaanite city of Arad arose on an earlier Chalcolithic occupation (Amiran, 1978). In the valley of the Dead Sea, the "cities of the plain" (Genesis 13:12), including Sodom and Gommorah, probably flourished in the southern Ghor region of the Dead Sea. To the eastern were the settlements of Bab edh-Dhra, Numeira, Khanazir, and e-Safi (Stern, 1993). Near Mizpe Shalem, a small temple was erected to serve both a local and a transient population (Bar-Adon, 1989).

The only occupation known from the Intermediate Bronze period (MBI) is a small settlement in Jericho, whereas during the Late Bronze period, the only settlement is indicated by a meager, short-lived, occupation layer, also in Jericho. Apparently, during a period of about a thousand years, from about 2,200 to 1,250 B.C.E., the Dead Sea region, unlike the earlier periods, was not inhabited by a population numerous enough to leave remains recognizable today.

Additional settlement occurred during Iron Age II, the period of the Judean kingdom. The central settlement was at Tel Goren, near the western edge of the Dead Sea (Mazar et al., 1966). Along the shore, several settlements and military outposts were established: at Qumran near the northwestern shore, at Rujm el Bahr at the north shore of the lake, at Qasr el-Yahud and at Ein el-Turba located at the western edge of the lake (Bar-Adon, 1989). At the same time, settlements developed in the eastern Negev, which borders on the west side of the Dead Sea. However, in this area most of the new settlements were military outposts, reflecting political developments that occurred during the seventh century B.C.E at the end of the Israelite period.

During the late Iron Age, Assyria was declining as the dominant power of the Near East, while the countervailing rise of Babylon had begun. The kingdom of Judah, under threat from Edom, built a series of fortified military outposts along its remote eastern frontier parallel to the Dead Sea and also established a series of outposts on the eastern Negev plateau (Beit-Arieh and Cresson, 1991). However, around 600 B.C.E, the time of the destruction of Jerusalem by the Babylonians, the Edomites apparently succeeded in penetrating the Negev and dominating areas and trade routes leading to the Mediterranean ports. An important corroborative site in this regard is the Edomite shrine, which my team and I discovered and subsequently excavated at Horvat Qitmit (Fig. 23-1; Beit-Arieh, 1995).

During the return of Zion, some of the returning population settled at Tel Goren, which continued to exist into the Hellenistic period. Also, during this period, the En Boqeq oasis was settled for the first time (Gichon, 1993), and military outposts that were originally founded during the Israelite period were reestablished at Rujm el-Bahr, Qasr el-Yahud, and En el-Turba (Fig. 23-1). Perhaps these outposts were built by the Hasmonian king Alexander Janneaus (Bar-Adon, 1989).

At Horvat Qumran, originally an Iron Age settlement, renewed occupation in the second to first century B.C.E. by a Judean desert sect identified with the Essenes was discovered. In the vicinity, a large cemetery used by the sect was found, and in nearby caves, scrolls of the sect were uncovered (Stern, 1993, p. 1135-1241). The scrolls include original transcriptions of Old Testament books, which antedated by more than 1,000 years the earliest hitherto extant Old Testament manuscripts (Yadin, 1984). During this period (the second temple period), a Jewish community flourished at En Gedi, as has been recently proven by the large tombs uncovered in the area (Hadas, 1994).

In the Early Roman or Herodian period, Herod built himself a great palace at Masada, west of the Dead Sea. The site was captured by Jewish rebels after the destruction of Jerusalem in 70 C.E. by the Romans. But, its capture after a long siege by the Romans in 76 C.E., brought to a close the great Jewish rebellion against imperial Rome. The excavation of Masada by Yigael Yadin and his expedition is considered one of the greatest

Table 23-1 Chronology of archaeological periods

Period	Approximate Dates
Pre-Pottery Neolithic	8,300–5,500 B.C.E.
Pottery Neolithic	5,500–4,500 B.C.E.
Chalcolithic	4,500–3,300 B.C.E.

HISTORICAL ARCHAEOLOGICAL PERIODS

Bronze Age (Canaanite period)	
Early Bronze Age IA-B	3,300–3,000 B.C.E.
Early Bronze Age II	3,000–2,700 B.C.E.
Early Bronze Age III	2,700–2,200 B.C.E.
Middle Bronze Age I (EB IV)	2,200–2,000 B.C.E.
Middle Bronze Age IIA	2,000–1,750 B.C.E.
Middle Bronze Age IIB	1,750–1,550 B.C.E.
Late Bronze Age I	1,550–1,400 B.C.E.
Late Bronze Age IIA	1,400–1,300 B.C.E.
Late Bronze Age IIB	1,300–1,200 B.C.E.
Iron Age (Israelite period)	
Iron Age IA	1,200–1,150 B.C.E.
Iron Age IB	1,150–1,000 B.C.E.
Iron Age IIA	1,000–900 B.C.E.
Iron Age IIB	900–700 B.C.E.
Iron Age IIC	700–586 B.C.E.
Babylonian and Persian periods	586–332 B.C.E.
Hellenistic period	
Early Hellenistic period	332–167 B.C.E.
Late Hellenistic period	167–37 B.C.E.
Roman and Byzantine periods	
Early Roman period	37 B.C.E.–132 C.E.
Herodian period	(37 B.C.E.–70 C.E.)
Late Roman period	132–324 C.E.
Byzantine period	324–638 C.E.
Early Arab to Ottoman periods	
Early Arab period	638–1099 C.E.
(Umayyad and Abbasid)	
Crusader and Ayyubid period	1099–1291C.E.
Late Arab period	1291–1516 C.E.
(Fatimid and Mamluk)	
Ottoman period	1516–1917 C.E.

C.E., Common Era (A.D.).
B.C.E., Before the Common Era (B.C).

archaeological enterprises ever undertaken in Israel. The expedition's scientific reports have recently begun to be published (Yadin and Naveh, 1989; Cotton and Geiger, 1989; Netzer, 1991; Foerster, 1995).

During the Early Roman Period, a large "industrial" installation was constructed at En Boqeq oasis, apparently for the compounding of medicinal and cosmetic products (Gichon, 1993). During and after the Bar-Kochba revolt (in the second century

C.E.), a Jewish population settled in the area, exploiting the En Gedi oasis for living and the caves for escaping from the Roman army.

A characteristic feature of the Byzantine period (fourth to seventh century C.E.) is the dense settlement of the entire area, including the Dead Sea region. From Horvat Qumran in the north to En Boqeq in the south, Byzantine building remains have been discovered, mostly on the ruins of earlier sites. In the fifth century C.E., a Jewish community flourished in an En Gedi oasis, where a synagogue was uncovered (Stern, 1993).

SPECIAL RESOURCES OF THE DEAD SEA

The long record of settlement in the Dead Sea region, an unequivocal desert environment, requires an explanation. Two factors seem to account for this record. One is the very terrain of this desert: Its deep gorges and many caves provided an ideal environment for anyone seeking isolation, whether for ideological reasons or for escape from a pursuing enemy. The second, and certainly most significant factor, for the "permanent" settlements was undoubtedly the Dead Sea's natural resources, which served as a locally unique source for raw materials needed in the ancient world.

One of these materials was asphalt, or bitumen, the earliest use of which occurred in the preceramic site of Gilgal. Wooden figurines coated with bitumen, as well as bitumen-coated skulls from the Neolithic period, were found in the Nahal Hemar Cave. A lump of asphalt was found in stratum II of Canaanite Arad from the Early Bronze period. At more distant sites like Tel Eirani and Tel Dalit, jars coated on the inside with bitumen have been found. Moreover, the discovery of asphalt, which evidently originated in the Dead Sea, in a stratum dating to the third millennium B.C.E. in the Nile Delta region proves that this material was exported to Egypt, where it found widespread use, for example, in caulking clay pots, in mummification, and so forth (Connan et al., 1992). Bitumen has also been found from later periods, for example, at the Intermediate Bronze site of Ein Zik in the Negev highlands and at the Iron Age II sites of Tel Ira and Tel Malhata in the eastern Negev.

A second natural resource of the Dead Sea was undoubtedly salt, which was as much in demand during earlier periods as it is today. Apparently, it was produced from the briny water of the lake itself or mined from the solid salt deposits that existed nearby. In truth, there is no archaeological evidence for this, since salt is readily dissolved in rain and flood waters, leaving no physical trace. Some indication for the exploitation of salt may be present in the name Tel Malhata. The original (Arabic) name of this site in the eastern Negev, *Tel el-Milh* (mound of salt), was given during an unknown period (probably in the Middle Ages), apparently because this site served as a trading center for salt, the source of which was probably the Dead Sea.

Additional economic resources of the Dead Sea region known to us from literary and archaeological sources were fruit crops, such as *apharsimon* (almost certainly not the persimmon known by this Hebrew name today) and dates, as well as various spice plants, for the cultivation of which the dry desert climate was ideal.

The region of the Dead Sea has a rich cultural history which is evident in the extensive archaeological remains from prehistoric time to the present. Contrary to the idea that the Dead Sea is an uninhabitable wasteland, the area's unique resources, including asphalt, salt, and freshwater, and an abundance of caves and topographic havens made it a very habitable site. There is still much to be learned about the cultural history of this fascinating enironment.

REFERENCES

Amiran, R., 1978, *Early Arad I*: Jerusalem, The Israel Exploration Society, 138 p. 193 pl.

Bar-Adon, P., 1980, *The cave of the treasure*: Jerusalem, The Israel Exploration Society, 298 p..

Bar-Adon, P., 1989, Excavations in the Judean Desert: *Atiqot*, v. 9, p. 1–91 (in Hebrew).

Bar-Yosef, O., and Alon, D., 1988, Nahal Hemar: *Atiqot*, v. 18, p. 1–81 (in Hebrew).

Beit-Arieh, I., 1995, Horvat Qitmit: An Edomite shrine in the Biblical Negev: Tel Aviv, Institute of Archaeology, Tel Aviv University, 318 p.

Beit-Arieh, I., and Cresson, B., 1991,'Uza a Judean outpost on the eastern Negev border: *Biblical Archaeologist*, v. 54, p. 126–135.

Connan, J., Nissenbaum, A., and Dessort, D., 1992, Molecular archaeology—Export of Dead Sea asphalt to Canaan and Egypt in the Chalcolithic-Early Bronze Age (4th–3rd Mill. B.C): *Geochimica et Cosmochimica Acta*, v. 56, p. 2743–2759.

Cotton, H. M., and Geiger, J., 1989, *Masada: The Latin and Greek documents*: Jerusalem, Israel Exploration Society, 300 p.

Foerster, G., 1995, *Masada: Art and architecture*: Jerusalem, Israel Exploration Society, 264 p.

Gichon, M., 1993, *En Boqeq Ausgrabungen In Einer Oase Am Toten Meer*: Mainz, Philipp von Zabern, 461 p.

Hadas, G., 1994, Nine tombs of the Second Temple Period at En Gedi: *Atiqot*, v. 24, p. 75 (in Hebrew).

Kenyon, K. M., 1957, *Digging up Jericho*: New York, Praeger Publisher, 272 p.

Levy, T. E., 1986, The Chalcolithic Period: *Biblical Archaeologist*, v. 49, p. 82–108.

Mallon, R., Koppel, R., et al., 1940, *Teleilat Ghassul I–II*.

Mazar, B., Dothan, T., and Dunayevski, I., 1966, En Gedi—The first and second seasons of excavations, 1961–1962: *Atiqot*, v. 5, p. 1–100.

Netzer, E., 1991, *Masada: The buildings—Stratigraphy and architecture*: Jerusalem, Israel Exploration Society, 680 p.

Stern, E., ed., 1993, The new archaeological encyclopedia of archaeological excavations in the Holy Land (4 vols.): Jerusalem, Israel Exploration Society and Carta.

Ussishkin, D., 1980, The Ghassulian shrine at En Gedi: *Tel Aviv*, v. 7, p. 1–44.

Yadin, Y., 1984, *The Temple Scrolls (3 vols.)*: Jerusalem, Israel Exploration Society.

Yadin, Y., and Naveh, J., 1989, *Masada: The Aramaic and Hebrew ostraca and jar inscriptions*: Jerusalem, Israel Exploration Society, 232 p.

24. GEOCHEMICAL AND HYDROLOGICAL PROCESSES IN THE COASTAL ENVIRONMENT OF THE DEAD SEA

Yoseph Yechieli and Joel R. Gat

Lakes in general and saline lakes in particular exchange material with their environment. The watershed and the immediate coastal areas contribute the salt load to the lake; in a terminal lake, such as the Dead Sea, this salt input accumulates over the lifetime of the system (Bentor, 1961). However, the reverse process, namely, the transfer of material from a lake to its environment, may also be a significant factor in the geochemical balance of the lake. This reverse process operates by means of water spray and coastal wave splashing on one hand, and through leakage of lake waters into the adjacent banks on the other. Part of the salt may later be flushed back into the lake, but diagenetic processes on land may alter the chemical composition. Further, the loss of salt by aeolian transport from the lake and its coastal environment is believed to be a significant factor in the geochemical balance and to limit the salinity buildup of saline lakes (Langbein, 1961).

These exchange processes between a lake and its environment assume increased significance when lake levels change on a seasonal or longer time scale. During high water stand, brines invade the coastal region and are then drained or flushed back into the lake as the waters recede. The resultant to-and-fro flux of salt between the lake and land may overshadow the net import or export of the salts. The study of these coastal processes is thus of interest for the geochemistry of saline lakes.

The Dead Sea, the terminal lake of the Jordan River system, is an extremely saline lake located in an extremely arid environment (annual precipitation of only 70 mm). As has been often described, large amounts of the freshwater input to the lake have been diverted for irrigation purposes since the middle of this century, resulting in a lowering of the lake at an average rate of 0.5 m/year (Klein, 1985). The last 10 years are characterized by the steepest decline in lake level, averaging about 0.8 m/year (Anati and Shasha, 1989). In addition, the water level fluctuates annually (seasonally), with large changes following very wet years, for example, the rise of the water level by 2 m during the rainy season of 1992. The changes in the water level are in phase with the salinity changes of the surface waters (Anati, chapter 8, this volume). Since 1978, the Dead Sea surface waters have reached saturation with respect to halite at the extremes of the low water stand (Stiller et al., chapter 15, this volume).

With the lowering of the lake waters, sediments are exposed along the receding shorelines. These sediments were former bottom sediments of the Dead Sea, impregnated by the Dead Sea brine. This new land, the *terra nova*, is then partially drained but also undergoes processes resulting from exposure to its new atmospheric and hydrological environment. The elevation relative to recorded variations of the Dead Sea water level in recent years enables determination of the length of exposure. As an example, the area that was exposed between 1980 (when the Dead Sea elevation was -400 m) and 1992 (elevation -407 m) is shown in Figure 24-1c. Obviously, the extent of that area depends on the steepness of the land slope (Fig. 24-1c). This slope is steeper along the western margin than along the axis of the rift in the north-south direction.

The lowering of the Dead Sea level and its effect on the surroundings mimics, on a shortened time scale, the occurrences induced by climatic changes on the coastal areas of other lakes and the ocean. As such a model, the *terra nova* is a rewarding study site.

In this chapter, we describe observations and preliminary studies on the evolution of the newly exposed coastal areas carried out in a few specific locations (Fig. 24-1) on the western shore of the lake by us and by Michal Shatkay. Obviously much more detailed and extensive work will be needed to quantify these processes throughout the coastal region of the Dead Sea area.

THE GEOHYDROLOGICAL BACKGROUND

The areas we studied are located between the Dead Sea and the western borders of the rift valley. This area is mainly composed of continental sediments of Quaternary age (Fig. 24-2). The sediments are clastic (clay, sand, and gravel), deposited in fan deltas, with some intercalations of lacustrine sediments (clay, gypsum, and aragonite), belonging to the Lisan Formation (Sneh, 1979) or older units, such as the Samra Formation. In general, the sediments are of finer grain size farther from the mountains and closer to the sea. This is evident in the delta of Wadi Zeelim, where mainly gravel layers can be seen 4.5 km from the Dead Sea (with some silty and clayey layers), whereas near the Dead Sea (at a distance of less than 0.5 km from the shoreline), the layers are mainly silt and clay. The area occupied by the Quaternary sediments varies greatly (the distance from the Dead Sea to the rift margin is about 5 km in Wadi Zeelim, but less than 1 km at En Gedi and the Turiebe area). The slope of the exposed shore also varies significantly between the different parts of the Dead Sea coast (Fig. 24-1). West of the border of the rift, the sediments are of Cretaceous age, consisting mainly of limestone and dolomite with some marly layers (Fig. 24-2).

Information about the nature of the newly exposed sediments before their exposure is available from the works of Neev and Emery (1967), Garber (1980), and Levy (1987), who studied the sedimentology of the Dead Sea itself; of special relevance are those studies dealing with the shallow parts of the Dead Sea, which were later exposed. The Dead Sea sediments are composed of alternations of thin laminae of dark layers (mainly detritic, such as calcite, clays, and quartz) and light layers (mainly lacustrine, such as aragonite and gypsum). The aragonite concentration was found to be very high in the white laminae (85-99%; Garber, 1980) and much lower in the dark laminae (less than 25%; Garber, 1980).

During 1959–1960, the shallow parts of the Dead Sea bottom were reported to be covered with a hard crust of gypsum (Neev and Emery, 1967). Similar gypsum crusts were reported to have occurred also in the 1930s, and Neev and Emery (1967) claimed that gypsum had precipitated from the Dead Sea water at least during the century before 1960. This gypsum crust was not found in the bottom of the Dead Sea in studies conducted in the 1980s (Levy, 1987). Although the Dead Sea brine was supposedly close to halite saturation, no substantial halite precipitation was observed before the turnover in 1979. Formation of halite crystals at the Dead Sea surface was observed in 1979 (Steinhorn, 1983), 1982 (Levy, 1987), and occasionally since that time. Some of this halite precipitated to the Dead Sea bottom,

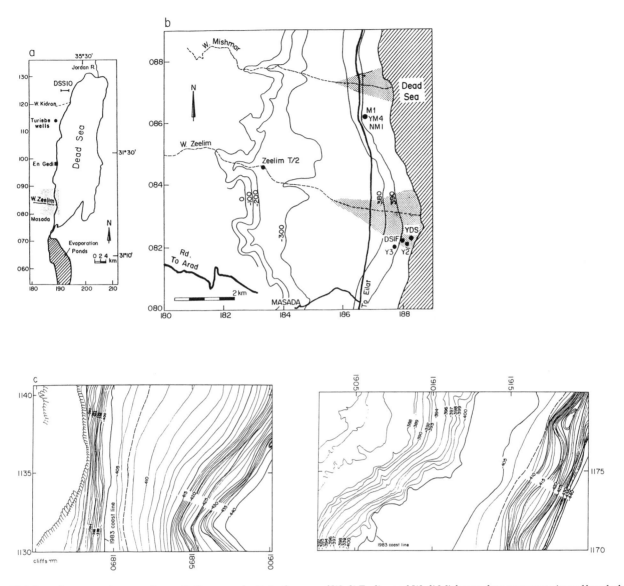

Figure 24-1 Location maps: A, general map; B, the research site in the area of Wadi Zeelim and Wadi Mishmar; dots represent sites of boreholes; c, detailed topographic and bathymetric maps of the Dead Sea coast (topography by Mehish Ltd. and bathymetry by the Institute for Limnological Research, 1984). The 1983 coastline is also shown.

especially in the shallow margins (Gavrieli, chapter 14, and Stiller et al., chapter 15, this volume).

Because of its location below sea level, the Dead Sea basin is the hydrological outlet for the eastern slopes of the Judean mountains. The runoff pattern is due to flash floods that occur in the larger wadis a few times a year during the winter months and to subsurface drainage. The latter manifests itself in the aquifers found in the region and in the springs that discharge both above and below the lake's surface.

Three main aquifers have been identified in the area to the west of the Dead Sea (Naor et al., 1987). The deepest aquifer is built of sandstone layers of the Kurnub Group of Lower Cretaceous age; the aquifer above it is built of limestone and dolomite layers of the Judea Group of Upper Cretaceous age; and the third aquifer is the alluvial aquifer of Quaternary rocks. The alluvial aquifer is generally separated from the other aquifers by faults on the western margin of the Dead Sea rift. The main source of freshwater into the Quaternary aquifer appears to be precipitation in the mountain area, 10–30 km to the west. These rains recharge the Judean aquifer in the mountain region, which later discharges into the Quaternary aquifer (Fig. 24-2); some of the water arrives as floods into the rift valley. Because of the small amount of rain that falls directly on the Dead Sea shores and the high evaporation rate, the direct local recharge to groundwater is expected to be very small.

Many springs discharge close to the Dead Sea shore. A survey conducted along the west coast of the Dead Sea revealed the discharge of these springs into the Dead Sea to be about 80 x 10^6 m^3 of water (Gutman and Simon, 1984). The discharges, as well as the locations of the springs, change in response to the changes in the Dead Sea level (Gutman and Simon, 1984). Most spring waters are of the Na-Cl and Na-Mg-Cl type, with salinity ranging from that of relatively fresh water (~1,000 mg/l Cl) to brines of the Dead Sea type. The chemical and isotopic compositions of some of these springs have been studied by Mazor et al. (1969), Gat et al. (1969), and Starinsky (1974) to trace their

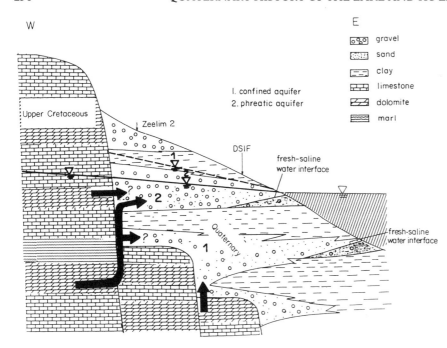

Figure 24-2 Schematic cross section showing the field relations between the different aquifers and the Dead Sea. The western part is built of limestone, dolomite, and some marl belonging to the Judea Group aquifer, whereas the area between the cliff and the Dead Sea is built of gravel and clay belonging to the Quaternary aquifer.

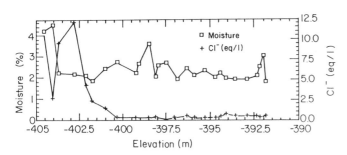

Figure 24-3 Moisture content and chlorinity versus elevation (or distance) in cross section (DSS10) along the Dead Sea coast (from Shatkay et al., 1988). A constant value of about 0.3 eq/l is reached at a distance of about 25 m (elevation -401 m) from the shoreline.

origin and explain the chemical processes that determine their composition. A summary is given also by Mazor (chapter 25, this volume). Springs are also known to flow from the bottom of the Dead Sea basin (Ben-Avraham and Ballard, 1984), contributing groundwater of varying compositions. The existence of such sublacustrine springs was suggested by Stiller and Chung (1984) based on the radium balance in the Dead Sea. Because of the lowering of the Dead Sea level, some of the outlets are now located above the present Dead Sea water level.

The depth of groundwater level varies according to the distance from the Dead Sea. Near the shore (distance of up to 100 m), the depth of the water table is less than 5 m, whereas farther from the shore toward the mountains (distance of 0.5–1.5 km), the water level depth is of the order of several tens of meters. The unsaturated zone, above the water table, consists of unconsolidated sediments as described previously. Some of the sediments, which were filled with the Dead Sea solution until not long ago, are now dry, whereas others still have a high moisture content.

A notable feature observed near the Dead Sea shore is that of "wet spots" (Shatkay and Magaritz, 1987). These wet spots are areas of about 5–10 m in diameter with a high moisture content

of about 10% gravimetric moisture at the surface. Wet spots are characterized by fine-grained sediments (mainly clay and silt) and are separated from each other by drier areas (moisture content less than 5%). Many of the wet spots are located outside the flood plains of the major wadis, which excludes the possibility of their originating from runoff water. It had been postulated that these wet spots are groundwater discharge zones (proto sabkhas).

SALINITY BUILDUP AND DISSIPATION ALONG THE DEAD SEA SHORES

An amazing sight along the Dead Sea shores on many occasions are thick crusts of halite crystals that accumulate on boulders along the shores, rapidly changing form and texture only to disappear a few days later. This phenomenon, which results from the splash of waves, is prominent when the Dead Sea waters approach halite saturation (at low water stand) but barely apparent when the waters are more diluted.

The effect of spray and splash from the Dead Sea on the coast has not been studied in detail. The only work on the subject is that of Shatkay et al. (1988), who investigated the salinity content in the topsoil of the *terra nova* as a function of distance from the shore. Their work included sampling surface sediments along three cross-sectional transects from the Dead Sea shore westward (shown in Fig. 24-3). These represent an elevation span of 12.5 m, corresponding to an exposure time of up to 50 years. Figure 24-3, taken from Shatkay et al. (1988), shows a salinity buildup in the first 10–20 m from the shore and a decline to lower levels farther inland. This area of salinity buildup near the shore is the spray zone. They also found evidence for removal of the original Dead Sea salts in several areas. According to Shatkay et al. (1988), the removal of salts along one of the cross sections (Fig. 24-3) resulted from local rain only, whereas in the other two cross sections, the pattern of decrease in salinity with distance from the Dead Sea shore was apparently perturbed by upwelling groundwater. More thorough research is needed to exclude the possibility that, in some of those areas, surface floodwater was also involved in the removal of salt. If rain were indeed the main flushing factor, the salinity would have been expected to decrease continually with time of exposure; in fact, a constant value of about 0.3 eq/l is reached at a distance of about 25 m from the shoreline (Fig. 24-3). Shatkay et al. (1988) attributed this phenomenon to the addition of salt by aeolian transport from the sea, or from the immediate coastal area, to the more distant area.

Indeed, the aeolian salt transport (removal or accumulation) is a mechanism to be considered. Although removal of salt could be effective in the dry areas, in the wet spots (discussed subsequently), the high moisture content and the nature of the salts can be expected to result in the attachment of the salt to the surface, preventing the salts being picked up by the wind.

FIELD OBSERVATIONS ON WATER AND SALT IN THE SEDIMENTARY SEQUENCE

A number of boreholes were drilled in the years 1989–1992 (Yechieli, 1993) in the newly exposed coast, at location shown in Fig. 24-1B. Some of these boreholes were drilled down to the water table region (NM1, Y2, and YDS) and, in one case, to 24 m below the water table (DSIF). Others penetrated only to the upper part of the unsaturated zone (M1, Y1, YM4, and Y3).

All boreholes were located outside the present main streams. Borehole YDS was drilled in a wet spot in Wadi Zeelim, near the Dead Sea shoreline, to the water table at a depth of 6 m (sampled only in the upper 3.5 m). It is located at an elevation of -403 m (exposed in 1980) comprised of recent clay sediment (near the surface, sediments are younger than 400 yr; Yechieli, 1993). This area is quite flat, with several steep creeks crossing toward the Dead Sea. Boreholes M1, YM4, NM1, and Y1 were all dug in wet spots located south of the alluvial fan of Wadi Mishmar (Fig. 24-1B) and north of Wadi Zeelim, in an elevated area outside the main stream. These boreholes are located in a very flat area with a very shallow slope toward the Dead Sea, which was at a distance of about 1.1 km. Borehole Y2, located in the southern part of the alluvial fan of Wadi Zeelim (Fig. 24-1b), was designed to study the whole unsaturated zone down to the water table. This drilling was done at a dry site. Borehole Y3 was also drilled in a dry area and sampled to a depth of 12 m (the water table at the site is at a depth of 14 m).

In addition to measurements in these boreholes, measurements were also conducted in the Turiebe area (see location in Fig. 24-1a) in boreholes that were drilled in 1971 (Baida and Goldschtoff, 1972). The chemistry of the groundwaters was found to vary widely, ranging from relatively fresh water (3.5 and 1 g/l of Cl in wells NM1 and Turiebe 4s, respectively) to a very saline brine (210 g of Cl/l in well DSIF).

The groundwater levels in the Dead Sea coastal area are changing in accordance with the changes in the Dead Sea level. For example, the water level has decreased by about 6 m in well Turiebe 4s in the last 20 years in response to the 9 m decrease in Dead Sea level (Fig. 24-4, data from the archives of the Israel Hydrological Service). The response of the groundwater level to the 1.8-m water level rise of the Dead Sea during the winter of 1992, an extremely rainy year with exceptionally large input into the Dead Sea, was rapid (on the order of days, both in the Turiebe wells and in well DSIF), implying a good hydraulic connection between the well area and the Dead Sea shore (Yechieli, 1993).

Observations in the unsaturated zone

During the drilling, sediment samples were obtained from the unsaturated zone. The interstitial solutions were extracted in the laboratory and the total ion content, including that of solid evaporites, was determined by flushing the sediments (for details of the extraction methods, see Yechieli et al., 1996).

Profiles of ion concentrations, of $\delta^{18}O$ values of the interstitial waters, and of the moisture content, are shown in Figures 24-5, 24-6, and 24-7. In general, the concentrations of most ions in the interstitial solutions in profile YM4, which is located in a wet spot about 1 km from the Dead Sea shore, increase toward the surface (for example, Mg increases from 300 meq/l at a depth of 3 cm to 7,500 meq/l near the surface; Fig. 24-5). Peak concentration for Na occurs at a depth of 25 cm (Fig. 24-5). Sulfate content shows the opposite trend: the highest concentration (75 meq/l) is found at the bottom of the profile, with concentrations decreasing toward the surface (about 10 meq/l; Fig. 24-2). The interstitial solutions near the surface are more saline that the Dead Sea water, and the Na/Cl ratio in these solutions (less that 0.1) is lower than the Na/Cl ratio of Dead Sea water (0.27).

A nearly uniform profile, different from that in YM4, is seen in borehole YDS, located in a wet spot about 70 m from the Dead Sea shore, where the ion concentrations of interstitial waters are more than half that of the Dead Sea and the composition is similar to that of the Dead Sea (Fig. 24-6). Only the sample taken near the surface is more saline than the Dead Sea water and has a lower Na/Cl ratio (0.12).

A more complex picture is seen in the deeper Y2 profile, where the lowest chloride concentrations (300 meq/l) were detected at depths of 3–4 m (Fig. 24-7). This zone of minimal Cl content coincides with a lithological change from a gravel layer to a clay layer.

In most profiles, a maximum enrichment in the concentrations of ^{18}O and deuterium is apparent at a certain depth below the surface. The depth at which these maxima occur in shallow boreholes, drilled in the wet spots, such as YM4 and M1, is 20 to 25 cm (Figs. 24-5, 24-7). In boreholes Y2 and Y3, drilled in drier areas (Fig. 24-7), the maximum enrichment in both isotopes was found at a depth of 1 m.

Tritium levels in the deep boreholes Y1, Y2, and Y3 decrease from concentrations ranging from 20 to 40 TU at the surface to

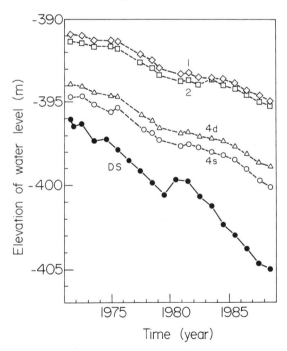

Figure 24-4 Variations in water level in the Turiebe wells and in the Dead Sea in the years 1972–1989 (data from the archives of the Israel Hydrological Service).

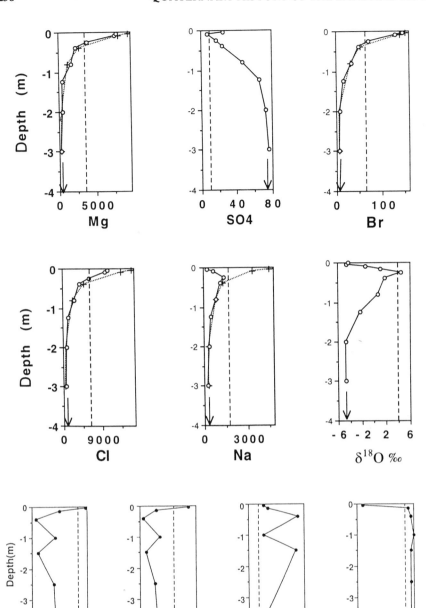

Figure 24-5 Chemical profiles of the interstitial solution from borehole YM4 (in meq/l). Also shown is the $\delta^{18}O$ profile in this borehole. Open circles denote real concentrations obtained by the DEIS method, whereas X's denote total concentration by WAE(a) method (connected by dotted line); these concentrations are higher than the real concentrations due to the presence of solid salts in the sample (Yechieli, 1993). The broken lines denote concentration of Dead Sea solution. Arrows denote the composition of groundwater in that area (borehole NM1), which was found at a depth of 7m.

Figure 24-6 Chemical profiles of the interstitial solution from borehole YDS (data in meq/l). Also shown are profiles of the ionic ratio of Na/Cl. Broken lines denote concentration of Dead Sea solution.

near zero at depth (Fig. 24-7). The surface concentrations are unusually high compared to those found in present precipitation, which is about 10 TU (Yechieli et al., 1993).

Optical microscopic examination of sediments from some of the profiles reveals gypsum and halite crystals in the samples. Scanning electron microscope (SEM) analyses were conducted on samples from boreholes M1 by Shatkay (1985 and pers. com., 1994). These analyses show that gypsum is present throughout the profiles (Fig. 24-8a). In the upper part of the unsaturated zone (upper 30–40 cm), halite is also found (Fig. 24-8). Carnallite ($KMgCl_3 \cdot 6H_2O$), sylvite (KCl), and bischofite ($MgCl_2 \cdot 6H_2O$) were also detected by the SEM in the upper 5 cm of some samples. Some of the carnallite and bischofite are well crystallized (bischofite crystals, about 5 μm in size), and their spectrum matches that of standard minerals, suggesting that their nature is authigenic (Shatkay, 1985). Sylvite is also present in this layer.

PROCESSES IN THE NEWLY EXPOSED SHORES

After a new land area along the new Dead Sea shore is exposed, certain processes take place consecutively (Fig. 24-9). In the beginning, in response to the lowering of the coastal saltwater-freshwater interface, the main processes are drainage of some of the original Dead Sea solution and penetration of groundwater of composition different from Dead Sea water by lateral inflow or by upflow of water through fault planes, commonly found in the Dead Sea rift.

The following processes may further modify the interstitial water composition:

(1) aeolian transport of spray and salt;
(2) percolation of surface runoff or local rains; the combination of aeolian transport and percolation could contribute salts to the sediment profile;
(3) lateral water movement in the unsaturated zone as inter-

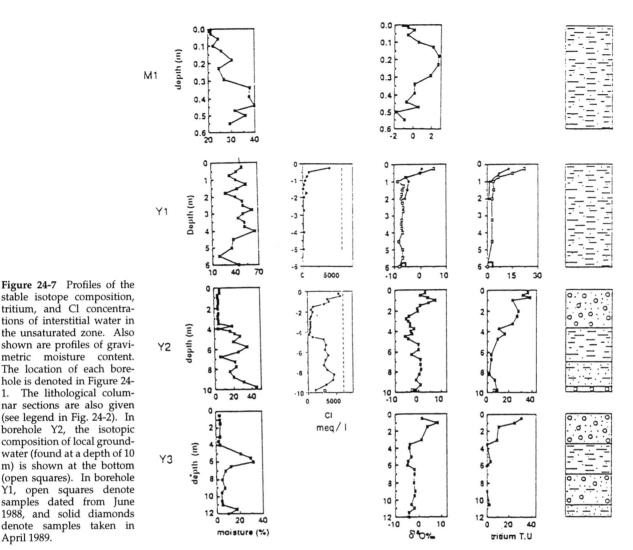

Figure 24-7 Profiles of the stable isotope composition, tritium, and Cl concentrations of interstitial water in the unsaturated zone. Also shown are profiles of gravimetric moisture content. The location of each borehole is denoted in Figure 24-1. The lithological columnar sections are also given (see legend in Fig. 24-2). In borehole Y2, the isotopic composition of local groundwater (found at a depth of 10 m) is shown at the bottom (open squares). In borehole Y1, open squares denote samples dated from June 1988, and solid diamonds denote samples taken in April 1989.

flow or as perched horizons; this water could be fresh or saline and have a composition different from that of the Dead Sea;

(4) capillary rise of water from the saturated zone or from perched water horizons;

(5) evaporation from within the sediment column leading to high concentration of solutes; and

(6) hygroscopic water adsorption at the surface of highly saline sediments.

The adsorption of vapor by the (extremely) saline top sediments is noted especially in environments such as the Dead Sea area with an abundance of hygroscopic salts like $MgCl_2$, $CaCl_2$, etc. Furthermore, the high salinities in the interstitial solution may impose severe osmotic gradients along the sedimentary column and thus provide further impetus for water movement.

Although evaporation and aeolian transport would further increase the salinity, the flushing mechanisms tend to decrease it. Evidently, the latter finally gain the upper hand, but the rate at which reclamation occurs depends critically on the hydrological setting of the area. Along the newly exposed western coast (along which the presently described research in concentrated), the following areas with different hydrological settings can be distinguished:

(1) areas where spring discharge is dominant, found mainly in the northern section of the coastline, for example, the En Fes-

hkha and Turiebe (Kane-Samar) areas;

(2) Wadi discharge zones where both surface and subsurface flow is prominent on a wide area basis; and

(3) the elevated areas between the major wadis.

Obviously, the rates of the various processes are quite different under these different circumstances.

The flushing processes

The reduction of salinity in the unsaturated, as well as in the saturated zone, of the sediment column depends on flushing with freshwater. Downward, upward, or horizontal flushing is possible. The chemical and isotopic compositions of the interstitial solutions can be used as tracers to study the rate and direction of flushing.

In more humid environments, the downward percolation of precipitation is the dominant transport mode in the unsaturated zone. This is not expected to occur in an arid environment where evaporation exceeds rain depth by many orders of magnitude (Gat, 1988). Only where surface flows marshal a considerable water depth at any site can we expect to see downward percolation, which could express itself in the form of a fully flushed soil profile. The sampling was, however, conducted outside the floodplains, and the inverse salinity profile (with higher salinity at the surface), which was the rule throughout

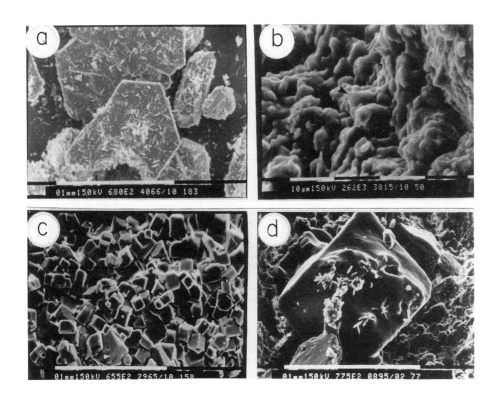

Figure 24-8 SEM analysis of minerals found in the sediments in the unsaturated zone on the Dead Sea coast (M. Shatkay, pers. com.): A, gypsum, covered by aragonite (needles); B, carnallite; C, halite, perfect cubic crystals; d, halite, some aragonite needles.

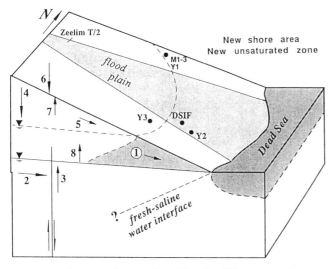

Figure 24-9 Conceptual model describing the different water transport mechanisms in the study area (broken line denotes the former groundwater and Dead Sea levels): 1, drainage of residual Dead Sea solutions as a result of the receding of the Dead Sea in the last few decades; 2, transport of groundwater originating in the west; 3, upflow of water through fault plane; 4, percolation of rainwater and floodwater; 5, lateral flow in the unsaturated zone; 6, adsorption of air moisture by the saline surface sediments; 7, evaporation; and 8, capillary rise.

The fact that the stable isotope composition in the deeper part of borehole Y1 (below 1 m) resembles that of local groundwater, such as found in well NM1 (Fig. 24-7), indicates that the original Dead Sea solution has been flushed completely. This resemblance is further supported by the low salinity found at this depth in boreholes YM4 (Fig. 24-5) and by the fact that the isotopic values are those of the meteoric waters, being situated on the East Mediterranean Meteoric Water Line (Gat and Dansgaard, 1972). The isotopic composition of fresh interstitial water in Y1, which is located only about 1 km west of the Dead Sea, is also similar (though not quite identical) to that of the fresh groundwater in well Zeelim T/2, which is located 4 km west of the Dead Sea (Fig. 24-1).

The location of borehole Y1 outside the alluvial fan of W. Zeelim, which precludes massive infiltration of floodwater, and the high moisture content found throughout the profile support the hypothesis that floodwater infiltrated along the wadi bed farther upstream and that this water then flowed laterally to the site. Furthermore, the low tritium concentrations of 3 TU and less (Fig. 24-7), which are measured below a depth of 1 m, imply that the fresh component at this site is neither water found at the water table in well Zeelim 2 (7.9 TU) nor recent floodwater (5-13 TU). Although some of the flushing might have started earlier, when the area was still covered by the Dead Sea, by relatively fresh groundwater that discharged as an underwater spring in the bottom of the Dead Sea, it is reasonable to assume that most of the flushing (which has diluted the original solution about 10- to 20-fold) has occurred since the retreat of the Dead Sea. In borehole NM1, water was encountered in a confined aquifer at a depth of 6–7 m (after drilling, water level rose by 3.5 m). The similarity between the chemical composition of groundwater in well NM1 and interstitial solution at a depth of 6 m in that area supports the hypothesis that this type of groundwater is the same as the ascending water that completely flushed the profile (e.g., the Na/Cl ratio is about 0.9 in both solutions, whereas in the Dead Sea solution, the ratio is 0.27).

the *terra nova*, attests to the inefficiency of this flushing mode. Yet we note the gradual decrease in surface salinity found by Shatkay et al. (1988), even in the absence of any agent other than the local rain. A relatively low tritium concentration, which was measured at the surface in borehole NM1 in the summer of 1992 and which is similar to that of most recent precipitation, also indicates percolation of some rain to shallow depths in the unsaturated zone.

Much less flushing has occurred in the area of YDS, which is located less than 100 m from the present-day shoreline and was exposed only about 10 years ago. Here the interstitial solutions have a salt content of more than half that of the Dead Sea brine (Fig. 24-6), and no indication of precipitation or dissolution of salts (besides dilution) was observed; for example, the Na/Cl ratio (0.29) remained similar to that of the original Dead Sea solution.

The isotopic composition of water found in the deeper part of profile Y2 (at a depth greater than 4 m) indicates mixing between freshwater and Dead Sea water. Since borehole Y2 is located in an area that was covered by the Dead Sea until 30 years ago, it is reasonable to assume that the sediments were saturated with the Dead Sea solution until that time. The present isotopic composition suggests that these sediments were flushed later by fresher groundwater and subsequently became part of the unsaturated zone. The tritium content and the stable isotope composition of interstitial water in the unsaturated zone in borehole Y2 (between depths of 5 and 10 m) could also be explained by a mixture of groundwater and a residual Dead Sea solution.

The amount of flushing (the duration and volumes of flushing water) of each layer by groundwater depends on the rate of decrease in the water level of the Dead Sea and on the hydraulic conductivity of each layer. As a result of the low hydraulic conductivity of the clay and silt layers (below a depth of 4 m), the flushing of the Dead Sea brine from these sediments is not complete. Conversely, in the gravel and sand layer (2–4 m deep) of relatively high hydraulic conductivity, the sequences have been efficiently leached of the Dead Sea solutions by meteoric water, with isotopic values typical of flood- and rainwater ($\delta^{18}O$ = - 5‰). This differential leaching process is clearly reflected by the change in salinity of the interstitial water at the gravel-clay contact zone (Fig. 24-7). The floodwater penetrated into the gravel and sand horizon but not into the clay layers below, because of their low permeability. This geological setup is conducive to the formation of a perched horizon on top of the clay layer. The recharge of the freshwater could have been from local rains or from floods penetrating at a distance of up to a few km upstream to the west. The perched horizon is probably temporary, depending on the amount of rain- and floodwater in that area.

Profile Y3 is located in an area that had not been submerged (under the Dead Sea) since at least 1900. The isotopic composition of most of the water samples (except those obtained near the surface), including those obtained from the clay layer, is quite similar to that of flood water, indicating that the profile has been flushed by freshwater. The interstitial solutions, even in the clay layers of Y3, are more depleted in ^{18}O and deuterium compared to profile Y2, reflecting the longer period of exposure and flushing of by fresh water. The absence of tritium (less than 2 TU, Fig. 24-7) in the lower part of profile Y3 (below 4 m) indicates that the flushed interstitial water is older than 50 years. The lack of interaction of the sediments with more recent water probably results from the location of borehole Y3 outside the main stream of the alluvial fan of Wadi Zeelim (Fig. 24-1). We propose that the Y3 section has been outside the alluvial fan for at least the last 50 years. Furthermore, recent Dead Sea water could not have been found in profile Y3, because the elevation of borehole Y3 is -389 m, greater than any elevation reached by the Dead Sea in the last century (Klein, 1985).

Unlike the situation described previously, the rate of flushing by floodwaters is very rapid in areas located in the main streams of the alluvial fans. The depth of flushing depends on the sequence of sediments in each specific are; that is, a section consisting of gravel may be flushed completely, whereas a

clayey layer will prevent the fresh floodwater from penetrating downward and flushing the sediment at depth.

The formation of "wet spots"

The phenomenon of the "wet spots" is manifested in the high moisture content of near-surface sediments in the unsaturated zone (Fig. 24-10). The facts that (1) most of the wet spots are located outside the alluvial fans of the major wadis and (2) no correlation was found between the location of the wet spots and areas of low topographic elevation (where surface runoff could accumulate) indicate that the occurrence of wet spots is not to be related to surface runoff. In the wet spot profiles, the moisture content near the surface remains high even after 3–4.5 months without rain.

There are two main factors controlling the moisture content in the surface sediments: the percentage of fine sediment and the high salt concentrations. These two factors are usually interrelated in the Dead Sea region because the more clayey sediments are less flushed and thus retain more of the original saline Dead Sea solution. A higher fraction of fine sediment and higher salt content will enhance adsorption of air moisture. This is evidenced by the fact that boreholes Y1 and M1, which are located in silt and clay layers, are part of the wet spot (θ_g = 10–20% near the surface; Fig. 24-7), whereas boreholes Y2 and Y3, which are located in an area consisting mainly of gravel, are dry (θ_g = 3–4%; Fig. 24-7).

To test the significance of the salt content on adsorption of moisture from the air, laboratory experiments were conducted by Yechieli (1993) on original Dead Sea shore sediments and on the same sediments after removal of their salt by leaching with distilled water. The leached and salt-containing samples were introduced into a drying oven at 105°C. After 24 hours, the samples were exposed to the free air in the lab at a temperature of 19–23°C and a relative humidity of 60%. The amounts of water adsorbed by the samples were found to differ significantly (Fig. 24-11): The salt-rich sediments (Cl- content of 0.10–0.14 $g/g_{sediment+salt}$) reached a gravimetric moisture content of over 10%, whereas the leached sediments (less that 1 mg Cl- in 1 g sediment) absorbed less than 2% water (Fig. 24-11). We suggest that, despite the difference in moisture content of the air in the lab (~60%) and in the field (~35%), the adsorption mechanism is probably similar. Preliminary experiments, conducted

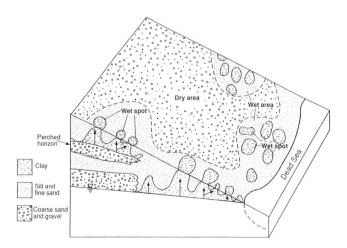

Figure 24-10 A schematic diagram showing the location of the wet spots and the mechanism of capillary rise.

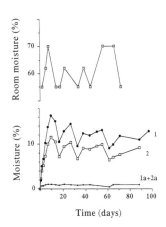

Figure 24-11 Laboratory determination of hygroscopic water adsorbed by Dead Sea shore sediments. The original salt content of sample 1 is higher (143 mg Cl in 1 g of sediment) than that of sample 2 (102 mg Cl in 1 g of sediment). Samples 1a and 2a have been leached, and their salt content is only 1 mg Cl in 1 g of sediments. The fluctuations in moisture content of all the samples reflect the changes in moisture content of the air in the room.

at the Dead Sea coast, indeed showed a considerable amount of moisture adsorption (about 5% in the same saline samples as described previously). These experiments show the salt content, rather than the clay content, to be the main factor that controls adsorption of water.

Shatkay and Magaritz (1987) explained the location of the

wet spots by mechanism of ascending saline groundwaters. Although this may be valid in some places, it cannot explain the existence of wet spots in areas where there is a gravel layer at shallow depth ($\approx$ 1 m) that would interrupt the upward water movement. At least in those places, it is clear that the wet spots are indeed surface phenomena based on the makeup of the sediments.

The effect of evaporation

In the shallow part (depth less than 3 m) of boreholes Y2 and Y3 and in the shallow borehole M1, the profiles of stable isotopes of the interstitial water clearly indicate that evaporation has occurred. Thus the isotopic composition of these interstitial waters depicts a line on a $\delta^{18}O$–δD plot whose slope is generally less steep than that of the Dead Sea-meteoric groundwater mixing line and does fit an isotopic "evaporation line" (Gat, 1980). Moreover, some of these interstitial waters are more enriched in $\delta^{18}O$ than is the Dead Sea end-member (Y2 and Y3; Fig. 24-12), which excludes the possibility of their being a mixed product of the Dead Sea water and freshwater.

In the shallow borehole M1, where the surface zone was sampled in detail (Fig. 24-7), there is a clear trend of decreasing moisture content toward the surface, indicating that evaporation from the top of the unsaturated zone is significant. This is reflected also by the higher salinity and the enrichment in ^{18}O and deuterium in borehole M1.

The stable isotope profiles of the interstitial waters (Figs. 24-5, 24-7) are consistent with an upward flux of water into an evaporation front. Using the (steady rate) model of Allison and Barnes (1985), Yechieli et al. (1993) estimated evaporation rates of up to tens of mm H_2O per year. Because of the uncertainty in

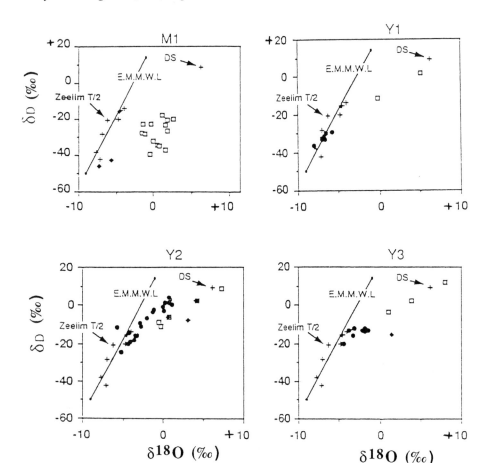

Figure 24-12 $\delta^{18}O$ versus δD of the interstitial water in the unsaturated zone. The concentrations of the stable isotopes of the Dead Sea water are from Gat (1984) and Horita and Gat (1989). Open squares denote water samples from the upper part of the profiles in the unsaturated zone, which have undergone evaporation. Filled diamonds denote water samples from the very top that have interacted with the atmosphere. The remaining water samples are denoted by filled circles. Other water samples from fresh groundwater, floods, rains, and the Dead Sea are denoted by crosses. EMMWL is the East Mediterranean Meteoric Water Line (Gat and Dansgaard, 1972) whose equation is $\delta D = 8$ $\delta^{18}O + 22‰$

assessing some of the parameters needed for the calculation, these evaporation rates should be considered only as order-of-magnitude estimates for the area studied. As expected, evaporation rates from the surface of the unsaturated zone are much lower than the evaporation rate from the surface of the Dead Sea (1,380 mm/yr—Stanhill, 1983; 1,460 mm/yr—Carmi et al., 1984).

NEW SALT PHASES IN THE SEDIMENTARY PROFILE

The high evaporation in this arid area and the very saline original solutions result in a concentration profile with maximal values near the surface. The high concentrations finally reached cause precipitation of several salts in the unsaturated zone. The two extraction methods employed (direct extraction of interstitial solution—DEIS—in which the ion concentration in the interstitial solution is obtained, and water addition extraction—WAE—in which the total ion content in the sample, both in the liquid and the solid, is obtained; Fig. 24-13a) enable detection of solid phases of soluble salts in the sediments (Yechieli, 1993). Note the unusual pattern of the expected concentration profiles in which there is a deflection point (denoted as β), defined as the depth above which the specific ion concentration in the interstitial solution begins to decrease with increasing salinity while the total ion content continues to increase (Fig. 24-13A).

This pattern is explained by the fact that the solid phase begins to precipitate at the depth at which the solubility product is exceeded. A schematic model describing the sequence of precipitation of the various salts is depicted in Figure 24-13b. Each salt appears to precipitate at a different depth, depending on its solubility. For example, sodium salt (halite) is expected to be found at a greater depth, where the interstitial solutions have a lower salinity, than the K-salt (carnallite), which is found closer to the surface.

The profiles in borehole YM4, depicted in Figure 24-5, indeed resemble those of the schematic model. Near the surface, for a depth less than β, the increase in ion activity of the solution and the common ion effect cause a decrease in the solubility product of NaCl, which precipitates at a lower Na concentration than that expected in a pure NaCl solution. This phenomenon is reflected in the deflection of the concentration versus depth curve toward lower Na values for depth less than ß. At the same time, the total Na concentration (obtained by the WAE method) increases, reflecting halite accumulation in the sediments. This is also indicated by the decrease in the Na/Cl ratio (between the β point and the surface) in the interstitial solution relative to that of the Dead Sea brine, implying that halite precipitation takes place. The presence of halite as perfect cubic crystals, which were seen using the SEM method (Fig. 24-8), supports this conclusion.

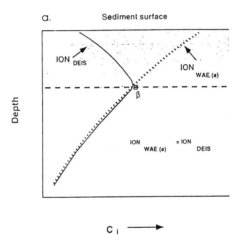

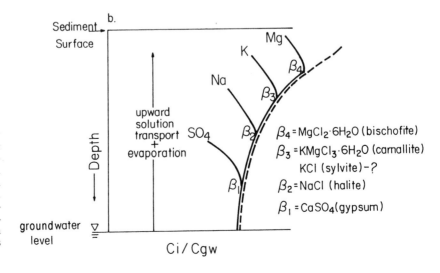

Figure 24-13 A, schematic representation of the distribution of a specific ion between solid and liquid phases in the unsaturated zone. The area above β is the zone where the ions will exist in both interstitial solution and solid minerals (shaded area). The solid line denotes the concentrations in the interstitial solutions obtained by the direct extraction method (DEIS) and the dashed line denotes the total concentrations obtained by the extraction involving water addition (WAE); B, schematic model describing the location in the sedimentary sequence where precipitation of the various salts takes place. Broken line denotes increasing evaporation and thus increasing salinity. C_i is the concentration of an ion in the sample, and C_{gw} is its concentration in groundwater in that area. The inverted triangle denotes the groundwater level.

As we get closer to the surface, saturation with respect to other (more soluble) minerals is reached, and the minerals are precipitated according to the sequence halite-carnallite-bischofite (Fig. 24-13b). As for the potassium, SEM analysis suggests that its salts (such as sylvite and carnallite) begin to precipitate only close to the soil surface (β = 5–10 cm). In the case of Mg, no β point is observed at all (Fig. 24-5). However, in the near-surface sample, the concentrations obtained by the WAE methods are somewhat higher than those obtained by the DEIS method (Fig. 24-5). Thus, it is possible that Mg salts do precipitate near the surface. Both carnallite and bischofite were indeed detected in this zone by SEM. We must stress, however, that the presence of bischofite (a very soluble mineral) may be a biased result of sample preparation for SEM analysis. Preliminary results of the *PhrqPitz* code calculation also show saturation with respect to halite and carnallite for the near-surface samples. These samples are also very close to saturation with respect to bischofite.

As for sulfate minerals, $CaSO_4$ was observed in the SEM pictures in all samples. The decrease of SO_4 concentration in the interstitial solution toward the top of the profile (Fig. 24-5) is controlled by the $CaSO_4$ saturation values. Indeed, it was found using the *PhrqPitz* code that the whole section is saturated with respect to gypsum. As the solution becomes more saline and the Ca concentration increases, $CaSO_4$ precipitates at a lower SO_4 concentration.

Shatkay and Magaritz (1987) reported the formation of dolomite. In their research, conducted in the area between Wadi Mishmar and Wadi Zeelim (denoted in Fig. 24-1 as M1), evidence was found for recent dolomitization and sulfate reduction at a depth of 6–12 cm. At that depth, the Mg/Ca ratio is relatively high, and the SO_4/Cl ratio and gypsum content are low. The dolomite was found to be euhedral and close to ideal dolomite in composition (by SEM, EDS, and X-ray diffractometer analyses; Shatkay and Magaritz, 1987); this differs from the anhedral detritic dolomite that was derived from the Upper Cretaceous rocks. Shatkay and Magaritz (1987) explain the dolomite formation by using a mixing zone model whereby the Dead Sea residual solution is postulated to mix with relatively fresh groundwater in the newly exposed coast.

SUMMARY: THE CONCEPTUAL EVOLUTIONARY MODEL

The processes described previously can be summarized in a chronological evolution model of events occurring following exposure of the sediments, as depicted in Figure 24-14. During the period when the Dead Sea covered the entire area under study (at time t_1 in Fig. 24-14), these sediments were still covered by the Dead Sea and were impregnated by a saline interstitial solution of the Dead Sea type. This interstitial solution is known to vary somewhat with depth because of previous changes in the Dead Sea salinity (Stiller et al., 1983). As the bot-

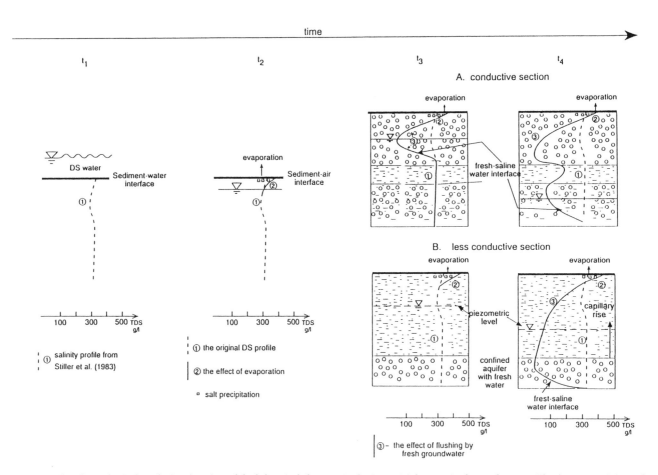

Figure 24-14 The chronological evolution (t_1–t_4) model of chemical changes in the interstitial water in the study area. The first stage (t_1) is when the area is still covered by the sea, and the second stage is a short time after exposure. "A" denotes the situation in a conductive layer, whereas "B" denotes the situation in a layer with low conductance.

tom sediments became exposed to the atmosphere, drainage of the residual Dead Sea solution to the Dead Sea occurred. Evaporation from the unsaturated zone resulted in salt precipitation within the sedimentary column (at time t_2 in Fig. 24-14). Flushing, resulting from the horizontal movement of fresh groundwater, could lead to two different scenarios according to the composition of the lithological profile (at time t_3 in Fig. 24-14). In one case, where the profile is composed mainly of gravels (at times t_3 and t_4 "A" in Fig. 24-14), the upper part is flushed rapidly by fresh groundwater, which intrudes laterally concomitant with the drainage of Dead Sea water. In places where the section is mainly composed of clay of very low hydraulic conductivity (at time t_3 "B" in Fig. 24-14), the freshwater penetrates at a much lower rate, leaving the original saline interstitial solution unflushed for a longer time. This situation will prevail until freshwater enters a more conductive deep (e.g., gravel) layer (at time t_4 "B" in Fig. 24-14). At this stage, the freshwater will continue to flow laterally either as a perched horizon or in a confined aquifer, where it may develop a high hydrostatic pressure. In both cases, the upward movement of water is enhanced by capillary forces, with pore radii determining the magnitude of water rise.

If fresh groundwater were involved in one of these mechanisms (lateral flow or capillary rise), the net result would be a profile of lower salinity in the unsaturated zone compared to the original Dead Sea water (at time t_4 in Fig. 24-14). However, the effect of evaporation in the vicinity of the sediment-atmosphere interface will lead to a renewed increase in the salinity and, eventually, to salt deposition. When the latter occurs, the chemistry of the aqueous phase is altered fundamentally.

All the diagenetic processes described previously have not yet formed a normal soil with several horizons in the studied area. The deposition of salt crust at several depths can probably be regarded as an embryonic stage of soil formation. It would probably take hundreds or thousands of years to form normal soils, such as the reg soil described by Amit and Gerson (1986) in the floodplains of Wadi Zeelim, whose age was estimated to be less than 14,000 years.

The data summarized herein show that, in the newly formed coastal zone, the replacement of the saline water of the Dead Sea brine by relatively fresh meteoric water can be quite rapid because of the runoff of surface and groundwater from the contiguous mountain region. However, the aridity of the area expresses itself by the combined action of surface evaporation and aeolian salt transport from the Dead Sea shore, which counteracts the effect of the flushing by freshwater.

Acknowledgment

This work has greatly benefited from the help of the late Prof. M. Magaritz, who was the thesis advisor of both Y. Yechieli and M. Shatkay in their work on the Dead Sea coast.

REFERENCES

Allison, G. B., and Barnes, C. J., 1985, Estimation of evaporation from the normally "dry" Lake Frome in South Australia: *Journal of Hydrology*, v. 78, p. 229–242.

Amit, R., and Gerson, R., 1986, The evolution of Holocene Reg (gravelly) soils in deserts—An example from the Dead Sea region: *Catena*, v. 13, p. 59–79.

Anati, D. A., and Shasha, S., 1989, Dead Sea surface-level changes: *Israel Journal of Earth Science*, v. 38, p. 29–32.

Baida, A., and Goldschtoff, Y., 1972, Groundwater survey in En Awar–En Tureibe area, Dead Sea Coast: Israel Water Inst. Ltd. (Tahal), Hydrology Section, Pub. HR/72/86 (in Hebrew).

Ben-Avraham, Z., and Ballard, R. D., 1984, Near bottom temper-

ature anomalies in the Dead Sea: *Earth and Planetary Science Letters*, v. 71, p. 356–360.

Bentor, Y. K., 1961, Some geochemical aspects of the Dead Sea and the question of its age: *Geochimica et Cosmochimica Acta*, v. 25, p. 239–260.

Carmi, I., Gat, J. R., and Stiller, M., 1984, Tritium in the Dead Sea: *Earth and Planetary Science Letters*, v. 71, p. 377–389.

Garber, R. A., 1980, The sedimentology of the Dead Sea [Ph.D. thesis]: Troy, New York, Rensselaer Polytechnic Institute, 169 p.

Gat, J. R., 1980, The isotopes of hydrogen and oxygen in precipitation, *in* Fritz and Fontes, eds., Handbook of environmental isotope geochemistry: Amsterdam, Elsevier, p. 21–47.

Gat, J. R., 1984, The stable isotope composition of Dead Sea water: *Earth and Planetary Science Letters*, v. 71, p. 361–376.

Gat, J. R., 1988, Groundwater recharge under arid conditions, *in* Yu-Si Fok, Ed., Infiltration principles and practices: Post conference proceeding ICIDA, WRRC, University of Hawaii, p. 245–257.

Gat, J. R., and Dansgaard, W., 1972, Stable isotope survey of the fresh water occurrences in Israel and the northern Jordan rift valley: *Journal of Hydrology*, v. 16, p. 177–212.

Gat, J. R., Mazor, E., and Tzur, Y., 1969, The stable isotope composition of mineral waters in the Jordan Rift Valley, Israel: *Journal of Hydrology*, v. 7, p. 334–352.

Gutman, Y., and Simon, A., 1984, Hydrological survey of the western shore of the Dead Sea, Mediterranean–Dead Sea project: Summary of Research Reports, 5:304–318 (in Hebrew).

Horita, J., and Gat, J. R., 1989, Deuterium in the Dead Sea— Remeasurement and implications for the isotopic activity correction in brines: *Geochimica et Cosmochimica Acta*, v. 53, p. 131–133.

Klein, C., 1985, Fluctuation of the levels of the Dead Sea and climatic fluctuation in the country during historical times: International Association of Hydrological Science International symposium on scientific basis for water resources management, Jerusalem, 19–23 Sept., v. 2, Additional paper and posters summaries 2, p. 197–224.

Langbein, W. B., 1961, Salinity and hydrology of closed lakes: U.S. Geological Survey Professional Paper 412, 20 p.

Levy, Y., 1987, The Dead Sea—Hydrographic, geochemical, and sedimentological changes during the last 25 years (1959–1984): Geological Survey of Israel (in Hebrew, English abstract).

Mazor, E., Rosenthal, E., and Ekstein, J., 1969, Geochemical tracing of mineral water sources in the southwestern Dead Sea basin, Israel: *Journal of Hydrology*, v. 7, p. 246–275.

Naor, H., Katz, D., and Harash, Y., 1987, Hydrogeological study to locate production wells in Wadi Zohar area: Tel Aviv, Tahal, Report 04/87/17.

Neev, D., and Emery, K. O., 1967, The Dead Sea — Depositional processes and environments of evaporites: Geological Survey of Israel Bulletin 41, 147 p.

Shatkay, M., 1985, The newly exposed sediments on the western shores of the Dead Sea — Chemical and mineralogical analysis [M.Sc. thesis]: Rehovot, Weizmann Institute of Science, 95 p.

Shatkay, M., and Magaritz, M., 1987, Dolomitization and sulfate reduction in the mixing zone between brine and meteoric water in the newly exposed shores of the Dead Sea: *Geochimica et Cosmochimica Acta*, v. 51, p. 1–7.

Shatkay, M., Magaritz, M., and Gat, J. R., 1988, Salt leaching in the new exposed shores of the Dead Sea: *Journal of Coastal Research*, v. 4, p. 257–272.

Sneh, A., 1979, Late Pleistocene fan deltas along the Dead Sea rift: *Journal of Sedimentary Petrology*, v. 49, p. 541–552.

Sofer, Z., and Gat, J. R., 1972, Activities and concentrations of Oxygen-18 in concentrated aqueous salt solutions—Analytical and geophysical implications: *Earth and Planetary Science Letters*, v. 15, p. 232–238.

Stanhill, G., 1983, Evaporation from the Dead Sea—A summary of research to date: Bet Dagan, Volcani Center, Report 6, 19 p.

Starinsky, a., 1974, Relation between Ca-chloride brines and sedimentary rocks in Israel [Ph.D. thesis]: Jerusalem, The Hebrew University, 176 p. (in Hebrew, English summary).

Steinhorn, I., 1983, In-situ salt precipitation at the Dead Sea: *Limnology and Oceanography*, v. 28, p. 580–583.

Steinhorn, I., and Gat, J. R., 1983, The Dead Sea: *Scientific American*, v. 249, p. 102–109.

Stiller, M., and Chung, Y. C., 1984, Radium in the Dead Sea — A possible tracer for the duration of meromixis: *Limnology and Oceanography*, v. 29, p. 574–586.

Stiller, M., Kaushansky, P., and Carmi, I., 1983, Recent climatic changes recorded by the salinity of pore water in the Dead Sea sediments: *Hydrobiologica*, v. 103, p. 75–79.

Yechieli, Y., 1993, The effects of water level changes in closed lakes (Dead Sea) on the surrounding groundwater and country rocks [Ph.D. thesis]: Rehovot, Israel, Weizmann Institute of Science, p.

Yechieli, Y., Magaritz, M., Shatkay, M., and Ronen, D., 1996, Early diagenesis of highly saline sediments after exposure. (submitted).

Yechieli, Y., Magaritz, M., Shatkay, M., Ronen, D., and Carmi, I., 1993, Processes affecting interstitial water in the unsaturated zone at the newly exposed shore of the Dead Sea, Israel: *Chemical Geology (Isotope Geoscience)*, v. 103, p. 207–225.

25. GROUNDWATERS ALONG THE WESTERN DEAD SEA SHORE

Emanuel Mazor

DIVERSITY AND COMPLEXITY OF GROUNDWATER SYSTEMS ALONG THE WESTERN SHORE OF THE DEAD SEA

The western shore of the Dead Sea is rich in groundwater occurrences (Fig. 25-1, Table 25-1), with an exceptionally wide range of properties, for example, composition of dissolved ions (Fig. 25-2), concentration of radioactive elements, water age, isotopic composition, and temperature. Thus, a large number of different water types exists in this segment of the Dead Sea rift. The groundwater system along the shores of the Dead Sea has been complicated in the past three decades by the lowering of the Dead Sea level and by intensive groundwater exploitation. The present discussion avoids these complications because it is largely based on data collected in the 1960s, when the level of the Dead Sea was high and rather stable and when exploitation was still limited. Effects of the recent lowering of the Dead Sea level on water movement in the exposed banks are addressed by Yechieli et al. (1993) and Yechieli and Gat (chapter 24, this volume).

The Dead Sea shore is bounded by the fault system of the rift, marked by complex geological structures and a large variety of lithologies. The geological complexity and variability of the groundwater properties point to the existence of countless distinct aquifers.

The term *aquifer* is applied to a permeable rock body that contains water in all its voids and can sustain producing wells or springs. All parts of an aquifer are hydraulically interconnected. Thus, adjacent wells or springs that tap groundwater with a similar composition, age, and temperature are likely to tap the same aquifer, whereas adjacent wells or springs that significantly differ in the properties of their waters belong to different aquifers (Mazor et al., 1992). This definition leaves room for two major types of aquifers: through-flow and stagnant. In through-flow aquifers, groundwater flows through the aquifer. This implies the existence of active recharge and discharge, the best examples of which are phreatic aquifers. In contrast, in stagnant aquifers, water is stagnant (static) with no through-flow and no recharge and discharge; they resemble traps of oil, gas, or brines. The continental sediments filling the valley, as well as the deeper rock strata in the marginal faulted blocks, seem to host stagnant aquifers next to through-flow aquifers. The stagnant nature of an aquifer may be established with entrapment between rocks of low permeability, burial below the active base of drainage, high water age, and stacking distinct aquifers that differ in pressure, composition, and age.

Preferred flow in karstic and fractured systems takes place toward the Dead Sea rift, in addition to the flow in through-flow aquifers and the entrapment in stagnant aquifers (Mazor and Kroitoru, 1990; Kroitoru et al., 1992).

Saline springs and seepages that have compositional similarities to the Dead Sea issue at the western shore of the Dead Sea, at a distance of a few meters from the waterline. Such water was also encountered in nearshore wells. Brines of peculiar composition are found in seepages and boreholes around the salt dome of Mt. Sedom (Mazor, 1962). Bentor (1961) examined the possible formation of the Dead Sea salts by partial evapora-

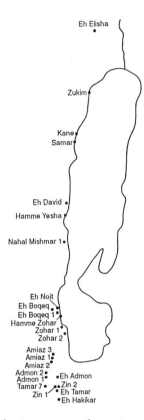

Figure 25-1 Groundwater sources on the western shore of the Dead Sea that were studied.

tion of seawater, entrapped in the Dead Sea rift, and concluded that another source is needed to explain the origin of two thirds of the observed salt inventory. He suggested that the salts of the Dead Sea originate from two main sources: about one third from the Jordan River and two thirds from the discharge of highly saline springs.

Starinsky (1974) examined the possibility that the saline springs on the Dead Sea shore originate by mixing of Dead Sea water with various freshwaters, but he pointed out that, in the few cases he examined, different elements observed in the same spring reflected different mixing ratios. Hence, he preferred the theory that the springs are of an independent source and contribute to the salt inventory of the Dead Sea. In contrast, Mazor et al. (1969, 1980) advocated that the springs on the shore of the Dead Sea recycle Dead Sea water mixed with water emerging from various aquifers. The discrepancy in some of the mixing ratios raised by Starinsky may be resolved by invoking local water-rock interactions before and after mixing with the Dead Sea brines. Arguments in favor of the mixing model are discussed in the following sections.

265

Table 25-1 Major ions (mg/l) in groundwater of the Dead Sea.

No	Name	Notes	Co. E.	Co. N.	Type	T °C	K	Na	Ca	Mg	Cl	HCO$_3$	SO$_4$
1	Ein Elisha				F	21	2	18	48	23	28	299	12
2	Nahal Arugot		1868	961	F		3	63	44	30	113	195	53
3	Nahal David		1869	978	F		3	22	44	29	77	161	35
4	Amiaz 1	32 m			F			54	45	21	85	91	107
5	Amiaz 1	62 m			KN		37	467	132	82	922	158	294
6	Amiaz 1	70 m			KN			321	81	31	469	105	313
7	Amiaz 1	85 m			KN		165	2698	830	519	6578	146	788
8	Amiaz 1				KN		214	2993	1002	613	7505	232	922
9	Ein Boqeq		1837	637	F-KN			337	143	78	581	183	420
10	Ein Boqeq		1837	637	F-KN		9	314	100	70	550	180	351
11	Tamar 5		1785	423	KN-F	39	19	383	163	65	629	286	392
12	Tamar 5	9 m	1785	423	KN-F			508	206	111	915	287	522
13	Tamar 5		1785	423	KN-F			228	165	68	603	280	
14	Ein Hakikar		1883	413	KN			563	276	137	1099	250	574
15	Ein Hakikar		1883	413	KN			575	240	138	1120	250	570
16	Neot Kikar 1		1863	388	KN		30	474	243	148	986	278	639
17	Neot Kikar 1		1863	388	KN		24	457	247	137	969	232	616
18	Neot Kikar 1		1863	388	KN		20	338	167	110	674	183	498
19	Ein Tamar		1835	438	KN			513	207	127	957	255	574
20	Ein Tamar		1835	438	KN			457	244	135	940	280	582
21	Ein Tamar		1835	438	KN			460	245	134	940	280	580
22	Tamar 3	51 m	1837	433	KN		31	386	187	80	667	299	465
23	Tamar 3	72 m	1837	433	KN			508	207	102	915	238	527
24	Tamar 3		1925	1233	KN	34	37	594	248	143	1124	262	622
25	Tamar 3		1837	433	KN		35	581	267	142	1114	286	679
26	Tamar 3		1837	433	KN		32	640	250	152	1115	286	663
27	Tamar 1	229 m	1794	439	KN	35		613	1370	112	1127	226	3582
28	Tamar 6	144 m	1797	951	KN			488	181	65	766	305	429
29	Tamar 6		1797	951	KN		51	853	245	89	1303	292	784
30	Tamar 7		1805	458	KN		81	816	248	153	1485	274	770
31	Tamar 7	232 m	1805	458	KN		106	1043	144	125	1489	338	864
32	Tamar 4	82 m			KN			565	213	127	1057	277	537
33	Tamar 4	175 m			KN			495	215	88	794	500	426
34	Tamar 4	200 m, start pumping			KN			473	169	62	718	259	464
35	Tamar 4	200 m, stop pumping			KN			491	178	85	830	210	497
36	Tamar 4				KN	36	34	506	215	125	980	219	601
37	Zin 1	30 m	1831	447	KN			595	198	123	1042	232	609
38	Zin 1	30 m, pumping	1831	447	KN	30		583	201	119	1010	268	587
39	Zin 3	pumping	1835	447	KN			552	206	124	933	265	579
40	Zin 3		1835	447	KN		63	899	292	207	1392	262	644
41	Zin 3		1835	447	KN		65	881	279	192	1855	250	642
42	Zin 4	artesian	1825	454	KN			737	132	119	1120	244	615
43	Zin 4		1825	454	KN	33		1336	218	171	2322	286	619
44	Zin 5				KN			635	196	110	1081	238	577
45	Zin 5				KN		58	725	255	173	1514	280	604
46	Zin 5				KN		57		246	162	1456	280	609
47	Zin 6				KN			551	207	137	1028	293	562
48	Zin 6	artesian			KN	28	34	520	190	140	1061	290	555
49	Admon 1	2 m, pumping	1806	488	KN	33	36	681	214	86	1062	335	613
50	Admon 1		1806	488	KN		43	756	221	97	1195	304	681
51	Admon 1		1806	488	KN		44	709	191	82	1039	299	668
52	Admon 2	120 m	1805	496	KN			735	177	104	1081	329	642
53	Admon 2	252 m	1805	496	KN			984	275	120	1500	305	912
54	Admon 2	275 m	1805	496	KN			1533	272	102	2305	317	877
55	Admon 2		1805	496	KN		81	1534	261	104	2422	323	844
56	Admon 2		1805	496	KN		77	1341	224	104	2064	183	895
57	Amiaz 2		1822	528	KN		195	3318	508	1094	8744	219	830
58	Amiaz 2		1822	528	KN		230	3537	1309	784	9440	171	905
59	Amiaz 1	32 m			KN		37	467	132	82	922	158	294
60	Amiaz 1	62 m			KN			321	81	31	469	105	313
61	Amiaz 1	70 m			KN		165	2698	830	519	6578	146	788
62	Amiaz 1	85 m			KN		214	2993	1002	613	7505	232	922
63	Amiaz 3	306 m	1824	548	KN		298	1757	874	247	4090	91	1214
64	Amiaz 3		1824	548	KN		83	1500	966	288	3735	216	1424
65	Zohar 3	50 m	1853	605	KN		123	1329	438	428	3361	256	905
66	Zohar 1		1840	620	KN		106	1288	465	357	3739	231	
67	Zohar 1	158 m	1840	620	KN		157	2203	585	511	5389	247	755
68	Zohar 1	169 m	1840	620	KN		108	1610	452	354	3840	219	719
69	Nahal Boqeq 1	38 m	1840	675	KN		141	1917	566	294	3430	256	1835
70	Nahal Boqeq 1	42 m	1840	675	KN		144	1902	577	288	3395	280	1835
71	Nahal Boqeq 1	59 m	1840	675	KN		123	1833	605	234	3149	256	1877
72	Nahal Boqeq 1	66 m	1840	675	KN		111	1810	623	193	3000	286	1860
73	Nahal Boqeq 1		1840	675	KN		91	1643	644	186	2867	280	1710

Table 25-1 (Continued) Major ions (mg/l) in groundwater of the Dead Sea.

No	Name	Notes	Co. E.	Co. N.	Type	T°C	K	Na	Ca	Mg	Cl	HCO3	SO4
74	Nahal Boqeq 1	97 m	1840	675	KN		94	1010	612	193	2867	256	362
75	Kane-Samar	Well 3 (shallow)			F		7	42	62	38	118	233	10
76	Kane-Samar	Well 1 (shallow)			F		8	43	53	44	132	274	8
77	Kane-Samar	Well 2 (shallow)			F		19	43	63	44	151	277	13
78	Kane-Samar	Small spring.	1888	1128	F	26	14	77	86	70	285	319	8
79	Kane-Samar	Shallow well.	1887	1138	F-Z-Y	26	31	107	80	82	385	290	53
80	Kane-Samar	Spring	1888	1142	F-Z-Y	26	76	350	175	290	1390	294	116
81	Kane-Samar	Spring	1890	1146	F-Z-Y	26	79	368	180	276	1405	329	98
82	Kane-Samar	Seepages	1885	1124	Z-Y	27	68	370	190	302	1540	293	105
83	Kane-Samar	Spring	1892	1150	Z-Y	25	87	460	260	375	2218	322	122
84	Kane-Samar	Spring	1892	1148	Z-Y		107	625	270	412	2299	322	141
85	Kane-Samar	Spring	1893	1151	Z-Y	28	79	437	280	450	2470	335	118
86	Kane-Samar	Seepages	1885	1122	Z-Y	26	120	781	300	562	2971	276	125
87	Kane-Samar	Spring	1896	1153	Z-Y	26	92	687	300	550	3099	340	175
88	Kane-Samar	Spring	1896	1152	Z-Y	29	137	750	340	625	3484	353	130
89	Kane-Samar	Spring	1888	1142	Z-Y	27	124	937	440	875	4833	284	98
90	Kane-Samar	Well	1887	1139	Z-Y	27	250	1250	590	1150	6428	279	166
91	Kane-Samar	Riverlets	1888	1140	Z-Y	27	225	1500	675	1375	7577	284	156
92	Zukim	Canal	1934	1261	T-N		134	1240	457	498	3902	263	96
93	Zukim	Canals	1932	1253	T-N		92	790	323	314	2706	245	71
94	Zukim	Pool	1927	1246	T-N	26	57	495	228	263	1738	275	45
95	Zukim	Small stream	1928	1244	T-N	27	66	530	238	272	1811	293	52
96	Zukim	Stream	1928	1244		30							
97	Zukim	Small stream.	1928	1244									
98	Zukim	Canal.	1927	1244		28							
99	Zukim	Stream, large.	1928	1244		28							
100	Zukim	Stream, medium.	1926	1244	T-N	28	84	670	292	329	2326	298	70
101	Zukim	Stream, large, smell.	1926	1243		28							
102	Zukim	Stream	1927	1240	T-N	29	224	805	410	484	3214	281	84
103	Zukim	Spring medium.	1923	1233		29							
104	Zukim	Saline spring, smell.	1925	1233	Z-Y	31	815	5125	2032	4000	23845	227	147
105	Zukim	Spring, smell.	1925	1233	Z-Y	31	1350	8557	3546	6895	40178	298	225
106	Nahal Boqeq 1	97 m, pumping	1840	675	KN-ZY		255	1873	890	1136	7671	259	171
107	Nahal Boqeq 1	97 m, pumping	1840	675	ZY		672	4906	1973	4148	23387	216	76
108	Nahal Boqeq 1	97 m, pumping	1840	675	ZY		880	6619	2530	5572	31317	201	74
109	Zohar 1		1840	620	KN-ZY(?)		342	4831	1453	1457	13943	183	658
110	Zohar 1		1840	620	KN-ZY(?)		564	3278	2290	2887	24922	146	757
111	Hamme Zohar	4 m, drill 1	1848	640	ZY		2080	14560	6112	12085	69922	195	726
112	Hamme Zohar	5 m, drill 8	1848	640	ZY		1630	11702	5879	9529	57441	168	597
113	Hamme Zohar	4 m, drill 3	1848	640	ZY		1048	9493	3446	6948	41555	152	675
114	Hamme Zohar	7 m, drill 1	1848	640	ZY		1028	10395	3497	6577	41839	177	704
115	Hamme Zohar	1 m, drill 4	1848	640	ZY		780	8892	2686	5173	34464	49	634
116	Hamme Zohar	a N. spring	1848	640	ZY		860	7800	3140	6300	36400	181	711
117	Hamme Zohar	Spr. 10 m off shore	1848	640	ZY	28	1450	11880	5100	9880	56700		675
118	Hamme Zohar	Spr. 20 m off shore	1848	640	ZY	28	2120	15850	6660	14200	79400	179	713
119	Hamme Zohar	Spr.30 m off shore	1848	640	ZY	28	1980	14760	6270	13000	74200	177	727
120	Ein Noit		1840	677	KN		167	2060	630	303	2700	279	1841
121	Ein Noit		1840	677	KN		62	1340	600	122	2030	245	1710
122	Nahal Mishmar	18 m	1867	870	ZY		955	4700	1650	2321	14250	4760	
123	Nahal Mishmar	30 m	1867	870	ZY		755	3830	1400	1921	12150	4380	735
124	Nahal Mishmar	33 m	1867	870	ZY		1067	9630	3000	4625	31300	4390	512
125	Nahal Mishmar	33 m, before pumping	1867	870	ZY		3070	32026	8499	18655	121158		79
126	Nahal Mishmar	33 m, after pumping	1867	870	ZY		2905	32400	9000	18250	118200	1698	119
127	Hamme Yesha	7 m	1859	920	ZY		2905	25690	10800	15550	106900	256	1152
128	Hamme Yesha	17 m	1859	920	ZY		3940	27400	14300	20990	134000	378	835
129	Hamme Yesha	24 m	1859	920	ZY		3656	28651	10180	19518	121677	86	1029
130	Hamme Yesha	24 m	1859	920	ZY		4024	29076	10648	20910	127644	24	967
131	Hamme Yesha	pumping	1859	920	ZY	39	3520	27361	9813	18086	114510	107	1078
132	Hamme Yesha	pumping	1859	920	ZY	39	3630	27700	11000	18850	116500	129	1150
133	Hamme Yesha	pumping	1859	920	ZY	39	3525	26900	12000	18250	116500	124	1158
134	Hamme Yesha	pumping	1859	920	ZY	39	3630	27700	11000	18850	118000	107	1180
135	Hamme Yesha		1859	920	ZY	39	3525	27700	12000	18250	120000	115	1171
136	Hamme Yesha		1859	920	ZY	39	3525	27700	11000	18850	118000	110	1174
137	Hamme Yesha		1859	920	ZY	39	3455	27700	11000	18850	118000	107	1182
138	Hamme Yesha		1859	920	ZY	39	3320	27700	11000	18850	118000	105	1194
139	Hamme Yesha		1859	920	ZY		3500	27800	10300	18700	117530	142	1060
140	Dead Sea	Surface					6600	38000	15150	39600	218130	233	578
141	Dead Sea	Surface					7260	36570	13780	37000	194440		600
142	Dead Sea	310 m					8540	38600	16080	44540	221950		711

Data from Mazor et al. (1969, 1972, 1973) and Kroitoru et al. (1989).

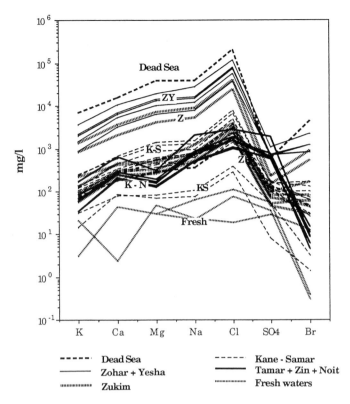

Figure 25-2 Fingerprint diagrams of springs and wells along the western Dead Sea shore, analyzed in the early 1960s (prior to the lowering of the Dead Sea level and the intensive groundwater removal). Ion concentrations vary over a range of four orders of magnitude. Composition and genesis of the various waters are discussed in the text.

FRESHWATER RECHARGE

Several freshwater springs issue along the western margins of the Dead Sea rift, creating oases, for example, the Elisha Spring, Jericho (28 mg Cl/l), En David (77 mg Cl/l), and En Boqeq (550 mg Cl/l). Freshwater was also encountered in shallow wells, for example, Amiaz I (85 mg Cl/l). The spring groups along the Dead Sea shore include freshwater seepages (Table 25-1), for example, at the Kane-Samar group (118 mg Cl/l). The main geochemical feature is the low concentration of dissolved ions. In the composition graphs that show mixing lines, described in the following sections, the freshwater end member always plots near the origin of the axes, that is, its salt load is negligible compared to the salts contributed by the different saline end members.

The Judean Desert, west of the Dead Sea, cannot be the recharge region of this fresh water because its rains are scarce (less than 100 mm/yr), whereas the Judean Mountains enjoy about 550 mm/yr and have an eastward flow of groundwater (Kroitoru et al., 1985, 1992; Mazor and Kroitoru, 1990). The high-discharge freshwater Elisha Spring at Jericho has the stable isotope composition of the rain of the Judean Mountains, indicating that the recharge takes place 22 km from the spring. Bomb-tritium (25 TU) was observed in the Elisha Spring in 1968, that is, 15 years after the concentration of atmospheric tritium began to increase. This observation was interpreted by Kroitoru et al. (1985, 1992) and Mazor and Kroitoru (1990) as indicating rapid flow (22:15 = 1.4 km/yr or faster), which sug-

gests flow in karstic (fossil) channels. Thus, fresh groundwater recharged in the Judean Mountains reaches the Dead Sea basin.

THE KIKAR-NOIT WATER GROUP

Several tens of producing wells and a few small springs bearing the names Kikar, Tamar, Zin, Admon, Amiaz, and Noit (Fig. 25-1) belong to one composition group, named the Kikar-Noit Water Group (Mazor et al., 1969). These waters occur 1 to 7 km southwest of the former shallow part of the Dead Sea (Fig. 25-1, Table 25-1). For the present discussion, 33 analyses of repeatedly collected samples from the 11 Tamar and Zin wells and the Noit Spring are presented in Fig. 25-3 (data points of more wells would clog the graphs). The fingerprint diagram (Fig. 25-3a) reveals that the relative ionic abundances are rather uniform and differ significantly from the Dead Sea values. The composition diagrams (Fig. 25-3b) reveal that the Tamar, Zin, and Noit waters constitute a mixing or dilution line in the plots of Na, K, Mg, and Br as a function of Cl, but the Dead Sea water is not an end member. The 20 repeatedly collected samples of the Admon, Amiaz, and Hakikar wells and spring (Table 25-1) reveal the same compositional characteristics.

Thus, the Kikar-Noit waters constitute an independent groundwater group that is not related to the Dead Sea composition. Groundwaters of the Kikar-Noit type occur in northern segments of the Dead Sea rift, especially around Lake Tiberias. The name Tiberias-Noit Water Group was suggested for the whole suite of the Dead Sea rift waters of similar compositional features. The Tiberias-Noit waters have been suggested to originate from seawater that was entrapped in local aquifers when the Mediterranean Sea invaded the Dead Sea rift via the Esderlon Valley, forming the precursor of Lake Lisan (Mazor and Mero, 1969). However, the Tiberias-Noit waters differ from seawater composition and therefore water-rock exchange reactions were suggested that caused depletion of Na and proportional enrichment of Ca and in cases also Mg. This hypothesis was supported by (1) observation of Tiberias-Noit type waters also on the shores of the Gulf of Suez (Mazor, 1968a; Mazor et al., 1973b), where interaction of seawater with aquifer rocks is plausible; (2) laboratory experiments in which seawater was stirred with Dead Sea rift rocks (Mazor, 1968b; Mazor et al., 1973b); and (3) a large-scale release of NaCl-rich waste disposal of an abattoir in which local wells were enriched in Cl that was accompanied by Ca and Mg rather than Na (Mazor et al., 1981).

The Kikar-Noit waters have the following characteristic features:

(1) Ionic composition (by weight) of Na>Ca>Mg>K and Cl>SO_4>HCO_3>Br.

(2) Total dissolved ion concentrations are observed to vary from 1.5 to 17 g/l. The cause of the large range in ion concentrations can be either dilution by recent recharge from the Judean Mountains or original dilution at the time of seawater encroachment into existing aquifers.

(3) Observed temperatures range from 27 to 42°C, reflecting different depths of circulation or entrapment.

(4) Often H_2S can be smelled, and in the few cases it was measured, the concentrations varied from 2 mg/l to a maximum of 40 mg/l. The latter value was found at the early stages of water abstraction from the Tamar 3 well (Mazor et al., 1969). Mazor and Rosenthal (1967) postulate that the H_2S stems from bacteriological decay of SO_4, which occurs in the range of 400 to 700 mg/l in the Kikar-Noit Water Group.

(5) Radium concentrations are 5 to 10 pc/l (picoCuries per liter), and radon concentrations are 1,000 to 3,000 pc/l (Mazor, 1962). These are medium values that lie between the very low concentrations in the freshwaters and the higher concentra-

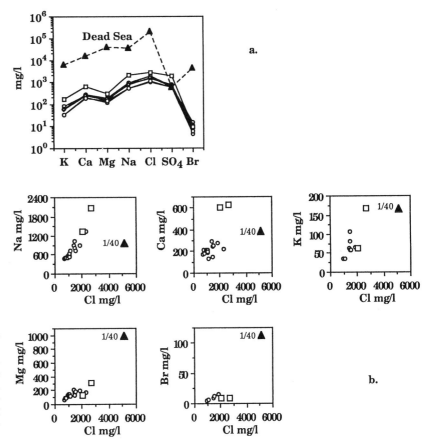

Figure 25-3 Repeated measurements in 11 Tamar and Zin wells (circles) and Noit spring (square) that belong to the Kikar-Noit Water Group: A, fingerprint diagram showing that the waters have similar relative dissolved ionic abundances, and they significantly differ from the Dead Sea composition; B, composition diagrams that show mixing lines for which the Dead Sea (triangle) is not an end member.

tions found in the Dead Sea-related water groups, discussed in the following sections.

(6) Low carbon-14 concentrations have been found in the following wells (Table 25-2): Tamar 3 (5.2 pmc), Tamar 11 (0 pmc), Tamar 7 (4.7 pmc), Zin 6 (4.2 pmc), and Neot Hakikar (37 pmc). The accompanying $\delta^{13}C$ values were -8 to -11% (Carmi, 1987; Carmi et al., 1971), in the range expected for water that interacted with limestone or dolomite. Thus, taking 65 pmc as the initial ^{14}C concentration (following Kroitoru et al., 1989), water ages in the range of 5 ka to more than 20 ka years are obtained for the analyzed Kikar-Noit groundwaters.

(7) Kikar-Noit waters were encountered in boreholes that tapped a large variety of members and formations of the Kurnub Group and the Judea Group (Mazor et al., 1969).

After examining the general features in common to the waters of the Kikar-Noit Water Group, we propose a common origin from an ancient episode of seawater intrusion into the Dead Sea rift. These waters are stored in a large number of distinct aquifers, mainly found in the basin of Lake Tiberias, the Dead Sea basin, and the northern Arava valley. The waters emerge under pressures that may stem from compaction by the overlying rocks, coupled with tectonic compression.

GROUNDWATERS WITH DIFFERENT SALINITY VALUES ENCOUNTERED DURING DRILLING

In several boreholes, different groundwaters were encountered during drilling (see Table 25-1). For example, the samples from depths of 32, 62, 70, and 85 meters at the Amiaz 1 borehole belong to the local Kikar-Noit Water Group and reveal a general trend of increased salinity with depth (Fig. 25-4). An ion

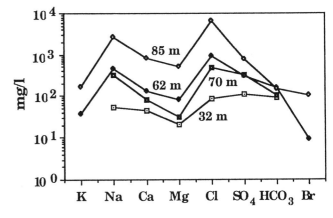

Figure 25-4 Fingerprint diagrams of water samples collected from the Amiaz-1 borehole during drilling. Kikar-Noit type water was encountered with a trend of increasing salinity with depth.

concentration increase with depth is also seen in the Tamar 3 and Admon 2 data (Table 25-1), whereas a decrease in the ion concentration is observed in the data of Tamar 4 and the nonpumped samples of Nahal Boqeq 1 (Table 25-1). Because of the lack of necessary auxiliary data, for example, water heads, it is difficult to determine whether the reported water samples represent separated overlying aquifers or mixing of freshwater with Kikar-Noit water, in proportions that vary with the depth of each borehole.

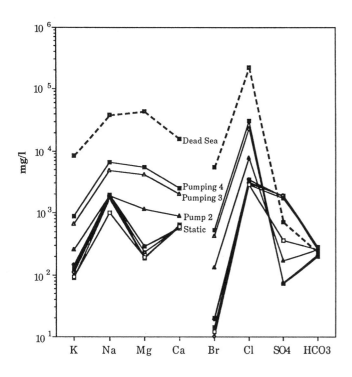

Figure 25-5 Fingerprint diagrams of water samples obtained from the Nahal Boqeq-1 borehole during drilling and at stages of a pumping test. As pumping advanced, the water composition indicated Dead Sea water encroachment.

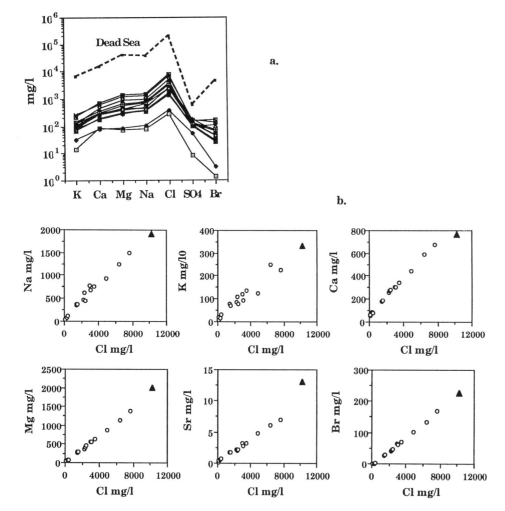

Figure 25-6 Data of springs of the Kane-Samar complex: A, fingerprint diagram, revealing composition similarity to the Dead Sea; B, composition diagrams, revealing different degrees of mixing of very fresh water (recharged from the Judean Mountains) and Dead Sea water (triangle, divided by 20 to fit into the diagram).

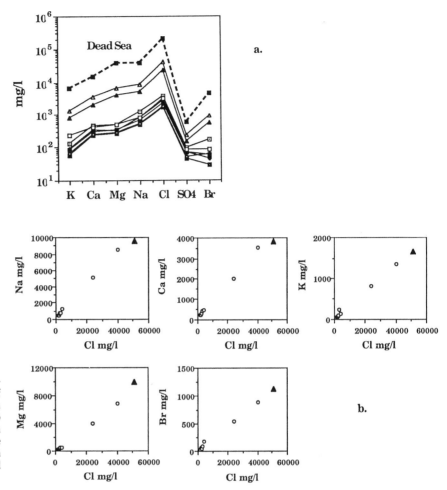

Figure 25-7 Data of springs from the Zukim complex: a, fingerprint diagram, revealing a group of rather fresh waters, that differ from the Dead Sea composition (best seen by the Br values), and two samples that are highly mixed with Dead Sea water; b, composition diagrams, revealing that the two more saline samples are mixtures of Dead Sea water (divided by 4 to fit into the graph) and rather fresh water.

DEAD SEA AFFINITY (ZOHAR-YESHA) WATER GROUP

Several saline waters, issuing in springs and occurring in wells that are located on the Dead Sea shore, reveal relative ion abundances similar to those found in the Dead Sea.

Encroachment of Dead Sea water as a result of pumping tests

Pumping tests performed in the Nahal Boqeq 1 borehole (600 meters from the Dead Sea water line, 97 meters deep, 4,000 m^3 pumped during 76 hours) and the Nahal Mishmar 1 borehole (100 meters from the Dead Sea, 33 meters deep) caused encroachment of Dead Sea type water (Table 25-1 and Fig. 25-5). These observations provide some idea of the local shape and dimension of the Dead Sea–freshwater interface.

Kane-Samar complex of slightly saline springs

The Kane and Samar spring groups occur at the Dead Sea waterfront along a stretch that is a few kilometers long. Table 25-1 and the fingerprint diagram (Fig. 25-6a) reveal a water salinity range of more than an order of magnitude, yet with an ion abundance pattern resembling that of the Dead Sea. Thus, mixing of freshwater and Dead Sea water has been postulated (Mazor et al., 1973a). This point is further checked in the detailed composition diagrams of Na, K, Ca, Mg, Sr, and Br as a function of Cl (Fig. 25-6b), in which the concentrations of the

Kane-Samar springs lie on lines that extrapolate to zero values at the lower end, and to the Dead Sea values at the higher end (triangle, divided by 20 for accommodation in the figures). Thus, in the Kane-Samar spring complex, freshwater recharged from the Judean Mountains seems to be mixed with up to 3% Dead Sea water, with the mixing occurring right at the water's edge. In this case, the bulk of the water comes with the freshwater component, whereas the bulk of the dissolved ions comes from the Dead Sea.

Tritium in the Kane-Samar springs was found in 1972 to be from less than 3 to 17 TU (Table 25-2), indicating a post-1954 component of the freshwater (Mazor et al., 1973).

Radium was less than 50 pc/l, and radon was in the range of 200 to 5,900 pc/l (Table 25-2). Elevated radon concentrations characterize all Dead Sea rift water groups (Mazor, 1962), and the origin of the radon is still an enigma.

Zukim (Feshkha) springs complex

The Zukim spring complex is situated on the northwestern shore of the Dead Sea (Fig. 25-1), at the waterline. In 1970, the complex encompassed about 20 springs, situated along a 3-km stretch of the Dead Sea shore. The following features characterize this group of springs (Fig. 25-7):

(1) There is a wide range of salinity, for example, Cl ranged from 1.7 to 40 g/l (Table 25-1).

Table 25-2 Minor ions (mg/l), isotopes (‰), tritium (TU), carbon-14 (pmc), radium and radon (pC/l) from ground water along the western shore of the Dead Sea.

No.	Name	Li	Sr	Br	D	±	O-18	±	Tritium	±	C-14	±	Radon	Radium
4	Amiaz 1				-42		-6.2							
5	Amiaz 1			9										
7	Amiaz 1			103										
8	Amiaz 1			106										
9	Ein Boqeq				-31		-5.7							
16	Neot Kikar 1				-25		-5.1							
22	Tamar 3				-3.3									
28	Tamar 6				-50		-7.1							
31	Tamar 7				-36		-6.6							
32	Tamar 4				-29		-6.2							
49	Admon 1				-37		-6.8							
52	Admon 2				-43		-6.7							
57	Amiaz 2				-40		-6							
75	Kane-Samar	<0.001	0.4	<0.001										<50
76	Kane-Samar	<0.001	0.1	<0.001										<50
77	Kane-Samar	<0.001	0.5	<0.001										<50
78	Kane-Samar	0.0	0.6	1					13	4			4200	<50
79	Kane-Samar	0.0	0.7	3					16	5			1600	<50
80	Kane-Samar	0.1	1.7	26					8	5			3900	<50
81	Kane-Samar	0.1	1.7	26					5	3			310	<50
82	Kane-Samar	0.1	1.8	30					3	4			<200	<50
83	Kane-Samar	0.1	2.2	42					15	5			200	<50
84	Kane-Samar	0.1	2.0	44					5	3			370	<50
85	Kane-Samar	0.2	2.2	46					17	5			5900	<50
86	Kane-Samar	0.2	3.2	63					3	3				<50
87	Kane-Samar	0.2	2.8	60					7	3			<200	<50
88	Kane-Samar	0.2	3.2	71					14	3			8700	<50
89	Kane-Samar	0.3	4.8	103					11	4			1440	<50
90	Kane-Samar	0.4	6.1	134					13	5			<200	<50
91	Kane-Samar	0.4	7.0	168					5	4				<50
92	Zukim	0.1	7.6	169	-25.0	1.5	-5.73	0.12					20540	
93	Zukim	0.1	4.8	44	-27.3	1.5	-5.68	0.12	0	6			4000	
94	Zukim	0.1	2.2	29	29.5	1.5					35.6	4	30000	
95	Zukim	0.1	3.9	61	-24.3	1.0	-5.48	0.70					23000	
96	Zukim				-24.3	2.0	-6.06	0.12	31	7			7530	
97	Zukim				-25.3	1.0	-5.68	0.12					20000	
98	Zukim				-22.7	1.0	-5.80	0.07	34	8	33.1	1.2	8160	
99	Zukim				-29.6	2.0	-5.95	0.12					10940	
100	Zukim	0.1	5.1	58	-31.0	1.0	-5.67	0.05	36	9			3386	
101	Zukim				-31.0	1.0	-5.65	0.12					14130	
102	Zukim	0.1	6.5	83	-27.8	1.0	-5.57	0.12	11	7			3820	
103	Zukim				-31.1	1.0	-6.00	0.12					10440	
104	Zukim	0.8	46.0	541	-27.3	1.5	-5.00	0.12					2130	
105	Zukim	1.5	31.0	886	-20.0	1.0	-4.11	0.12	13	6			2836	
111	Hamme Zohar				-36		-4.5							
120	Ein Noit			9	-50		-7							
121	Ein Noit			8			-7.1							
127	Hamme Yesha				-12		0.4							
140	Dead Sea			3570										
141	Dead Sea			4650										
142	Dead Sea			5450										

Data from Gat et al. (1969), Mazor (1962), Mazor et al. (1969, 1972).

(2) The fresher waters constitute a group of uniform composition resembling that of the Kikar-Noit Water Group.

(3) The two saline waters have a composition similar to that of the Dead Sea water (Fig. 25-7a), and the plots of Na, Ca, K, Mg, and Br as a function of Cl (Fig. 25-7b) lie on mixing, lines that extrapolate to freshwater at the lower end, and to the Dead Sea value (triangle, divided by 4) at the upper end.

(4) The temperature ranged from 20 to 31°C, that is, the springs are slightly thermal, as are all the Kikar-Noit waters.

(5) Tritium varied from 0 to 36 TU (Mazor and Molcho, 1972)

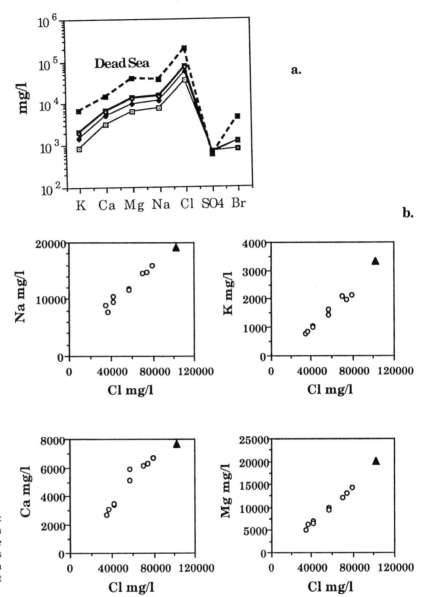

Figure 25-8 Data of the Hame Zohar Springs: a, fingerprint diagram revealing waters with different ion concentrations but all having the Dead Sea pattern; b, composition diagrams revealing almost perfect mixing lines with Dead Sea water (triangle, divided by 2 to fit into the plot).

with no correlation to the salinity, indicating a recent freshwater component.

(6) A carbon-14 concentration of 33 and 35 pmc was observed in the two freshest water samples (Mazor and Molcho, 1972), indicating presence of an old-water end member - that of the Kikar-Noit Water Group _ that has been observed to be old all over the Dead Sea rift.

(7) In 1970, radon ranged from 2,100 to 30,000 pC/l (pC/l = 10^{-12} Curie/l; Table 25-2), compared to only 12 pc/l in the fresh En David spring (Mazor, 1962; Mazor and Molcho, 1972). No correlation was observed between the Rn concentration and Cl or any other parameter. A similar range of elevated Rn concentration was previously found in the Tiberias-Dead Sea rift waters (Mazor, 1962). The occurrence of high Rn concentrations was already reported by Rosenberg (unpublished report to the Palestinian Potash Co., 1944), who calculated that the Rn flux from the Zukim springs area is a world record. It seems that Rn emanation to the surface is locally enhanced because of the intensive faulting rather than because of uranium-rich rocks.

Thus, three water groups seem to mix in the Zukim spring complex: freshwater recharged at the Judean Mountains, Kikar-Noit type water, and Dead Sea water.

Hamme Zohar complex of springs and wells

The Hamme Zohar springs (Fig. 25-1) were developed as a spa in the early 1960s. Measurements were conducted in springs issuing a few meters from the waterline, in three offshore springs covered by up to two meters of Dead Sea water, and in a number of partially self-flowing shallow wells, all concentrated in a 100-m-long strip of shoreline (Mazor et al., 1969). The following features were observed:

(1) The ion concentrations in the various samples varied significantly, but all resembled the Dead Sea composition, as seen in the fingerprint and composition diagrams (Fig. 25-8).

(2) All the waters contained H$_2$S. The highest value measured was 20 mg/l.

(3) Temperatures ranged from 28 to 33°C, with the latter value stemming from a 65-m-deep borehole.

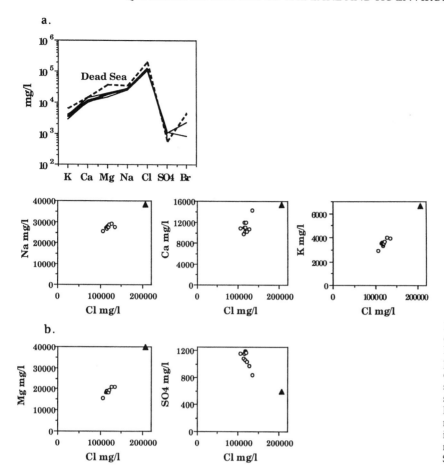

Figure 25-9 Data from samples at different depths and pumping stages of the Hamme Yesha well (Table 25-1): a, fingerprint diagrams that reflect high ion concentrations with an abundance pattern similar to that of the Dead Sea; b, composition diagrams that reveal positive correlation lines for Na, K, and Mg, as well as a negative correlation line for SO_4, indicating mixing with Dead Sea water; one end member contains more SO_4 than the Dead Sea

(4) Values of Ra were in the range of 13 to 32 pC/l, and Rn concentrations were in the range of 500 to 21,000 pC/l (Mazor et al., 1969).

Thus, the Hamme Zohar waters are mixtures of Dead Sea water, which provides the bulk of the salts, and Kikar-Noit waters, which contribute the slightly elevated temperature, the H_2S, the slightly elevated radium, and the conspicuous concentration of radon.

Hamme Yesha (En Gedi 3) shallow well

Hamme Yesha is a flourishing spa based on shallow wells and located 3 km south of the En Gedi settlement. The first well, En Gedi 3, is located at the narrow coast bounded between the Dead Sea waterline and the Dead Sea rift escarpment. Early measurements were taken in the 1960s at drilling sites down to 30 meters and at various pumping stages (Mazor et al., 1969). The following features emerged:

(1) All water samples had a high salinity, about half the Cl concentration of the Dead Sea water (Table 25-1).
(2) The relative ion abundances were close to those of the Dead Sea (Fig. 25-9a).
(3) Positively correlated mixing lines were present on the graphs of Na, K, and Mg as a function of Cl concentration (Fig. 25-9B), with the Dead Sea as one end member. The mixing with Dead Sea water is especially evident by the negative correlation line of SO_4 as a function of Cl (Fig. 25-9b), which shows that the other end member contains more SO_4 than the Dead Sea.
(4) Additional parameters, measured at the Hamme Yesha waters, revealed that the other mixing end member belongs to

the Kikar-Noit Dead Sea rift waters; these parameters included elevated temperature (39°C), high H_2S (40 mg/l), and elevated radon (1,500 pC/l).

Thus, the Hamme Yesha is an extreme manifestation of mixing two types of water: Kikar-Noit water (contributing elevated temperature and high concentrations of H_2S, radium, and radon, as well as SO_4 and a relatively lower Br abundance) and a nearly equal amount of Dead Sea water (contributing the major ions and Br).

ISOTOPIC COMPOSITIONS

Studies of deuterium and oxygen-18 concentrations conducted by Gat et al. (1969) and Mazor and Molcho (1972) revealed the following features (see Fig. 25-10 and Table 25-2): two end member groups are evident - the freshwaters flowing from Judean Mountains recharge and the relatively fresh Kikar-Noit Water Group (both have relatively low Cl concentrations, low deuterium, and oxygen-18 values) and the Dead Sea water (high concentrations of the three parameters). The most conspicuous examples of the mixing are the Hamme Yesha and Hamme Zohar waters.

The isotopic evidence for mixing of Dead Sea water in the saline springs at the Dead Sea shore is crucial for determining whether the springs contributed a substantial part of the ions dissolved in the Dead Sea or whether Dead Sea water is recycled in the springs. The latter hypothesis is in full accord with the observations, whereas the former implies huge underground reservoirs with an extremely high coincidence that their ionic and isotopic compositions fall on the observed mixing lines.

CONCLUSIONS

Three major water groups were identified in the study area: fresh waters, Kikar-Noit waters typical of the Dead Sea rift, and mixtures of these waters with the Dead Sea water in the immediate neighborhood of the Dead Sea shore. The fresh waters are recharged at the Judean Mountains and flow toward the Dead Sea drainage basin through various rock formations of the Judea Group and, in part, in old karstic channels. They are distinct by their very low concentration of dissolved ions.

The Kikar-Noit waters, found also in northern and southern segments of the Dead Sea rift, have the following characteristics: (1) an ionic composition that resembles seawater that exchanged about 10% of its Na for Ca and Mg; (2) ^{14}C ages of 5 ka to more than 20 ka years, which favor storage since a paleo-recharge event; (3) elevated temperatures of up to 40°C, indicating storage at a depth of several hundred meters; (4) high SO_4 concentrations, partly gained by dissolution of gypsum occurring as veins in the host rocks; (5) high H_2S concentrations, probably from biogenic decomposition of SO_4; (6) slightly enriched in radium and significantly enriched in radon; and (7) issue from springs or tapped in boreholes in a large variety of rock formations and structures.

The fresh waters and the Kikar-Noit waters occur in a large number of separated aquifers, defined by tectonic and lithological structural combinations that typify the Dead Sea rift, especially along its margins. The nearshore groundwater plumbing must be complicated because springs of fresh cold water issue tens of meters from springs that are saline and warm, for example, at the Zukim and Hamme Zohar sites.

Mixing with Dead Sea water is advocated for the Dead Sea Affinity Water Group based on (1) the perfect linear correlation observed in plots of the concentration of dissolved ions as a function of Cl concentration, with the Dead Sea values lying on these correlation lines (at the upper end, and, in the case of SO_4 at the Hamme Yesha water, at the lower end), and (2) the isotopic compositions that reveal Dead Sea mixing ratios similar to those calculated from the dissolved ions.

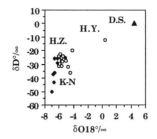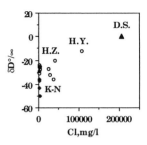

Figure 25-10 Stable isotopes and Cl concentrations. The Hamme Yesha (H.Y.), Hamme Zohar (H.Z.), and the most saline members of the Zukim complex lie on distinct mixing lines, with the Dead Sea as one end member and the Kikar-Noit Water Group (K.N.) and freshwater as the other end members (following Gat et al., 1969). The scatter of the values of the K-N waters reflects their paleoclimatic origins (Mazor, 1993).

REFERENCES

Bentor, K. B., 1961, Some geochemical aspects of the Dead Sea and the question of its age: *Geochimica et Cosmochimica Acta*, v. 25, p. 239–260.

Carmi, I., 1987, Rehovot radiocarbon measurement III: *Radiocarbon*, v. 29, p. 100–114.

Carmi, I., Noter, Y., and Schlesinger, R., 1971, Rehovot radiocarbon measurement I: *Radiocarbon*, v. 13, p. 412–419.

Gat, J., Mazor, E., and Tzur, Y., 1969, The stable isotope composition of mineral waters in the Jordan Rift Valley, Israel: *Journal of Hydrology*, v. 7, p. 334–352.

Kroitoru, L., Carmi, I., and Mazor, E., 1989, Groundwater ^{14}C activity as affected by initial water-rock interactions in the Judean Mountains, Israel: *Chemical Geology* (Isotope Geoscience Section), v. 79, p. 259–274.

Kroitoru, L., Mazor, E., and Gilad, D., 1985, Hydrological characteristics of the Wadi Kelt and Elisha springs: Proceedings of the Jerusalem Symposium, Sept. 1985, Scientific Basis for Water Resources Management, IAHS, publication no. 153, p. 207–218.

Kroitoru, L., Mazor, E., and Issar, A., 1992, Flow regimes in karstic systems: The Judean anticlinorium, central Israel, *in* Pallock, H., and Back, W., eds., *Hydrogeology of Selected Karst Regions*: International Association of Hydrogeologists, Verlag H. GMBH, Hannover, p. 339–354.

Mazor, E., 1962, Radon and radium content of some Israeli water sources and a hypothesis on underground reservoirs of brines, oils, and gases in the Rift Valley: *Geochimica et Cosmochimica Acta*, v. 26, p. 765–786.

Mazor, E., 1968a, Compositional similarities between hot mineral springs in the Jordan and Suez Rift Valleys: *Nature*, v. 219, p. 477–478.

Mazor, E., 1968b, Genesis of mineral waters in the Tiberias-Dead Sea Rift Valley, Israel: Proceedings of the 24th International Congress of Geology, v. 7, Genesis of Mineral and Thermal Waters, Prague, p. 65–80.

Mazor, E., 1993, Interrelations between groundwater dating, paleoclimate and paleohydrology, *in* Applications of Isotope Techniques in Studying Past and Current Environmental Changes in the Hydrosphere and the Atmosphere: Vienna, International Atomic Energy Agency, p. 249–257.

Mazor, E., Drever, J. I., Finely, J., Huntoon, P. W., and Lundy, D. A., 1992, Hydrochemical implications of groundwater mixing—An example from the southern Laramie Basin, Wyoming: *Water Resources Research*, v. 29, p. 193–205.

Mazor, E., and Kroitoru, L., 1990, Boundary conditions needed for groundwater modeling, derived from isotopic and physical measurements—Mediterranean-Dead Sea transect: Fifth Annual Canadian/American Conference on Hydrogeology, p. 443–452.

Mazor, E., Levitte, D., Truesdell, A. H., Healy, J., and Nissenbaum, A., 1980, Mixing models and ionic geothermometers applied to warm (up to 60°C) springs - Jordan Rift Valley, Israel; *Journal of Hydrology*, v. 45, p. 1–19.

Mazor, E., and Mero, F., 1969, Geochemical tracing of mineral and fresh water sources in the Lake Tiberias basin, Israel: *Journal of Hydrology*, v. 7, p. 318–333.

Mazor, E., and Molcho, M., 1972, Geochemical studies on the Feshkha springs, Dead Sea basin: *Journal of Hydrology*, v. 15, p. 37–47.

Mazor, E., Nadler, A., and Harpaz, Y., 1973a, Notes on the geochemical tracing of the Kane-Samar spring complex, Dead Sea basin: Israel *Journal of Earth Sciences*, v. 22, p. 255–262.

Mazor, E., Nadler, A., and Molcho, M., 1973b, Mineral springs in the Suez Rift Valley—Comparison with water in the Jordan Rift Valley and postulation of a marine origin: *Journal of Hydrology*, v. 20, p. 289–309.

Mazor, E., and Rosenthal, E., 1967, Notes on the sulfur cycle in the mineral waters and rocks of the Tiberias-Dead Sea Rift

Valley, Israel: *Israel Bulletin of Earth Sciences*, v. 16, p. 197–204.

Mazor, E., Rosenthal, E., and Eckstein, J., 1969, Geochemical tracing of mineral water sources in the southwestern Dead Sea basin, Israel: *Journal of Hydrology*, v. 7, p. 246-275.

Mazor, E., Verhagen, B. T., Sellschop, J. P. F., Jones, M. T., and Hutton, L. G., 1981, Sodium exchange in a NaCl waste disposal case (Lobatse, Botswana)—Implications to mineral water studies: *Environmental Geology*, v. 3, p. 195–199.

Starinsky, A., 1974, Relationship between Ca-rich brines and sedimentary rocks in Israel [Ph.D. thesis]: The Hebrew University, Jerusalem, 176 p. (in Hebrew, English abstract).

Yechieli, Y., Magaritz, M., Shatkay, R., Ronen, D., and Carmi, I., 1993, Processes affecting terrestrial water in the unsaturated zone at the newly exposed shore of the Dead Sea, Israel: *Chemical Geology* (Isotope Geoscience Section), v. 103, p. 207–225.

26. THE BOTANICAL CONQUEST OF THE NEWLY EXPOSED SHORES OF THE DEAD SEA

Erga Aloni, Amram Eshel, and Yoav Waisel

Several years of below-average rainfall, combined with the concomitant increase in demand for irrigation water, have led to the intensive exploitaion of the Jordan River water, reduced the influx of replenishing water into the Dead Sea, and caused a considerable drop in the Dead Sea water level (compare to Klein, 1982, 1993). As a result, substantial areas of hypersaline substrates have emerged from the sea to form the new coast.

The conquest of such new sites by terrestrial plants, the establishment of new plant communities, the momentum of their development, and the processes of their succession depend on complex biotic and abiotic factors (Aloni and Waisel 1990). The ability to invade such new habitats depends on the physiological tolerance of the individual plant species, the dispersal mechanisms of their seeds, and the rate of salt leaching by rain and by runoff water. In this chapter, we describe our investigation of the processes involved in plant succession on this *terra nova*.

MATERIALS AND METHODS

For this investigation, we selected seven different sites along the northwestern coast of the Dead Sea. At each site, several line transects were made, starting at the beach and ending at the first stands of the permanent plant communities. Each of the transects was between 200 and 900 m long, depending on the local topography. Plant species composition and plant cover were recorded along each of the transects from 1989 through 1992. Most of the data were collected during the spring, when the annual plants were fully developed.

Records of the water level were obtained from the Dead Sea Works. The area is extremely arid and remote, and no permanent meteorological station exists in the immediate vicinity. An estimate of the mean annual precipitation in the area between Nahal Kidron and En Gedi, was therefore, extrapolated from data from the following stations: (1) Sedom, 47 mm; (2) En Gedi, 84 mm; and (3) Kalia, 88 mm (Yaffe, 1972). We estimate precipitation to be 50 mm.

The Dead Sea is usually a calm sea. Nonetheless, during periods of easterly winds, it becomes stormy, and the coast is sprayed with fine sea water droplets (for example, Waisel, 1972; Hecht et al., 1984, chapter 10, this volume; Sirkes et al., chapter 9, this volume).

Samples of the substrate were taken, and the particle size, electrical conductivity of the saturated soil extracts, and Na^+, K^+, Ca^{+2}, Mg^{+2}, and Cl^- concentrations were determined (Table 26-1).

Soil and substrate particles were sieved through wire nets with holes of 0.2, 4, 6, and 20 mm. Particles of each size class were collected in 10-liter buckets with perforated bottoms, with 10 kg in each bucket. A gentle spray of distilled water was used to simulate the leaching rates of the newly exposed substrates by rain and runoff water.

RESULTS AND DISCUSSION

The rate of the decline of the lake level varied slightly between years; the average rate was 40 cm/yr during the 1960s and 60 cm/yr between 1970 and 1990. The size of the coastal belt, which emerged annually, varied with the local topography. An example of the investigated area and the changes of sea level that occurred during the last 20 years in the delta of Nahal Kidron is presented in Figure 26-1.

Two types of coast were distinguished, according to the texture of the substrate: shingle beaches and fine-textured beaches. The coarse pebbles of the shingle beaches were covered by several fine interchanging layers of sodium chloride, gypsum, and

Figure 26-1 Panoramic view of the Dead Sea coast near Nahal Kidron, May 1990.

Table 26-1 Changes in the content of Na⁺, K⁺, (Ca²⁺Mg²⁺), and Cl⁻ (meq/l) and electrical conductivity (EC) induced by leaching.

	Before Leaching					After Leaching				
	Na^+	K^+	$Ca^{2+}Mg^{2+}$	Cl^-	EC	Na^+	K^+	Ca^{2+}	Cl^-	EC
Sand	125	44	327.3	462.7	23	92.5	11	215	281	22.8
Pebbles 4 mm	30	4.7	74.7	1039.3	10.9	24	4.2	58.6	78.6	8.51
Pebbles 6 mm	31.2	4.8	75	105	11.3	23.7	4.4	70	56	7.9
Pebbles 20 mm	32	4.8	75.9	118.4	11.73	16.3	3.5	48.3	64	6.89

Original soil sample was taken from the sterile belt.

aragonite. The silt and clay particles of the fine-textured clay substrates were also loaded with NaCl and adsorbed by sodium. These substrates form a sodic-alkaline soil.

Rainfall and runoff water wash the substrate of its excess salts, and enable the germination and establishment of terrestrial plants. To obtain an estimate of the amount of water required for such leaching for each soil type, samples of the substrate were separated into size classes and exposed to gentle sprays of distilled water. Results indicate that the leaching rates varied with particle size and were faster for large pebbles. Nevertheless, approximately 30 mm of water is necessary to reduce the salinity of the saturated substrate extract by 75% (Fig. 26-2).

We do not exactly know what the leaching rate in situ is (compare to Carmi et al., 1984). However, on the average, leaching by 200 mm of rain (4–5 years of rainfall) is required to reduce the salt content of the substrate to a level that will enable the establishment of the first pioneer halophytes (salt tolerant plants). The leaching requirement of the fine-textured substrates is an order of magnitude higher (Table 26-1). Nevertheless, the fine-textured clays characterize the drainage basins of the wadis, where they form thick underground pans with low water infiltration capabilities. Thus, because of the topography and structure of the soil profile, such fine-textured substrates accumulate additional water from adjacent sites (for example, Shatkay, 1985; Yechieli and Gat, chapter 24, this volume) and, therefore,

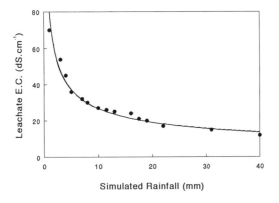

Figure 26-2 Electrical conductivity of the extract of coastal substrate samples as a function of the amount of spray of distilled water.

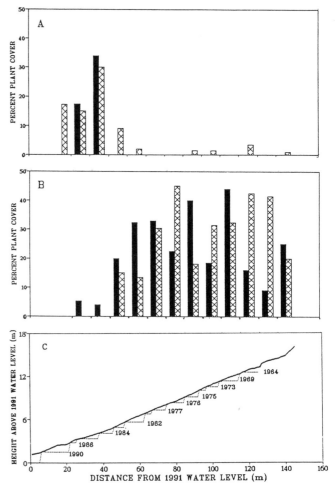

Figure 26-3 Distribution of plants (compare to Table 26-2) across the Dead Sea coast at the outlet of Nahal Kidron: A, distribution of pioneer halophytes: Full bars—annuals; hatched bars—perennials; B, distribution of desert plants: Full bars—annuals; hatched bars—perennials; and C, the topography of the coast and the water level change during the last 20 years.

Table 26-2 Plant species that constitute the main plant cover on the emerging coast of the Dead Sea.

First belt[a]	Life form	Second belt[b]	Life form
Aizoon hispanicum	A[c]	*Aaronsohnia faktorovskyi*	A
Anabasis setifera	P[d]	*Amberboa crupinoides*	P
Anthemis maris mortui	A	*Anthemis maris-mortui*	A
Atriplex leucoclada	P	*Anvillea garcinii*	P
Atriplex halimus	P	*Asteriscus hierochunticus*	A
Atriplex holocarpa	A	*Atriplex halimus*	P
Linaria haelava	A	*Capparis aegyptiaca*	P
Mesembryanthemum nodiflorum	A	*Fagonia mollis*	P
Mesembryanthemum forsskallii	A	*Forsskaolea tenacissima*	P
Mathiola aspera	A	*Heliotropium maris-mortui*	P
Rumex cyprius	A	*Kickxia acerbiana*	P
Salsola baryosma	A	*Limonium lobatum*	A
Salsola inermis	A	*Medicago laciniata*	A
Suaeda fruticosa	P	*Ochradenus baccatus*	P
Suaeda aegyptiaca	A	*Papaver sp.*	A
		Pulicaria dysenterica	P
		Rumex cyprius	A
		Salsola inermis	A
		Salsola baryosma	P
		Silene linearis	A
		Tamarix nilotica	P
		Trichodesma africana	P
		Zygophyllum dumosum	P

[a]First belt: constituents of the pioneer communities.
[b]Second belt: constituents of the subsequent communities.
[c]A: annual.
[d]P: perennial.

are leached by larger quantities of water than the coarse-textured substrates. In spite of this, the time that elapsed between the water retreat and the establishment of the first plants is very similar on shingle beaches and on fine-textured ones.

The establishment of plants on the shores of the Dead Sea follows a lag period of approximately 4 years between the time of emergence of the coast and the time of appearance of the first higher plants. During that time, the area next to the water line remains sterile. The average width of this sterile belt in the Kidron area was 12 ±5 m during 1990 (Fig. 26-3). It took 4 to 5 years for the first pioneer plants to germinate and establish seedlings (Fig. 26-4).

Figure 26-4 Young seedlings of the pioneer halophyte *Atriplex holocarpa* forming the first belt of plants.

Table 26-3 Some characteristics of saturated substrate extracts of the Kidron area, at various distances from the 1992 water line.

Distance from the sea	Year of emergence	Substrate type	Habitat	Salinity pH	EC dS/cm	Na⁺ (mM)
8 m	1990	Sandy loam	Sterile	9.2	99.2	703
35 m	1987	Sandy	First belt of halophytes	7.6	87.0	7.5
80 m	1976	Sandy	Second belt of desert plants	8.0	5.2	11.0

Table 26-4 Plant cover on the emerging coast of the Dead Sea near En Gedi.

	First belt[a] (% of area cover)		Second belt[b] (% of area cover)	
	1990	1992	1990	1992
Annual plants	4.8 ± 2.3	9.5 ± 3.4	10.0 ± 4.6	20.4 ± 9.3
Perennial plants	8.2 ± 3.2	7.5 ± 5.0	18.5 ± 6.9	15.0 ± 2.3
Total cover	14.0 ± 5.3	17.0 ± 5.4	28.5 ±12.5	25.4 ± 5.8

[a]First belt is characterized by the pioneer communities.
[b]Second belt is made up of subsquent communities.

The first stage of plant succession started with a small number of local halophytes (Waisel, 1972). These included annual species, such as *Mesembryanthemum forsskalii*, *M. nodiflorum*, *Aaronsohnia faktorovskyi*, *Rumex cyprius*, *Linaria haelava*, *Anthemis maris mortui*, and *Salsola inermis*. The perennial species that followed included chenopods, such as *Anabasis setifera*, *Suaeda fruticosa*, *Salsola baryosma*, *Atriplex halimus*, etc. This pioneer plant community, which also included the neophyte *Atriplex holocarpa*, constituted the first vegetation belt of the retreating coast (Aloni and Waisel, 1990). Because it was a pioneer community, it characterized the area for only a few years (Fig. 26-3).

After additional leaching occurred during subsequent winters, the quality of the substrate was modified and plant succession progressed. New plants replaced the pioneer species and, in the course of 8–10 years, became dominant, forming a second belt of vegetation (Table 26-2, Fig. 26-3).

The distribution of the pioneer plant species of the first belt of terrestrial vegetation has gradually declined with time and with increasing distance from the coast. Most of the pioneers were replaced in due time by other species of the surrounding desert, a process that proceeded for almost 10 years. This second stage of succession was enabled by the sharp decline in the electrical conductivity of the soil solution (down to 5 dS/cm²) and by a drop in the pH of the soil extract (Table 26-3).

There is no clear establishment of distinct plant communities at each belt, but new plants gradually penetrate into the area until the prevailing climax community take over. Such a community constitutes the dominant vegetation unit of the surrounding area, and represents the vegetation of areas that were not affected by the Dead Sea water for millennia. The prevailing climax communities are dominated by *Zygophyllum dumosum* and by trees of *Acacia* sp. (Fig. 26-5) on the nearby mountainsides and by *Ziziphus spina-christi* in wadi beds.

The establishment of pioneer plant communities on this *terra nova* and their gradual development to the higher stages of the successional sere depend primarily on the residual salinity of the substrate, that is, on the rates of salt leaching from the substrate. The success of these plant communities is also dependent on the proximity of sources of viable seeds, on the salt tolerance of the new immigrants, and probably also on the presence of other organisms in such a site. These factors do not seem to be limiting (Danin, 1976), and the diversity of the species increased considerably within a short time (Table 26-2).

The distribution of plants and the diversity of the developing communities depend in part on the rate of seed supply. Seeds of certain species, for example, *Mesembryanthemum* sp. and *Suaeda* sp., are distributed mainly by runoff water, which flows toward the shore of the Dead Sea from neighboring higher

Figure 26-5 A stand of *Acacia* trees on a site that constituted the sea bottom 30 years ago.

ground. Such species appear in dense stands near water courses or in small ditches. Other plant species, for example, *Salsola* sp., *Atriplex holocarpa*, *Rumex cyprius*, and *Tamarix* sp., are distributed mainly by the wind. Such plants may disperse seeds over large distances and thus reach the emerging coast faster and in great numbers. Nevertheless, the establishment of plants on each belt depends on the seed availability as well as on the tolerance of the young seedlings to the presence of residual salts.

The appearance of *Zygophyllum dumosum*, the dominant species of the prevailing climax (Waisel et al., 1987), started on the shingle beach 8–15 years after emergence and on fine clay substrates after 15–20 years.

The total plant cover of both belts remained rather constant (Table 26-4) within the period 1990–1992, in spite of a noticeable change in species composition.

The rate of plant succession in this habitat is relatively fast, and the establishment of plants is driven, in the first place, by the rate of leaching of the substrate. The surrounding desert soil is rocky and shallow, only slightly developed, and moderately saline. The permanent vegetation on such soils constitute sparse shrub communities dominated by *Zygophyllum dumosum*. According to our estimates, a stable community of this type is established within 15–20 years after the retreat of the Dead Sea, depending on the local water regime.

SUMMARY

The water level of the Dead Sea dropped gradually during the 30 years between 1962 and 1992. The substrate that was exposed along the receding shore was extremely saline and remained sterile for 4–5 years. During that period, even the low precipitation (average of 50 mm per annum for this area) was enough to leach the substrate. Indeed, it reduced the salinity of the soil to a level that allowed the establishment of a few species of extreme halophytes. Those halophytes constituted the first stage of succession of higher plants on the newly exposed coast. This stage of plant development remained dominant on the examined sites for an additional 4–5 years. When salinity was further reduced, the conditions enabled the penetration of various desert species and, thus, the establishment of the first stable

community. After 8–10 years, a third stage of succession was reached, that is, a plant community dominated by *Zygophyllum dumosum*. Such a community is usually the permanent community on the neighboring hillsides, away from the coast of the Dead Sea. Thus, succession in the Dead Sea area is dictated mainly by two abiotic factors: salinity and precipitation.

REFERENCES

Aloni, E., and Waisel, Y., 1990, Recuperation of plant cover after destruction by man: Patterns and rates, *in* Frey, V., and Kurschner, H., eds., Proceedings Third Plant Life of South-West Asia symposium: Berlin, September 1990, p. 15.

Carmi, I., Gat, J. R., and Stiller, M., 1984, Tritium in the Dead Sea: *Earth and Planetary Science Letters*, v. 71, p. 377–389.

Danin, A., 1976, Plant species diversity in the Dead Sea valley: *Oecologia*, v. 22, p. 251–259.

Hecht, A., Ezer, T., and Mandelzweig, R., 1984, Currents, waves and meteorology in the Dead Sea: Israel Oceanography and Limnology Research Report, No. 46., p. 84.

Klein, C., 1982, Morphological evidence of lake level changes on the western shore of the Dead Sea: *Israel Journal of Earth Science*, v. 31, p. 67–94.

Klein, C., 1993, The effects of the last 4000 years' climate fluctuations in Eretz Israel on the Dead Sea level and Man, *in* Graber, M., Cohen, A., and Magaritz, M., eds., *Regional implications of future climate change*: Jerusalem, The Israel Academy of Science and The Ministry of the Environment, p. 52–56.

Shatkay, M., 1985, The newly exposed sediments on the westren shores of the Dead Sea. Chemical and mineralogical analysis [M.S. thesis]: Rehovot, Israel, The Weizmann Institute of Science, 96 p.

Waisel, Y., 1972, *Biology of halophytes*: New York, Academic Press, 395 p.

Waisel, Y., Agami, M., and Eshel, A., 1987, Human interference and vegetation dynamics in historical time. *in* Greuter, W., Zimmer, B., and Behnke, H. D., eds., Proceeding XIV International Botanical Congress, Berlin, August 1987, p. 6–09–1.

Yaffe, S., 1972, Climate of the Dead Sea: Beth Dagan, Meteorological Service, Instructive Papers no. 12 (in Hebrew).

27. DEAD SEA RESEARCH: SYNOPSIS AND FUTURE

Joel R. Gat, Tina M. Niemi, and Zvi Ben-Avraham

The work summarized in this book records an impressive research effort by a varied group of investigators from many disciplines and institutions. These researchers were attracted to this unique system mostly because of its scientific and intellectual challenge, rather than by practical and applied considerations, even though the latter did play some role, foremost because of the exploitation of the mineral wealth of the Dead Sea and the development schemes that crop up now and then.

This collection highlights the achievements and the knowledge gained, but it also manifests the gaps in our knowledge and understanding of this system. Some of these shortcomings can be attributed to the political divisions in the region, which enable the collection of only a fragmentary set of data from the lake and its coastal and atmospheric environment. Moreover, the ever-changing nature of this system (one of its most fascinating features) dictates an uninterrupted program of observations, since data missed at any particular point in time and space cannot be repeated at a later time (as would be the case for a steady-state system). The amount of research invested, however, whether measured in the number of people engaged, the observation period, or the budget invested, falls short by orders of magnitude relative to the size and complexity of the system and its changeability as compared to other lakes, such as the Sea of Galilee. The absence of a permanent coastal facility and of a properly equipped research vessel, for example, is painfully evident.

As a result, we know very little about basic properties of the Dead Sea waters, such as wave structure, optical properties, and surface slicks, to name a few. The meteorological regime is but sketchily recorded, and the thermohaline model of the lake is based essentially on a single station in the center of the lake, without a clue about its three-dimensional structure. A basic parameter of this evaporitic system, namely, the rate of evaporation, is still a bone of contention.

The geological structure of the Dead Sea basin, unlike its waters, is relatively well known. In fact, this is one of the best studied strike-slip basins on Earth. Yet, as more is known about the subbottom structure of the basin, new questions arise. Therefore, new studies are being planned to investigate such topics as the nature and rate of neotectonic activity, the style of faulting along the boundary faults, and the deep structure. Improved monitoring of seismicity should help to better constrain earthquake epicentral data. Coring in the northern Dead Sea basin has the potential for revealing a long and continuous sedimentary record for the region shedding new light on the history of climate, lake levels, and changes in the environment. With peace between the Dead Sea area nations, studying these problems will be easier because the entire width of the basin will be available for geological and geophysical research.

With the encouraging change in the political climate of the region and the promise of cooperation between scientists from the opposing shores of the lake, it is hoped that some of these gaps will be closed and that the sporadic forays by the concerned scientists will be replaced by a systematic limnological research program. This is especially desirable in view of the uncertain future of this unique body of water and the need to be able to foresee possible developments and to guide the activities.

Several scenarios for the future of the Dead Sea can be envisioned. Each of these is one of profound change and unparalleled research challenges. It is quite unlikely that the freshwater presently diverted from the Jordan–Yarmouk river system would be restored to the Dead Sea. On the contrary, even more of the water inflows will probably be used to quench the thirst of these arid lands, including the possible desalination of some of the brackish waters that are as yet unutilized. If so, the shrinkage of the Dead Sea is destined to continue at the present, or even an accelerated, pace. A very preliminary estimate by Z. Sirkes (pers. com., 1994) puts the final new equilibrium level of the Dead Sea 150 m below the present one within the next century and a half. By that time, the waters would be much saltier. This factor, together with the reduced area and more shielded situation, is expected to reduce the evaporative water loss to an extent that would balance the remnant freshwater inflows by occasional floods.

Obviously, far-reaching geochemical changes in the water can be expected if this process is allowed to take its course. Also, the lake can then be expected to become completely abiotic. Moreover, the large, newly exposed shorelines will provide challenging research opportunities as they are desalinated and reclaimed. Exposure of the lake bottom sediments and structures would permit geological study of the depositional and geomorphological systems.

If, however, one of the many development schemes to bring seawater into the area were to be implemented, then the chemical changes would be still more dramatic because of the introduction of extraneous salts. Massive precipitation of gypsum, for example, would affect not only the sedimentation pattern, but because of the increase in suspended matter and a possible permanent whitening of its surface, it would affect the whole thermohaline and energy balance of the system.

It is quite unthinkable that the research community will not rise to the occasion of such an important and challenging issue. It is only to be hoped that the relevant authorities will also recognize these needs in time and provide the financial support necessary for this endeavor. We hope that this book has contributed to an increased awareness of the scientific challenges and practical needs that reside in this strange but fascinating body of salt and water.

INDEX